Dynamics of Engineered Artificial Membranes and Biosensors

This state-of-the-art guide provides a powerful toolkit for building artificial membranes, combining techniques for synthesis with mathematical modeling.

- Drawing on the most recent advances in bioengineering, biochemistry, and computational biology, it describes how to precisely construct synthetic lipid bilayer membranes to mimic the remarkable properties of biological membranes, and shows how they can be used to develop biosensors and diagnostic devices.
- Experimental and modeling case studies provide insight into how artificial cell membranes actually operate at the molecular level, and molecular dynamics simulation code accompanying the book online enables readers to reproduce the key experimental results presented throughout.
- Multi-physics models for predicting membrane performance and improving design are developed, with coverage including molecular dynamics, coarse-grained molecular dynamics, Brownian dynamics, continuum theory, and reaction-rate theory.

This book is essential reading for researchers, students and professionals in bioengineering, biophysics, and electrical engineering.

William Hoiles is a Research Fellow in the Department of Electrical and Computer Engineering at the University of British Columbia.

Vikram Krishnamurthy is a Professor at Cornell Tech and the School of Electrical and Computer Engineering at Cornell University. He is a Fellow of the IEEE and the author of *Partially Observed Markov Decision Processes* (Cambridge, 2016).

Bruce Cornell is an Adjunct Professor in the School of Life Sciences at the University of Technology, Sydney, and at Western Sydney University. He is also the Principal Scientist at Surgical Diagnostics Pty Ltd and SDx Tethered Membranes Pty Ltd.

Dynamics of Engineered Artificial Membranes and Biosensors

WILLIAM HOILES
University of British Columbia, Vancouver

VIKRAM KRISHNAMURTHY
Cornell University, New York

BRUCE CORNELL
University of Technology, Sydney, and Surgical Diagnostics Pty Ltd., Sydney

CAMBRIDGE
UNIVERSITY PRESS

University Printing House, Cambridge CB2 8BS, United Kingdom

One Liberty Plaza, 20th Floor, New York, NY 10006, USA

477 Williamstown Road, Port Melbourne, VIC 3207, Australia

314–321, 3rd Floor, Plot 3, Splendor Forum, Jasola District Centre, New Delhi - 110025, India

79 Anson Road, #06-04/06, Singapore 079906

Cambridge University Press is part of the University of Cambridge.

It furthers the University's mission by disseminating knowledge in the pursuit of education, learning and research at the highest international levels of excellence.

www.cambridge.org
Information on this title: www.cambridge.org/9781108423502
DOI: 10.1017/9781108526227

© Cambridge University Press 2018

This publication is in copyright. Subject to statutory exception and to the provisions of relevant collective licensing agreements, no reproduction of any part may take place without the written permission of Cambridge University Press.

First published 2018

Printed in the United Kingdom by TJ International Ltd. Padstow Cornwall

A catalogue record for this publication is available from the British Library

Library of Congress Cataloging-in-Publication data
Names: Hoiles, William, author. | Krishnamurthy, V. (Vikram), author. | Cornell, Bruce, author.
Title: Dynamics of engineered artificial membranes and biosensors / William Hoiles
   (University of British Columbia, Vancouver), Vikram Krishnamurthy (Cornell University,
   New York), Bruce Cornell (University of Technology, Sydney).
Description: Cambridge : Cambridge University Press, [2018] | Includes bibliographical
   references and index.
Identifiers: LCCN 2017057899 | ISBN 9781108423502 (hardback : alk. paper)
Subjects: LCSH: Membranes (Biology) | Membranes (Technology) | Lipid membranes. | Biosensors.
Classification: LCC QH601 .H65 2018 | DDC 572/.577–dc23 LC record available
   at https://lccn.loc.gov/2017057899
ISBN 978-1-108-42350-2 Hardback

Additional resources for this publication at www.cambridge.org/engineered-artificial-membranes

Cambridge University Press has no responsibility for the persistence or accuracy of URLs for external or third-party internet websites referred to in this publication, and does not guarantee that any content on such websites is, or will remain, accurate or appropriate.

Contents

Preface		*page* xv
List of Abbreviations		xviii

Part I Introduction and Background

1 Motivation and Outline — 3
   1.1 Why Membranes? — 3
   1.2 Guided Tour of the Book — 4

2 Biochemistry for Engineers: A Short Primer — 9
   2.1 Bonded and Nonbonded Molecular Interactions — 9
   2.2 Lipids, Vesicles, and Bilayers — 12
   2.3 Lipid Bilayers — 14
      2.3.1 Archaebacteria and DphPC Bilayers — 14
      2.3.2 Energetics of Lipid Bilayers — 15
      2.3.3 Structure of Lipid Bilayers — 16
   2.4 Peptides and Proteins — 16
   2.5 Ion Channels — 21
   2.6 Tethers, Spacers, and the Bioelectronic Interface — 24
      2.6.1 Tethers — 25
      2.6.2 Spacers — 25
      2.6.3 Bioelectronic Interface — 26
   2.7 How to Visualize Macromolecules — 27
   2.8 Closing Remarks — 29

3 Engineered Artificial Membranes — 30
   3.1 Membranes — 30
   3.2 Artificial Membrane Architectures — 32
   3.3 Engineered Artificial Tethered Membranes — 35
   3.4 Sensing with Engineered Tethered Membranes — 39
      3.4.1 Device 1: Ion-Channel Switch (ICS) Biosensor — 41
      3.4.2 Device 2: Pore Formation Measurement Platform (PFMP) — 42

		3.4.3 Device 3: Electroporation Measurement Platform (EMP)	43
		3.4.4 Device 4: Electrophysiological Response Platform (ERP)	44
	3.5	Multiphysics Dynamic Models of Engineered Tethered Membranes	45
		3.5.1 Ab Initio Molecular Dynamics	46
		3.5.2 Molecular Dynamics	47
		3.5.3 Coarse-Grained Molecular Dynamics	47
		3.5.4 Continuum Theories	48
		3.5.5 Reaction-Rate Theory	48
	3.6	Electrolyte Dynamics: Steric Effects and Double-Layer Charging	49
	3.7	Future Technologies: Implantable Medical Devices, Diagnostics, and Therapeutics	51
		3.7.1 Cochlear and Retinal Implants	51
		3.7.2 In Vitro Medical Diagnostics (IVDs)	52
		3.7.3 Molecular Therapeutics	54
		3.7.4 Biological Neural Networks	54
		3.7.5 Microeletrodes and Single-Cell Measurements	55
		3.7.6 Summary	58
	3.8	Closing Remarks	58

Part II Building Engineered Membranes, Devices, and Experimental Results

4 Formation of Engineered Tethered Membranes — 61

	4.1	Introduction	61
		4.1.1 Engineered Tethered Membrane: Structure	62
		4.1.2 Overview of Tethered Device	63
	4.2	Building an Engineered Artificial Membrane	64
		4.2.1 Solvent-Exchange Technique	64
		4.2.2 Evaluating the Quality of the Engineered Membrane	70
	4.3	Inserting Proteins and Ion Channels into Engineered Artificial Membranes	71
		4.3.1 Spontaneous Insertion Method	72
		4.3.2 Electrochemical Insertion Method	72
		4.3.3 Proteoliposomal Insertion Method	74
	4.4	Laboratory Exercise: Tethered Membranes and Spontaneous Insertion of Gramicidin Channels	76
		4.4.1 Prepare the Engineered Tethered Membrane for Spontaneous gA Ion-Channel Insertion	77
		4.4.2 Spontaneous Insertion of gA Ion Channels	79
		4.4.3 Measuring Membrane Conductance Response	79
	4.5	Complements and Sources	81
	4.6	Closing Remarks	82

5 Ion-Channel Switch (ICS) Biosensor — 83

- 5.1 Introduction — 83
- 5.2 ICS Biosensor: Construction and Formation — 85
- 5.3 Operation of the ICS Biosensor — 87
  - 5.3.1 Large and Small Analyte Detection — 87
  - 5.3.2 Impedance Response of ICS Biosensor for Digoxin and b-F_{ab} — 87
- 5.4 ICS Biosensor: Flow Velocity, Binding-Site Density, and Specificity — 90
  - 5.4.1 Flow Velocity and Binding-Site Density — 91
  - 5.4.2 Specificity in Complex Environments — 94
- 5.5 Detection of Influenza A in Clinical Samples — 95
  - 5.5.1 ICS Biosensor Preparation and Clinical Trials for Rapid Influenza A Diagnosis — 95
  - 5.5.2 Influenza A Clinical Samples — 96
  - 5.5.3 Results of Influenza A Clinical Trial — 96
- 5.6 ICS for Multianalyte Detection — 97
  - 5.6.1 Biosensor Arrays — 98
  - 5.6.2 Multi-Analyte Detection — 99
- 5.7 Complements and Sources — 101
- 5.8 Closing Remarks — 102

6 Physiochemical Membrane Platforms — 103

- 6.1 Introduction — 103
- 6.2 Device 1: Pore Formation Measurement Platform (PFMP) — 104
  - 6.2.1 Pore Formation Measurement Platform: Introduction — 104
  - 6.2.2 Pore Formation Measurement Platform: Construction — 105
  - 6.2.3 Pore Formation Measurement Platform: Operation and Experimental Measurements — 106
- 6.3 Device 2: Electroporation Measurement Platform (EMP) — 107
  - 6.3.1 Electroporation Measurement Platform: Introduction — 107
  - 6.3.2 Electroporation Measurement Platform: Formation — 108
  - 6.3.3 Electroporation Measurement Platform: Operation and Experimental Measurements — 109
- 6.4 Device 3: Electrophysiological Response Platform (ERP) — 110
  - 6.4.1 Electrophysiological Response Platform: Overview — 110
  - 6.4.2 Electrophysiological Response Platform: Formation — 112
  - 6.4.3 Electrophysiological Response Platform: Operation and Experimental Measurements — 114
- 6.5 Complements and Sources — 115
- 6.6 Closing Remarks — 117

7 Experimental Measurement Methods for Engineered Membranes — 118

- 7.1 Introduction — 118

7.2 Electrical Response of Engineered Membranes ... 118
    7.2.1 Electrical Impedance Measurements ... 120
    7.2.2 Time-Dependent Electrical Measurements ... 122
    7.2.3 Interpretation of Measured Current Response ... 125
7.3 Spectroscopy and Imaging Techniques for Engineered
    Tethered Membranes ... 127
    7.3.1 X-Ray Reflectometry for Measuring Area per Lipid ... 128
    7.3.2 Nuclear Magnetic Resonance Measurements of the
         Conformation and Orientation of Gramicidin A ... 129
    7.3.3 Fluorescence Recovery after Photobleaching for Measuring
         Lipid Diffusion ... 129
    7.3.4 Neutron Reflectometry for Measuring Membrane Thickness and
         Reservoir Thickness ... 131
    7.3.5 Summary ... 133
7.4 Complements and Sources ... 134
7.5 Closing Remarks ... 135

Part III Dynamic Models for Artificial Membranes: From Atoms to Device

8 Reaction-Rate-Constrained Models for Engineered Membranes ... 139

8.1 Introduction ... 139
8.2 Fractional-Order Macroscopic Model ... 140
    8.2.1 Fractional-Order Derivatives: Double-Layer Capacitance and
         Charging Dynamics ... 144
    8.2.2 Fractional-Order Macroscopic Model: Sinusoidal and
         Time-Varying Excitation Potential ... 147
    8.2.3 Determining the Quality of an Engineered Membrane Using the
         Fractional-Order Macroscopic Model ... 148
8.3 Experimental Measurements: Fractional-Order Macroscopic Model ... 150
    8.3.1 Spacer Surface and Electrolyte Concentration ... 151
    8.3.2 Variation in Membrane Types and Tether Density ... 152
    8.3.3 Estimating the Dielectric Constant of the Membrane ... 152
8.4 Modeling Membranes with Sterol Components ... 154
    8.4.1 Fractional-Order Model for Cholesterol in Engineered
         Membranes ... 154
    8.4.2 Impedance Analysis of Engineered Membranes Containing
         Sterol Molecules ... 156
8.5 Complements and Sources ... 157
8.6 Closing Remarks ... 158

9 Reaction-Rate-Constrained Models for the ICS Biosensor ... 159

9.1 Introduction ... 159
9.2 Detection of Analyte Species in the Reaction-Rate Regime ... 161

		9.2.1	Aside: From Chemical Equations to Reaction-Rate Differential Equations	161
		9.2.2	Reaction-Rate Model of the ICS Biosensor	162
		9.2.3	Singular Perturbation Analysis of Dimer Concentration	165
		9.2.4	Detection of Human Chorionic Gonadotropin (hCG)	166
	9.3	Microelectrode ICS (mICS) Biosensor and Hidden Markov Model (HMM)		167
		9.3.1	Hidden Markov Model for mICS Biosensor	168
		9.3.2	Hidden Markov Model Statistical Signal Processing	170
		9.3.3	Detection of Monoterpene Oxidation Product (MTOP)	171
	9.4	Complements and Sources		172
	9.5	Closing Remarks		172

10 Diffusion-Constrained Continuum Models of Engineered Membranes — 174

10.1	Introduction		174
10.2	Mass Transport versus Reaction-Rate-Limited Kinetics		176
	10.2.1	Damköhler and Péclet Numbers	176
	10.2.2	Characterization of Operating Regime	177
10.3	Mass-Transport-Limited Model of the ICS Biosensor Dynamics		178
	10.3.1	Poisson's Equation: Electrostatics	180
	10.3.2	Nernst–Planck Equation: Advection and Diffusion	181
	10.3.3	Poisson–Nernst–Planck Equation	182
	10.3.4	Estimating the Reaction Rates in the ICS Biosensor	185
	10.3.5	Experimental Results: Streptavidin, TSH, Ferritin, and hCG	186
10.4	Biosensor Arrays: Numerical Case Study		187
	10.4.1	Biosensor Array Model	189
	10.4.2	Mass-Transport Phase Diagram	190
	10.4.3	Sensor Array Can Mitigate Mass-Transport Limits	193
10.5	Pore Formation Dynamics: Models for PGLa Antimicrobial Peptides		195
	10.5.1	Generalized Reaction-Diffusion Equation	197
	10.5.2	Analyte and Surface Reaction Mechanism of PGLa	197
	10.5.3	Dynamic Model of Electrolyte and Surface Diffusion of PGLa	199
	10.5.4	Experimental Results: Reaction Dynamics of PGLa	201
10.6	Asymptotic Poisson–Nernst–Planck Model and Lumped Circuit Parameters		203
	10.6.1	Double-Layer Capacitance and Electrolyte Resistance for Blocking Electrode	204
	10.6.2	Double-Layer Capacitance for Reaction-Limited Electrode	206
10.7	Complements and Sources		208
	10.7.1	Poisson–Nernst–Planck (PNP) Model	208
	10.7.2	ICS Biosensor Arrays and Multicompartment Models	208
	10.7.3	Parameter Estimation and System Identification	209
10.8	Closing Remarks		211

11	**Electroporation Models in Engineered Artificial Membranes**	212
	11.1 Introduction	212
	11.1.1 Applications of Electroporation	212
	11.1.2 What Is Electroporation?	213
	11.1.3 Mesoscopic Model of Electroporation	214
	11.1.4 Organization of This Chapter	216
	11.2 Smoluchowski–Einstein Equation	217
	11.2.1 Source Term and Energy Term of the Smoluchowski–Einstein Equation	219
	11.2.2 Summary	222
	11.3 Multiphysics (Mesoscopic) Model of Electroporation	222
	11.3.1 Equivalent Circuit Model of Electroporation	222
	11.3.2 Singular Perturbation Approximation and Electrical Dynamics	224
	11.4 Continuum Model of Electroporation: Aqueous Pore Conductance and Double-Layer Capacitance	227
	11.4.1 Continuum Model 1: Generalized Poisson–Nernst–Planck (GPNP) Equation	229
	11.4.2 Continuum Model 2: Poisson–Fermi–Nernst–Planck (PFNP) Equation	232
	11.5 Computing Engineered Tethered-Membrane Parameters from Continuum Theory	235
	11.5.1 Computing Pore Conductance	235
	11.5.2 Electrical Potential Energy for Pore Formation	236
	11.5.3 Computing Pore Capacitance	237
	11.5.4 Double-Layer Capacitance	238
	11.5.5 Detection Tests for Ionic Correlation Effects	238
	11.6 Faradic Reactions at the Bioelectronic Interface	242
	11.6.1 Faradic Reactions and Double-Layer Charging at the Bioelectronic Interface	242
	11.6.2 Faradic Reaction Boundary Conditions for the PFNP Continuum Model	244
	11.7 Complements and Sources	247
	11.8 Closing Remarks	248
12	**Electroporation Measurements in Engineered Membranes**	250
	12.1 Introduction	250
	12.2 Aqueous Pore Conductance, Capacitance, and Electrical Energy	253
	12.2.1 Aqueous Pore Conductance	254
	12.2.2 Aqueous Pore Electrical Energy	256
	12.2.3 Aqueous Pore Capacitance	259
	12.3 Pore Radii and Membrane Conductance Dynamics	260
	12.4 Sensitivity of Current Response to Model Parameters	260
	12.5 Effect of Tether Density of Membrane Electroporation Dynamics	262

12.6		Heterogeneous Membrane Mixtures	265
12.7		Membranes with Sterol Inclusions	267
12.8		Estimating Hydration Ion Size and Faradic Reaction Rates	269
12.9		Electrical Double-Layer Charging Dynamics	271
	12.9.1	Spatially Dependent Dielectric Constant at the Bioelectronic Interface	272
	12.9.2	Voltage-Dependent Double-Layer Capacitance	275
12.10		Large Excitation Potentials and Double-Layer Charging Dynamics	276
12.11		Complements and Sources	279
12.12		Closing Remarks	281

13 Electrophysiological Response of Ion Channels and Cells — 282

13.1		Introduction	282
13.2		Dynamic Model of Embedded Ion Channels	284
13.3		Electrophysiological Response of a Voltage-Gated Ion Channel	285
13.4		Dynamic Model of Electrophysiological Response of Cells	286
	13.4.1	Macroscopic Model of the Electrophysiological Response Platform	287
	13.4.2	Cellular Membrane Conductance and Charging Dynamics	289
13.5		Electrophysiological Response of Skeletal Myoblasts	291
13.6		Complements and Sources	292
13.7		Closing Remarks	293

14 Coarse-Grained Molecular Dynamics — 295

14.1		Introduction	295
14.2		Basics of Coarse-Grained Molecular Dynamics	299
	14.2.1	From an Atomistic to a Mesoscopic Coarse-Grained Description of Engineered Membranes	300
14.3		Atomistic-to-Observable Model of Tethered Membranes	301
14.4		Aside: The Fokker–Planck Equation	305
	14.4.1	Kolmogorov and Fokker–Planck Equations	306
	14.4.2	First-Passage Time and the Arrhenius Equation	308
14.5		Coarse-Grained Molecular Dynamics Model for the Bioelectronic Interface and Water	310
	14.5.1	Percus–Yevick Equation and Water Density at the Bioelectronic Interface	311
	14.5.2	Density Profile of Water at the Bioelectronic Interface	315
	14.5.3	Fokker–Planck Equation: Spatially Dependent Water Diffusion Coefficient	317
	14.5.4	Diffusion Tensor of Water in Tethering Reservoir	320
	14.5.5	Summary	321
14.6		Tethered Membrane Dynamics and Energetics	322
	14.6.1	Lipid Energetics and Pore Density	322

		14.6.2	Line Tension and Surface Tension	325
		14.6.3	Deuterium Order Parameter	328
		14.6.4	Lipid Lateral Diffusion	328
		14.6.5	Geometric Properties of Tethered Membranes	330
		14.6.6	Summary	332
	14.7	Control of Tethered-Membrane Properties by Sterol Inclusions		332
		14.7.1	Lateral Diffusion Dynamics of Lipids and Cholesterol	333
		14.7.2	Biomechanics of Lipids and Cholesterol	334
	14.8	Molecular Diffusion and Langevin's Equation		336
		14.8.1	Langevin's Equation and Diffusion of Molecules	336
		14.8.2	Nonstationary Lipid Diffusion with Sterol Inclusions	339
	14.9	Case Study: Atomistic-to-Observable Model PGLa Pore Formation in Tethered Membranes		341
		14.9.1	Coarse-Grained Molecular Dynamics Simulation of Tethered Membrane Containing PGLa	342
		14.9.2	Diffusion of PGLa and Membrane Properties from Coarse-Grained Molecular Dynamics	345
		14.9.3	Surface Binding and Oligomerization of PGLa from Coarse Grained Molecular Dynamics	346
	14.10	Complements and Sources		348
	14.11	Closing Remarks		351
15	**All-Atom Molecular Dynamics Simulation Models**			**353**
	15.1	Introduction		353
	15.2	Basics of Molecular Dynamics		353
		15.2.1	Potential Energy Functions	355
		15.2.2	Macroscopic Parameters and Statistical Ensembles	358
		15.2.3	Numerical Methods for Molecular Dynamics	365
	15.3	MD Simulations for the Dynamics of Engineered Membranes		371
	15.4	Aqueous Pore Formation Dynamics in Tethered Membranes		374
	15.5	Capacitance and Dipole Potential of Tethered Membranes		376
	15.6	Modeling Ion Permeation and Channel Conductance		378
		15.6.1	Models for Ion Permeation: From Ab Initio to Reaction Rate	378
		15.6.2	Gramicidin Channel Conductance Estimation Using Distributional Molecular Dynamics	381
	15.7	Gramicidin A (gA) Dimer Dissociation and Reaction-Rate Estimation		383
		15.7.1	Molecular Reaction Dynamics of Gramicidin Channel Dissociation	384
		15.7.2	Gramicidin A Reaction Rates	385
	15.8	Complements and Sources		387
	15.9	Closing Remarks		389
16	**Closing Summary for Part III: From Atoms to Device**			**390**

Appendices

Appendix A   Elementary Primer on Partial Differential Equations (PDEs)   395
A.1   Linear, Semilinear, and Nonlinear Partial Differential Equations   395
A.2   Linear Partial Differential Equations and Boundary Conditions   396
A.3   Nondimensionalization of Partial Differential Equations   397
A.4   Solutions of Partial Differential Equations   401

Appendix B   Tutorial on Coarse-Grained Molecular Dynamics with Peptides   404
B.1   Constructing the All-Atom and Coarse-Grained Structure of a Peptide   404
B.2   Construction of Coarse-Grained Lipid Bilayer   406
B.3   How to Insert PGLa Peptides in the Transmembrane State   409
B.4   Note on Publication-Quality Figures   410

Appendix C   Experimental Setup and Numerical Methods   412
C.1   Ion-Channel Switch Biosensor   412
C.2   Pore Formation Measurement Platform: PGLa   414
C.3   Tethered-Membrane Parameters: Pore Conductance
and Electrical Energy   415
C.4   Coarse-Grained Molecular Dynamics (CGMD) Simulations   416
C.5   CGMD Simulation Setup for PGLa   419
   C.5.1   Simulation Setup of All-Atom Molecular Dynamics   420

Bibliography   421
Index   447

Preface

Biological membranes such as the cell membrane play an essential and ubiquitous role in cellular communication, energy storage, structural support, and protecting the contents of living cells. Biological membranes are dynamic structures with remarkable properties: they have moving parts, they act as electrical circuits with long-range memory, they contain protein macromolecules (ion channels) that selectively open and close to admit ions, thereby regulating the electrical activities of cells, and they spontaneously form pores (electroporation) which is crucial for the operation of antimicrobial drugs.

This book focuses on building and mathematically modeling artificial membranes that mimic biological membranes. Recent advances in bioengineering and biochemistry allow us to build artificial membranes precisely to mimic the remarkable properties of biological membranes and also to build synthetic biological devices out of these membranes. Also advances in computational molecular biology allow us to construct large-scale computer simulation models at the atomic-spatial scale to gain deep insight into the dynamic properties and structure-function relationship of such artificial membranes and devices. This dual approach of synthesis (building membranes and novel devices) and analysis (mathematical modeling) is vital for the future development of sensing devices, drug delivery mechanisms, and synthetic biological devices.

Summary. This book provides a comprehensive description of artificial membranes aimed at advanced undergraduate and graduate students, and researchers in electrical engineering, biological and chemical engineering, biophysics, and applied mathematics. Construction of artificial membranes and devices is intimately linked with careful mathematical modeling so as to predict the performance and improve the design. As a result this book is organized into three interrelated parts:

Part I gives an overview of the book along with an elementary primer on biochemistry for engineers and applied mathematicians.

Part II deals precisely with engineering and building artificial membranes, building novel synthetic biological devices out of these artificial membranes, and evaluating how such devices perform in experimental and clinical studies. The devices studied include a super-resolution biosensor (which can be viewed as a fully operational nanomachine), an electroporation platform (for studying how membranes spontaneously form water-filled pores), and an electrophysiological platform (for noninvasive measurements of cells).

Part III develops mathematical models that operate at multiple spatial and temporal scales to capture the dynamics of artificial membranes and devices, starting from atoms and ending in the macroscopic device. Several levels of modeling abstractions are studied: molecular dynamics (at the atomic scale resolution), mesoscopic models (Poisson–Nernst–Planck models and generalizations) that take into account fluid flow dynamics, and reaction-rate models at the macroscopic device level.

The combination of engineered membranes (synthetic biology hardware discussed in Part II) together with mathematical models (software discussed in Part III) yields a powerful set of engineering tools to go from structure to function. The hardware allows us to consider various components in their natural state (as they interact with other components), whereas the software then zooms into specific subparts in isolation. This combination of hardware (ex vivo and in vitro) and software (in silico) is a unique feature of our book.

A challenge encountered with synthetic biological devices such as artificial cell membranes is the bioelectronic interface: charged ions carry information in biological membranes, while electrons carry information in electrical devices. Several sections of this book are devoted to building and detailed mathematical modeling of the bioelectronic interface at several levels of abstraction.

Readership. This book is written for advanced undergraduate students, graduate students, and researchers in electrical engineering, biological and chemical engineering, biophysics, and applied mathematics. The prerequisites are modest. A student or researcher with an applied mathematics, physics, bioengineering, or electrical engineering background (signal processing, circuits and systems, control theory) can follow the material in this book without a detailed knowledge of membrane biology and biochemistry. An undergraduate knowledge of cellular biology, signals and systems, and physics involving diffusion-type partial differential equations is adequate to understand the material. Also, the mathematical models proposed in this book for the dynamics of artificial membranes involve computer simulations using molecular dynamics, finite element, boundary element, and finite difference numerical methods. These are typically covered in computational physics and engineering mathematics undergraduate courses. Parts of this book have been class-tested in graduate and senior-level undergraduate courses at the University of British Columbia and the University of Technology Sydney.

This is not a book on the biology of membranes; instead we are interested in engineering membranes, namely, synthesizing (building membranes and devices) and analyzing (mathematically modeling) them. Given the multidisciplinary nature of this book, it is natural that several topics have been omitted which some readers might deem as being important. We do not cover molecular pumps, ionophores, or endocytosis/exocytosis. Also our discussion of ion channels focuses on engineering – how to use them to design and build synthetic biological devices and how to model these devices; we do not present detailed discussions on ion channel experiments (patch clamping) or modeling (Hodgkin-Huxley models, permeation).

Appendices and Internet supplement. The appendices discuss elementary partial differential equations and nondimensionalization, how to perform both coarse-grained molecular dynamics and all-atom molecular dynamics simulations, and the typical parameter values used in the mathematical models of engineered artificial membranes. Using the molecular dynamics simulation code provided, the reader can repeat the computer simulation results documented in this book. Actual engineered membrane devices can be obtained by contacting the authors. Using these devices, the reader can repeat the experiments documented in this book.

To keep the cost manageable, all figures are printed in black and white. Color figures can be downloaded for free from the website of the book hosted at www.cambridge.org/engineered-artificial-membranes. The website also contains molecular dynamics simulation code and other pedagogical material.

Context. This book is truly a multidisciplinary effort between academia and industry. In the 1990s, Cornell (a biochemist) invented a remarkable biosensor built out of ion channels and synthetic membranes. In 2005, Krishnamurthy (with a background in electrical engineering and applied mathematics) and Cornell started collaborating on modeling the dynamics of this biosensor. The aim was to understand how a biosensor device comprising moving ion channels and complex structural components embedded in the membrane together with electrochemical reactions could be mathematically modeled to explain its remarkable sensitivity. In 2012, Hoiles (with an engineering physics background) joined Krishnamurthy's group as a PhD student and worked on molecular dynamics interpretations of various aspects of engineered membranes and devices. With more than a decade of fruitful collaboration, the three authors now have a useful understanding of artificial membranes and novel devices built out of these membranes – from a physics, biochemistry, engineering and applied mathematics viewpoint. We hope that this book conveys our multifaceted understanding of this fascinating area.

Acknowledgments. The authors are grateful to collaborators and family for their support. Krishnamurthy gratefully acknowledges the supportive academic environment at Cornell Tech and School of Electrical and Computer Engineering at Cornell University.

Abbreviations

AMBER	assisted model building and energy refinement.
BCC	body-centered cubic.
BLM	bilayer lipid membrane.
CGMD	coarse-grained molecular dynamics.
CHARMM	Chemistry at Harvard Molecular Mechanics.
CM	cushioned membrane.
CPE	constant-phase element.
DLP	half-membrane-spanning tethered lipid.
DMPC	1,2-dimyristoyl-sn-glycero-3-phosphocholine.
DMPG	1,2-dimyristoyl-sn-glycero-3-phosphorylglycerol.
DOPC	1,2-dioleoyl-sn-glycero-3-phosphocholine.
DphPC	zwittrionic C20 diphytanyl-ether-glycero-phosphatidylcholine.
DPPC	1,2-dipalmitoyl-sn-glycero-3-phosphocholine.
DSPC	1,2-distearoyl-sn-glycero-3-phosphocholine.
EMP	electroporation measurement platform.
ERP	electrophysiological response platform.
FCC	face-centered cubic.
FEA	finite-element analysis.
FLB	freestanding lipid layer.
FRAP	fluorescence recovery after photobleaching.
gA	Gramicidin.
gA-dig	Gramicidin digoxin.
GDPE	C20 diphytanyl-diglyceride ether.
GPNP	generalized Poisson–Nernst–Planck.
GROMOS	Groningen Molecular Simulation.
HBL	hybrid bilayer lipid membranes.
hCG	human Chorionic Gonadotrophin.
HMM	hidden Markov model.
ICS	ion-channel switch biosensor.
IVDs	in vitro medical diagnostics.
MARTINI	A coarse-grained molecular dynamics force field.
MCT	mode-coupling theory.
MD	molecular dynamics.
MRD	molecular reaction dynamics.

List of Abbreviations

MRSA	methicillin-resistant *Staphylococcus aureus*.
MSD	mean-square displacement.
MSE	mean-squared error.
MSL	membrane-spanning lipid.
MSLOH	synthetic archaebacterial membrane-spanning lipid.
MTOP	monoterpene oxidation product.
NMR	nuclear magnetic resonance.
NPV	negative predictive value.
NR	neutron reflectometry.
ODE	ordinary differential equation.
PBS	phosphate-buffered saline.
PDE	partial differential equation.
PEG	Polyethylene Glycol.
PFMP	pore formation measurement platform.
PFNP	Poisson–Fermi–Nernst–Planck.
PGLa	peptidyl-glycyl-leucine-carboxyamide.
POPC	1-palmitoyl-2-oleoyl-sn-glycero-3-phosphocholine.
POPG	palmitoyloleoyl-phosphatidylglycerol.
PPV	positive predictive value.
SAPC	1-stearoyl-2-arachidonoyl-sn-glycero-3-phosphocholine.
SDPC	1-stearoyl-2-docosahexaenoyl-sn-glyerco-3-phosphocholine.
SLB	supported lipid bilayer.
SLPC	1-stearoyl-2-linoleoyl-sn-glycero-3-phosphocholine.
SOPC	1-stearoyl-2-oleoyl-sn-glycero-3-phosphocholine.
SP	spacer.
tBLM	tethered bilayer lipid membrane.
TRP	transient receptor potential.
TSH	thyroid stimulating hormone.
VDAC	voltage-dependent anion channel.

Part I
Introduction and Background

1 Motivation and Outline

In simple terms a biological membrane consists of two layers of fat that slide on each other – hence membranes are called lipid bilayers. Biological membranes include more than just the cell membrane. Within the cell, there are several structures that are bound by membranes; these structures are called organelles and include the nucleus, mitochondria, endoplasmic reticulum, Gogli apparatus, and lysomes.

1.1 Why Membranes?

Why biological membranes? Membranes and the macromolecules embedded in them perform extremely important biological functions. They form selectively permeable barriers that enclose cells and organelles within the cell. The cell membrane is the target of physical, chemical, and biological agents such as thermal and mechanical stress, toxins, hormones, viruses, and microbes.

Biological membranes exhibit remarkable properties. Embedded in the membrane are protein macromolecules that perform crucial functions for a living cell. For example, ion channels are subnanosized pores formed out of proteins in the membrane that selectively open and close and allow ions to flow into the cell. It is known that almost 25 percent of genes code for membrane proteins. Also, more than 50 percent of available drugs target membrane proteins [369]. Cytoskeletal filaments and sterols, such as cholesterol, give structural stability to the membrane. Antimicrobial drugs bind to specific sites in the membrane of bacterial cells and induce pores in the membrane that compromise the integrity of the membrane and lead to bacterial cell death. Applying a voltage across a membrane causes the membrane to spontaneously form pores – this process is called electroporation and is crucial for drug delivery mechanisms. A membrane has several moving parts that comprise lipids and macromolecules that perform a variety of biological tasks. For example, the assembly of new cellular membranes commonly results from old membranes in which membrane-bound enzymes construct new lipid molecules. These new lipid molecules then either diffuse into the old membrane or form vesicles which can merge with other membranes via vesicle fusion. Vesicle fusion requires the coordination of several proteins and macromolecules as biological membranes do not spontaneously fuse. The process of vesicle fusion and cellular membrane assembly is still an active area of research.

Why artificial membranes? Given the importance and remarkable properties of biological membranes, there is strong motivation to build artificial membranes that mimic them. An artificial membrane allows us to target specific structures in biological membranes and study their properties to elucidate the structural and dynamic properties of cell membranes in a controlled environment. More importantly, one can build synthetic biological devices out of precisely engineered artificial membranes that achieve remarkable properties. For example, the moving parts of the artificial membrane can be precisely engineered to design a supersensitive biosensor that can detect as low as femtomolar concentrations of target molecules. Apart from the challenge of mimicking a biological membrane, another challenge encountered with artificial cell membranes is the bioelectronic interface. Charged ions carry information in biological membranes, whereas electrons carry information in electrical devices. So how should a bioelectronic interface be built to interface ions and electrons?

1.2 Guided Tour of the Book

This book provides a comprehensive treatment of how to engineer artificial membranes: their design, construction, experimental studies, and mathematical models. The book consists of three interrelated parts. The schematic organization and interdependencies of the three parts are illustrated in Figure 1.1.

Part I is an elementary primer on biochemistry for applied mathematics and engineering readers. The background is important for understanding how engineered artificial membranes are constructed and mathematically modeled. The description includes lipids, peptides and proteins, ion channels, chemical components of engineered artificial membranes, and the bioelectronic interface. Additionally, the relation between cell membranes and the types of engineered artificial membranes present in the literature is described.

Part II discusses methods for constructing engineered artificial membranes and devices built out of these engineered artificial membranes. In Chapter 4 the formation process of engineered artificial membranes constructed using the solvent-exchange technique is presented. Additionally, information is provided on how to insert peptide and protein macromolecules of interest, and how to grow cells on the membrane surface. Chapter 5 provides a comprehensive presentation of the ion-channel switch (ICS) biosensor which is composed of an engineered membrane and is designed to detect specific analyte species.[1] A schematic of the ICS biosensor is provided in Figure 1.2. The ICS biosensor is composed of millions of fully functioning nanomachines embedded in an engineered tethered membrane. The nanomachines are designed to bind with specific analyte species such as proteins, hormones, polypeptides, microorganisms, oligonucleotides, DNA segments, polymers, and viruses in cluttered electrolyte

---

[1] Throughout this book, the term *analyte* denotes the target molecules that we wish to detect. The analyte is typically administered in an electrolyte solution; this is called the analyte solution. We are interested in estimating the concentration of analyte in the solution.

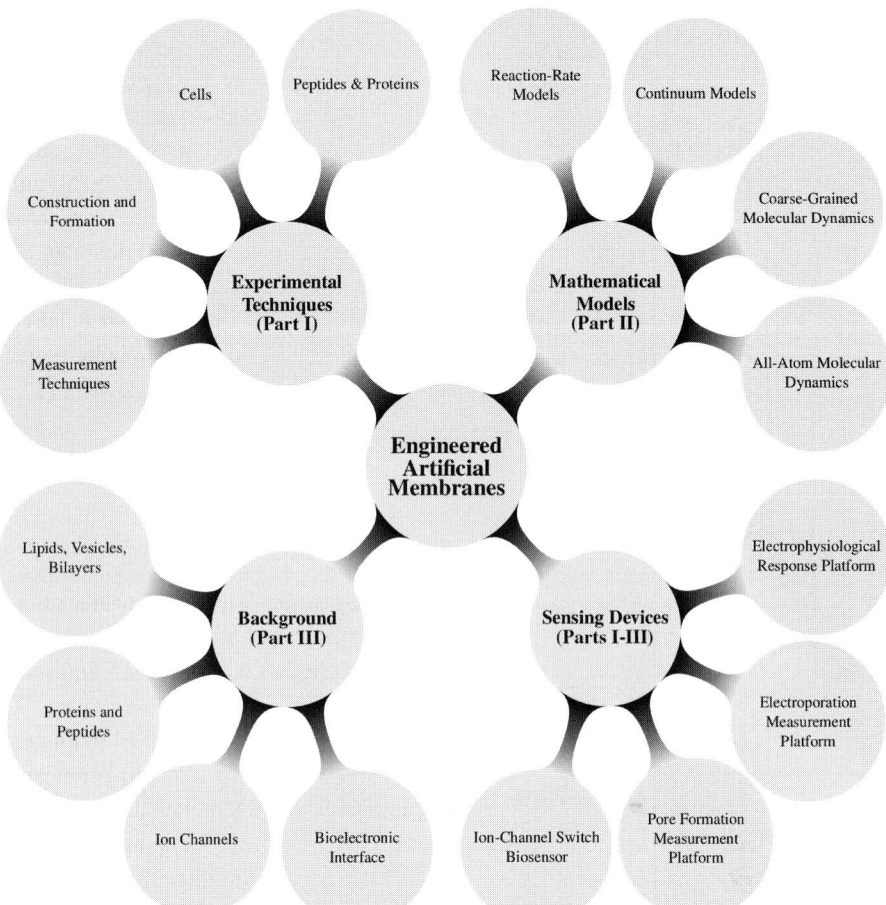

Figure 1.1 Concept chart of the topics presented in this book. Part I presents a biochemistry primer for engineered artificial membranes. Part II focuses on experimental techniques and Part III describes mathematical models of engineered artificial membranes. Parts II and III discuss the four sensing devices, namely, the ion-channel switch biosensor, the pore formation measurement platform, the electroporation measurement platform, and the electrophysiological response platform.

environments. If the analyte is present, these nanomachines will bind with the analyte, causing a decrease in the conductance of the membrane. At a macroscopic level, the conductance change allows the ICS biosensor to detect femtomolar (1×10^{-15} molar) concentrations of target analyte species. Chapter 6 presents the formation and operation of the physiochemical engineered membrane platforms (pore formation measurement platform [PFMP], electroporation measurement platform [EMP], and electrophysiological response platform [ERP]). Chapter 7 presents several experimental techniques for measuring the structural and functional properties of engineered membranes using electrochemical and optical techniques.

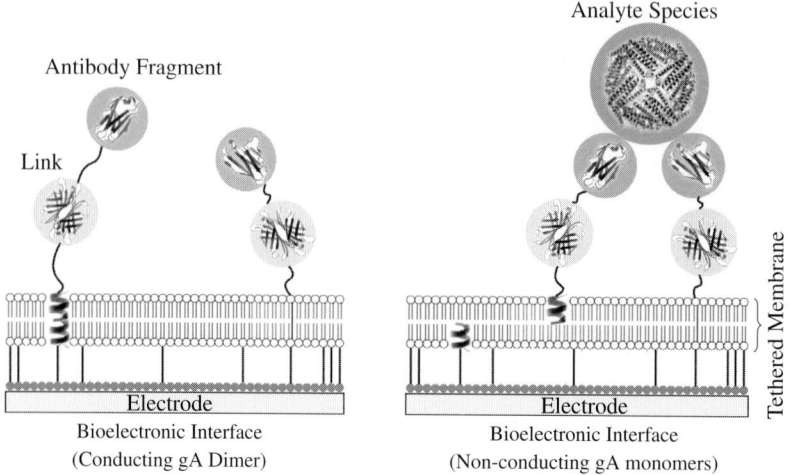

Figure 1.2 Schematic of the ion-channel switch (ICS) biosensor. The ICS biosensor is composed of mobile gramicidin (gA) monomers attached to antibody fragment receptors through a streptavidin-biotin linkage, tethered gA monomers, and tethered-membrane-spanning lipids which are connected to antibody fragment receptors through a streptavidin-biotin linkage. The antibody fragment receptor is designed to bind to a specific analyte species in the analyte solution. When the antibody fragment receptors bind with the analyte species, this causes the population of conducting gA dimers in the ICS biosensor to decrease, which results in an overall decrease in the conductance of the tethered membrane.

Part III deals with mathematical and computer simulation models of engineered membranes. The aim is to use dynamic models to relate the structure of an engineered membrane to its function. The mathematical and simulation models that we develop account for the moving components in membranes, the behavior of macromolecules in the membrane, transport phenomena in the analyte solution, and the bioelectronic interface. We use engineered membranes (hardware) in combination with mathematical models (software) to go from structure to function. The hardware (Part II) allows us to consider various components in their natural state (as they interact with other components), whereas the software (Part III) then zooms into specific subparts in isolation. This combination of hardware (ex vivo and in vitro) and software (in silico) is a unique feature of this book. Conceptually, Part II of the book can be viewed as a biomimetic hardware approximation to real-life membranes. Part III then constructs mathematical models for the devices in Part II and can be viewed as a second-level approximation to real-life membranes.

The models used in Part III include all-atom molecular dynamics, coarse-grained molecular dynamics, continuum theory, and reaction-rate theory. The time and spatial scales associated with each of these models are illustrated in Figure 1.3. These multi-time-scale models allow us to predict the response of engineered artificial membranes from the microscopic (molecular scale) to the macroscopic (experimentally measurable device) scale. Throughout Part III, discussion of the numerical methods used to evaluate all the models are provided to ensure that the reader can repeat all the

1.2 Guided Tour of the Book 7

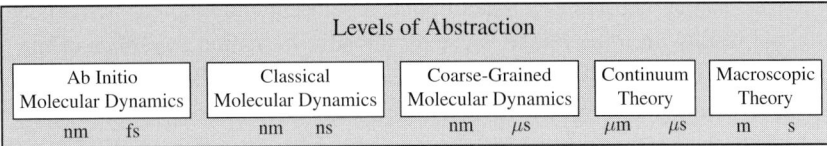

Figure 1.3 Time and spatial scales associated with engineered tethered-membrane models presented in this book.

computations in this book. Additionally, methods to visualize all-atom molecular dynamics and coarse-grained molecular dynamics models of engineered artificial membranes are also provided. Chapter 8 begins with a detailed study of the application of reaction-rate theory to study important membrane properties. This includes electrochemical impedance spectroscopy, the most widely used method for measuring membrane properties, and surface reaction dynamics for the interaction of molecules with the membrane surface. Chapter 9 introduces reaction-rate models for the ICS biosensor, and black-box models of the ICS biosensor. Chapter 10 provides continuum theories useful for modeling the advection-diffusion processes in ICS biosensors and the PFMP. These allow important biological parameters to be estimated such as analyte concentration and diffusion from the measured current response from the engineered membrane platforms. Chapter 11 presents a detailed analysis of the process of electroporation in engineered membranes. Additionally, using extensive experimental measurements, we illustrate how the dynamic models can be used to measure the dynamics of embedded ion channels and cells grown on the surface of the engineered membrane. Chapter 13 illustrates how the ERP with mesoscopic models can be used to estimate the electrophysiological response of ion channels and cells grown on the surface of engineered tethered membranes. Chapter 14 introduces how coarse-grained molecular dynamics can be used to compute important biological parameters of engineered membranes, and how these can be related to experimental measurements. Chapter 14 also presents a case study of how coarse-grained molecular dynamics together with a continuum model can be used to understand the effect of antimicrobial drugs on the membrane. At the most fundamental level is all-atom molecular dynamics. Chapter 15 introduces how molecular dynamics can be used to gain important insight into the statics and dynamics of engineered membranes. Molecular dynamics simulations are used to estimate essential parameters of artificial membranes, understand aqueous pore formation in membranes, and model switching mechanisms of gramicidin channels (which is a crucial part of the ICS biosensor). An interesting aspect of Chapters 14 and 15 is that they use molecular dynamics simulation to understand the structure-function relationship of synthetic biological devices; traditionally molecular dynamics has been used for studying biological molecules. Chapter 16 closes with a complete overview of the atom-to-device model of engineered tethered membranes constructed in Part III of this book.

Case studies. This book contains several important experimental and modeling case studies that provide a wealth of insight on how artificial cell membranes operate. For example, using measurements from the ICS and dynamic models, we estimate the concentration of the molecule species streptavidin thyroid stimulating hormone, ferritin, and human chorionic gonadotrophin in solution and whole blood. Using experimental results from the PFMP we study the pore-formation dynamics of the protein toxin α-hemolysin [397] from *Staphylococcus aureus*, and the pore-formation dynamics of the antimicrobial peptide peptidyl-glycylleucine-carboxyamide (PGLa). Using results from the EMP important insights are gained into the effect the bioelectronic interface, archaebacterial, *Escherichia coli*, and *Saccharomyces cerevisiae* lipids, electrolyte concentration, and cholesterol content have on the process of electroporation. Using the dynamics and experimental measurements from the ERP we study the electrophysiological response of the voltage-gated sodium ion channel, and skeletal myoblast cells.

Finally, this book has three appendices. Appendix A is a short elementary review of partial differential equations (including construction of nondimensionalized models) that are used to model membranes. Appendix B provides details for performing coarse-grained molecular dynamics and all-atom molecular dynamics simulations of engineered artificial membranes. Appendix C provides numerical methods for solving the continuum models, including the physical constants, of engineered artificial membranes presented in the book.

Color Figures. High-resolution colored figures for the book can be downloaded from the website of the book at www.cambridge.org/engineered-artificial-membranes

2 Biochemistry for Engineers: A Short Primer[1]

Biochemistry deals with chemical processes occurring in living organisms. An elementary understanding of the biochemistry of cell membranes is important for understanding the key ideas behind engineered artificial membranes and biosensing. This chapter gives a brief description of the key chemical components in cell membranes, namely, lipids, proteins, and peptides. Lipids are the primary component of all cell membranes; and peptides and proteins are the primary component of signaling in living organisms. Additionally, we discuss tethers and spacers, which provide the basis for connecting engineered membranes to a bioelectronic interface; this bioelectronic interface is connected to electrical measurement equipment that allow signals to be acquired from the device.

With the availability of powerful computer-based visualization tools (discussed at the end of this chapter), it is strongly recommended that the reader use these tools to get a better feel for the various types of biochemical molecules described in this chapter. These software tools are also a first step in more sophisticated molecular dynamics simulations that are discussed in Part III of the book.

2.1 Bonded and Nonbonded Molecular Interactions

A chemical bond is an attraction between two or more atoms that facilitates the formation of multiatom chemical compounds. There are two main types of chemical bonds: ionic bonds and covalent bonds. In addition there are also van der Waals forces, a type of nonbonded interaction between atoms, which include attraction or repulsion between atoms that do not arise from covalent or ionic bonds. Specifically, van der Waals forces are a consequence of quantum dynamics in which repulsive or attractive forces between atoms result from correlations in the fluctuating polarizations of atoms. In this section we define the covalent and ionic bonds as well as the van der Waals interaction between atoms and their importance in the dynamics of cell membranes. Additionally, we describe their importance when constructing molecular dynamics simulations.

Covalent bond. A covalent bond is a chemical bond formed by sharing electrons between atoms. Covalent bonds are key to the structure of nearly all chemical

---

[1] This chapter can be omitted by readers who are familiar with chemical bonds, lipid bilayers, peptides and proteins, and cell membranes.

Covalent Bond: Electron shared between atoms

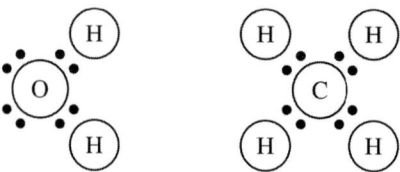

Ionic Bond: Electron not shared between atoms

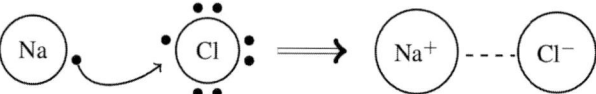

Hydrogen Bond: Dipole-dipole interactions

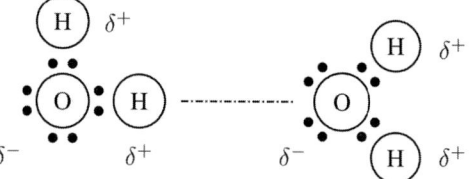

Figure 2.1 Examples of the covalent bond, ionic bond, and hydrogen bond. Na, H, C, and O represent sodium, hydrogen, carbon, and oxygen, respectively, with the black dots indicating electrons. In the ionic bond, the electron from the sodium ion is transferred to the chloride ion to create NaCl. In water (H_2O), the electrons of the hydrogen atom are pulled toward the oxygen atom, creating a slightly positively charged region (δ^+). The other side of the oxygen atom also pulls electrons, creating a slightly negatively charged region δ^-. Since the charge distribution in water is not equal it will have a dipole moment. The interaction of these positively and negatively charged regions between water is an example of hydrogen bonding.

compounds such as water, lipids, peptides, and proteins. The covalent bond is one of the strongest bonds compared with ionic bonds and van der Waals interactions. Typically the covalent bond has a bond energy in the range of 120 to 1000 kJ/mol (120–500 kJ/mol for single bonds, and up to 1000 kJ/mol for triple bonds). In Figure 2.1 two examples of covalent bonds are provided. The first is between the hydrogen atoms and the oxygen atom in water, and the second is between the hydrogen atoms and carbon atom. A particularly important covalent bond in artificial membranes is the coordinate covalent bond, which occurs between sulfur and gold. A coordinate covalent bond is a two-electron covalent bond in which the two electrons derive from the same atom. Another important covalent bond is the disulfide bond (or S-S bond), which is typically found in protein molecules and is used to regulate the chemical structure of the protein.

In molecular dynamics simulations (discussed in Part III of the book), these types of strong-bonded interactions are modeled as a spring where the spring constant is proportional to the bond energy.

Ionic bonds. An ionic bond is formed when an electron is transferred from one atom to another, causing the formation of positive and negative ions.[2] The typical bond energy of an ionic bond is in the range of 20 to 50 kJ/mol. Figure 2.1 provides an example of an ionic bond for sodium chloride. In sodium chloride (NaCl), the electron from the sodium atom (Na) is completely transferred to the chlorine atom (Cl). This transfer of the electron causes the sodium to become positively charged (i.e., to become Na^+) and the chloride to become negatively charged (i.e., to become Cl^-). The sodium ion Na^+ is known as a cation, which has a positive charge, and the chloride ion Cl^- is known as an anion, which has a negative charge. The ionic bond involves the attraction of the positively charged cation with the negatively charged anion. Note that, in aqueous solutions, once the ions have formed then they will dissociate because the ionic bond is weak compared with the formation of nonbonded interactions of the ions with water. That is, in aqueous solutions, simple ions of biological significance, such as Na^+, K^+, Ca^{2+}, Mg^{2+}, and Cl^-, do not exist as free, isolated entities. Instead, each is surrounded by a stable, tightly held shell of water molecules.

In molecular dynamics simulations, models of ions typically include a hydration shell (a shell of water molecules). This is a reasonable approximation for ion interactions in electrolyte[3] solutions as any interaction between ions typically involves water molecules. Note that this is a nonbonded interaction that involves van der Waals forces and electrostatic interactions. For ion interactions that are devoid of water, specialized models are necessary to account for specific ion-to-atom interactions.

Van der Waals forces. The attractive or repulsive forces between atoms or molecules other than those that result from covalent or ionic bonds are known as van der Waals forces. These include electrostatic, dipole-dipole, dipole-induced dipole, and London (instantaneous dipole-induced dipole) forces. Dipole-dipole interactions result from attractive forces between the positive end of one polar molecule (difference in charge distribution) and the negative end of another polar molecule. Dipole-dipole forces have strengths that range from 8 to 40 kJ/mol.

An important type of van der Waals force is the *hydrogen bond*, illustrated in Figure 2.1. The hydrogen bond is crucial to all life on earth; it plays an important role in the DNA double helix, the structure of proteins, and the properties of water. A hydrogen bond is the electromagnetic attraction between polar molecules in which hydrogen is bonded to a larger atom, such as oxygen or nitrogen. The attraction between hydrogen and the other polar molecule is a result of dipole-dipole interactions of the hydrogen atom and the other polar molecule. The most common example of a hydrogen bond is between water molecules, as illustrated in Figure 2.1. Hydrogen bonds play an important role in the process of electroporation in cell membranes, and they are also partly responsible for the secondary, tertiary, and quaternary structures of peptides and proteins.

In molecular dynamics simulations, nonbonded interactions are modeled using the Lennard-Jones potential, and Coulomb's law. The parameters of the Lennard-Jones potential are selected to match the results of ab initio molecular dynamics simulations.

---

[2] An ion is an atom that has a total electric charge which results from the loss or gain of an electron.
[3] An electrolyte comprises positive ions (cations) and negative ions (anions) in liquid water (H_2O).

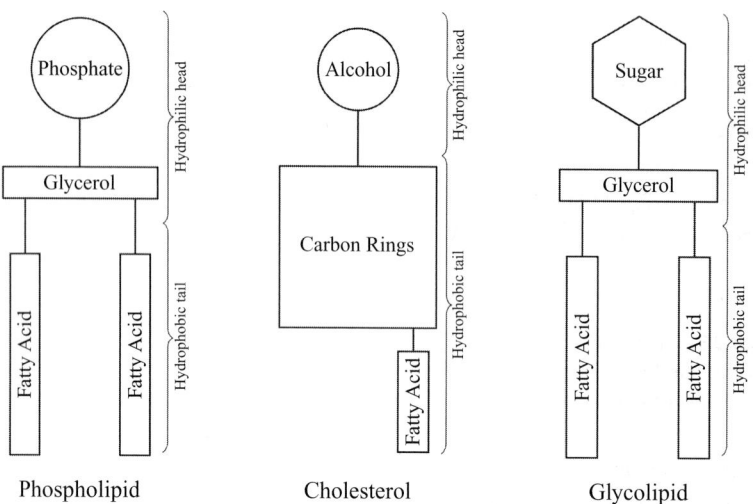

Figure 2.2 Schematic of the chemical components of phospholipids, cholesterol, and glycolipids.

2.2 Lipids, Vesicles, and Bilayers

In simplistic terms, the cell membrane comprises two layers of fat that slide on each other, hence the name *lipid bilayer*. The fundamental building block of cell membranes are lipids.[4] A lipid is a special type of biomolecule that interacts weakly with water (hydrophobic) or that contains sections that both weakly interact with water and strongly interact with water (amphiphilic). Hydrophobic molecules tend to be nonpolar and interact strongly with other neutral and nonpolar molecules. Hydrophilic molecules, also known as water-loving molecules, tend to be polar and strongly interact with water, which is also a polar molecule. Amphiphilic molecules contain both hydrophilic and hydrophobic sections. A common misconception is that hydrophobic molecules are water repulsive. This is not the case, however; it is merely that hydrophobic molecules tend to aggregate with other hydrophobic molecules when contained in a polar solvent such as water.

Cell membranes are primarily composed of the following three types of lipids: phospholipids, cholesterol, and glycolipids. A schematic of these lipids is illustrated in Figure 2.2. Each of these lipids plays an important role in cell membranes.

Phospholipids. All cell membranes contain phospholipids. As illustrated in Figure 2.2, the phospholipid contains both a hydrophilic[5] head group, typically composed of phosphate and glycerol molecules, and hydrophobic tails which are predominantly composed of fatty acid chains. When phospholipids are placed in a solution of water their fatty acid tails aggregate to minimize the interaction of these molecules with water. The aggregation of phospholipids in water can form different structures, including *micelles*, *vesicles* or *liposomes*, and *lipid bilayers*, which are illustrated in Figure 2.3. Micelles are used to transport insoluble chemicals between cells, whereas

---

[4] Lipids include fats, waxes, and fat soluble vitamins (A, D, E, and K).
[5] A molecule is hydrophilic if it is attracted to water; a molecule is hydrophobic if it is not attracted to water.

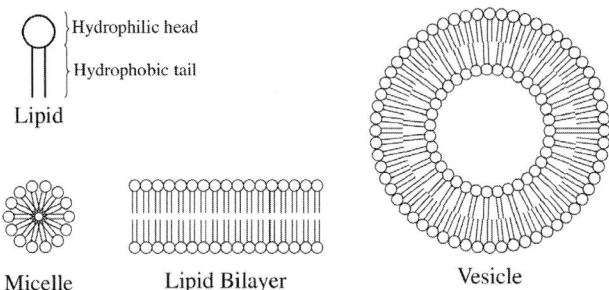

Figure 2.3 Schematic of a lipid, and possible lipid aggregate structures (micelle, lipid bilayer, and vesicle) that form when lipids are placed in water. Micelles and vesicles are used as the building blocks for forming engineered artificial membranes which are composed of a lipid bilayer.

vesicles and liposomes are used to transport large proteins between cells. Vesicles are commonly used for bilayer mechanistic studies and high-throughput screening for bilayer-perturbing drugs [177]. The lipid bilayer is the fundamental structure of both cell membranes and engineered artificial membranes. Note that, although there are numerous types of phospholipids, each contains the basic building block of a phosphate and glycerol head group, and a tail group composed of fatty acids.

Cholesterol. Cholesterol is a type of lipid. It is a steroid (the molecule is not classified as a fat or an oil) that is found in eukaryotic[6] (e.g., animal) cell membranes and is illustrated in Figure 2.2. Cholesterol is composed of a hydrophilic hydroxyl (or alcohol) head group, and a hydrophobic tail composed of carbon rings and a fatty acid. As such, the hydrophilic hydroxyl head group of cholesterol is attracted to the hydrophilic head group phosphate and glycerol of phospholipids. Similarly, the hydrophobic carbon rings and the fatty acid of cholesterol are attracted to the hydrophilic fatty acids of phospholipids. Therefore, cholesterol typically resides in the interior of cell membranes. (Recall the inside of a cell membrane is hydrophilic.)

Cholesterol controls the shape and mobility of lipids in cell membranes. Increasing the concentration of cholesterol in the cell membrane causes changes in membrane thickness, lipid diffusion, and the permeability of the cell membrane. A cell membrane also has large patches of lipids with high concentrations of cholesterol called *lipid rafts*. That is, at high cholesterol concentrations, phase separation[7] occurs, where cholesterol promotes the formation of lipid rafts in cell membranes. Lipid rafts are important for communication between cells. For example, in eukaryotes cholesterol aids in the formation of lipid rafts with a specific thickness and lipid mobility to promote the operation of specialized signaling proteins and peptides [175].

Glycolipids. Glycolipids are present in eukaryotic cell membranes. They are lipids that contain a carbohydrate (also known as sugar) receptor connected to the head group of a lipid. Note that the major constituents of the head group are the sugars. A schematic

---

[6] Eukaryotes contain cells with a nucleus and organelles, and possess a complex cell-division process. Eukaryotes include all multicellular organisms. Single-celled organisms such as bacteria are prokaryotes.

[7] Phase separation is the separation of fluid phases that contain different concentrations of common components. In our context phase separation refers to the separation of lipids and cholesterol domains in the membrane.

of a glycolipid is provided in Figure 2.2. The primary function of glycolipids is to use the carbohydrate receptor to bond specific chemicals to the cellular membrane surface, or to promote the interaction between cell membranes. This type of lipid is used in engineered membranes that are designed to interact with cells grown on the surface of the engineered membrane.

Experimental measurements of the atomic structure of lipids can be performed using nuclear magnetic resonance (NMR) techniques in combination with molecular dynamics simulations for equilibration. A large repository of lipid bilayer structures that have been estimated using these techniques can be found in the Lipidbook repository.[8] This is an excellent resource for visualizing the atomic structure of lipids and lipid bilayer membranes.

2.3 Lipid Bilayers

Cell membranes of living organisms and some viruses consist of a lipid bilayer.[9] The lipid bilayer plays a crucial role in cells as it provides a barrier that prevents the passage of chemicals (ions, proteins, and peptides) from the interior of the cell to the exterior. This is important for cellular communication, which involves controlling the release of ions, signaling proteins, and signaling peptides from the interior of the cell to the exterior. Below we discuss the DphPC bilayer, as well as the energetics and structure of lipid bilayers. Engineered tethered membranes can be constructed with several types of lipid bilayers, discussed in Chapter 3; however, the DphPC bilayer plays a crucial role in the construction of biosensors and biomimetic devices because of its stability.

2.3.1 Archaebacteria and DphPC Bilayers

In this book, we focus a lot on DphPC membranes. DphPC bilayer membranes are constructed from the lipids extracted from archaebacteria.[10] Archaebacteria have existed on earth for more than 3 billion years and live in a wide variety of environments including volcanic vents, the bottom of the sea, and even space. Specifically, archaebacteria can survive in very acidic or alkaline conditions,[11] high-temperature environments of up to 120°C, and in space where the organisms are exposed to high ultraviolet and gamma radiation levels in a vacuum. The remarkable resistance of archaebacteria to these harsh environments results from their membranes being composed of DphPC lipids – also referred to as archaebacteria lipids. Figure 2.4 illustrates one example of a lipid bilayer composed of an ester-DphPC (chemical name: 1,2-diphytanoyl-sn-glycero-3-phosphocholine) lipid from archaebacteria.

---

[8] https://lipidbook.bioch.ox.ac.uk/
[9] Biological membranes include more than just the cell membrane. Within the cell, there are several structures that are bound by membranes; these include the nucleus, mitochondria, endoplasmic reticulum, Gogli apparatus, and lysomes.
[10] Archaebacteria are single-celled organisms that have no cell nucleus or any other membrane-bound organelles in their cells.
[11] pH is a logarithmic scale used to specify the acidity or basicity (alkalinity) of an electrolyte solution. Formally $pH = \log_{10}(M_H)$ where M_H is the molar concentration of hydrogen ions. The pH scale ranges from 0 for the most acidic to 14 for the most basic. Pure water is neutral and has a $pH = 7$.

2.3 Lipid Bilayers

DphPC Lipid
(Atomic Structure)

DphPC Bilayer Lipid
Membrane

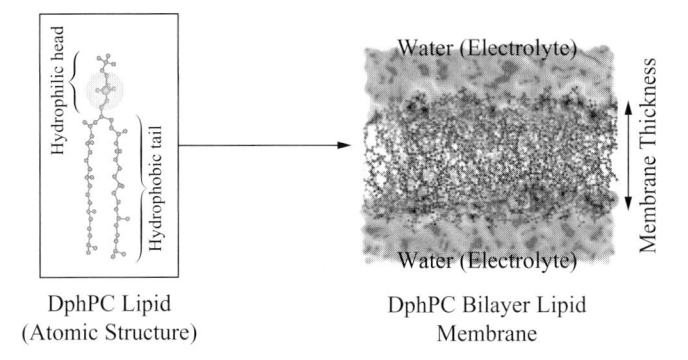

Figure 2.4 Schematic of an ester-DphPC lipid bilayer in an electrolyte solution. The phosphate group of the ester-DphPC lipid is illustrated by the large gray sphere on the atomic structure. The atomic structure of the ester-DphPC lipid bilayer was estimated using all-atom molecular dynamics. As seen, the hydrophilic head group of the ester-DphPC lipid is adjacent to the water, and the hydrophobic tail groups of the ester-DphPC lipids contain no water. This is the equilibrium structure of the lipid bilayer.

It is important to note that the lipids in archaea are chemically different from other organisms. The tails of archaea lipids contain isoprenoid side chains; that is, the lipids do not contain straight fatty acid chains found in other organisms. This difference in molecular structure of the lipid tails allows the archaea cell membrane, and therefore DphPC bilayers, to remain intact in harsh environments. Given the extremely stable bilayers that can be formed from archaebacteria lipids, these lipids are used extensively in the design of engineered artificial membrane sensing devices.

2.3.2 Energetics of Lipid Bilayers

The energetics and physical properties of lipid bilayers are of importance in Parts II and III of the book, where devices are constructed and modeled using artificial lipid bilayers.

The equilibrium structure of the lipid bilayer is a result of the *hydrophobic effect* of lipids. The hydrophobic effect is the observed tendency of hydrophobic molecules to aggregate in aqueous solution to maximize the interaction of hydrophilic molecules with water. An everyday example of the hydrophobic effect is the interaction of oil (which is hydrophobic) with water: the oil will aggregate to minimize its interaction with water. In water, hydrogen-bonding[12] networks exist between water molecules. However, the introduction of hydrophobic molecules disrupts this hydrogen-bonding network. Hydrophobic molecules aggregate to minimize this disruption to the hydrogen-bonding network of water. This minimizes the free energy of the system containing water and lipids. Therefore, the aggregation of lipids to form a lipid bilayer in water is an entropy-driven

---

[12] A hydrogen bond is the electrostatic attractive force between the hydrogen attached to an electronegative atom of one molecule and an electronegative atom of a different molecule. An example is between the hydrogen atom of one water molecule and the oxygen atom of another water molecule.

reaction that minimizes the free energy of the water-lipid system. This bilayer lipid structure is illustrated in Figure 2.4.

2.3.3 Structure of Lipid Bilayers

Two important structural questions regarding the lipid bilayer are as follows:

> How do proteins and peptides penetrate and move through the membrane?
> How resistant is the membrane to deformations?

A typical lipid bilayer is between 3 and 5 nm thick with each lipid molecule having an average surface area of 0.6 to 0.9 nm^2. Therefore, a cell membrane is composed of billions of such lipid components. The lipid bilayer exhibits mechanical properties of both solids and liquids. Important properties of lipid bilayers include lipid mobility; the thickness of the membrane; its ability to compress, expand, or bend; its propensity to curve; and the dielectric constant of the membrane. These play a crucial role in the response of the lipid bilayer to external excitations which may result from a difference in the population of charged ions on either side of the membrane, temperature variations which introduce the thermal fluctuations of the lipids, or the reaction dynamics of proteins and peptides that interact with the membrane. The membrane exhibits both viscous and elastic characteristics when undergoing deformations that can result from ion gradients and thermal fluctuations. Two important properties of the lipid bilayer are how it responds to external stress and how it responds to externally applied electric fields. The *viscoelastic response* of the cell membrane measures the ability of the membrane to resist shear strain when a shear stress is applied, and the ability of the membrane to return to its original shape once the stress is removed. An *electrophysiological response* is any response of the membrane that is caused by the applied electric field, such as mechanical deformations, charge accumulation at the surface, and the formation of aqueous pores in the membrane.

2.4 Peptides and Proteins

Proteins and peptides are polymers of amino acids; that is, they are composed of long chains of amino acids. Proteins and peptides can be found in cell membranes and perform a variety of tasks, including transporting chemical species across the membrane (i.e., acting as ion channels), acting as receptors or binding cites for chemical species on the surface of the membrane, and controlling the properties of the lipid membrane.

A simple example of the formation of a peptide from amino acids is the creation of a dipeptide – a peptide composed of two amino acids as illustrated in Figure 2.5. The amine group covalently bonds with the carboxyl group of the other amino acid to form the dipeptide. This process can repeat several times with the different amino acids in Table 2.1 to form a specific peptide. The main difference between proteins and peptides is that peptides are composed of a continuous chain of 50 or fewer amino acids. A continuous chain of amino acids is known as a polypeptide. Proteins contain

2.4 Peptides and Proteins 17

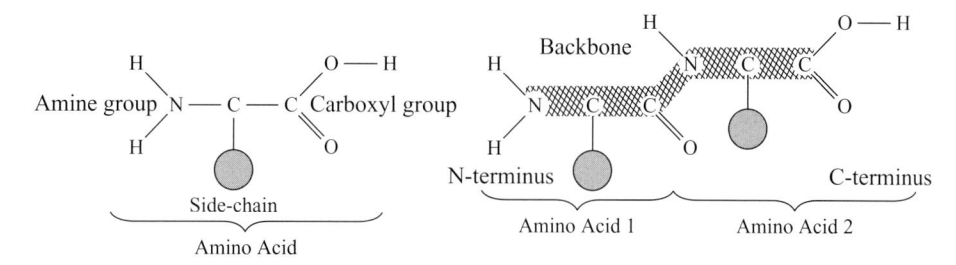

Figure 2.5 Chemical structure of an amino acid and a peptide composed of two amino acids. H is hydrogen, N is nitrogen, C is carbon, and O is oxygen. The side-chain molecule governs if the amino acid is a weak acid, a weak base, hydrophilic, or hydrophobic. The peptide is formed by the covalent bond between the carbon of the carboxyl group of one amino acid with the nitrogen in the amine group of the other amino acid. The backbone atoms of the dipeptide are indicated with the crosshatch pattern. The orientation of the backbone atoms defines the secondary structure of the peptide. The N-terminus (terminus amine group) is the start of the peptide and the C-terminus (terminus carboxyl group) is the end of the peptide. Note that if the location of the side chain is on the left-hand side of the α-carbon (carbon connected to the carboxyl group and side chain) of the amino acids that comprise a peptide or protein, the peptide or protein is referred to as L (levorotatory, or left handed), otherwise it is D (dextrorotatory, right handed). Peptides and proteins primarily consist of left-handed amino acids.

more than 50 amino acids and can be composed of several polypeptides. Given the size difference between peptides and proteins, each plays a different role in the cell membrane as discussed below.

Peptides. Peptides are composed of a continuous chain of amino acids in which the carboxyl end of each amino acid is linked to the amine[13] end of another amino acid. This type of covalent link between the nitrogen and carbon between amino acids is known as the backbone of the peptide and is illustrated in Figure 2.5. Amino acids that reside in a peptide are commonly referred to as residues in the literature. The side-chain molecules of the amino acid can make an amino acid a weak acid or a weak base, and a hydrophilic molecule if the side chain is polar[14] or hydrophobic if it is nonpolar. There are 20 different proteinogenic amino acids that can be used to create peptides. These 20 proteinogenic amino acids are defined in Table 2.1 with their chemical name, three-letter code, and one-letter code. Each of these identifiers is used throughout the literature on engineered artificial membranes. Each peptide is composed of an N-terminus and a C-terminus, as illustrated in Figure 2.5. The N-terminus (terminus amine group) is associated with the start of the peptide and the C-terminus (terminus carboxyl group) is associated with the end of the peptide. There are thousands of different types of peptides that operate in cell membranes with each associated with a specific task that is dependent on the sequence of amino acids contained in the peptide. Note that the

---

[13] In organic chemistry, amines are compounds and functional groups that contain a basic nitrogen atom with a lone pair.

[14] A polar molecule has a net dipole as a result of the opposing charges; a nonpolar molecule has a negligible net dipole.

Table 2.1 Proteinogenic[a] amino acids.

Amino Acid	3-Letter	1-Letter
Alanine	Ala	A
Arginine	Arg	R
Asparagine	Asn	N
Aspartic acid	Asp	D
Cysteine	Cys	C
Glutamic acid	Glu	E
Glutamine	Gln	Q
Glycine	Gly	G
Histidine	His	H
Isoleucine	Ile	I
Leucine	Leu	L
Lysine	Lys	K
Methionine	Met	M
Phenylalanine	Phe	F
Proline	Pro	P
Serine	Ser	S
Threonine	Thr	T
Tryptophan	Trp	W
Tyrosine	Tyr	Y
Valine	Val	V

[a] A proteinogenic amino acid is a molecule that can covalently bond with the backbone of other amino acids in a peptide or protein. Although there are numerous amino acids, only proteinogenic amino acids can be used to construct peptides and proteins.

molecular structure of most peptides is time dependent: the distance between the amino acids (or residues) is not a constant over time in peptides. These changes in structure can result from the binding of chemical molecules to the peptide, from thermal vibrations, and from collisions with other molecules.

Proteins. A protein is a large macromolecule that consists of one or more long chains of amino acids. Since a peptide is composed of a single continuous chain of amino acids, a protein can be viewed as a polypeptide or combination of several peptides. The number of amino acids (or residues) in a protein ranges from 50 for small proteins to approximately 33,000 amino acids for the largest known protein. Recall that the interaction of an amino acid with other molecules is governed by the amino acid's side chain; this side chain can be weakly acid, weakly basic, polar, or nonpolar. The structural and chemical reaction dynamics of the protein is therefore governed by the combined effect of all the amino acid side chains in the protein.

The atomic structure (spatial coordinates of atoms in three-dimensional space) of a protein molecule is determined by its amino acids. The equilibrium structure of a protein is known as its *native conformation*. However, as proteins interact with other molecules, they may change shape – this change in shape is known as a conformation change.[15] These changes can result from the binding of chemical molecules to the protein, and from thermal vibrations and collisions with other molecules.

In this book, we focus on proteins in the membrane. Such membrane proteins play an important role in molecular recognition such as in the mammalian immune system. They perform a variety of tasks beyond cell surface recognition: transporting molecules and ions across the membrane, promoting the adhesion of specific cells and interaction between cells, and accelerating, or catalyzing chemical reactions between molecules in the membrane. Membrane proteins can be classified into three groups: integral polytopic proteins, integral monotopic proteins, and peripheral membrane proteins. Integral polytopic proteins are permanently attached to the membrane and are proteins that cross the membrane at least once – these are also known as transmembrane proteins. Integral monotopic proteins are permanently attached to the membrane and are proteins that are attached to only one side of the membrane. Peripheral membrane proteins are proteins that are temporarily attached to either the membrane or to integral proteins by a combination of hydrophobic, electrostatic, and other noncovalent chemical interactions. All these membrane proteins play an important role in the chemical reactions in engineered artificial membrane biosensors.

Structure of peptides and proteins. The molecular structure (coordinates of the atoms) of peptides and proteins is defined by the primary structure, secondary structure, tertiary structure, and quaternary structure of the amino acids. The *primary structure* of the peptides and proteins is defined by the sequence of covalently linked amino acids in the peptide or protein. The *secondary structure* is related to the backbone structure of the peptide and protein and is determined by hydrogen bonding (see Footnote 7). Additionally, disulfide bonds, a type of covalent bond which forms between different amino acids of a protein chain (for example between the thiol groups of cysteine), also contributes to the secondary structure of proteins. The backbone is the structure that is produced by the covalent bond between the nitrogen and carbon molecules in the amino acids. Any molecules that are not associated with the backbone are called side chains. There are three common shapes for the backbone structure: α-helix, β-sheet, and random (also known as coil structure). In the α-helix backbone structure the amino acids are arranged in a right-handed helical structure where each amino acid residue corresponds to approximately a $100°$ turn in the helix. In the β-sheet backbone structure the amino acid chains are arranged adjacent to other chains and form an extensive hydrogen bond network with their neighbors. Specifically the N–H groups in the backbone of one amino acid chain establish hydrogen bonds with the C=O groups in the backbone of the adjacent amino acid chain. A peptide typically only contains a single backbone structure. However, proteins tend to contain a combination of α-helix, β-sheet, and coiled backbone structures. The backbone structure of peptides and proteins is dependent on the local environment. For example, the common secondary structure for a

---

[15] Proteins are flexible molecules which have several different shapes (or conformations). The shape of the protein is a result of the amino acids contained in the protein and their interaction with other proteins, lipids, and environmental factors such as temperature and ionic concentration. Typically within the membrane it is assumed that proteins do not change conformation.

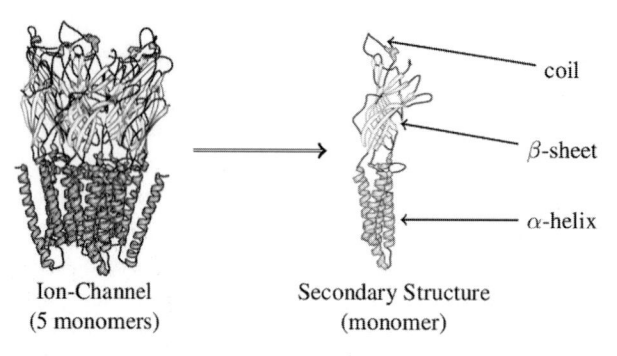

Figure 2.6 Schematic of the secondary structure of the ligand-gated ion channel from *Escherichia coli*.

membrane-bound peptide is an α-helix, whereas the secondary structure of a peptide in water is a coil. Figure 2.6 illustrates the three types of secondary structures (coil, β-sheet, and α-helix) for a ligand-gated ion channel found in *Escherichia coli*. The *tertiary structure* is associated with the complete three-dimensional shape of the peptide or protein – that is, the coordinates in three-dimensional space of each atom in the peptide and protein. The protein molecule will bend and twist in such a way as to achieve the lowest energy state. The *quaternary structure* involves the clustering of several individual peptides or repeated amino acid groups into a specific shape. Each amino acid group is defined as monomer, and a collection of these monomers defines the protein structure. An example of this is the peptide gramicidin A (gA), an important molecule used in biosensors, which is composed of two gramicidin monomers as illustrated in Figure 2.7.

Chirality peptides and proteins. Chirality is a geometric asymmetry property of amino acids and molecules. A molecule is chiral if the mirror image of the molecule results in a pair of nonsuperimposable shapes. The chirality of an amino acid is either L (levorotatory, or left handed) or D (dextrorotatory, or right handed). If the location of the side chain is on the left-hand side of the α-carbon (carbon connected to the carboxyl group and side chain) of the amino acid, then the amino acid has left-handed chirality;

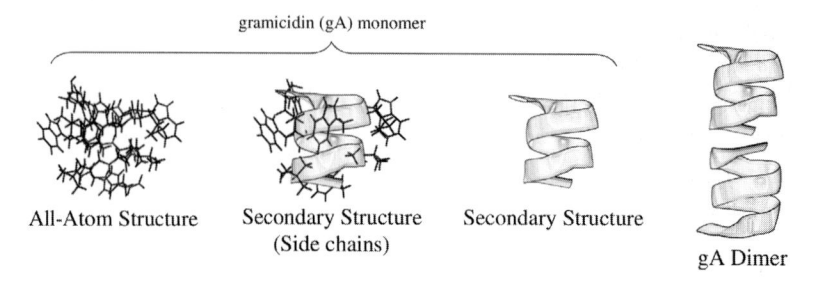

Figure 2.7 Schematic of the secondary structure of the gramicidin A peptide. The three representations of the gA monomer (all-atom structure, secondary structure with side chains, and secondary structure with no side chains) and the gA dimer secondary structure. The gA dimer is a conducting ion channel that allows the permeation of ions through the biological membrane.

otherwise the amino acid has right-handed chirality. Proteins and peptides are composed of amino-acid sequences with each amino acid containing an identical chirality. That is, the peptide or protein is a homochiral molecule where the chirality is derived from the chirality of each amino acid. In mammalian proteins only L-chirality proteins are found; in bacteria such as gramicidin (gA) both L- and D-chirality proteins are found, which permits a very efficient structure of low molecular weight with an internal hydrogen-bonding pattern that forms a channel. A channel with a similar conductance formed exclusively from L amino acids possesses a molecular weight of 150 kDa, such as the KcsA potassium channel. Both of these channels are discussed in more detail below.

Experimental measurements of the structure of proteins and peptides can be performed using NMR or X-ray crystallographic techniques. A large repository of protein structures that have been determined using these techniques can be found at the RCSB Protein Data Bank.[16] This is an excellent resource for visualizing the structure of proteins. However, note that these chemical structures are not static. Peptides and proteins are dynamic and can have different shapes depending on the environmental conditions they are in.

2.5 Ion Channels

Ion channels are water-filled aqueous pores composed of protein molecules that span the cell membrane and facilitate the diffusion of electrolyte ions[17] across the membrane. Ion channels are responsible for most electrical activities in the nervous system, including communication between cells and the influence of hormones and drugs on cell function.

Why are ion channels necessary in cell membranes? The interiors of cell membranes are hydrophobic and also have a low dielectric permittivity, which prevents ions from crossing the cell membrane. The cell membrane can be viewed as an electrical insulator that prevents the flow of ions. Ion channels provide a highly conductive pathway that allows ions to traverse the cell membrane. The conductance of a single ion channel ranges from 1 to 150 pS. Since ion channels are embedded in cellular membranes they contain both hydrophobic regions (water resistant) and hydrophilic regions (water seeking). Ion channels are composed of three basic parts: a central conduction pathway (opening) for ions to pass through, an ion recognition site to allow passage of specific ions (selectivity filter), and one or more gates that may open or close. Additionally, some ion channels have sensor molecules (attractor sites) that can control the operation (opening or closing) of the gates.[18] These receptor sites can act as sensors for a large variety of stimuli, including changes in cell-membrane potential, chemicals, temperature, and

---

[16] www.rcsb.org/
[17] Common electrolyte ions are sodium (Na^+), potassium (K^+), calcium (Ca^{2+}), magnesium (Mg^{2+}), chloride (Cl^-), hydrogen phosphate (HPO_4^{2-}), and bicarbonate (HCO_3^-).
[18] Gating in an ion channel is actually a conformational change in the protein structure. Such dynamics have been studied extensively in biophysics.

Nonconducting gA monomers ⟷ Conducting gA dimer ⟷ Nonconducting gA monomers

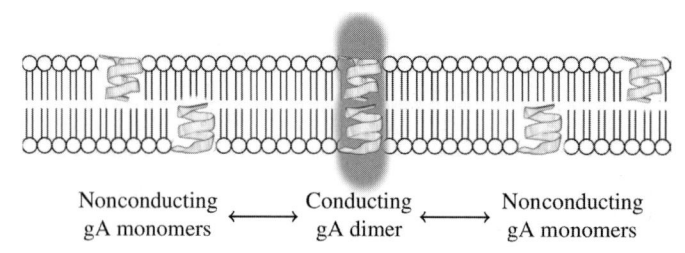

Figure 2.8 Schematic of the nonconducting gA monomers and conducting gA dimer. The gA monomers can react through a reversible dimerization step to form a conducting gA dimer ion channel. gA is an example of a peptide ion channel as the ion channel only exists when the two peptides bond to form the conducting dimer. The full atomic structure of gA is provided in Figure 2.7.

the mechanical stretch of the cell membrane. There are six main classes of ion channels: peptide, voltage gated, ligand gated, light gated, mechanosensitive, and transient receptor potential (TRP). Each of these classes of ion channels is discussed below.

Gramicidin channels. Peptide ion channels do not exhibit gating (opening and closing). Instead the entire channel forms and dissociates over time; this formation and disassociation of the channel occurs multiple times. When the channel forms due to aggregation of specific peptides, it conducts, and when the peptides or channel disassociate, the conduction pathway disappears. This is in contrast to other ion channels which have a gating mechanism to allow the flow of ions through the membrane. Note that protomers[19] can also be used to construct peptide ion channels.

An important example of a peptide ion channel is the gA channel illustrated in Figure 2.8. gA channels are arguably the best understood ion channels and have been studied for more than five decades. gA channels play a significant role in the construction of biosensors built out of artificial membranes (as discussed in subsequent chapters). A gA channel comprises two symmetric halves: each half is called a monomer, and the combination of the two halves together is called a dimer. Each gA monomer consists of 15 amino acids. When two gA monomers combine to form a gA dimer this reaction is known as dimerization. The gA monomers are nonconducting; however, the gA dimer forms a conducting pathway for ions through the lipid bilayer, so a gA dimer can be viewed as a monovalent-cation-selective conducting ion channel.

The conducting pore in a gA dimer has an α-helix structure. In the gA dimer ions move through the α-helix structure of the gA dimer. Additionally, the gA dimer also allows water to move, in single file, through the α-helix structure of the channel. Note that the addition of methylbenethonium chloride, or similar cationic blockers, can obstruct the passage of ions through the dimer.

Voltage-gated ion channels. Voltage-gated ion channels play an important role in living organisms as they regulate the concentrations of ions between the interior and

---

[19] A protomer is composed of one or several polypeptide chains that are used to construct a protein composed of several protomers.

2.5 Ion Channels

Figure 2.9 Schematic of the KcsA channel top and side views, and a side view of one of the KcsA monomers with the location of the gating regions of the K^+ ions. The KcsA channel is a tetramer composed of four monomers.

exterior of the cell. This process is crucial for the propagation of neuronal and muscular action potentials throughout organisms. Voltage-gated ion channels open and close (this opening and closing constitutes the gating process) based on the concentration of electrolyte ions in proximity to the membrane surface. This difference in concentration of electrolyte ions causes a transmembrane voltage.[20] The voltage-gated ion channel is controlled by the transmembrane voltage. Voltage-gated ion channels are frequently ion specific; however, some may also conduct similarly sized and charged ions. The functionality of voltage-gated ion channels is attributed to three main units: the transmembrane voltage sensor, the channel, and the gate. The transmembrane voltage sensor is composed of a set of charged peptides. From Coulomb's law, the transmembrane potential will cause an electric force to act on these charged peptides, which can cause a conformational change of the peptides. This conformational change can either promote the flow of ions or restrict the flow of ions – that is, open or close the ion channel.

An example of a voltage-gated ion channel is the potassium (K^+) ion channel. The structure of the potassium channel was first measured using X-ray crystallography by the MacKinnon group in 1998. The potassium channel consists of four monomers that come together to form a symmetric tetramer as illustrated in Figure 2.9. The gating mechanism of the potassium channel is a result of the reorientation of amino-acid molecules in the filter region that only allow the flow of K^+. The potassium channel opening (also known as activation) is caused by changes in the membrane potential or changes in the concentration of specific ligands[21] and can occur at different rates, and over different voltage or ion concentration ranges. Channel closing (deactivation) occurs upon removal of the activating stimulus and can also proceed at different rates. Although the molecular events that underlie activation and deactivation are known in significant detail, the mechanisms that account for differences in activation and deactivation rates are not well understood.

Ligand-gated ion channels. These channels play an important role in the nervous system of living organisms and are essential for the detection of neurotransmitters. For

---

[20] The difference in electric voltage (potential) between two sides of the lipid bilayer.
[21] A ligand is an ion or molecule attached to a metal atom.

example in sensory transduction the ligand-gated ion channel is responsible for detecting certain odors, which allows organisms to have a sense of smell. The main objective of the ligand-gated ion channel is to act as a transducer that converts the chemical signal (ligand or other molecule) into an electrical signal (flow of ions across the membrane). Ligand-gated ion channels are typically composed of two chemical groups. The first is used to detect if the ligand is present, and the second is associated with the channel and gating mechanism.

Light-gated ion channel. Light-gated ion channels can be viewed as transducers that convert a light stimulus into an electrical signal. This electrical signal is caused by the passage of ions through the light-gated ion channel. Light-gated ion channels, such as channelrhodopsins, are used as sensory photoreceptors in unicellular organisms. Additionally, these light-gated ion channels are very useful for controlling the concentration of ions in proximity to lipid bilayers using light. The best-studied ligand-gated ion channel is bacteriorhodopsin, which uses the energy from photons to move protons across the membrane out of the cell. Note that light-gated ion channels are not the same as photoreceptor proteins, which do not conduct ions across the membrane. Light-gated ion channels work in a similar fashion to the other gated ion channels. When a light stimulus (a photon) reacts with the light-gated ion channel this can either open or close the channel. Therefore, light-gated ion channels can be viewed as a photoswitch that can be used to control the flow of ions across a membrane.

Mechanosensitive ion channels. In mechanosensitive channels the gating mechanism is controlled by the deformation (changes in tension, thickness, or curvature) of the membrane in which they are embedded. These ion channels play an important role in living organisms and are involved in several sensing mechanisms, including touch, hearing and balance, and thirst, as well as participating in cardiovascular regulation. Mechanosensitive channels can be nonselective between anions and cations, and selective. The gating mechanism of the mechanosensitive channels results from conformation changes between an open state and a closed state, which results from the mechanical stimulus.

Transient receptor potential ion channels. These ion channels are found in most organisms and are involved in several sensory processes including the sensations of pain, temperature, different kinds of tastes, pressure, smell, and vision. The TRP ion channel is cation selective and contains a very large diversity of activation mechanisms that can be used to open or close the channel. The unique property of TRP ion channels is that the duration of the channel being open or closed is not governed by the environmental conditions; that is, if the channel is activated by a stimulus, then the channel will remain open or closed for a duration of time.

2.6 Tethers, Spacers, and the Bioelectronic Interface

Artificial membranes are synthetic replicas of biological membranes. We will say a lot more about artificial membranes and how they can be precisely engineered in Chapter 3. Typically, artificial membranes contain chemical components that are not found

2.6 Tethers, Spacers, and the Bioelectronic Interface

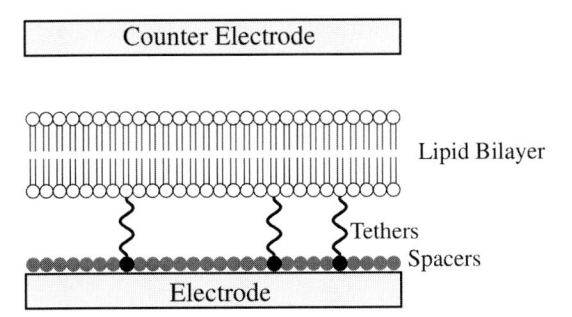

Figure 2.10 Schematic of an engineered (artificial) cell membrane containing tethers, spacers, a lipid bilayer, and the bioelectronic interface. The bioelectronic interface is composed of the counter electrode and electrode.

in biological membranes. The three key building blocks of artificial membrane devices are tethers, spacers, and a bioelectronic interface as illustrated schematically in Figure 2.10.

2.6.1 Tethers

As illustrated in Figure 2.10, the tethers are used to bind the engineered membrane to the bioelectronic interface. Any chemical component can be used for the tether as long as it can bind with the membrane and the bioelectronic interface. A commonly used chemical component for the tethers in artificial membranes is polyethylene glycol (PEG), which is a chain of ethylene glycol molecules. These tethers are then anchored to the bioelectronic interface (gold surface) via a benzyl disulfide group. The gold-sulfur anchoring requires formation of gold-thiolate bond(s). The specific process of the formation of these bonds has only recently been elucidated using advanced X-ray and scanning tunneling microscopy in combination with density-functional theory computations. The gold-thiolate coordinate covalent bond has a strength close to that of the gold-gold bond. Note that hydrophilic tethers act as a reservoir similar to the interior of a cell; that is, they mimic the cytoskeleton supports found in cells.

The fraction of the tethers per unit area in the membrane is called the tether density. Typical tether densities that we will study in this book range from 1 to 100 percent. Membranes with low tether densities (1 to 10 percent) allow more access for ions to the bioelectronic interface, and insertion of large macromolecules in the membrane, which is useful in the fabrication of biomimetic devices. Membranes with 100 percent tether density are useful for the constructing biomimetic devices for growing cells on the surface of the engineered membrane.

2.6.2 Spacers

The spacers, illustrated in Figure 2.10, are chemical structures in an engineered membrane that only bind to the bioelectronic interface and not the membrane. They are composed of a disulfide molecule connected to four polyethylene glycol molecules, and are terminated by a hydroxyl group. The disulfide molecule anchors the spacer to the gold bioelectronic interface. The spacers are used to fill in the space between the tethers on the bioelectronic interface. If no spacers are present, then the bioelectronic interface would be saturated with tethers, thereby causing all the lipids adjacent to the bioelectronic interface to be tethered.

2.6.3 Bioelectronic Interface

In biological systems and engineered membranes, signal communication is performed using ions (electrically charged particles), whereas in electronic systems, information is conveyed via electrons. The bioelectronic interface forms the interface between the ions and electrons. Measuring the electrical response of an engineered membrane requires building a bioelectronic interface that connects the membrane with electronic instrumentation. The bioelectronic interface is illustrated schematically in Figure 2.10. Building models of the biolelectronic interface has challenges, namely,

- accounting for the diffuse three-dimensional distribution of ions in solution relative to the delocalized property of electrons in metals[22] and
- constructing multiphysics models of the slow diffusion kinetics of ions in solution (approximately 1×10^{-9} m^2/s) compared to the relatively instantaneous diffusion kinetics of electrons in metals (approximately 1×10^{-2} m^2/s).

In Part III of the book, we construct mathematical and computer simulation models of the bioelectronic interface – from the macroscopic scale all the way down to the atomic scale. Here, in keeping with our biochemistry-for-engineers theme, we discuss the chemical composition of the bioelectronic interface of engineered membranes.

The bioelectronic interface is composed of an electrode and a counter electrode (also known as an auxiliary electrode). The bioelectronic interfaces can be composed of silicon, gold, platinum, or other metals that do not impede the function of the cellular membrane. The bioelectronic interface we consider in this book is composed of a gold electrode coated with polyethylene oxide spacers and tethers, and a counter electrode composed of gold. Gold electrodes are used because they are biologically inert – the gold electrode does not react with peptides, proteins, or the lipid bilayer. Another benefit of using the gold electrode is that the oxidation potential[23] is approximately 0.9 V. Beyond this electrode potential the gold will oxidize (lose electrons) and enter the electrolyte. For example, at an electrode potential of 0.93 V, the oxidation half-reaction $Au + 4Cl^- \rightarrow [AuCl_4]^- + 3q$ will result, forming chloroauric acid $[AuCl_4]^-$ with q denoting the electron charge. Therefore, electrode potentials below 0.9 V should be used to excite engineered tethered membranes to minimize the effects of oxidation reactions on the surface of the electrode. Additionally, at high voltages an enormous van der Waals

---

[22] Metals contain a large portion of delocalized electrons that are not associated with a specific atom or covalent bond. Application of an electric potential across the metal allows these delocalized electrons to move, producing a current.

[23] The oxidation potential (also known as the standard oxidation potential) is the potential difference across the electrode surface required to remove electrons from the electrode.

attraction results for any molecule near the bioelectronic interface surface. To prevent physisorption[24] some specific coating must be employed. Note that typical membrane potentials for animal cells are on the order of 0.1 V, which is significantly less than the oxidation potential of gold.

To perform experimental measurements, a time-dependent voltage (potential) is applied across the counter electrode and electrode and the response, in terms of an electrical current, is measured. The measured current provides information about the conductance of the lipid bilayer, and the accumulation of electrical double layer charges at the surface of the electrodes and lipid bilayer. At each electrode electrons interact with the electrolyte ions to induce a current in the wires connected to the electrodes. The counter electrode is designed to have a surface area much larger than that of the electrode to ensure that the electron to electrolyte ion reactions (charge transfer) occur fast enough so as not to limit the charge transfer process between electrons and electrolyte ions at the electrode surface. Note that only the electrode contains the engineered artificial membrane; the counter electrode is in direct contact with the electrolyte solution.

2.7 How to Visualize Macromolecules

We strongly recommend that the reader visualize various macromolecules in the membrane to get a better intuitive feel for them. VMD (Visual Molecular Dynamics) and PyMOL are two popular software packages that can be used to visualize macromolecules.[25] Most atom coordinate files in the Protein Data Bank are available in a format that either VMD or PyMOL can read.

Example 1: α-hemolysin. Let us use PyMOL to visualize the protein α-hemolysin which is a pore-forming toxin in membranes. α-hemolysin is obtained from the methicillin-resistant *Staphylococcus aureus* (MRSA) superbug.

The first step is to obtain the atom coordinate file for this protein. Searching the Protein Data Bank (PDB), we find that PDB ID 4P24 is the α-hemolysin we are searching for which can be obtained from the webpage www.rcsb.org/pdb/explore/explore.do?structureId=4P24. We see that the atom positions of this molecule are available in the file 4p24.pdb, which can be read by PyMOL. The second step is to load the 4p24.pdb file into PyMOL. From the initial visualization in PyMOL, we see the atomic structure of α-hemolysin is composed of 32,256 atoms. It is useful visualize the α-hemolysin protein using the secondary-structure representation. In PyMOL, this can be performed by issuing the command "show cartoon." The resulting image from PyMOL is given in Figure 2.11.

Example 2: Hydrogen bonds. In this example we study the hydrogen-bonding structure between the two gA monomers in the gA channel illustrated in Figure 2.7 using

---

[24] The adsorption of ions into the surface of the electrode.
[25] These visualization tools can be obtained from the webpages www.ks.uiuc.edu/Research/vmd/ and www.pymol.org/.

(Side View) (Top View)

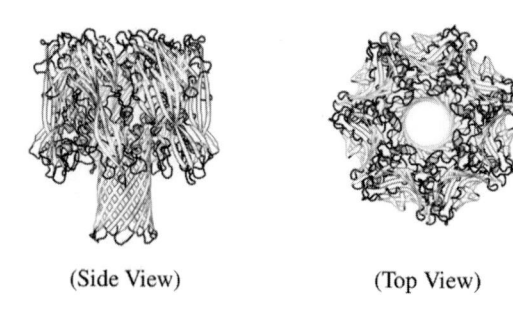

Figure 2.11 PyMOL secondary-structure representation of the α-hemolysin protein (PDB ID 4P24). The α-hemolysin protein contains 32,256 atoms; therefore, it is useful to represent this protein using the secondary-structure representation.

PyMOL. As will be discussed in subsequent chapters, gA is an integral component in biosensors built out of engineered membranes.

The first step is to obtain the atom coordinate file for the gA channel. Searching the Protein Data Bank, we find that PDB ID 1MAG is associated with the gA channel in a lipid bilayer. The coordinate file 1mag.pdb for the gA channel can be downloaded from the webpage www.rcsb.org/pdb/explore/explore.do?structureId=1MAG. Opening the 1mag.pdb file in PyMOL we initially see the atomic structure of the gA channel, which is composed of 552 atoms.

The next step is to estimate the positions of the hydrogen bonds between the two gA monomers. We follow the tutorial at https://pymolwiki.org/index.php/Displaying_Biochemical_Properties#Hydrogen_bonds_and_Polar_Contacts. The main commands to issue to PyMOL once the 1mag.pdb is loaded are the following:

```
hide everything
alter 1-16/,ss='H'
show cartoon
select /1mag//A/
[A]→polar contacts→to any atoms
```

The command "alter 1-16/,ss='H'" ensures that the gA monomer secondary structure is interpreted as an α-helix; the command "select /1mag//A/" selects one of the gA monomers, and "[A]→polar contacts→to any atoms" finds all the hydrogen bonds between the two gA monomers in the gA channel. The final result of the hydrogen bonding between the two gA monomers is illustrated in Figure 2.12 and is composed of six hydrogen bonds. Therefore, the dissociation energy of the gA channel is approximately equal to the energy required to break these six hydrogen bonds.

Summary. The outcome of the above two examples is a three-dimensional visualization of the secondary structure of α-hemolysin and the gramicidin A ion channel, as well as the hydrogen bonding between the two gA monomers in the gA channel. The visualization software packages VMD and PyMOL contain several options for creating detailed representations of proteins and membranes. Tutorials on constructing molecular

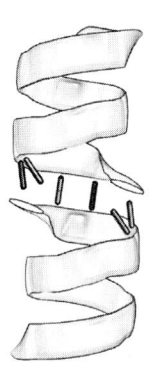

Figure 2.12 PyMOL secondary-structure representation of the gA channel (PDB ID 1MAG) and the six hydrogen bonds, illustrated by the black lines, between the two gA monomers generated using PyMOL.

visualizations for VMD can be found at www.ks.uiuc.edu/Research/vmd/current/docs.html#tutorials, and for that PyMOL can be found at https://pymolwiki.org/index.php/Category:Tutorials.

2.8 Closing Remarks

This chapter has provided an elementary primer on the important biochemical components of engineered membranes. Each of the topics outlined above is the subject of numerous books and papers. Although we have only scratched the surface of an enormous area, the background provided is sufficient for understanding the key aspects of Parts II and III of the book. More detailed studies can be found in any undergraduate-level textbook in biochemistry such as [10, 129, 296]. Chapter 3 gives a detailed account of artificial membranes.

3 Engineered Artificial Membranes

3.1 Membranes

Biological membranes are ubiquitous, ranging from the cell membrane which encompasses a cell, to membranes for organelles within the cell such as the nucleus and mitochondria. This book focuses exclusively on the construction and modeling of engineered artificial membranes. We use the term "engineered membranes" to emphasize the fact that we will precisely engineer various components of the artificial membrane to achieve specific structural properties. Additionally, we model the dynamics of these engineered membranes from atoms to device. The results from artificial membranes elucidate the structural and dynamic properties of biological membranes. The interaction of biological membranes with macromolecules is key to the fight against human disease (rational drug design and construction of therapeutic protocols). As such, engineered artificial membranes in combination with accurate mathematical models provide an important tool for innovation in the pharmaceutical industry.

Since artificial membranes are meant to mimic biological membranes, we start by discussing the key properties of one of the most important biological membranes, namely, the cell membrane. The cell membrane provides an interface between the cell and its environment as illustrated in Figure 3.1. The extracellular fluid (depicted in Figure 3.1) contains all components that are not contained in the cell, and the cytosol is composed of components that are contained in the cell. Cell membranes are dynamic and respond to their environment as the result of an external stimulus. To recap from Chapter 2, the cell membrane is composed of three primary components:

(i) **Lipids**. Lipids are primary components of the cell membrane. All lipids are composed of a hydrophilic head (attracted to water) and hydrophobic tails (water neutral). The lipids separate the extracellular fluid from the cytosol of the cell.
(ii) **Macromolecules**. These are responsible for cellular communication both in cells and between cells. Given the low permeability of biological membranes, the primary task of macromolecules is to transport molecules from the extracellular fluid to the interior of the cell. Additionally, there exist other macromolecules which act as molecular receptors, detect vesicles carrying cellular cargo, and facilitate the transport of other macromolecules across the membrane from the extracellular fluid to the cytosol, or from the cytosol to the extracellular fluid. Low-molecular-weight molecules also exist in the membrane that modulate membrane structure

3.1 Membranes   31

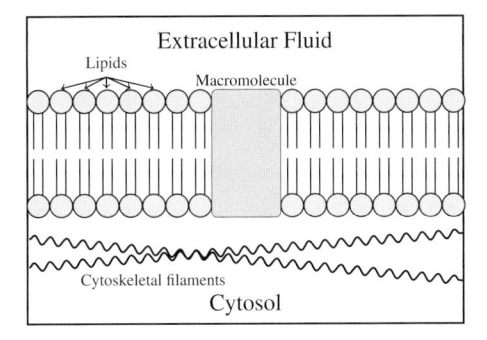

Figure 3.1 Schematic of a cell membrane. The extracellular fluid represents the content outside the cell and the cytosol is the interior of the cell, with the membrane separating the two domains. Note that cell membranes have several different types of lipids, macromolecules, and cytoskeletal filament components. Additionally, the cytosol contains a host of other organelles (specialized structures within a living cell) which are not illustrated. It is important to note that the cell membrane is a dynamic system: the individual lipid layers slide on each other.

  (e.g., sterols, progesterone, and testosterone), as well as a range of other peptides that act as toxins or defensive elements to the membrane.
(iii) **Cytoskeletal filaments**. These provide structural support for the membrane and determine or modulate the shape of the cell. Note that macromolecules, such as Band 3[1], act as tethers within natural membranes that link to the cytoskeleton for structural stability.

The development of novel drugs and therapeutic protocols for effective drug delivery require knowledge of dynamic processes present in cell membranes. These in turn depend on the properties of the lipids, macromolecules, and cytoskeletal filaments.

Why artificial membranes? Biological cells can be grown in the laboratory, so why do we need to construct artificial membranes to study the dynamics of cell membranes? The primary reason is *specificity*: an artificial membrane allows us to target specific components in a biological membrane and examine their functionality. Recall that cell membranes contain thousands of protein macromolecules, peptides, lipids, and cytoskeletal filaments which contribute to the dynamics of the cell membrane. So it is difficult to study the effect of a specific protein macromolecule on the dynamics of the cell membrane since we cannot isolate if changes in the membrane were due to this specific macromolecule or were the effect of other proteins and peptides present in the membrane. The aim of an engineered artificial membrane is to create a biomimetic[2] system that mimics the cell membrane and also allows us to study specific biological processes in an experimentally controllable environment.

This chapter discusses five important classes of engineered artificial membrane architectures found in the literature. Of these engineered membrane architectures, we focus

---

[1] Chemical structure can be obtained from the Protein Data Bank at www.rcsb.org/pdb/explore/explore.do?structureId=4KY9
[2] A biomimetic system is a physical device that mimics features of a biological system. This is similar to an in vivo system in which biological molecules are used outside their normal biological environment.

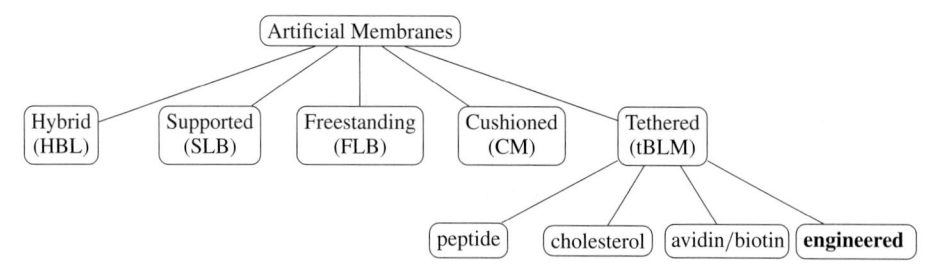

Figure 3.2 Classification of artificial membranes. This book focuses exclusively on engineered tethered bilayer lipid membranes (tBLMs).

on engineered tethered membranes (also known as tethered bilayer lipid membranes) as these can be used to study several cellular membrane processes and for the construction of biosensors. Biosensors are biomimetic systems that use biological components, such as protein receptors, to detect specific chemical species in the extracellular fluid. The key concepts the book aims to convey are how to perform experimental measurements using engineered artificial membranes and how to model the dynamics of engineered artificial membranes.

3.2 Artificial Membrane Architectures

The purpose of artificial membranes is to mimic the biological membranes of living organisms in an experimentally controllable environment.[3] This is equivalent to performing experimental measurements on membranes both ex vivo[4] and in vitro. For the reader's convenience, Figure 3.2 lists the various artificial membranes that are described below.

Five types of artificial membrane architectures have been widely studied (Figure 3.3 displays a schematic representation of their key features):

(i) **hybrid bilayer lipid membranes** (HBLs) [15, 58, 320, 326, 383],
(ii) **supported lipid bilayer** (SLB) [50, 414],
(iii) **freestanding lipid bilayer** (FLB) [306, 349, 350, 446],
(iv) **cushioned membranes** (CMs) [354, 355],
(v) **tethered bilayer lipid membranes** (tBLMs) [84, 103, 243, 294].

A common theme of all five artificial membrane architectures is that they are in proximity to a flat support. The reason this type of flat support is used is that it is easy to construct and eases the analysis of the experimental results from these devices. If an irregular surface was used then the formation of a homogeneous bilayer is difficult

---

[3] Parameters that are desirable to control include concentration of electrolyte, chemical molecules in the electrolyte, temperature, transmembrane potential, bilayer lipid membrane composition, and the tethering density.

[4] Ex vivo means taking place outside the normal membrane environment in a cell, but with minimal alterations from the natural conditions of the membrane.

3.2 Artificial Membrane Architectures

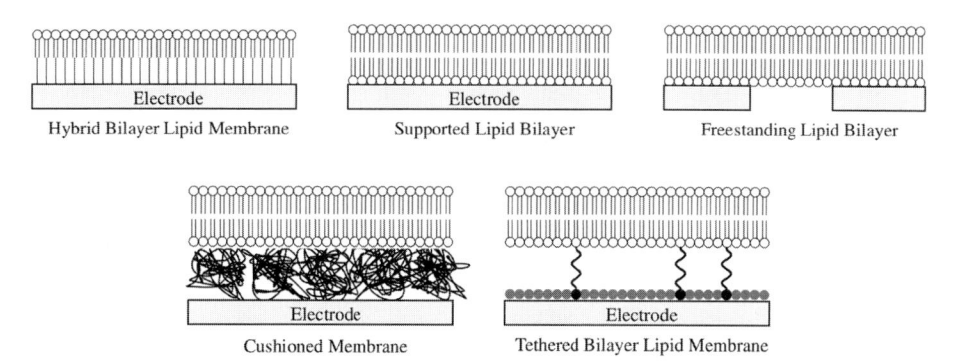

Figure 3.3 Schematic structure of five types of engineered artificial membrane architectures: hybrid bilayer lipid membrane, supported lipid bilayer, freestanding lipid bilayer, cushioned membranes, and tethered bilayer lipid membranes. The counter electrode is not illustrated but is present for all the engineered membranes.

as vesicles (clumps of lipids) may form because of the irregularities. Additionally, a disruption of the membrane seal at regions of high curvature can also occur.

Hybrid bilayer lipid membranes. HBLs provide a useful artificial membrane structure due to their stability – the lipids can only dissociate from the bilayer if the bond between the fatty acids and the tethering molecule that bonds the lipid to the bioelectronic surface is broken. This is typically a coordination bond, which is significantly stronger than the hydrophobic effect[5] which links the lipids in a lipid bilayer. This high stability results from the lipid tails being chemically bonded to the flat surface. An issue with the HBLs is that since only a lipid layer is formed, it does not mimic the dynamics of a bilayer lipid membrane. Additionally, since the lipid tails are chemically bonded to the flat surface, the dynamics of lipids are dramatically effected [15]. In cellular membranes the lipids move throughout the membrane and do not remain in a fixed position. HBLs cannot be used to study the dynamics of macromolecules that are inserted into the engineered membrane as only a single lipid layer is formed. HBLs can also not be used to study the dynamics of ion channels as there is no water reservoir into which ions can flow after crossing the membrane.

Supported lipid bilayer. An SLB is composed of an artificial membrane that rests directly on the bioelectronic interface.[6] The formation of the SLB is performed by bringing unilamellar vesicles of lipids into contact with the bioelectronic interface as described in [342]. A unilamellar vesicle is a spherical shell that encloses a small amount of water using a single lipid bilayer. This unilamellar vesicle, when brought into contact with the flat surface, unfolds and forms the flat bilayer lipid membrane. An issue with the SLB is that the engineered membrane is in close proximity to the surface, which causes the surface to interfere with the dynamics of the lipids. Formally, there is a *frictional coupling* between the flat surface and lipids that will impact the lipid dynamics. Also, the close contact can compress the bilayer by way of the bilayer being

---

[5] This is the tendency for hydrophobic molecules to aggregate in water.
[6] Recall from §2.6 that the bioelectronic interface of the engineered artificial membrane is used to allow electronic equipment to be connected to the engineered membrane to perform experimental measurements.

attracted to and compressed against the gold. The lack of a water reservoir means very little water is present between the membrane surface and gold bioelectronic interface. Another limitation with the SLB is that it is difficult to insert macromolecules into the engineered membrane as a result of the proximity of the formed bilayer to the flat surface. Large macromolecules require free space on both sides of the bilayer [391, 415].

Freestanding lipid bilayer. A benefit of the FLBs is that the engineered membrane is not in proximity to a planar electrode support, as illustrated in Figure 3.3. This allows nearly any size of macromolecule to be inserted into the engineered membrane. Additionally, since both sides of the engineered membrane are in contact with an aqueous solution, different macromolecules can be introduced to either side of the engineered membrane. These types of multimacromolecule reactions are important for studying biological processes such as the transfer of adenosine triphosphate[7] to an inserted macromolecule. Additionally, since the FLBs are not in proximity to the flat surface, there is no frictional coupling effect present in the engineered membrane [215]. An issue with FLBs is the difficulty in constructing the surface support of the engineered membrane. Typically a micrometer-size opening is used; then the formed engineered membrane will have a short lifetime [476]. This results because the engineered membrane does not have sufficient support in this micrometer-size opening. Recall that in cellular membranes the membrane is supported by cytoskeletal filaments. Smaller openings on the order of tens of nanometers can be constructed using advanced processing techniques [386]. An issue with this technique is that, since experimental measurements are being performed on such a small sample of engineered membrane, any minor heterogeneous lipid dynamics are dramatically enhanced. This presents an issue with creating experimentally reproducible measurements, as any small lipid defect can dramatically impact the results. Additionally, it is difficult to ensure the electrolyte solution does not penetrate the region between the solid support and the membrane surface. These effects are not present when measurements are performed using large patches of engineered membranes. It should be noted that the FLBs, although useful in studying larger ion channels, do not mimic biological membrane as they do not include any tethering to cytoskeleton supports.

Cushioned membranes. A benefit of CMs is that they use a cushion between the lipid bilayer and the flat surface. The cushion reduces the effects caused by frictional coupling between the flat surface and the engineered membrane. Additionally, the cushion also allows larger macromolecules to be inserted into the bilayer compared with SLBs, where the bilayer is not in direct proximity to the flat surface [415]. The cushion can be composed of several different types of chemical components such as polymer chains, polyelectrolyte layers, carbohydrates, and peptides. The selection of which chemical component to use is dependent on the desired properties of the cushion (firm, flexible, etc.). A limitation with the CMs is that when experimental measurements are performed it is difficult to separate whether the measured dynamics are a result of the cushion, the biological processes in the engineered membrane, or both [272].

---

[7] Adenosine triphosphate is a high-energy molecule present in all living organisms that is involved in nearly all cellular signaling processes.

Tethered bilayer lipid membranes. tBLMs use tethered lipids or tethered macromolecules to anchor the engineered membrane to the flat surface. This is particularly useful as the tethered lipids or macromolecules mimic the response of the cytoskeletal filaments found in cellular membranes. Therefore, the dynamics of the lipids in the engineered membrane do not suffer from the effects of frictional coupling. The key advantage of using tethers is that they allow the incorporation of a relatively large amount of water per lipid between the membrane and the supporting surface, and they promote the isolation of the membrane from the surface, minimizing any membrane distortion.

Additionally, only tethering some lipids or macromolecules in the engineered membrane can allow a wide range of macromolecules to be inserted into the engineered membrane. The stability of the engineered membrane in the tBLMs is high and typically has a lifetime of several months. This provides sufficient time to study several types of biological processes. Note that typically the lifetime of macromolecules is on the scale of several days. A limitation with tBLMs is that direct access to both sides of the engineered membrane is not possible. Recall that this feature is present for the FLBs.

The five artificial membrane architectures (HBL, SLB, FLB, CM, and tBLM) all provide a biomimetic system for constructing biosensors and for performing experimental measurements of certain peptides and proteins. Deciding which engineered membrane architecture to use depends on the specific application. All the architectures can be used to study the reaction dynamics of surface-bound proteins and peptides. However, for the study of ion channels and large transmembrane proteins, only the FLB and tBLM can be used. The reason the dynamics of ion channels can only be studied using the FLB and tBLM is that the other engineered membranes do not provide sufficient space for ions to flow into the electrolyte reservoir between the electrode and the bilayer lipid membrane surface. Additionally, large transmembrane proteins have chemical groups that are both contained in the membrane and exterior to the membrane. Since the HBL, SLB, and CM have negligible space between the electrode and lipid bilayer surface this will significantly affect the gating and conformational dynamics of these large transmembrane proteins. The FLB, CM, and tBLM can be used to study the lipid mobility and structural properties of the lipid bilayer (e.g., thickness, area per lipid). However, the HBL and SLB are not suitable for this as the electrode surface non-negligibly interacts with the lipids in the lipid bilayer. FLBs are useful for studying any chemical processes that depend on chemical species on both sides of the lipid bilayer. For example, an FLB is particularly useful for studying ligand-gated ion channels when different electrolyte concentrations are desired on either side of the membrane. tBLMs provide a useful platform for studying the dynamics of proteins, peptides, lipid mobility, and structural properties the membrane when access to both sides of the membrane is not required.

3.3 Engineered Artificial Tethered Membranes

Of the five artificial membrane architectures displayed in Figure 3.3, this book focuses exclusively on tBLMs. Although we focus on tBLMs, the experimental techniques and dynamic models of the HBL, SLB, FLB, and CM are similar.

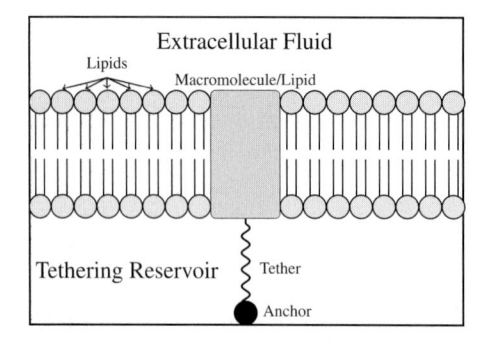

Figure 3.4 Schematic diagram illustrating a tethered bilayer lipid membrane (tBLM) architecture. The tBLM is composed of an engineered membrane comprised of lipids, embedded macromolecules, and tethered macromolecules or tethered lipids. The tethered molecule contains three components: an anchor which binds the tether to the surface, a tether, and either a macromolecule or a lipid that connects the engineered membrane to the tether. The extracellular fluid is any fluid outside the tethering reservoir. Note that the tether can be attached to molecules that span either half the lipid bilayer, or the full lipid bilayer (membrane-spanning lipid), or to a membrane-incorporated peptide.

In the rest of this section, to give the reader more perspective, we describe important properties of tBLMs. The atomic structure of tBLMs can be determined using several experimental techniques such as electrical measurements, X-ray reflectometry, and neutron reflectometry as will be discussed in Chapter 7. All tBLMs have a common structure: they use a surface tethered molecule that connects to a lipid or macromolecule in the engineered membrane as illustrated in Figure 3.4. The tether is anchored to the surface via anchoring molecules. Examples of anchoring chemistries include disulfide and lipoic acid. The surfaces (also known as the substrates) include gold, silicon dioxide, mercury, platinum, and glass. Note that silver is never used because it ablates off, and platinum is not typically used as a result of the difficulty of anchoring molecules to the platinum surface. The only requirement is that the anchor molecule can bind to the substrate. tBLMs are designed to provide sufficient space between the engineered membrane and surface support to eliminate the effects of frictional coupling between the surface and lipids, and to allow the insertion of large macromolecules while ensuring increased membrane stability, that is, to construct engineered membranes with a lifetime of several months.

As illustrated in Figure 3.5, there are several examples of tBLMs:

(i) peptide tethered membrane [56, 292, 293, 345, 390, 461, 472, 473],
(ii) protein tethered membrane [133],
(iii) cholesterol tethered membrane [199],
(iv) polymer tethered membrane [205, 379, 415, 446],
(v) avidin/biotin tethered membrane [416, 417], and
(vi) engineered tethered membrane [84].

Important properties that affect the selection of which type of tBLM to use are the stability of the tether and anchor, how the tether will affect the dynamics of the engineered membrane and inserted macromolecules, and the difficulty associated with formation of

3.3 Engineered Artificial Tethered Membranes

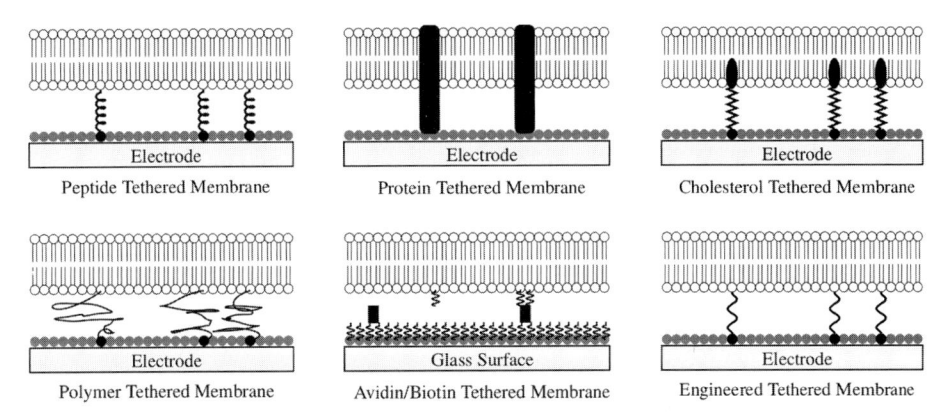

Figure 3.5 Schematic of the six classes of tethered bilayer lipid membranes (tBLMs): the peptide tethered membrane, protein tethered membrane, cholesterol tethered membrane, polymer tethered membrane, avidin/biotin tethered membrane, and engineered tethered membrane. The electrode/glass is covered with an anchoring chemical molecule (gray dots), and the line shapes connecting the anchoring molecule to the bilayer lipid membrane indicate the associated tether. For the protein tethered membrane the black rectangles indicate the protein which connects to the anchors and the lipid membrane. The cholesterol in the cholesterol tethered membrane is indicated by the black ellipsoids. For the avidin/biotin membrane the black rectangle is the avidin molecule, the anchors are composed of molecules connected to a biotin molecule, and some lipids in the bilayer are also connected to a biotin molecule.

the tBLM and performing experimental measurements. Below we outline some of the benefits and limitations of each of the tBLMs presented in Figure 3.5.

Peptide tethered membranes. Peptide tethered membranes are constructed by covalently bonding a peptide to the hydrophilic head or head group of a lipid, and then anchoring this peptide-lipid structure to the surface. Using the peptide-lipid structure to anchor the engineered membrane to the surface has the benefit that the tether binds directly to the same lipid that the engineered membrane is composed of. This reduces the impact of the tether on the lipid dynamics and dynamics of inserted macromolecules. The limitation with this method is that the peptide may interfere with the dynamics of charged chemicals in proximity to the protein and can also affect the dynamics of the lipids. If a positively charged peptide is used as a tether, this will attract negatively charged ions to the peptide [190]. Increasing the population of ions of a particular polarity can enhance the conduction of ion channels selective for that polarity of ion – for example, selecting cations to promote gramicidin conduction.

Protein tethered membranes. Protein tethered membranes are constructed by anchoring a protein to the surface and to the interior of the engineered membrane. Since we know that the interior of the membrane is composed of hydrophobic (water-repulsive) lipid tails, the section of the protein that anchors to the engineered artificial membrane must contain hydrophobic amino acid molecules. A crucial issue with the protein tethered membrane is to ensure sufficient spacing between the anchored proteins is maintained. If the anchored proteins are all clumped into a specific region then sections of the engineered membrane will collapse to the surface, introducing heterogeneous frictional coupling effects. An additional limitation of this method is that the protein may interfere with the lipid dynamics or the dynamics of charged chemicals in proximity to the protein.

Cholesterol tethered membranes. The cholesterol tethered membrane architecture is similar to that of the peptide tethered membrane, except that instead of the tether anchoring to a lipid, the tether is anchored to a cholesterol molecule. Note that cholesterol contains both hydrophilic and hydrophobic portions, allowing it to be used to anchor with the engineered membrane. Cholesterol is useful as an anchor to study the dynamics of cell membranes from animals which contain cholesterol. A limitation of the cholesterol tethered membrane is that cholesterol can cause significant changes in the lipid dynamics. Additionally, cholesterol is not present in the cellular membranes of prokaryotes (bacteria and archaea); therefore, this engineered membrane should not be used to study the dynamics of prokaryotic membranes.

Polymer tethered membranes. Polymer tethered membranes are engineered membrane architectures that use polyethylene glycol[8] as the tether. The polyethylene glycol is covalently bonded to the hydrophilic lipid head and anchors to an amine-coated surface via a carboxylic acid. A very useful property of the polymer tethered membrane is that the experimentalist can control the density of tethers compared to density of lipids. This is an important parameter to control as different cellular membranes contain different densities of cytoskeletal filaments. The dynamics of cellular membranes are dependent on the densities of cytoskeletal filaments. A limitation with the polymer tethered membrane is that the tether is contained in both lipid bilayers of the engineered membrane. This can interfere with biological processes that are dependent on the binding of macromolecules to the surface of the membrane. Additionally, the tethered lipids that are not connected to the surface will also impact the dynamics of the lipids.

Avidin/Biotin tethered membranes. The avidin/biotin tethered membrane is composed of a complex multimacromolecule tethering structure. The tether is composed of a biotinylated[9] bovine serum albumin that is linked to biotinylated lipids using either an avidin or a biotin linker. The tether is then anchored to an aldehyde-modified glass surface. This complex tethering structure provides significant separation of the formed engineered membrane from the surface. Given the large separation between the membrane and surface, the dynamics of large macromolecules can be studied using this tethered membrane architecture. A limitation of this method compared to the other tethered membrane architectures is that optical measurement techniques must be used to measure the dynamics of the membrane or inserted macromolecules.

Engineered tethered membranes. The engineered tethered membrane is composed of a polyethylene glycol (PEG) linker with one end linked to the electrode surface and the other end covalently linked to a hydrophobic group that becomes entangled in the lipid bilayer. Notice that this is a similar construction to the peptide tethered membrane. However, a key difference here is that PEG groups are used in the space between the membrane and sulfur-gold anchors at the gold electrode. These engineered membranes have lifetimes of several months. Remarkably, the engineered membrane is not damaged

---

[8] A description of polyethylene glycol can be found in Chapter 2.
[9] Biotinylation is the process of covalently attaching biotin to a peptide, protein, or lipid.

when cells are grown on the surface of the membrane. Additionally, the engineered tethered membrane can be constructed in approximately 20 minutes using techniques from first-year undergraduate chemistry. The major benefit of the engineered membrane is that all aspects of the engineered tethered membrane can be controlled, which include the tether density, lipid composition, tether length, and composition of the extracellular fluid in contact with the membrane surface. A limitation of the engineered tethered membrane is that the solution in contact with the tethered lipids cannot be controlled independently at the inner and outer surfaces. Recall that this limitation is present in all engineered membranes except for the FLBs, which have a significantly shorter lifetime.

In this book we focus on the experimental analysis and dynamic modeling of engineered tethered membranes. Though we focus on engineered tethered membranes the experimental techniques and dynamic models are useful for all the engineered membrane architectures presented in Figures 3.3 and 3.5.

3.4 Sensing with Engineered Tethered Membranes

In this book, we consider four important sensing devices built out of engineered tethered membranes:

(i) the ion-channel switch (ICS) biosensor,
(ii) the pore formation measurement platform (PFMP),
(iii) the electroporation measurement platform (EMP), and
(iv) the electrophysiological response platform (ERP).

These four devices employ an engineered membrane that mimics the electrophysiological properties of real cell membranes, and a gold electrode bioelectronic interface to which electrical instrumentation is connected. Recall from §2.6.3 that the bioelectronic interface is an important component of sensing with an engineered membrane. Experimental measurements of all four devices are performed by estimating the time-dependent conductance of the engineered membrane, which is dependent on the bioelectronic interface, and the ensemble of aqueous pores and conducting ion channels present.

The use of inert gold electrodes as the bioelectronic interface is superior to the use of redox active electrodes for two reasons. First, if redox active electrodes are used, the metal will ablate, causing the tethers to dissociate from the electrode surface, destroying the membrane [88]. Second, redox active electrodes release metal ions into solution which can interfere with the electrophysiological response of proteins and peptides. The inert gold electrode capacitively couples the electronic domain to the physiological domain without the issues associated with redox electrodes. However, the diffusion-limited effects of ions at the electrode surface must be accounted for when modeling the four tethered membrane measurement platforms.

Common molecular components used to construct engineered tethered membranes are illustrated in Figure 3.6. Using these molecular components it is possible to construct several unique membrane devices such as the ICS, PFMP, EMP, and ERP. Schematics

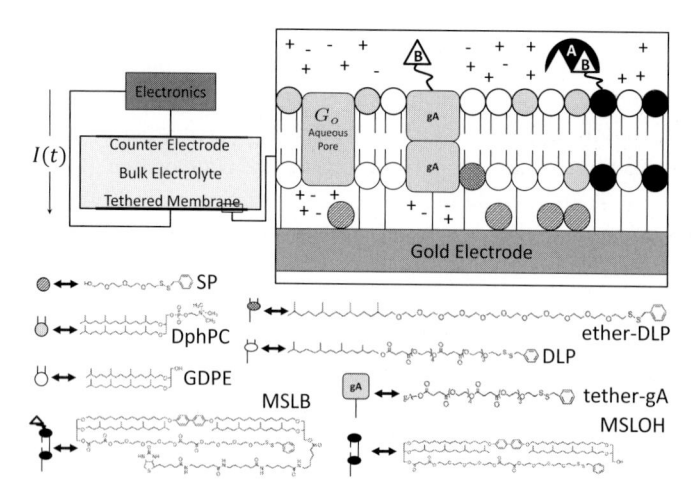

Figure 3.6 Overview of the engineered membrane and molecular components. The "Electronics" block represents the electronic instrumentation that generates the excitation potential between the gold electrode and gold counter electrode and records the current response $I(t)$. G_o is a transient aqueous pore. The conducting gramicidin (gA) dimer is shown and is composed of two gA monomers. A represents the analyte species, and B the analyte receptor. MSLOH denotes synthetic archaebacterial membrane-spanning lipids, DLP half-membrane-spanning tethered lipids, DphPC and GDPE mobile half-membrane-spanning lipids, MSLD membrane-spanning lipids, and SP a spacer.

of these four engineered membrane devices (namely, ICS, PFMP, EMP, and ERP) are provided in Figure 3.7. These four devices are representative of several important applications involving artificial cell membranes and will be studied in great detail in this book. A useful property of these engineered membrane devices is that the experimentalist can precisely select the density of tethers and the membrane composition. The constructed membrane has a lifetime of several months. The engineered membrane is composed of a self-assembled monolayer of mobile lipids and a self-assembled monolayer of tethered and mobile lipids. The tethered lipids are anchored to the gold electrode via a benzyl disulfide component which is connected to a polyethylene glycol chain. Spacer molecules are used to ensure the tethers are spread over the gold electrode. The intrinsic spacing between tethers and spacers is maintained by the benzyl disulfide moieties. A time-dependent voltage potential is applied between the electrodes to induce a transmembrane potential of electrophysiological interest; this results in a current $I(t)$ that is dependent on the charging of the electrical double layers and the conductance of the engineered membrane. The measurements of the membrane conductance of the ICS and PFMP are performed using impedance measurements, and measurements of the membrane conductance of the EMP and ERP are performed using current response measurements as illustrated in Figure 3.7. The reason these two different measurement techniques are used is that in the case of the ICS and PFMP the membrane conductance dynamics changes on a time scale of seconds, while for the EMP and ERP the conductance changes on a time scale of milliseconds.

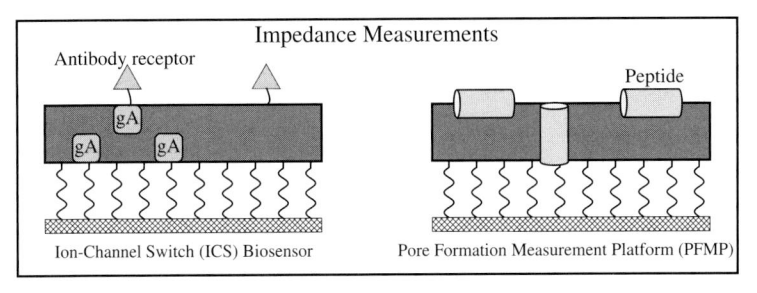

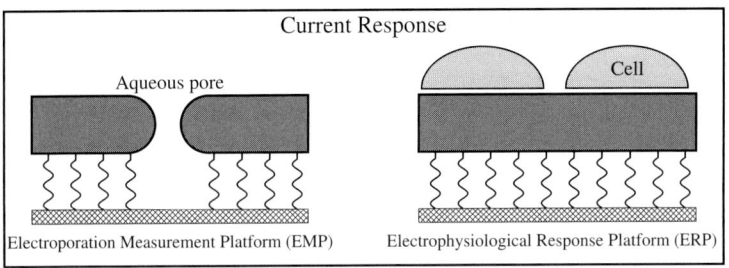

Figure 3.7 A schematic diagram of the four tethered membrane devices, namely, ICS, PFMP, EMP, and ERP. In each subfigure, the engineered tethered membrane is depicted by the gray rectangle, and the gold interface by the crosshatched rectangle. The unifying theme of all four devices is the use of an inert gold bioelectronic interface. Sensing using ICS and PFMP is performed by measuring the time-dependent impedance of the membrane as a result of changes in the concentration of conducting pores. The sensing mechanism of the EMP and ERP is performed by measuring the current response of the devices to a time-dependent excitation potential. The current response is dependent on the concentration of conducting pores, the polarization dynamics of the membrane surfaces, and the charging dynamics at the surface of the bioelectronic interface.

3.4.1 Device 1: Ion-Channel Switch (ICS) Biosensor

The first device that we briefly describe is the ICS biosensor [84]. By a biosensor, we mean a sensor built out of synthetic biological material – in our case, built out of an artificial membrane, ion channels, spacers, and tethers. The purpose of the biosensor is to detect the presence of analyte molecules. When it senses the presence of analyte molecules, the biosensor responds by changing its electrical impedance significantly. The amount the impedance changes depends on the concentration of the analyte molecules. Therefore, by examining the current flowing through the biosensor, one can detect the presence of analyte and estimate its concentration. The ICS biosensor can detect (sense) extremely small concentrations of analyte[10] molecules at femtomolar concentrations. Indeed when it encounters the analyte molecules specific to the antibody groups attached to the gramicidin and lipid molecules, the electrical impedance of the ICS biosensor increases dramatically – it is this property that results in high detection sensitivity requiring only minimal amounts of analyte and also the specificity

---

[10] Recall that analyte denotes the target molecules that we wish to detect.

to detect the presence of specific analyte molecules. We will have a lot more to say about ICS biosensor, in Chapter 4 (construction), Chapter 5 (clinical studies), and Chapters 8–15 (atomistic to macroscopic mathematical models for the dynamics of the ICS biosensor).

The ICS biosensor can be viewed as a fully functioning nanomachine constructed out of an engineered membrane with moving parts comprising gramicidin (gA) monomers and conducting gA dimer channels. The operation of the ICS biosensor can be viewed schematically at www.youtube.com/watch?v=6Ti83oO2ml4, which illustrates how a tethered gA monomer and a mobile gA monomer (purple cylinders) interact when a specific analyte molecule (green blob) binds to the receptor sites (red blobs) of the ICS. Gold electrodes constitute the bioelectronic interface between the electrical instrumentation and the electrolyte solution. The ICS biosensor has a lifetime of several months and can detect femtomolar (10×10^{-15} molar) concentrations of target species including proteins, hormones, polypeptides, microorganisms, oligonucleotides, DNA segments, and polymers in cluttered electrolyte environments. The ICS biosensor has also been used in clinical trials for the detection of influenza A.

In Part III we construct models of the ICS biosensor which utilize reaction-rate theory, and a combination of reaction-rate theory and the Nernst–Planck equations for advection and diffusion. Using the constructed model of the ICS biosensor it is possible to estimate important biological parameters from the experimentally measured impedance response of the ICS biosensor.

3.4.2 Device 2: Pore Formation Measurement Platform (PFMP)

The PFMP is the second device that we describe extensively in this book. The PFMP can be used to detect the presence of proteins and peptides that form pores in the membrane. Chapter 4 deals with the construction of the PFMP, Chapter 6 presents experimental measurements using the PFMP, and Chapters 8, 10, and 14 present atomistic to macroscopic mathematical models of the PFMP.

Since the artificial membrane surface can be precisely engineered to mimic prokaryotic, eukaryotic, and archaebacterial membranes, the PFMP can be used to measure the specificity of attack of protein and peptide toxins. Therefore, the PFMP can be used for rapid point-of-care detection of pore-forming toxins and for inexpensive pharmacology screening of novel antimicrobial peptides. Other techniques, which do not include tethered membranes, to study pore formation dynamics include lytic experiments, gel electrophoresis, site-directed mutagenesis, and cryoelectron microscopy [23, 110, 259, 309]. The benefit of using the PFMP compared to these methods is that the tethering density, electrolyte composition, membrane composition, and applied transmembrane potential can all be controlled by the experimentalist. Examples of tethered bilayer lipid membranes for the measurement of pore-forming toxins include that in which the membrane is composed of diphytanyl chains that are coupled via a glycerol to oligoethylene oxide spacers [442] and that which uses different lipid, anchoring, and spacer components than the PFMP but employs an identical solvent-exchange membrane formation protocol [274].

Regarding mathematical models for the pore formation dynamics, Part III presents a powerful suite of models ranging from macroscopic (phenomenological) reaction-rate models to microscopic (all-atom molecular dynamics). In Chapter 10 we construct a detailed model of the PFMP that allows experimental measurements from the PFMP to be used to estimate important biological parameters such as the reaction pathway of peptides that lead to pore formation. The model accounts for the diffusion of the proteins and peptides in solution and the reaction-diffusion processes present on the membrane surface. Finally, Chapter 14 illustrates how coarse-grained molecular dynamics simulations and continuum models can be used to explain the pore formation process at the atomistic scale, which includes membrane binding, and the processes necessary to create a peptide pore in the engineered membrane.

3.4.3 Device 3: Electroporation Measurement Platform (EMP)

We now discuss the third device built out of engineered membranes, namely, the EMP, that we study in this book. Electroporation is the process of aqueous pore formation in a membrane when a voltage is established across the membrane. Electroporation has applications in electrochemotherapy for antitumor treatment, protein insertion, cell fusion, debacterialization, and gene and drug delivery. Experimental platforms to study electroporation include synthetic bilayer lipid membranes and in vitro cells. However, synthetic nontethered bilayer lipid membranes do not provide a good representation of physiological systems since the effects caused by the cytoskeletal network are not present. All cells contain a cytoskeletal network that provides structural support for the biological membrane. Using cells provides a physiological system for validation; however, it is impossible to fully define the physiological environment that affects properties associated with electroporation. This motivates the need for an engineered membrane platform which gives the experimentalist control over tethering density, membrane composition, and physiological environment, unlike the synthetic lipid membrane and cell-based platforms.

The EMP is a tethered membrane platform that provides a fully controllable physiological environment with a tethered membrane. Construction of the EMP and experimental measurements are discussed in Chapter 6. The tethers in the EMP mimic the response of the cytoskeletal network in cells. Therefore, the EMP can be used to model the dynamics of electroporation; however, a dynamic model is required to relate the experimental measurements from the EMP to important biological parameters such as aqueous pore density and size.

Models of the electroporation process employ the Smoluchowski–Einstein equation derived from statistical mechanics, which requires specifying the energy of a pore. These models are discussed in detail in Chapter 11. The pore energy models are typically constructed by assuming the membrane can be modeled using continuum mechanics. However, the pore energy models can also be constructed from the results of coarse-grained molecular dynamics and all-atom molecular dynamics simulations. A key property that must be estimated is not only the pore energy, but also the conductance of an aqueous pore. Assuming symmetric electrolytes and electroneutrality, the aqueous pore conductance can be estimated via the Poisson–Nernst–Planck (PNP) system of equations. However, in the tethered membrane platform there exist electrical double layers at the surface of the membrane and electrode contacts. Using the PNP equations to estimate the pore conductance may not provide an accurate description of the actual pore conductance. Additionally, the Smoluchowski–Einstein equation (a nonlinear partial differential equation) is numerically prohibitive to solve, and the pore energy typically found in the literature does not include effects caused by asymmetric electrolytes, multiple ionic species, and the Stern and diffuse electrical double layers present. To address these issues, in this book we construct an electroporation model using asymptotic approximations to the Smoluchowski–Einstein equation coupled with the generalized Poisson–Nernst–Planck (GPNP) equation, or Poisson–Fermi–Nernst–Planck (PFNP) equation for modeling the electrodiffusive dynamics for estimating both the pore conductance and pore energy. These models account for the complex dynamics that are present in the EMP which results from asymmetric electrolytes, multiple ionic species, the Stern and diffuse electrical double layers, and the polarization of water. The GPNP is equivalent to the PNP system of equations if steric effects[11] are neglected. Additionally, the PFNP is equivalent to the GPNP if the polarizability of water is neglected. Note that near a pore entrance significant nonlinear potential gradients are present which restrict the current flowing through the pore; this effect is denoted as the "spreading conductance" and is dominant for pore radii significantly larger than the membrane thickness, causing the pore conductance to scale proportionally to the pore radius. The pore conductance scales proportionally to the pore radius for all pore radii, suggesting that the "spreading conductance" is dominant as a result of the electrode in proximity to the membrane surface and the nonlinear potential gradients present. This effect only becomes pronounced for large pores in free-floating membranes.

3.4.4 Device 4: Electrophysiological Response Platform (ERP)

The ERP is an engineered membrane platform designed to measure the electrophysiological response of ion channels and cells. Details of the construction and experimental measurements using the ERP are in Chapter 6. Using a fractional-order macroscopic model in Chapter 11, we illustrate how the electrophysiological response of ion channels and cells can be computed from the measured current response of the ERP.

Classical methods for the electrophysiological measurement of ion channels include patch-clamp electrophysiology in mammalian cells and two-electrode voltage clamp in *Xenopus oocytes* [107]. Both of these methods produce extremely information-rich data which can be used to validate ion-channel gating models. However, these approaches are labor intensive and require highly skilled staff to ensure reproducible results are obtained. Embedding ion channels in a controllable tethered membrane environment allows the electrophysiological measurement of ion channels in a platform that requires approximately 20 minutes to form using standard laboratory techniques. The membrane

---

[11] The steric effect results in a repulsive form between molecules as a result of the finite size of the molecules. A detailed description of the steric effect is provided in §3.6.

and electrolyte environment of the ERP can be controlled with embedded ion channels to study the voltage-gating dynamics in an environment that mimics the native environment of the ion channel. The ERP and dynamic model can therefore be used for high-throughput drug screening and the validation of ion-channel gating models.

There are two major methods for measuring the electrophysiological response of cells. The first is to use substrate-integrated microelectrode arrays, and the second is to use sharp or patch microelectrodes that puncture the cell [400]. A limitation of invasive cell measurement that employs sharp and patch microelectrodes is that a limited number of cells can be analyzed for a short period of time. Substrate-integrated microelectrode arrays provide a noninvasive method for measuring the electrophysiological response of cells; however, a major challenge when using these sensors is to ensure sufficient cell adhesion and coverage [46, 400]. Getting sufficient cell coverage to an extent that impedance measurements down to 10 Hz can be read is challenging. Leaks between cells provide membrane resistances that limit the power spectrum to 50–100 kHz, eliminating responses due to ion-channel currents which occur near the 10-Hz frequency regime. An emerging technology to ensure cell adhesion is to use a metal electrode coated with a polycationic film onto which an adhesion protein, such as Glycocalyx[12], is used to bind with the cell membrane [400]. Another popular glycoprotein, that can be purchased with a PEG linker and thiol attached, is Fibronectin, which can bind to membrane-spanning receptor proteins [338]. The ERP is designed to promote cell proliferation, allowing cell cultures to reach a sufficient coverage and adhesion to allow for their electrophysiological measurement. Remarkably this allows the response to be measured using a noninvasive technique in the proximity of a synthetic membrane that mimics the electrophysiological response of biological membranes.

3.5 Multiphysics Dynamic Models of Engineered Tethered Membranes

How can an engineered membrane consisting of a synthetic lipid bilayer, spacers, tethers, and a bioelectronic interface be modeled mathematically? Such models are crucial for predicting how devices built out of engineered membranes operate and are important for designing improved devices. In simple terms a membrane is a dynamic system (i.e., a system with memory) and it evolves over time according to the laws of physics. Indeed, the membrane has both movement dynamics (kinematics) and electrical dynamics. In this section we briefly introduce several levels of abstraction (atomistic to macroscopic) that can be used to model membrane dynamics and provide insight into the construction of the multiphysics models presented in this book. In Part III, we study these levels of abstraction in much more detail; see also [39] for a beautiful exposition on multiphysics models.

A multiphysics model comprises multiple physics-based models, each of which accounts for different physical phenomena at a different time scale. That is, each

---

[12] Glycocalyx is a macromolecule that the human body uses to distinguish between its own healthy cells and transplanted tissues.

Levels of Abstraction				
Ab Initio Molecular Dynamics	Classical Molecular Dynamics	Coarse-Grained Molecular Dynamics	Continuum Theory	Macroscopic Theory
nm fs	nm ns	nm μs	μm μs	m s

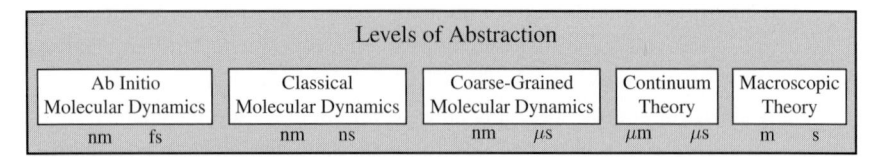

Figure 3.8 Schematic diagram illustrating the length and time scale achievable by the atomistic to macroscopic simulation methods in §3.5: nm is nanometers (1×10^{-9} m), μm is micrometers (1×10^{-6} m), fs is femtoseconds (1×10^{-15} s), ns is nanoseconds (1×10^{-9} s), and μs is microseconds (1×10^{-6} s).

physical model accounts for specific physical phenomena at a certain level of abstraction as illustrated in Figure 3.8. These levels of abstraction go from femtoseconds (time scale) and nanometers (spatial scale) to minutes and centimeters. To construct a multiphysics model, one must carefully interface the physical models at the various levels of abstraction, so that a model at a lower level of abstraction can be used to determine the parameters of the model at a higher level of abstraction.

Key to the development and improvement of multiphysics models of novel biosensing platforms is formulating accurate physical models of the membrane, macromolecules, and the bioelectronic interface.

To be of practical use, a multiphysics model must link the atomistic dynamics of water, ions, membrane lipids, proteins, and peptides to experimental measurements of the device at the macroscopic time scale. It is essential that the multiphysics model be based on physical principles and result in computationally tractable simulation algorithms. The computational tools of physics employed in this endeavor, from fundamental to phenomenological, are ab initio molecular dynamics, classical molecular dynamics, coarse-grained molecular dynamics, continuum theories, and reaction-rate theory. These approaches make various levels of abstractions in replacing the complex reality with tractable models. Each of these approaches has its strengths and limitations and involves a degree of approximation. Figure 3.8 provides an overview of the length scale and simulation time horizon achievable by general atomistic to macroscopic models.

3.5.1 Ab Initio Molecular Dynamics

At the lowest level of abstraction is ab initio (quantum-mechanics-based) molecular dynamics, which combines Newton's laws of classical physics with Schrödinger's equations of quantum physics. Ab initio molecular dynamics has no free parameters and therefore represents the ultimate tool for modeling tethered membrane platforms. However, performing ab initio molecular dynamics simulations is often computationally intractable since it involves solving multiparticle Schrödinger equations. Currently, the attainable length and simulation time horizons are on the order of a few nanometers and femtoseconds [203]. Ab initio models are not used in this book; however, the force field parameters in molecular dynamics (Chapter 15) are computed using quantum mechanics.

3.5.2 Molecular Dynamics

A simplification to ab initio molecular dynamics is to replace the potential energy function in the many-body quantum-mechanical Schrödinger equation by a phenomenological one. At this level of abstraction the force field is empirically parameterized to describe the pairwise interaction between ions with the dynamics of ions evaluated using Newton's equation of motion. This simulation method is known as molecular dynamics [237]. Molecular dynamics simulation models for engineered membranes are discussed in Chapter 15. Using molecular dynamics allows equilibrium thermodynamic and dynamic properties of a system at finite temperature to be computed assuming that the interatomic forces are known a priori.[13] Though molecular dynamics simulation only requires numerical evaluation of Newton's laws, the typical system size and simulation time achievable are on the order of nanometers and nanoseconds [267].

Computing important biological parameters in tethered membranes requires a simulation time horizon on the order of microseconds with a system size of tens of nanometers. In Chapter 15 we use molecular dynamics simulations to compute important parameters and processes, including the dynamics of aqueous pore formation, the electrostatic potential, intrinsic membrane dipole potential, and membrane capacitance. Additionally, we discuss the ion permeation dynamics of the gramicidin ion channel, and the dissociation dynamics of gA dimers.

3.5.3 Coarse-Grained Molecular Dynamics

One possible computational simplification of molecular dynamics[14] is to group atoms together into coarse-grained beads, with the bead-to-bead interactions empirically parameterized, allowing their dynamics to be evaluated using Newton's equation of motion. This method of abstraction is known as coarse-grained molecular dynamics (CGMD). Methods for constructing CGMD simulations for engineered membranes are discussed in Chapter 14. In CGMD, the force field is empirically parameterized by matching the bead dynamics to appropriate experimental data and all-atom molecular dynamics simulations. This allows the CGMD model to achieve simulation time horizons of microseconds with a system size of tens of nanometers.

In this book, the MARTINI force field [268, 321] is utilized for all CGMD simulations of the tethered membranes (see Chapter 14). The MARTINI model is based on taking four heavy atoms (e.g., four carbon atoms and hydrogen, or four water molecules) and representing this structure by a single coarse-grained bead. The MARTINI force field contains 18 possible beads which can be used to represent lipids and amino-acid sequences [267]. The MARTINI force field has been used for the study

---

[13] In ab initio molecular dynamics the interatomic forces are computed "on the fly" using quantum-mechanical electronic structure calculations. Quantum-mechanical effects play an important role in chemical bonding processes.

[14] Another possible simplification is to use Brownian dynamics. In Brownian dynamics water molecules are replaced by Brownian motion – explicit water is replaced by implicit water. Then the individual particles evolve according to Langevin's stochastic differential equation. We do not use Brownian dynamics in this book.

of tethered membranes [245, 451] and the oligomerization of peptides (refer to citations in [267]). In [245] the membrane consists of DOPC lipids and pegylated DOPC lipids with different tethering lengths. The simulation results show that decreasing the length of tethers increases the stability of the membrane. In [451] the membrane consists of DPPC lipids, pegylated DPPC lipids, and surfaces with hydroxyl-terminated β-mercaptoethanol, which is typically used to prepare the gold surface for tethered membrane assembly. The results in [451] provide the formation dynamics of the tethered membrane for different lipid concentrations and show that, once a sufficient concentration of lipids is present, a tethered bilayer lipid membrane will self-assemble.

The CGMD model of the ERP and ERP is composed of custom coarse-grained structures which model the bioelectronic interface, tethers, spacers, lipids, and antimicrobial peptide peptidyl-glycylleucine-carboxyamide. These structures are based on those reported in [245, 267, 451] and constructed using the methods provided in [268, 284]. The CGMD models of the tethered membranes are constructed with a focus on computing important biological parameters (e.g., diffusion, line tension, and surface tension) which can be used in continuum and macroscopic models. This allows the results from the CGMD simulations to be validated using experimental measurements from the PFMP and ERP.

3.5.4 Continuum Theories

A higher level of abstraction than CGMD is to apply the mean-field approximation [482], which allows the dynamics of ions to be modeled by a continuum theory. In mean-field theory, ions are treated not as discrete entities but as continuous charge densities that represent the space-time average of the microscopic motion of ions. This allows continuum models to achieve simulation time horizons of microseconds with a system size of micrometers. The most well-known continuum theory model for ion transport is certainly the Poisson–Nernst–Planck system of equations, which combines the Poisson equation from electrostatics and the Nernst–Planck equation for diffusion [176, 224, 253–255, 265, 280, 312, 353, 371, 483]. Primarily in a biological context, the PNP theory is used to model ion transport through ion channels and nanopores [80, 167, 176, 224, 265, 280, 371]. Continuum theories are employed in this book to relate the diffusion and chemical reactions of molecular species to changes in membrane conductance. When steric effects (effects that occur due to finite atom sizes) are significant, we utilize the generalized PNP continuum model to compute the conductance of aqueous pores and the electrical energy required to form a pore in the tethered membranes. These are necessary for computing the dynamic response of the tethered membrane to an elevated transmembrane potential. Continuum theories of engineered membranes are presented in Chapters 10 and 11.

3.5.5 Reaction-Rate Theory

The highest level of abstraction considered in this book is macroscopic reaction-rate theory. This is a phenomenological black-box input-output model for the macroscopic behavior of the engineered membrane, namely, how the current flowing through the membrane depends on the applied voltage and concentration of chemical species present. The key point is that the reaction-rate model is a system of nonlinear ordinary differential equations. The coefficients of the differential equations (model parameters) do not require a direct physical interpretation (hence the phrase black-box model). Since the parameters of the reaction rate do not have a direct physical interpretation, the length scale and time scale achievable are arbitrarily large (on the order of centimeters and hours). In this book, using a novel extension of the reaction-rate model – called the fractional order macroscopic model – and experimental measurements, we illustrate how important biological parameters can be estimated that include membrane conductance and electrode double-layer charging dynamics. Reaction-rate models are presented in Chapters 8 and 9.

3.6 Electrolyte Dynamics: Steric Effects and Double-Layer Charging

Several physical phenomena occur in the electrolyte solution that surrounds the tethered membrane and at the boundary of the bioelectronic interface. These physical phenomena, collectively called electrolyte dynamics, include steric effects, Coulomb correlations, polarization and screening effects, charge accumulation, diffusion-limited charge transfer, reaction-limited charge transfer, and ionic adsorption dynamics. The models constructed in this book can be used to account for all these electrolyte dynamics in engineered membranes.

Steric effects. In electrolytes, atoms occupy a certain amount of space since each atom has a finite size. The size of the atom is associated with the overlapping electron clouds (Pauli interaction or Born repulsion) between atoms. If two atoms are brought too close together, then a *steric* repulsive force occurs between the two atoms. Steric effects play an important role in the dynamics of electrolytes as they govern the maximum possible number of atoms that can fit in a specified volume of space. Additionally, steric effects also affect the diffusion dynamics of ions and molecules. In electrolytes, the movement of ions and molecules favors low-concentration regions which minimize the steric repulsive forces between atoms.

Coulomb correlations. Coulomb correlations (also known as electronic correlations) result from the interaction of electrons between atoms and molecules. In an atom, electrons occupy certain volumes of space as defined by the atomic orbital. The atomic orbital is a mathematical function from quantum mechanics which defines the probability of finding any electron of an atom in any volume of space around the atom. If the electrons move independently of the other electrons then the electrons are uncorrelated. In such cases the probability of finding the electrons in a certain volume of space can be evaluated using the Hartree product wave function. However, as we know from Coulomb's law, an electrostatic force (either attractive or repulsive) exists between charged atoms. This electrostatic force is a result of electrons interacting (i.e., the movement of electrons is correlated) and results in Coulomb correlations. Formally, to estimate the effects of Coulomb correlations on the electrostatic force between atoms requires solving the many-body Schrödinger equation to estimate the wave function as a function of space and time for all the electrons in the system. For engineered membranes this is a computationally prohibitive task. Therefore, Coulomb correlations are accounted for by assuming each atom has an associated charge and then using Coulomb's law to compute the electrostatic force between atoms. Coulomb's law states that the force between two point charges is proportional to the product of their charges, and inversely proportional to the square of the distance between them. If the molecules are interacting in a dielectric medium (an electrolyte that can be polarized, e.g., water), then to account for polarization effects the proportionality constant in Coulomb's law also includes the dielectric constant of the medium.

Polarization and screening effects. The polarization effect in electrolytes results from the ability of water to form dipoles[15] as a result of ionic gradients which form an electric field. Formally, the electric polarization effect results from the relative tendency of a charge of the electron cloud of an atom or molecule to have its charges displaced by any external electric field. The formed dipoles will then rotate to align with the externally applied electric field. This realignment of the formed dipoles will cause an overall reduction in the electric field. In addition to the realignment of the dipoles, ions can also move to reduce the electric field. The overall reduction in the electric field caused by the movement of dipoles and mobile charges is known as the screening effect (or electric-field screening effect). Note that the polarization effect and screening effect can be accounted for by the dielectric constant of the medium. These effects play an important role in engineered membranes when a large electric field is applied across the electrodes of the engineered membrane.

Charge accumulation. An electrostatic potential applied across the bioelectronic interface (equivalently, across the electrode to electrolyte interface; see Figure 3.7 on page 41), will promote the movement of oppositely charged molecules or ions to the electrode surface. The increase in charges at the electrode surface is known as charge accumulation. The movement of the molecules and ions to the electrode surface can be modeled using the Nernst–Planck diffusion equation, introduced in Chapter 10. However, if the movement of the ions and molecules cannot be modeled using the Nernst–Planck diffusion equation, then a combination of diffusion-limited charge transfer, reaction-limited charge transfer, and ionic adsorption dynamics may be present. Diffusion-limited charge transfer occurs when the diffusion coefficient of the ions or molecules is dependent on the concentration of these ions and molecules at the surface of the electrode. Reaction-limited charge transfer and ionic adsorption dynamics occur when the flux of ions and molecules to the surface is dependent on a surface reaction. An example of a reaction-limited charge transfer or ionic adsorption dynamic reaction is the faradic reaction that can occur at an electrode surface. In Chapter 11 we illustrate how faradic reactions at the surface of the electrode can be modeled using the generalized Frumkin–Butler–Volmer equation.

---

[15] An electric dipole results from the separation of positive and negative charges. In a molecule, the electric dipole moment is used as a measure of the separation of positive and negative charges.

3.7 Future Technologies: Implantable Medical Devices, Diagnostics, and Therapeutics

We conclude this chapter with a speculative discussion of key challenges in creating a new generation of active implantable medical devices using engineered tethered membranes. Although the rest of the book does not discuss such implantable devices, the engineered membranes we will build and mathematically model in Parts II and III are directly relevant to such devices. Therefore, this section serves as additional motivation for the rest of the book.

Implantable medical devices have applications in restoring or enhancing human capabilities and, more significantly, in redesigning the maintenance of human health. These advances will be achieved through nanometer-scale integration of external electronics with the human body. Specifically we describe below how engineered tethered membranes can be used in health care for the design of cochlear and retinal implants, in vitro medical diagnostics, molecular therapeutics, and the study of biological neural network communication. The construction and formation techniques of engineered membranes presented in Part II of the book provide the basis for the design of these medical devices.

3.7.1 Cochlear and Retinal Implants

Engineered tethered membrane devices can potentially enhance the performance of medical implants. Cochlear implants are now widely used. Although capable of restoring hearing for speech, they fall short of permitting an appreciation of music.[16] Retinal implants at present provide little more than assistance for navigation through brightly lit windows and doorways. It is becoming clear that simply increasing the electrode number and minimizing electrode size does not improve the performance of such implantable devices. The reason such high-resolution implant devices do not currently exist is due to major shortcomings in their electrode design. The key to the design of such high-resolution implants is to properly interface the electrodes to cells; this can be achieved using engineered tethered membranes.

A new generation of active implantable devices will permit the implementation of a new principle of health management – specifically, the use of neural arcs, which are naturally present in biology, for stimulation. An implantable device will participate in the information pattern through both reading and stimulating neuronal cells and other active tissue using these neural arcs. This participation will guide active cell populations away from deviant or diseased behavior back to normal function as gauged by reference to a previously recorded healthy average.

The starting point for communicating with neuronal cells and tissue is to construct an interface between the engineered tethered membrane and the biological membrane of an active cell line. The bilayer lipid membrane is essential for all sensing phenomena (sight, sound, taste, touch, and smell). Also, all signaling pathways in human physiology (action potentials, muscle innervation, and cell apoptosis) depend on lipid membranes.

---

[16] Refer to www.youtube.com/watch?v=SpKKYBkJ9Hw.

The engineered tethered membrane possesses the necessary skeleton to which novel chemistries can be added to provide the first iteration of correctly integrating electronic coupling with biological tissue. The major challenges that engineered tethered membranes overcome compared with the latest generation of implantable device electrodes is that they do not introduce toxins which can interfere with the normal operation of the cell, and they can be precisely positioned to be a few nanometers from the cell surface.

In more detail, signaling between cells and the electrode is conveyed by ion fluxes. All information between the cell and electrode is essentially differentiated with time as the electrode and cell surface double layers charge and discharge with ions. That is, if a constant voltage is applied then the current flowing between the cell and the electrode will decay to zero. This type of charging behavior is identical to that observed when charging an ideal capacitor. Ohmic contacts[17] (that act like an ideal resistor) cannot be used in implantable devices, owing to the rapid ablation of reactive metals in physiological salt solutions. In addition, silver, the most commonly used ohmic metal in implantable devices, is highly toxic and blocks monovalent ion channels. Interfacing with cells using engineered tethered membranes does not introduce any toxic molecules into the electrolyte. Additionally, the passage of ion fluxes between cells into the interstitial voids causes a major loss of read and excitation sensitivity.[18] Current approaches to electrode design attempt to achieve maximal efficiency by using large electrodes; however, these are still located at significant distances from the target cells. Another method is to use multielectrode arrays in the hope that at least a small number of electrodes are sufficiently close to a target cell to read or stimulate that cell. Distant excitation and overstimulation causes the stimulation of cells adjoining the target cell, causing a loss of resolution of information being conveyed from an electrode to a specific cell. Engineered tethered membranes overcome this challenge since the tethered membrane surface is only a few nanometers from the cells. Therefore, precise excitation and measurement of specific cells can be achieved.

The ultimate challenge for implantable devices is the design of microelectrode arrays of engineered tethered membranes that have a size on the scale of cells which are on the order of micrometers.

3.7.2 In Vitro Medical Diagnostics (IVDs)

Engineered tethered membranes can potentially replace the diagnostic therapeutic open-loop model of clinical assessment and treatment. At present medical care consists of a diagnostic assessment followed by phenomenological therapeutic treatment. The

---

[17] Ohmic contacts are composed of electrodes where oxidation and reduction reactions occur on the electrode surface, and where the current flowing through the electrodes is directly proportional to the voltage across the electrodes. The cathode is where reduction occurs, and the anode is where oxidation occurs. Cations (positively charged ions) move toward the cathode to be reduced, while anions (negatively charged ions) move away from the cathode. At the anode, anions react with the surface and release electrons which flow from the anode to cathode, resulting in an electric current between the electrodes. Note that the electrolyte ions exchange electrons directly with these ohmic contacts.

[18] A figurative example is transporting liquid in a colander. After a short time period, most of the liquid is lost.

process is both open loop and statistical. Drug trials provide, at best, only guidelines for dosage and outcomes. Also they require repeated clinical intervention, which can slow and mask the efficacy of positive results of the treatment.

In vitro medical diagnostics (IVDs) are devices that use physical techniques to report on biological circumstances, usually molecular marker concentrations, as evidence for subsequent treatment. These diagnostics can be categorized into four major areas:

(i) invader detection (bacterial, viral, prion, and metastasis cancer markers),
(ii) degradation marker detection (troponin I, CK-MB [creatine kinase for muscle type and brain type], myoglobin [cardiac markers], and amyloids [Alzheimer's disease],
(iii) regulatory marker detection (cholesterol, glucose, genetic, and gut microflora), and
(iv) physical metrics (repetition rates, temperature, and pressure).

These devices are complemented by clinical assessments of perception and stasis.

Modern IVDs arose from immunoassays first reported as radioimmunoassays and enzyme-linked immunosorbent assays more than 60 years ago. They have generated a major industry built around centralized pathology services. In the 1980s, a smaller point-of-care (POC) diagnostic industry emerged, focusing principally on enzyme electrodes for glucose testing. More recently POC diagnostics have experienced a significant growth of interest for identifying prohibited substances by law enforcement agencies. POC immunodiagnostic kits are almost exclusively based on an inexpensive, mass-produced, adaptation of paper chromatography known as the lateral flow assay (LFA). LFA kits are most commonly encountered as human chorionic gonadotrophin immunodiagnostic pregnancy test kits.

In the early 1980s the polymerase chain reaction process also bolstered interest in IVDs and created a new field of molecular diagnostics. Molecular diagnostics complemented immunodiagnostics with the detection of trace amounts of genetic sequences. Although greeted with enormous excitement in the fields of forensic science, genealogy, and in vitro fertilization screening applications, molecular diagnostics is of limited commercial interest as a mainstream medical diagnostic due to the relatively limited number of genetic diseases.

Molecular diagnostics had a further boost of interest ten years ago with the proposal that POC devices could perform rapid gene sequencing, based on a novel sequencing method using nanopores or ion channels. By passaging single-stranded DNA through nanopores or naturally occurring channels such as α-hemolysin, each of the four bases is found to occlude channel pores to a varying extent. This results in a sequence of current levels from which the base sequence of the DNA molecule may be determined.

Although much remains to be explored in improving the throughput and performance of in vitro diagnostic assays, the basic insights of using radioactivity or enzyme turnover for transducing a response from a primary detection mechanism are well known. Furthermore, despite recent developments such as quantum dots and single-molecule detection schemes appearing to have advanced IVD performance, the rate-limiting step remains the kinetics and affinity of the primary detection event. These invariably employ poly- or monoclonal antibodies which again are discoveries from over half a century ago.

Post 2000, in vitro medical diagnostics have become a commercial commodity. More than three million publications are currently listed under medical diagnostics on the Web of Science. It is now widely accepted that IVD science has matured both commercially and as a scholarly topic. It is unlikely that a major research investment into in vitro medical diagnostics will yield a profound scientific or technical advance. The major corporate laboratories such as those at Johnson & Johnson, Becton Dickinson, and Abbott have sophisticated research and development teams that are continually trawling the extensive academic research community, and advances will most likely be evolutionary and be shared with many others.

3.7.3 Molecular Therapeutics

The utility of molecular therapeutics is described in terms of empirical probabilities, with each new drug requiring extensive clinical trials before being released for general use. Therapeutics depend on the interaction of usually well-defined compounds with very complex and poorly defined biological systems. Engineered tethered membranes can be used to play a pivotal role in the design of molecular therapeutics as they provide a precisely controllable platform that mimics biological membranes of use for evaluating novel therapeutics.

The long-term utility of a new drug can be categorized into two major areas:

(i) repelling infection (antibiotics) and
(ii) function control (pain control [opioids, cannabinoids], metabolic modulation, ion-channel blockers [β-blockers, β-2 agonists, corticosteroids, fertility control], vaccines, and gamma globulins).

Therapeutics operate at both the multicellular system level and the level of individual cells. Operation at the single-cell level impacts the local cell population and subsequently elicits a secondary global response. Often the target will be a molecular interaction or supramolecular function. Typical examples range from the design of antibiotics to lower bacterial resistance, to designing antagonists, to cardiac ion channels such as the human ether-à-go-go-related gene (hERG) that modify cardiac muscle function. Active implantable medical devices constructed out of engineered tethered membranes will participate in the communication between cells and to combine the currently separate functions of diagnostics and therapeutics. This will be achieved by detecting ionic fluxes from individual cells and then stimulating the cellular array to reinforce desirable patterns of interaction between populations of cells. Notice that this is a very different approach than that currently employed by the pharmaceutical industry, in which an open-loop model is used where efficacy is gauged by clinical trials modulated only by the slow and noisy feedback from users.

3.7.4 Biological Neural Networks

Engineered tethered membrane arrays hold the promise of being able to observe and control the response of biological neural networks comprising interconnected neuron cells. Insight into how neurons communicate and how neural networks operate is critical for the design of electrophysiological neuronal therapies. An artificial neural network can be viewed as a parallel computational system with artificial neurons connected together to perform a specific task. However, since they comprise very crude approximations to biological neurons, they cannot be used to reproduce the complex dynamics that are present in biological neural networks.

Biological neural networks can be viewed as randomly interconnected Boolean elements that possess a delayed response. A fascinating example of the consequences of order being created from a randomly wired neural network may be seen at the following link: www.youtube.com/watch?v=h2ljAjI1GDA. The initial array of lights reflecting the status of each neuron within the neural network array starts out as a series of randomly distributed dots. Once one neuron of the neural network is activated, it takes only seconds to see a pattern of lights form that remains for several minutes. Such patterns are a consequence of large numbers of interconnected Boolean elements establishing a reinforced pathway of interaction. Such patterns may provide a clue to apparent complex and purposeful behavior by multicellular organisms.

A similar logic is applied to the ability of engineered tethered membranes to permit large populations of cells to be monitored through connection to a subset of the whole population. That subset would be active cells captured onto an engineered tethered membrane. Patterns of interconnected cellular ion currents of the whole population would be sampled by the cells on the electrode array. An example that could have a profound outcome would be to culture a population of pancreatic eyelet cells over an engineered tethered membrane array. As the cells detect glucose and secrete insulin, the tethered membrane array permits the detection of any patterns that indicate the correct functioning of the eyelet cells. The challenge is to devise electrical excitation algorithms, driven from the engineered tethered membrane, to reinforce the correct cell function should the eyelets start to malfunction.

3.7.5 Microeletrodes and Single-Cell Measurements

Novel engineered tethered membranes to measure and control the response of individual cells provide opportunities for the design of novel therapies. For example, physiological disorders (e.g., long Q-T syndrome in sudden cardiac death) are intrinsically multicellular communication failures rather than molecular functional faults within a single molecular population or within a single cell. Engineered membranes can be used to measure the source of such multicellular communication failures. Another example is the modulation of the behavior of eyelet cells, which would open the way to an implantable cell culture dialysis pouch for the treatment of type 2 diabetes. Experience in fabricating and characterizing these microelectrode tethered membranes requires that the leakage pathways between cells and the membrane surface are on the order of a few Ω/cm^2. This reinforces the fact that, in the absence of a well-sealed contact to a target cell, little or no ion gradient will be available to stimulate that cell. Similarly, the ion fluxes and gradients from a distant isolated cell from the electrode surface will soon dissipate and be lost, preventing any measurement of the cellular response.

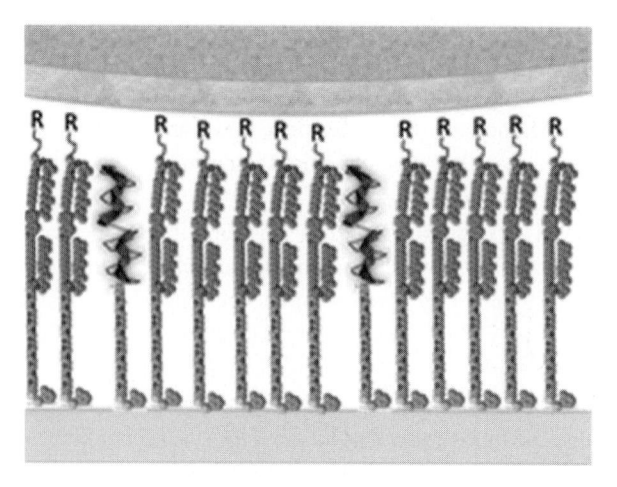

Figure 3.9 A color version of this figure can be found online at www.cambridge.org/engineered-artificial-membranes. For active cell adhesion, molecules "R" can be incorporated into the tethered membrane to facilitate the capture and close contact of the cells to the tethered membrane surface. Further elements included will be analogs of tethered bis-gramicidin A, illustrated in green, to permit ready access to ions between the tethering reservoir and the cell membrane. The interior of the cell is illustrated by the pink region, and the cell membrane by the light grey region. Generating electrode coatings without an engineered tethered membrane layer can cause nonspecific binding to the electrode and a loss of electrode efficiency.

Tethered membranes can be constructed with functioning assemblies of lipids and ion channels to make a molecular device with moving parts that can perform the task of a diagnostic device (e.g., the ICS biosensor). The ability to create such complex constructs demonstrates that similar engineered tethered membranes can be fabricated containing cell adhesion molecules (e.g., integrins and cadherins selectins) in their outer surface. An initial ex vivo approach that demonstrates the ability to capture cells at the surface of engineered tethered membranes has been demonstrated using the lectin conconavilin A. Figure 3.9 illustrates this method of fixing cells to the surface of an engineered tethered membrane. Cells attached in this manner were healthy and continued growing for hours. Achieving this quality of seal to the cell, with only the dimensions of the adhesion proteins between the surface of the tethered membrane and the cell surface, will permit both individual cell stimulation and cellular measurements.

Maintaining the longevity of cells chemically attached to the surface of tethered membranes is a challenge. However, it is well known from the function of pacing leads in cardiac pacemakers that, once the initial two-week encapsulation period is past, the leads continue stimulating and reading the ion currents from the cardiac pacing centers in the heart for typically more than eight years. Furthermore, a commercial venture launched in 2012 under the name of Axiom Corp. has a product for studying action potentials of cells grown over microelectrode arrays. A biocompatible polymer is coated over the array and the end user seeds cells onto the polymer surface and waits until the cells grow to confluence to provide a sufficiently sealed surface to permit the observation of ion current arising from the cell's action potential. The obvious shortcoming of this

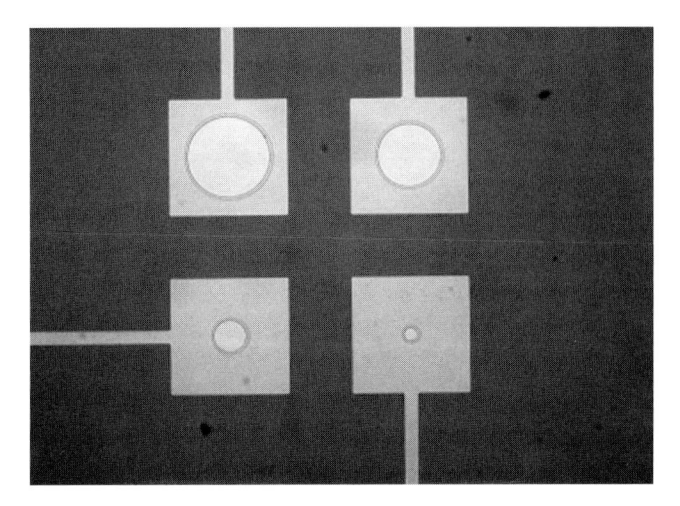

Figure 3.10 Example of a engineered tethered membrane array comprising electrodes with diameters in the range of 20 to 150 μm.

approach is that only a limited subset of cells, the epithelial cell lines in particular, readily forms well-sealed layers. Neuronal cell lines,[19] which are of the most interest in action potential detection, seldom form well-sealed layers, preventing their detection. Engineered tethered membranes designed with cell adhesion molecules provide a path to measuring the response of neuronal cells.

Performing single-cell measurements requires the design of microelectrode arrays. The key design challenge for such tethered membrane devices is to ensure the surface roughness and contamination of the electrode is minimized. Cleanliness of the gold surface is a particular challenge as classical photolithographic approaches to making microelectrodes require that the masking layers be removed chemically. Our recommended approach is to follow the photolithography with an ion milling step to eliminate residues from the wet chemistries. Further care must be taken not to roughen the surface by the ion milling. That is, the electrode surface roughness must be less than 1 nm to permit the formation of a tethered membrane with minimal defects. Engineered tethered membranes with electrodes of this quality will provide excellent coupling efficiency with the target sensitivity being one cell to each individually read electrode in a microelectrode array of tethered membranes. Figure 3.10 illustrates an example of a microelectrode array on which we have successfully formed engineered tethered membranes. To date, the smallest electrode diameter we have fabricated is 20 μm. The next goal is to design protein receptors to perform cellular measurements using these microelectrode arrays. Ultimately, electrode arrays will be required with individual elements of order of the dimensions of a cell (μm) with specialized methods for performing experimental measurements that can detect nano-Coulomb changes in charge that result from ions released by individual cells.

---

[19] A cell line is a cell culture that is formed from a single cell and therefore is composed of homogeneous cells.

3.7.6 Summary

In this section we discussed how engineered tethered membranes can potentially be applied to the design of molecular therapeutic protocols, in vitro and in vivo medical diagnostics, and implantable medical devices. The possible applications of engineered tethered membranes, although challenging, appear almost inevitable given the progress now made in the design of tethered membranes. At the most basic, the results from engineered tethered membranes can be used to construct rules of thumb for the rational design of antimicrobial peptides. These would be similar to Lipinski's rule of five[20] for the design of oral drugs [244]. At its most ambitious, it takes the open-loop diagnostic treatment of contemporary medicine and recreates an autonomous health process that closely parallels that which very likely evolved in nature. This also allows for the design of specific electrophysiological signals that can be used to inhibit cell growth and multiplication. A further outcome is that insights will be gained on ways of tailoring the activities of multicellular organisms to perform tasks more suited to our needs than their original evolved functions. The ultimate speculative goal is to work through a hierarchy of human health disorders, identify the dysfunctional metabolic cell populations, and then generate diagnostic and/or therapeutic protocols to provide autonomous treatments using engineered tethered membranes.

3.8 Closing Remarks

In this chapter we introduced several different types of engineered artificial membranes designed to mimic biological membranes, namely, the hybrid bilayer lipid membranes, the supported-lipid bilayer, the freestanding lipid bilayer, cushioned membranes, and tethered bilayer lipid membranes. Recall that a biological membrane comprises lipids, macromolecules, and cytoskeletal filaments. A common theme of all of these engineered artificial membranes is that the membrane is adjacent to a flat support to ease in the experimental measurement of membrane properties. We further discussed six classes of tethered bilayer lipid membranes: peptide tethered membranes, protein tethered membranes, cholesterol tethered membranes, polymer tethered membranes, avidin/biotin tethered membranes, and engineered tethered membranes. Throughout this book we focus on the experimental analysis and dynamic modeling of engineered tethered membranes. Note, however, that although we focus on engineered tethered membranes, the insights, experimental techniques, and dynamic models are useful for all the engineered membrane architectures presented in Figures 3.3 and 3.5. The chapter concluded with a speculative discussion of future applications of engineered tethered membranes.

---

[20] Lipinski's rule of five states that an oral drug should satisfy the following four conditions: molecular mass less than 500 Da, an octanol-water partition coefficient not greater than 5, at most five hydrogen bond donors, and fewer than ten hydrogen bond acceptors. The nomenclature "rule of five" is used because the quantities in the four rules are all multiples of 5.

Part II

Building Engineered Membranes, Devices, and Experimental Results

Part II of this book deals with the construction, formation, and operation of engineered tethered membrane devices. Detailed descriptions are provided on the molecular components of engineered membranes, methods for inserting peptides and proteins, and how to measure the structure and dynamics of these biomimetic devices.

Four biomimetic devices built out of engineered tethered membranes are discussed: the ion-channel switch (ICS) biosensor, the electroporation measurement platform (EMP), the electrophysiological response platform (ERP), and the pore formation measurement platform (PFMP). Several real-world examples are provided such as how the ICS biosensor can be used for the rapid detection of influenza A, how the PFMP can be used to infer the pore-formation dynamics of the antimicrobial peptide PGLa (peptidylglycylleucine-carboxyamide), how the EMP can be used to study the membrane conductance dynamics, and how the ERP can be used as a noninvasive method for measuring the response of ion channels and cells.

We also discuss how to perform experimental measurements using engineered tethered membranes to determine the structure and dynamics of the membrane and macromolecules in the membrane. The experimental techniques discussed include electrical measurements, and spectroscopy and imaging techniques (e.g., X-radiation refractometry, neutron reflectometry, fluorescence recovery after photobleaching, and nuclear magnetic resonance). These measurement methods not only yield important biological details about the membrane but also verify its structure. Since engineered tethered membranes mimic real biological membranes, the experimental studies reported involving antimicrobial peptides, electroporation, growth of cells, and other properties of the membrane give significant insight into how biological membranes function.

Parts II and III of the book together give a complete account of how to engineer artificial membranes: building them, mathematically modeling their dynamics, and interpreting and refining their design. The reader interested in mathematical modeling of engineered membranes can read §4.1, §§5.1–5.4, and parts of Chapter 6 before proceeding to Part III. For a laboratory exercise on building engineered tethered membranes, refer to §4.4.

4 Formation of Engineered Tethered Membranes

4.1 Introduction

Engineered membranes are synthetic lipid bilayers that are used to mimic biological cell membranes in biophysical and physiological applications. Recall from Chapter 3 that engineered membranes are a fundamental component in building high-resolution sensing devices (such as the ion-channel switch [ICS] biosensor), building stable platforms for measuring the electrophysiological response of cells, and also for using diagnostic tools and implantable devices. From a more fundamental point of view, engineered membranes allow us to study the dynamics of lipids, proteins, peptides, and cells in a controlled environment.

Recalling the various types of artificial membranes summarized in Figure 3.2 on page 32, this book focuses exclusively on artificial tethered bilayer lipid membranes (tBLMs). The atomic structure of these engineered tethered membranes has been determined using several experimental techniques such as electrical measurements, X-ray reflectometry, and neutron reflectometry as will be discussed in Chapter 7. The unique feature of such membranes is that they are tethered to a conductive solid surface, namely, the bioelectronic interface. Tethering is used to improve the mechanical stability of the engineered membrane. The conductive gold surface to which the membrane is tethered also provides a convenient electrode to generate localized ion fluxes.

This chapter discusses two important synthesis aspects of engineered tethered membranes:

(i) How does one build a tethered membrane?
(ii) How does one insert specific lipids, proteins, and peptides into the membrane?

We stress that engineered tethered membranes have several advantages compared to other types of artificial membranes. One can precisely control the membrane stability via the tether density. Also a tethered membrane can withstand electrochemical excitations in the range of -900 to $+900$ mV and convection and advection flows, and it has a shelf life on the order of months. This remarkable stability, compared to traditional whole cell voltage clamping experiments, for example, is a result of the disulfide chemistry, the nanometer hydrophilic submembrane space between the bioelectronic gold interface and the membrane, and the flexibility of lipid composition. For example, it is standard to use 100 percent tether density to ensure the membrane is sufficiently stable to allow the cells to grow on the surface of the membrane. In other cases, a 1:10 ratio of

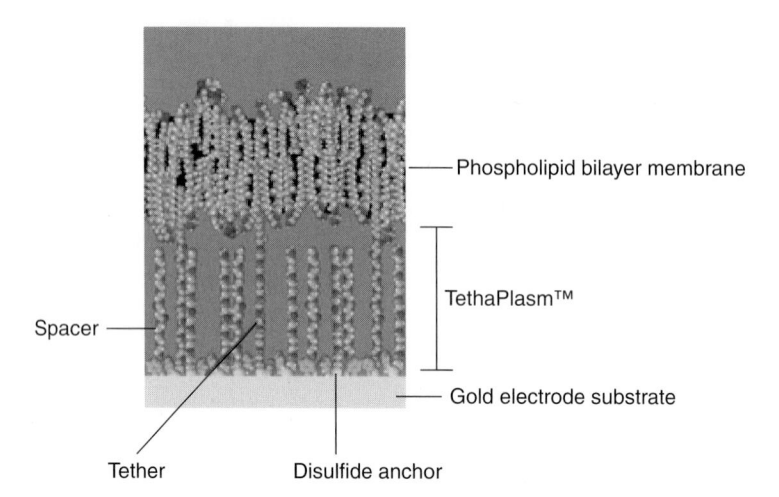

Figure 4.1 Building blocks of an engineered tethered membrane. The distance between the gold bioelectronic interface and the membrane surface is approximately 2–3 nm (measured from neutron reflectometry experiments) depending on the tethering chemistry (tether density, lipid type, and embedded macromolecules), and the membrane has an approximate thickness of 4 nm. Details on the experimental measurements of the atomic structure of engineered tethered membranes are provided in Chapter 7.

spacers to tethers provides the best compromise between membrane stability and flexibility.[1] Ratios as small as 1:100 have been used if greater flexibility is required for the insertion of large proteins, but less than this is not recommended as membrane integrity becomes unreliable. Conversely, ratios of up to 100 percent can be used if exceptional stability is required as needed for growing cells on the surface of the membrane.

Before proceeding with the synthesis procedure for an engineered membrane, in the rest of this section we briefly describe the structure of the resulting membrane and also describe a real-life membrane device.

4.1.1 Engineered Tethered Membrane: Structure

Engineered tethered membranes are planar tethered lipids, typically phospholipids, held above a gold electrode by a set of hydrophilic polyethylene glycol (PEG) chains covalently bonded to the gold surface by organic sulfur anchors, disulfides in particular being described here.[2] Lipophilic alkane phytanyl groups are bonded to the top of the PEG chain which act as a scaffold around which the membrane lipids spontaneously cluster, ultimately forming a continuous membrane. A schematic of the formed membrane is provided in Figure 4.1. Together the sulfur anchor, PEG chain, and phytanyl group are referred to as a molecular tether because it ties the membrane to the gold surface. In practice the tethers are separated from each other on the gold substrate by molecules called spacers. These spacers are similar in structure to tethers but they lack

---

[1] Flexibility means the ability to incorporate peptides or proteins of higher molecular weight than the lipids.
[2] Please see Chapter 2 for a short description of these chemical species.

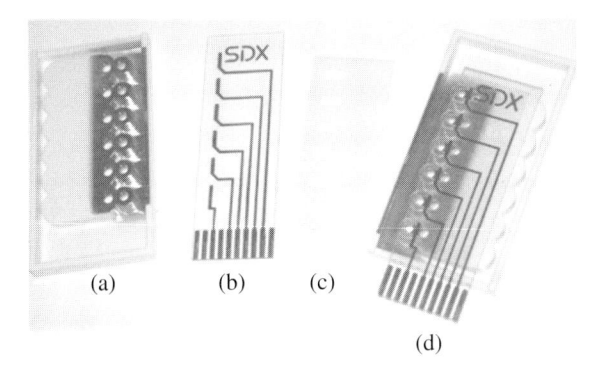

Figure 4.2 Components of the six-chamber flow-cell cartridge (tethaPlate™) for constructing engineered membranes. (a) An injected molded polycarbonate cartridge incorporating openings for six chambers and a sputtered gold return electrode on its underside. (b) A gold patterned $25 \times 75 \times 1$ mm³ polycarbonate slide with end connectors and six electrodes. (c) An adhesive laminate, 100 μm thick. (d) The assembled six-chamber flow-cell cartridge.

the lipophilic phytanyl group. These spacers are located at the surface of the gold bioelectronic interface and do not have any direct contact with the membrane surface. The tethers and spacers completely cover the gold surface but they are not closely packed as each contains a benzene molecule bonded to the disulfide anchor.

The key feature of the engineered tethered membrane is that it encloses an aqueous space between the membrane and the tethering surface. This space is essential to the function of membranes as it provides an aqueous ionic reservoir, equivalent to the aqueous ionic reservoir within the interior volume of all biological cells. Without this reservoir, the transport of ions across a tethered membrane would essentially be eliminated due to the absence of a sink into which transported ions can be absorbed or from which they can be sourced from the bioelectronic interface. Additionally, very few of the actual membrane-forming molecules are themselves tethered.

4.1.2 Overview of Tethered Device

As an example, Figure 4.2 illustrates the geometry of a six-chamber flow-cell cartridge for constructing engineered tethered membranes. The typical electrode size is of order mm². The geometry of the flow-cell chamber is important because any kinetic measurement of engineered membrane properties, in response to added molecular components, is influenced by mass transfer effects. Mass transfer effects arise from the membrane interactions of added solutes being slowed by inadequate mixing in the space above the surface of the engineered tethered membrane. The flow-chamber geometry is designed to minimize mass transfer effects above the surface of the engineered tethered membrane.

The self-assembly process of engineered tethered membranes commences with the creation of a self-assembled monolayer. This monolayer is created from ethanolic solutions typically comprising thiol- or disulfide-containing molecules brought into contact with a freshly coated gold surface. For example, a tethaPlate™ cartridge

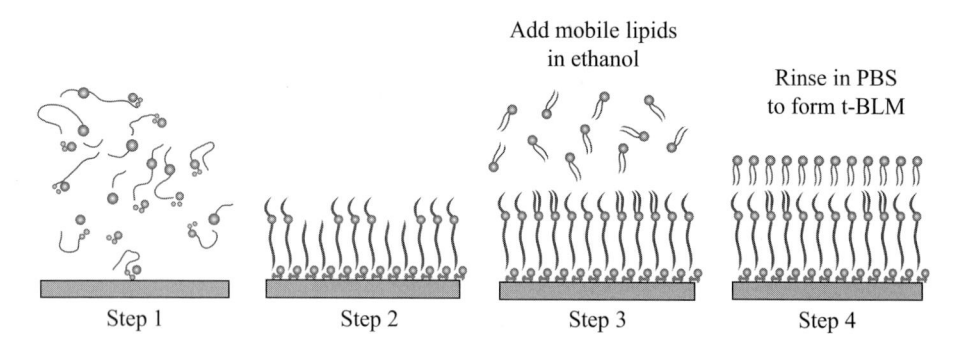

Figure 4.3 Schematic of the four-step procedure to form an engineered membrane described in §4.2.1.

(Figure 4.4) which comprises six chambers each with a gold electrode (surface area 2.1 mm^2 = 2.1 × 10^{-6} m^2 with dimensions 3.0 × 0.7 mm^2) that has been precoated with tether and spacer molecules. A second, uncoated, gold counter electrode sits 100 μm away from the membrane surface but does not physically interact with it. This geometric setup has proven to be exceptionally robust for engineered membrane preparations. The self-assembled monolayer is typically constructed by immersing a freshly gold-coated microscope slide into an ethanolic solution of the sulfur-based coating compound which contains the phospholipid mixture, resulting in the formation of the engineered membrane. Sulfur-gold chemistry is by far the most convenient, inexpensive, and popular approach used to assemble engineered tethered membranes. Chemistries other than sulfur interacting with gold may be used, but they all require more complex assembly procedures. This chapter is devoted exclusively to aspects of the sulfur-gold assembly approach.

4.2 Building an Engineered Artificial Membrane

This section gives the detailed procedure for building a tethered membrane; the actual laboratory procedure is described in §4.4.

4.2.1 Solvent-Exchange Technique

The construction of the engineered artificial membrane is performed using a four-step solvent-exchange technique [84] as illustrated in Figure 4.3 and discussed below.

Initial: Constructing the Electrode Surface

The geometry of the gold electrode can range from a few square microns, as used in multiplexed chip arrays, to 1–2 mm^2 for typical laboratory measurements (as illustrated in Figure 4.4), up to 50 cm^2, which is the size used for neutron reflectometry measurements. A commonly used and convenient format is polycarbonate microscope slides. Glass can also be used but this usually requires coating the gold with a bonding layer

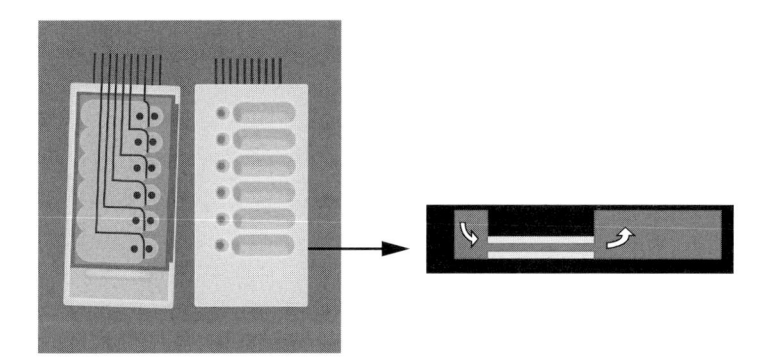

4.2 Building an Engineered Artificial Membrane 65

(a) Six-chamber flow-cell cartridge (tethaPlate™) for constructing engineered membranes. The electrolyte passes from the circular addition port to the oblong storage reservoir through a 100 μm flow-cell chamber sandwiched between the electrode with the engineered tethered membrane attached and the gold counter electrode on the reverse side of the white cartridge.

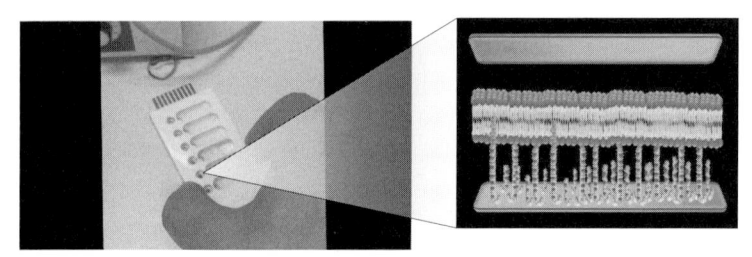

(b) The engineered tethered membrane (surface represented by red beads) is located in the space between the circular sample addition port and the oblong reservoir of the flow-cell cartridge. It is protected from damage by pipette tips at the top of the addition port. The gold counter electrode is on the reverse side of the cartridge top, instead of being 100 μm away from the membrane surface.

Figure 4.4 Six-chamber flow-cell cartridge (tethaPlate™) that is constructed using the components illustrated in Figure 4.2.

such as chromium or tungsten in order to secure the gold to the glass surface. Gold adheres to polycarbonate surfaces without needing a chromium or tungsten coating. Pinhole defects in the sputtered 100-nm gold layer are inevitable and result in access to the underlying electrochemically active chromium or tungsten, which can result in a distortion of electrical measurements and membrane properties.

One of the important requirements is that the surface on which the engineered tethered membrane is to be assembled must be very smooth. Typically, a surface roughness of less than 1 nm over an area of 1 μm^2 is required. The reason for this limitation is quite obvious. If one is attempting to assemble a 4-nm-thick lipid bilayer over a metal surface it is necessary that the surface does not have sharp edges or steps that prevent the surface coverage being continuous. An analogy is paint flaking off a poorly prepared rough surface.

Evaporation or preferably sputtering is usually employed with a gold thickness of no more than 100 nm in order to achieve a smooth surface. Following the deposition of the

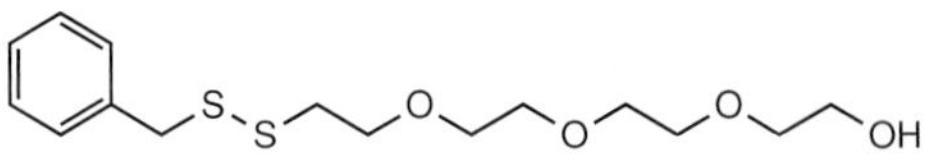

(a) Spacer molecule (Benzyldisulfide-TEG-OH).

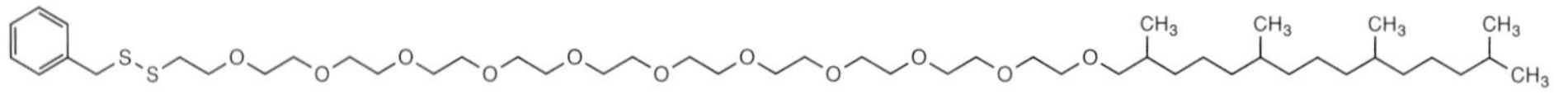

(b) Tether molecule (Ester free, Benzyl disulfide polyethylene glycol phytanyl).

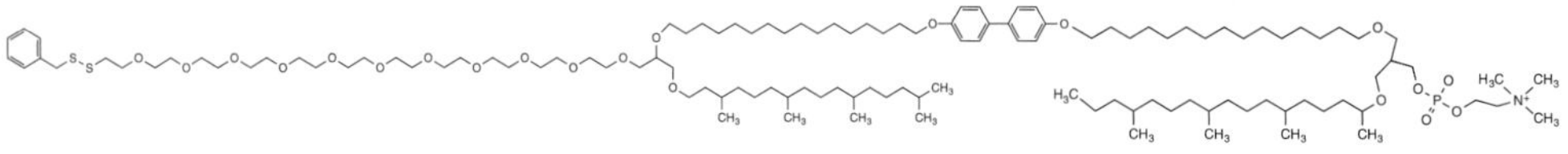

(c) Tether molecule (Phosphoryl choline terminated membrane spanning tethers).

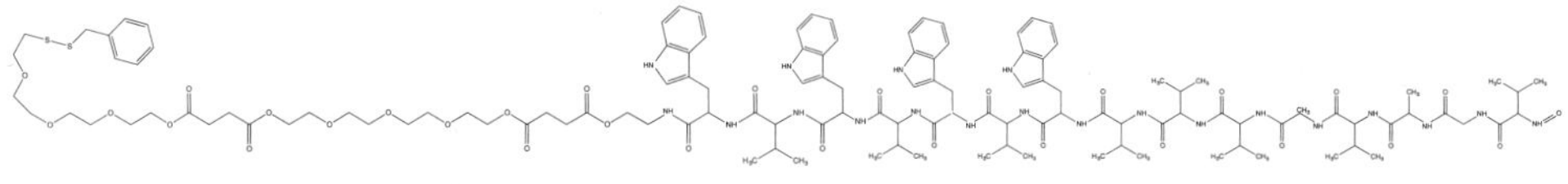

(d) Tether molecule (Tethered gramicidin A).

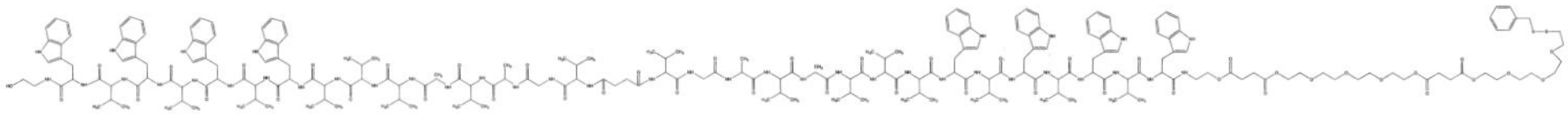

(e) Tether molecule (Tethered bis gramicidin A).

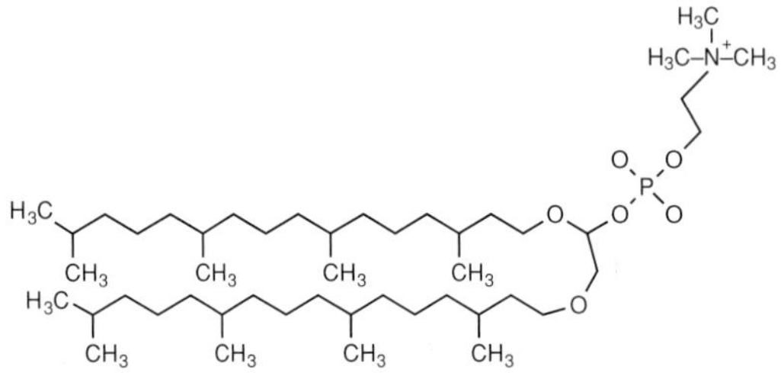

(f) Lipid molecule (C20 Diphytanyl di-ether phosphatidyl choline).

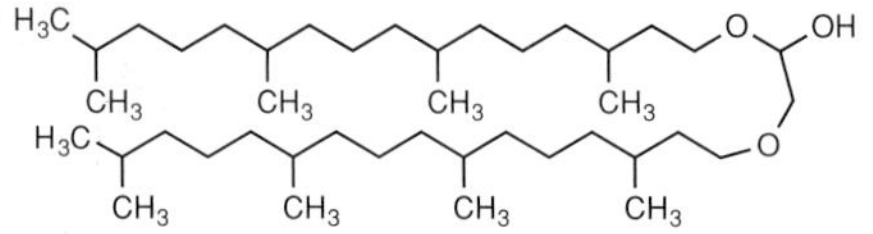

(g) Lipid molecule (C20 Glycerol diphytanyl ether).

Figure 4.5 Examples of tether, spacer, and lipid molecules used to construct engineered tethered membranes.

gold, the substrates are brought up to ambient air pressure and transferred to a sulfur-based coating solution. As seen in Figure 4.1, disulfides are preferred as they have a longer shelf life. The coating solution can contain a variety of compounds depending on the structure to be assembled.

Each constructed electrode is in an isolated flow cell with a common gold return electrode. The process to form an engineered membrane on the gold electrode is performed in four steps.

4.2 Building an Engineered Artificial Membrane

Tethers, Spacers, and Lipid Chemistries

A variety of molecules can be used for the tethers, spacers, and lipids in engineered tethered membranes. Examples of tether, spacer, and lipid molecules are illustrated in Figure 4.5. As mentioned above, disulfides are the preferred sulfur binding groups to participate in the coordinating sulfur-gold anchoring chemistry. Figure 4.5(a) shows a typical short, polar spacer molecule that is interspersed between the tethering molecules shown in Figure 4.5(b). A mix of tethers and spacers is used to produce an engineered tethered membrane. Engineered tethered membranes allow for the incorporation of larger peptide or protein molecules among the membrane lipids. Additionally, tethered membranes ensure space within the reservoir region is maintained to facilitate ion flow within the reservoir and to prevent the reservoir capacity from limiting the apparent ion flow across the membrane. This is particularly significant, as will be seen later, when studying the conduction of ion channels within tethered membranes.

The spacer molecule shown in Figure 4.5(a) is composed of an anchoring benzyl disulfide group connected to a four-oxygen ethylene glyco group terminated by an alcohol group. The tether molecules shown in Figure 4.5(b) are similarly composed of benzyl disulfide groups but connected to an 11-oxygen ethylene glycol group terminated by a C20 hydrophobic phytanyl chain. Figure 4.5(c) illustrates a tether molecule that spans the entire lipid membrane. The tethering molecule in Figure 4.5(d) illustrates a tether that is terminated by a gramicidin A molecule and is used in the ICS biosensor. Additionally, the tethering molecule in Figure 4.5(e) is used in the electrophysiological response platform (ERP) for measuring the electrophysiological response of cells. The tethering species functions by the terminating nonpolar, hydrophobic phytanyl or other group becoming entangled in the subsequently formed lipid bilayer membrane.

A formed engineered tethered membrane can be constructed from any bilayer-forming lipids. Figures 4.5(f) and 4.5(g) illustrate two such lipids that were chosen because of their chemical stability and because the phytanyl groups do not undergo a first-order phase transition from rigid to fluid liquid crystalline phase transitions at ambient temperatures. This is due to the steric interference of the methyl groups along the phytanyl chains. The chains are unable to pack as closely as straight methylene chains, preventing phase transitions. Different mixtures of the lipids illustrated in Figures 4.5(f) and 4.5(g) can be used to produce membranes with different properties. For example, a 30:70 mixture of these two lipids yields a minimum of the membrane permeation. A possible mechanism for this effect is the achievement of an optimal packing structure between these two lipids.

Step 1: Binding of Tethers and Spacers

Step 1 of the fabrication of the tethered membrane involves anchoring the tethers and spacers to the gold surface. The tethers provide structural integrity to the membrane and mimic the physiological response of the cytoskeletal supports of real cell membranes. The lipids and spacers are bound to the gold substrate using a benzyl disulfide, which is more stable against oxidation than a simple thiol and also provides optimal spacing of the spacers to minimize the conduction limitation of ionic species within the reservoir region between the engineered membrane and the gold electrode. The experimentalist

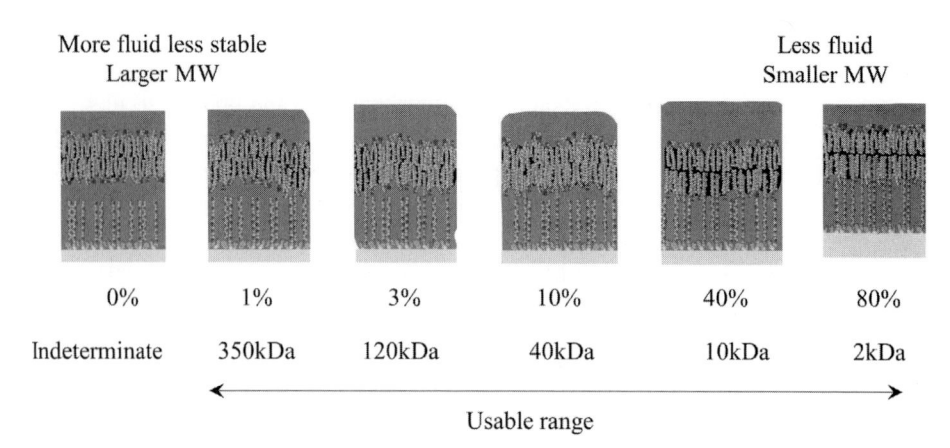

| More fluid less stable | | | | Less fluid | |
Larger MW				Smaller MW	
0%	1%	3%	10%	40%	80%
Indeterminate	350kDa	120kDa	40kDa	10kDa	2kDa

←——————————— Usable range ———————————→

Figure 4.6 Schematic diagrams of engineered tethered membranes containing tether densities in the range of 0 to 80 percent with the associated size of proteins and peptides that can be accommodated within the engineered membrane. The size of the proteins and peptides can be approximated by the molecular weight of the proteins and peptides (e.g., kDa).

can select the desired density of tethers by changing the concentrations of tethers and spacers for coating the gold electrode. The ratio of spacer molecules to the tethered lipids is denoted as the percent tethering density (e.g., for a 10 percent tether density, for every one tethered lipid there are nine spacer molecules). The advantage of a range of percent tethers within the inner leaflet of the engineered membrane is that it permits experiments to be designed where ion channels or polypeptides can be optimally incorporated into the bilayer. The average spacing of the tethered lipids can be used as an approximate guide to the molecular weight of molecules that can be accommodated within the engineered membrane. The basis of this estimate is a simple volume calculation based on the molecular weight of the membrane-associated component and a measured thickness of 4 nm for the engineered membrane [152]. Figure 4.6 schematically illustrates the relation between the molecular weights of components that can be accommodated within the engineered membrane as a function of the tether density. As Figure 4.6 illustrates, for a 10 percent tethering density, moieties of up to 40 kDa could potentially be incorporated into the membrane. The possible tethering densities range from 1 to 100 percent; however, a trade-off exists between the stability and ionic leakage of the membrane and the volume of lipid bilayer available to incorporate additional membrane-associated species such as an ion channel. Components ranging up to 350 kDa represent the limit of potential inclusions within the membrane. An additional consideration is the limited space available within the reservoir region between the membrane and the gold electrode. Bulky groups that project beyond the engineered tethered membrane will not be accommodated within the reservoir. However, many ion channels have extramembrane components that project beyond the membrane on only one side of the bilayer. These can be accommodated by the molecule orienting with the bulky group projecting into the bulk solution above the membrane. As seen later, the spontaneously membrane-inserting ion channel α-hemolysin is one example. Approximately 50 percent of the α-hemolysin protein projects beyond the membrane, and yet through its

4.2 Building an Engineered Artificial Membrane

preferred orientation with the extramembrane fraction in the aqueous phase the protein inserts into the membrane and conducts. Note that in the special case of 100 percent tethering, the engineered membrane is composed of a tethered archaebacterial-based monolayer with no spacer molecules. As experimentally illustrated in [152], it is possible to construct a 0 percent tether density membrane; however, the inner membrane leaflet adjacent to the gold electrode interacts strongly with the gold surface through physisorption, perturbing the membrane geometry and mobility, and providing little improvement to the membrane stability. The stability of 0 percent tether density is far less than even a 1 percent tether density membrane. As the electrolyte reservoir separating the membrane and electrode surface is required for the normal physiological function of the membrane, and noting that all prokaryotic and eukaryotic cell membranes contain cytoskeletal supports with a 1–10 percent tether density, the inability to construct a stable 0 percent tethered membrane is not relevant to constructing membranes that mimic membranes in vivo. When designing an experiment it is recommended that the highest tether density possible for a particular molecular-weight species being incorporated into the engineered membrane to be used. Components ranging from 1 to 350 kDa can be incorporated into the engineered membrane.

The spacer is composed of a benzyl disulfide connected to a four-ethylene glycol group terminated by an OH, and the tethers are composed of a benzyl disulfide connected to an eight-oxygen ethylene glycol group terminated by a C20 hydrophobic phytanyl chain.[3] To form the anchoring layer, an ethanolic solution containing 1 mM of engineered ratios of benzyl disulfide components is prepared. Note that in the special case of 100 percent tethering, the engineered membrane is composed of a tethered archaebacterial-based monolayer with no spacer molecules.

Step 2: Equilibration of the Tether and Spacer Surface Binding

To ensure that the tether and spacer solution has time to sufficiently bind with the gold surface, the solution is exposed to the gold surface for approximately 1 hour. The deposition process of the tethers and spacers can be measured using impedance measurements; however, this is not required for the formation process, only to check the quality of the deposition process. To remove any excess tethers and spacers, the surface is flushed with ethanol and stored either under ethanol or in an ethanol atmosphere for later use.

Step 3: Binding of Membrane Lipids

Step 3 involves the formation of a bilayer membrane by incorporating the membrane tethers. Note that these subsequently added lipids are not tethered and are therefore mobile in two dimensions perpendicular to the membrane surface. A variety of lipids can be used for this including C20 diphytanyl-diglyceride ether (GDPE), zwittrionic C20 diphytanyl-ether-glycero-phosphatidylcholine lipid (DphPC), and, palmitoyloleoyl-phosphatidylglycerol (POPG); however, in most cases the lipids

---

[3] Details on the use of benzyl disulfide and polyethylene glycol in engineered tethered membranes can be found in Chapter 2. These are common chemical components used for the design of biodevices as a result of their useful coordination bonding characteristics with the gold bioelectronic interface and low cost.

selected are soluble in ethanol [84, 273]. Ethanol is preferred when using a polymer substrate since a polycarbonate slide and cartridge are used. The advantage of the polycarbonate slide and cartridge is that the gold adheres to the polycarbonate without needing a chromium or tungsten coating. Pinhole defects in the sputtered 100-nm gold layer are inevitable and result in access to the underlying electrochemically active chromium or tungsten, which can result in a distortion of the impedance measures. Should studies of lipids be employed that require other solvents such as chloroform, which attacks polycarbonate, dilute mixtures of chloroform with ethanol may be used. The tethered membrane mobile lipids can be selected to provide an optimal impedance seal. Empirically it has been found that mixtures of 70 percent DphPC and 30 percent GDPE lipids are optimal. The solution containing a mixture of mobile lipids is brought into contact with the gold-bonded components from step 1. As an example, let us consider the formation of a 70 percent DphPC and 30 percent GDPE mixed tethered membrane: 8 μL of 3 mM of the 70 percent DphPC and 30 percent GDPE ethanolic solution is added to the flow chamber.

Step 4: Equilibration of the Tethered Membrane

To ensure that the tethered membrane has sufficient time to form, the solution in step 3 is incubated for 2 min at 20°C. Following the 2 minutes of incubation, 300 μL of phosphate-buffered saline is flushed through each flow chamber to remove any excess mobile lipids. The tethered membrane is equilibrated for 20–30 minutes prior to any experimental measurements. Throughout this process the impedance of the tethered membrane can be measured to infer the quality of the formed membrane and to determine when the transients of the chemical processes leading to membrane formation are complete.

The procedure for constructing an engineered membrane with different tethering density and lipid composition follows a similar procedure. For example, if only a DphPC membrane was desired, then in step 3 only 100 percent DphPC lipids would be used (with 0 percent GDPE). Note that a different lipid entirely could be used as well; for example, a 100 percent 1,2-dipalmitoyl-sn-glycero-3-phosphocholine (DPPC) lipid membrane could be constructed by replacing the DphPC and GDPE mixture in step 3 with only DPPC lipids. If a different tethering density was desired, this would involve simply changing the proportion of tethers and spacers used in step 1.

4.2.2 Evaluating the Quality of the Engineered Membrane

Thus far we have described the key steps for constructing a tethered artificial membrane. To evaluate the quality of the formed membrane, we could use neutron reflectometry (NR), which allows the determination of the atomic structure of the engineered membrane (e.g., membrane and tethering reservoir thickness, density, and roughness of the electrode surface). However, as will be discussed in Chapter 7, although NR can be used it requires labor-intensive experimental setup and high-cost equipment. A convenient way of evaluating the quality of the engineered membrane is to measure its impedance response.[4] When the engineered membrane is excited by an

---

[4] This is also commonly referred to as the electric impedance spectrum response.

external voltage, then the ions in the electrolyte solution migrate to the gold surface and insulating barrier of the membrane. Additionally, ions can also pass through aqueous pores in the membrane. These phenomena can be modeled by an equivalent electric circuit model composed of capacitors and resistors. The capacitors model the ion accumulation at the bioelectronic interface (gold surface) and membrane surface, while the resistor models the conduction of ions through the engineered membrane. To test the quality of the engineered membrane, we compare the estimated capacitance and resistance values to the expected capacitance and resistance values resulting from a high-quality membrane. Though any nonzero voltage excitation can be used, it is convenient to use impedance measurements to estimate the values of these capacitors and resistors.

Several methods can be used to perform impedance measurements using standard laboratory equipment such as an oscilloscope and a function generator. The quality of the tethered membrane is measured continuously using electrical impedance spectroscopy. A commercially available spectrometer was supplied by SDx tethered membranes. The tethaPod™ swept frequency impedance reader was used to measure the impedance at frequencies of 10,000, 5000, 2000, 1000, 500, 200, 100, 50, 20, 10, 5, 2, 1, 0.5, 0.2, 0.1, 0.05, 0.02, and 0.01 Hz with an excitation potential of 20 mV. Custom excitation potentials were produced and the resulting current was recorded using an eDAQ™ ER466 potentiostat (eDAQ, Doig Avenue, Denistone East) and an SDx tethered membrane tethaPlate™ adapter to connect to the assembled electrode and cartridge. The defect density in the formed membrane can be estimated from the impedance measurements using the protocol presented in [435]. A high-quality membrane will have a low conductance in the range of 1 to 100 $\mu S/cm^2$, and an associated capacitance in the range of 0.8 to 1.2 $\mu F/cm^2$ depending on the electrolyte pH, lipid type, and tether density of the engineered membrane. For all experimental measurements, the membrane contained negligible defects. To detect the presence of electrodesorption and release of portions of tethered membrane into solution, the capacitance of the gold electrode and membrane is monitored.

4.3 Inserting Proteins and Ion Channels into Engineered Artificial Membranes

There is increasing interest in the study of membrane-active peptides and proteins in engineered membranes. Specifically, it is of interest to study the dynamics of these species with respect to different transmembrane potentials, lipid compositions, mechanical stress, and electrolyte compounds. In this section we provide experimental methods for inserting various peptides and proteins into engineered membranes.

There are three experimental methods for inserting peptides and proteins into a tethered membrane:

(i) spontaneous insertion,
(ii) electrochemical insertion, and
(iii) proteoliposome insertion.

For compounds that spontaneously insert into the membrane (e.g., valinomycin, gramicidin A, alamethicin, α-hemolysin, chloride intracellular protein [CLIC1], and peptidylglycylleucine-carboxyamide [PGLa]) the membrane can be prepared using the method presented in §4.2.1. If the ions do not spontaneously insert, then it may be possible to use an electrochemical excitation to promote the insertion into the membrane surface. The application of an external electric potential across the membrane causes the formation of transient aqueous pores. The peptides and proteins can then insert into these formed pores; when the pores close, the peptides and proteins will reside in the membrane. For the insertion of water-insoluble peptides and proteins (e.g., voltage-dependent anion channels), the proteoliposome insertion method can be utilized. How each of these three methods is applied to inserting channels into an engineered membrane is provided below with recommendations on which method to use for which type of peptides and proteins.

4.3.1 Spontaneous Insertion Method

The spontaneous insertion method involves adding the peptides and proteins into the electrolyte in contact with the surface of the tethered membrane. The peptides and proteins diffuse to the surface of the membrane and then spontaneously insert. There are three requirements for the spontaneous insertion method to be used for studying peptides and proteins:

(i) The peptides and proteins must be soluble in the electrolyte solution.
(ii) The peptides and proteins must spontaneously insert into the membrane. This typically occurs if the peptides and proteins are amphiphilic molecules that contain both hydrophobic regions and hydrophilic regions. The hydrophobic regions of the peptide will associate with the hydrophobic regions of the lipid tails, and the hydrophilic regions of the peptide will associate with the hydrophilic regions of the lipid head group.
(iii) An important consideration for the spontaneous insertion method is the tethering density, as depicted in Figure 4.7. If the tethering density is too high this will inhibit the spontaneous insertion of the peptides and proteins. However, if the tethering density is too low then for high concentrations of peptides and proteins the molecules may irreversibly damage the stability of the membrane, causing patches of membrane to detach from the surface.

If all these requirements are satisfied then the spontaneous insertion method can be used to insert peptides and proteins into the engineered tethered membrane. Examples of some peptides and proteins that spontaneously insert into the membrane include valinomycin, gramicidin A, alamethicin, α-hemolysin, CLIC1, and PGLa.

4.3.2 Electrochemical Insertion Method

In the spontaneous insertion method, the insertion of peptides and proteins into the membrane is dependent on the concentration of these molecules in the electrolyte. If an

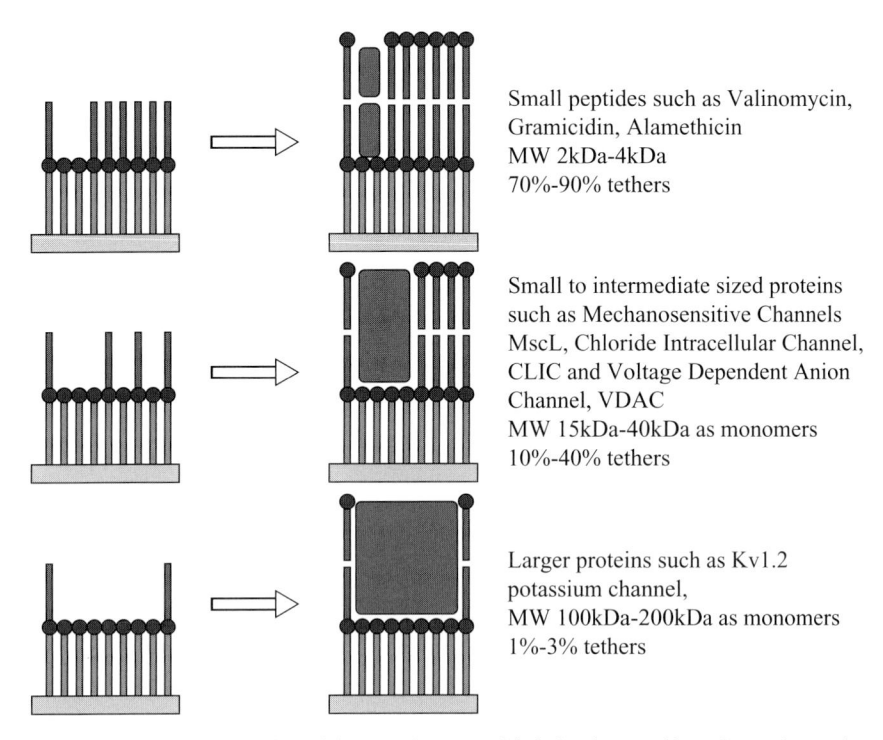

4.3 Inserting Proteins and Ion Channels into Engineered Artificial Membranes

Small peptides such as Valinomycin, Gramicidin, Alamethicin
MW 2kDa-4kDa
70%-90% tethers

Small to intermediate sized proteins such as Mechanosensitive Channels MscL, Chloride Intracellular Channel, CLIC and Voltage Dependent Anion Channel, VDAC
MW 15kDa-40kDa as monomers
10%-40% tethers

Larger proteins such as Kv1.2 potassium channel,
MW 100kDa-200kDa as monomers
1%-3% tethers

Figure 4.7 The molecular size of the protein or peptide being inserted into the engineered membrane must be considered because too high a tether density will prevent the spontaneous insertion of the molecule.

electrochemical potential (transmembrane potential) is applied to the engineered tethered membrane it is possible to promote the insertion of peptides and proteins. Note that all the conditions of the spontaneous insertion method must hold to apply the electrochemical insertion method for inserting peptides and proteins. The electrochemical insertion method is particularly useful if the peptides and proteins are charged because the application of an external voltage potential causes the charged molecules to diffuse in a direction parallel to the electric field.

How does the electrochemical insertion method work? When an electrochemical potential is applied across the engineered tethered membrane this causes an increase in the number and/or size of conducting aqueous pores in the membrane as a result of the process of electroporation. The number and size of the pores are dependent on the transmembrane potential. Once these aqueous pores have formed then proteins and peptides can diffuse into them. When the transmembrane potential is removed, these pores close and the peptides and proteins remain in the membrane. An example of this procedure is illustrated in Figure 4.8 for a peptide. Initially the peptide remains on the surface of the engineered tethered membrane. When a transmembrane potential is applied this causes the formation of an aqueous pore that the peptide can diffuse into. Once the peptide has diffused into the pore then the transmembrane potential is removed and the pore closes, leaving the peptide inserted in the membrane.

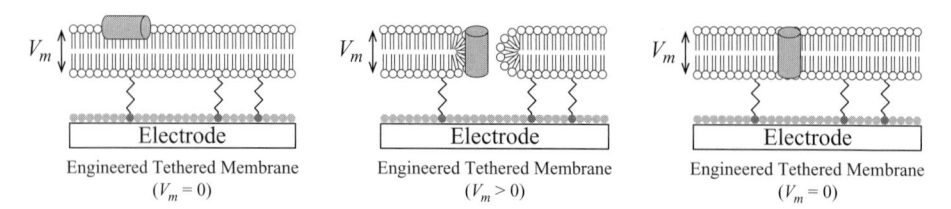

Figure 4.8 Schematic of the electrochemical insertion of a peptide (gray cylinder). Initially the peptide is bound to the surface of the engineered tethered membrane with no aqueous pores present. Then an electrochemical potential (transmembrane potential) V_m is applied to promote the creation of an aqueous pore. The peptide then diffuses into this aqueous pore. When the electrochemical potential is switched off ($V_m = 0$) then this aqueous pore closes, inserting the peptide into the membrane.

4.3.3 Proteoliposomal Insertion Method

The proteoliposomal insertion method can be used to insert insoluble peptides and proteins into an engineered tethered membrane directly from a biological membrane. This method of insertion involves four primary steps. The first step is to extract the insoluble protein from a biological membrane to form a detergent-lipid-protein complex that is soluble. Note that the removal of proteins and peptides from biological membranes is known as cell lysis. The formation of the detergent-lipid-protein complex is achieved using detergent molecules that are specifically designed to form hydrophobic-hydrophilic interactions that allow the extraction of insoluble peptides and proteins from biological membranes. A detergent-lipid-protein complex is illustrated in Figure 4.9. The second step is the formation of the tethered lipid surface which can be formed using steps 1 and 2 of the construction of an engineered tethered membrane outlined in §4.2.1. In the third step an insoluble protein contained in a detergent-lipid-protein complex is inserted into a liposome to form a proteoliposome – a lipid vesicle that incorporates specific proteins into the vesicular membrane as illustrated in Figure 4.9. The fourth step involves the proteoliposome bonding with the tethered lipid surface from the first step. The entire proteoliposomal insertion method can be viewed in Figure 4.9.

As an example of the proteoliposomal insertion method, consider the insertion of the voltage-dependent anion channel (VDAC) from the mitochondrial membrane of a eukaryote. The insertion protocol is provided below [101]:

(i) The first step of the process involves extracting the protein from the biological membrane to form the detergent-lipid-protein complex. To extract the VDAC from the mitochondrial membrane, we lysed the membrane by osmotic shock in a 10 mM Tris-HCl buffer for 45 min at 4°C. After centrifugation at 50,000g (a relative centrifugal force of 50,000 times that of Earth's gravitational pull [$g = 9.8$ m/s^2]) for 60 minutes, the mitochondrial pellet was solubilized in a 10 mM Tris-HCl buffer containing 3 percent Triton X-100 detergent for 45 minutes. Another centrifugation at 50,000g for 60 minutes was performed, and then the contents were placed in a dry hydroxyapatite/Celite (2:1) column (1.3 × 10 cm^2). The porin was

4.3 Inserting Proteins and Ion Channels into Engineered Artificial Membranes

Figure 4.9 Schematic of the proteoliposomal insertion method for inserting proteins and peptides into engineered tethered membranes. The first step is the creation of a detergent-lipid-protein complex from a biological cell membrane (e.g., cell lysis). The second step is the formation of a tethered lipid monolayer. The third step is the construction of a proteoliposome from the detergent-lipid-protein complex. The fourth step involves the binding of the proteoliposome with the tethered lipid monolayer to form the engineered tethered membrane with the inserted protein.

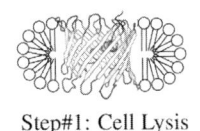

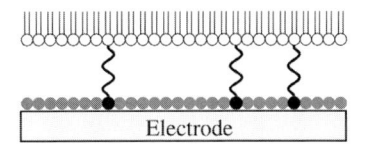

collected from the flow through of the column. The result is the 1-palmitoyl-2-oleoyl-sn-glycero-3-phosphocholine (POPC)- and Triton-X100-coated VDAC (detergent-lipid-protein complex) illustrated in Figure 4.9.

(ii) The second step is to construct the tethered lipid surface following steps 1 and 2 in §4.2.1. This creates a lipid monolayer to which lipid vesicles (or liposomes) can aggregate with to form a lipid bilayer.

(iii) The third step involves the formation of the proteoliposomes.[5] In the case of VDAC a POPC liposome is utilized, which can incorporate the VDAC from the second step.

(iv) The fourth step involves binding the formed proteoliposomes from the third step with the tethered lipid surface from the first step. Since the POPC-based proteoliposomes self-assemble with the POPC tethered lipids, sufficient time was given for the formed proteoliposomes to diffuse to the tethered lipid surface. Once sufficient time has been given to form the membrane, then all remaining proteoliposomes are rinsed from the electrolyte solution above the surface of the engineered tethered membrane.

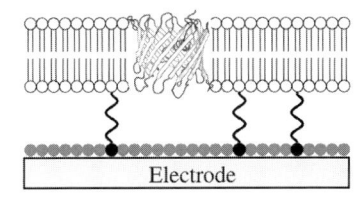

Notice that the proteoliposomal fusion insertion method requires several complex experimental steps which have to be refined for each type of peptide and protein of interest. Careful consideration should be given to ensure that the spontaneous insertion, and electrochemical insertion methods, cannot be applied to insert the molecule prior to applying this method. To increase the probability of a proteoliposome binding to the

---

[5] A proteoliposome is a lipid vesicle that contains proteins.

tethered lipid surface, detergent electrolyte solutions can be added at the fourth step above.

4.4 Laboratory Exercise: Tethered Membranes and Spontaneous Insertion of Gramicidin Channels

In this section we provide a laboratory description of how to construct an engineered tethered membrane containing different concentrations of the gramicidin (gA) channel.[6] The key learning objectives are to provide hands-on experience in building engineered tethered membranes, and performing impedance measurements to study the conductance dynamics of gA. Recall that gA plays a crucial role in the construction of biosensors (e.g., the ICS biosensor discussed in Chapter 5 and the ERP discussed in Chapter 6). A video demonstration of the laboratory exercise in this section can be viewed on YouTube at www.youtube.com/watch?v=F5NYLPcJDOs for the cartridge assembly and www.youtube.com/watch?v=lvmbwY-ek0E for the membrane formation.

The equipment required to construct the tethered membrane is illustrated in Figure 4.10. This includes a polycarbonate slide that contains six gold electrodes with a 10 percent tether density of spacers and tethers (chemical structure illustrated in Figure 4.5) and a polycarbonate cartridge that contains six flow-cell chambers with a common gold-coated counter electrode. These components are used to construct the flow-cell chambers illustrated in Figure 4.2. In addition to these components, the experimentalist will also require the following:

(i) 10- and 100-μL pipettes with tips to deliver the phospholipid solution (10 μL) and rinse with phosphate-buffered saline (100 μL),
(ii) tweezers to remove the slide from the sealed pack,
(iii) PBS (100 mL), and
(iv) a timer to measure 2-minute incubation times for forming the membrane and a 1-minute delay for the adhesive to seal.

To construct the engineered tethered membranes with different concentrations of gA ion channels, six 10-μL methanol solutions will be prepared that contain different concentrations of gA ion channels as follows:

Vial 1, methanol control (no gA);
Vial 2, 34 nM gA in methanol;
Vial 3, 68 nM gA in methanol;
Vial 4, 103 nM gA in methanol;
Vial 5, 137 nM gA in methanol; and
Vial 6, 171 nM gA in methanol.

Additionally, a vial of 3 mM DphPC/GDPE phospholipid solution with a ratio of 70:30, and a vial containing PBS are required to construct the tethered membranes.

---

[6] A short description of the gA ion channel is provided in Chapter 2.

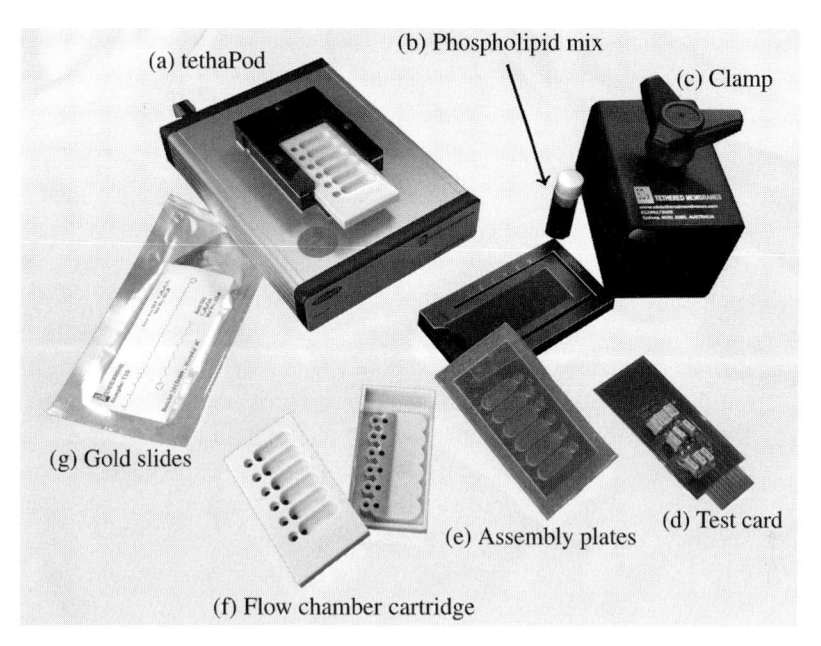

Figure 4.10 The equipment needed to build and measure the conductance of a tethered membrane. The components, clockwise from top left, include (a) a "tethaPod" tethered membrane impedance reader, (b) a vial of phospholipid mix (70 percent DphPC and 30 percent GDPE in 99.9 percent ethanol) to create the tethered membrane, (c) a clamp to compress the electrode onto the adhesive-coated cartridge top to make a flow-cell test element, and alignment jig to ensure the slide with the patterned gold electrode containing the tethered elements of the tethered membrane are correctly aligned with the cartridge top, (d) a test card to demonstrate the impedance reader is functioning correctly, (e) a silicon rubber pressure spreader and aluminum pressure plate to distribute the clamp pressure across the slide and cartridge during the adhesive sealing step, (f) flow chamber cartridge tops with gold counter electrode coated on the reverse face, also including adhesive coating on the rear face to permit assembly of the flow-cell cartridge, and (g) individual gold patterned slides coated with a monolayer of the tethers and spacers (90 percent benzyldisulfide bis-tetraethylene glycol spacers and 10% benzyldisulfide bis-tetraethylene glycol C16 phytanyl half membrane spanning tethers) stored in metallized sealed packs in a solution of 99.9 percent ethanol. The storage of the spacers, tethers, and lipids in the ethanol solution allows these components to have a shelf life of up to 12 months.

Prior to beginning the construction of the engineered tethered membrane, ensure that all equipment, instrumentation, and chemicals are available. Timing is critical for proper membrane formation. Read the entire experiment through before commencing this exercise.

4.4.1 Prepare the Engineered Tethered Membrane for Spontaneous gA Ion-Channel Insertion

In the first part of this exercise we will prepare the tethaPlate™ cartridge, illustrated in Figure 4.2, that contains six flow chambers. Each flow-cell chamber contains a gold-coated electrode with the tethers and spacers adhered to the electrode surface with a 10 percent tether density. A common gold counter electrode is positioned 100 μm above each electrode surface in the flow-cell chambers.

The tethaPlate™ cartridge is constructed using the following procedure:

(1) Cut open the silver foil pack and, using tweezers, remove the slide. Do not touch the gold electrode bioelectronic interface with fingers or any contaminants as this may damage the surface of the electrode and interfere with membrane formation.
(2) The electrode is stored in ethanol and you need to stand it on a tissue to dry. It takes approximately 2 minutes to dry the electrode surface.
(3) Align the dry slide over the alignment jig, ensuring electrode tracks and the SDX logo on the slide overlay each other. Using tweezers gently push the electrode into the slot. Ensure that you do not make contact with the surface of the gold electrode.
(4) Remove the top thin protective layer of plastic from the cartridge. Ensure that you remove only the thin protective layer and not the entire adhesive laminate. If the adhesive laminate is removed, this will cause the gold electrode to bind with the sticky surface used to bind the adhesive laminate to the cartridge.
(5) Position the white cartridge over the top and push into position. Note that one end has a cutout to accommodate the slide. Position this end toward the ten gold contact tracks. Once the two surfaces meet, do not peel them apart or attempt to relocate them as this will damage the electrode surface.
(6) Place the cartridge and electrode into the clamp and tighten. Allow to stand for at least 30 s before loosening pressure.

After completion of the above steps, the tethaPlate™ cartridge is ready for membrane formation. Since the tethers and spacers have a 10 percent tether density on the electrode surface, the formed membrane will have a 10 percent tether density.

The procedure to construct the 10 percent tether density engineered tethered membrane in each of the six flow-cell chambers in the tethaPlate™ cartridge is detailed below. Precise timing is required for proper formation of the membrane.

(1) Start stopwatch.
(2) Add 8 μL DphPC/GDPE phospholipid solution to flow chamber 1.
(3) At 15 s, add 8 μL DphPC/GDPE phospholipid solution to flow chamber 2.
(4) At 30 s, add 8 μL DphPC/GDPE phospholipid solution to flow chamber 3.
(5) At 45 s, add 8 μL DphPC/GDPE phospholipid solution to flow chamber 4.
(6) At 60 s, add 8 μL DphPC/GDPE phospholipid solution to flow chamber 5.
(7) At 75 s, add 8 μL DphPC/GDPE phospholipid solution to flow chamber 6.
(8) At 120 s, add 100 μL PBS to flow chamber 1.
(9) At 135 s, add 100 μL PBS to flow chamber 2.
(10) At 150 s, add 100 μL PBS to flow chamber 3.
(11) At 165 s, add 100 μL PBS to flow chamber 4.
(12) At 180 s, add 100 μL PBS to flow chamber 5.
(13) At 195 s, add 100 μL PBS to flow chamber 6.

Once the above is complete, this allows the lipids in the DphPC/GDPE phospholipid solution to bind to the tethers on the electrode surface. To remove any excess

DphPC/GDPE phospholipid solution, we rinse the flow-cell chambers with PBS solution. Specifically, add 100 μL PBS to each flow-cell chamber, and then remove 100 μL of solution from the outlet of the flow-cell chamber. Repeat this process three times to ensure that no excess DphPC/GDPE phospholipid solution is present in the flow-cell chamber. Following the above procedure, the tethered membrane will self-repair, becoming better sealed with a progressively lower conductance over the initial 30 minutes to an hour. Before performing any quantitative experiments a period of equilibration is recommended. To evaluate the quality of the tethered membranes, we measure the impedance response of the tethered membranes as discussed in §4.2.2. Recall that a high-quality membrane will have a membrane conductance in the range of 0.04 to 1.00 μS/mm^2, and an associated membrane capacitance in the range of 4 to 10 nF/mm^2.

4.4.2 Spontaneous Insertion of gA Ion Channels

Given the tethaPlate™ cartridge with the formed membranes in each of the six flow-cell chambers, we now use the spontaneous insertion method to insert different concentrations of gA ion channels into each of the tethered membranes.

Since gA spontaneously inserts into the engineered membrane, all that is required to insert these ion channels is to bring them into contact with the surface of the tethered membrane. The gA ion channels are inserted using the following procedure:

(1) Start stopwatch.
(2) Add 3 μL of methanol solution to flow chamber 1.
(3) At 15 s, add 3 μL of 34 nM gA in methanol solution to flow chamber 2.
(4) At 30 s, add 3 μL of 68 nM gA in methanol solution to flow chamber 3.
(5) At 45 s, add 3 μL of 103 nM gA in methanol solution to flow chamber 4.
(6) At 60 s, add 3 μL of 103 nM gA in methanol solution to flow chamber 5.
(7) At 75 s, add 3 μL of 171 nM gA in methanol solution to flow chamber 6.
(8) At 120 s, add 100 μL PBS to flow chamber 1, then aspirate using 100-μL pipette.
(9) At 135 s, add 100 μL PBS to flow chamber 2, then aspirate using 100-μL pipette.
(10) At 150 s, add 100 μL PBS to flow chamber 3, then aspirate using 100-μL pipette.
(11) At 165 s, add 100 μL PBS to flow chamber 4, then aspirate using 100-μL pipette.
(12) At 180 s, add 100 μL PBS to flow chamber 5, then aspirate using 100-μL pipette.
(13) At 195 s, add 100 μL PBS to flow chamber 6, then aspirate using 100-μL pipette.

After completion, each of the flow-cell chambers will contain a 3 percent methanol solution with different concentrations of gA ion channels embedded in the tethered membranes. When performing the above procedure, be careful not to incorporate air bubbles into the system as this will damage the tethered membranes.

4.4.3 Measuring Membrane Conductance Response

Given the engineered tethered membranes with different concentrations of gA ion channels, we now perform impedance measurements to determine how the membrane conductance changes as a result of gA ion-channel concentration. The interpretation of

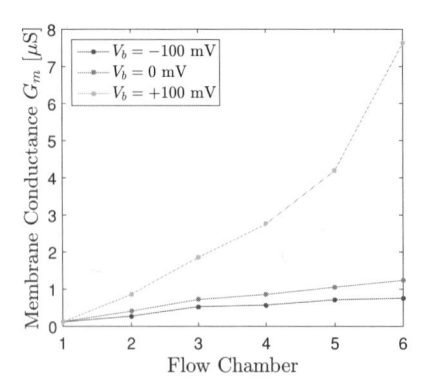

Figure 4.11 Conductance response of the six 10 percent tether density DphPC engineered membranes containing different concentrations of gA ion channels. The concentration of gA ion channels increases from flow chamber 1 to flow chamber 6 in the tethaPlate™ cartridge. V_b indicates the voltage bias used when estimating the membrane conductance G_m.

the impedance response to estimate membrane conductance G_m is discussed in Chapter 7 and Part III of the book; here we merely provide the experimental results with discussion.

The impedance measurements are performed using a sinusoidal voltage excitation $V_s(t) = V_0 \sin(2\pi f t) + V_b$ with a magnitude of $V_0 = 25$ mV at frequencies f of 1000, 500, 200, 100, 40, 20, 10, 5, 2, 1, 0.5, and 0.1 Hz and voltage biases V_b of -100, 0, and 100 mV. Typically $V_b = 0$ mV to measure the membrane conductance G_m; however, here we are interested in measuring how the membrane conductance varies depending on the transmembrane potential (e.g., the concentration of anions and cations adjacent to the tethered membrane surface) and concentration of gA ion channels present. Using the impedance measurements and dynamic model, the estimated membrane conductance of the tethered membranes in each flow-cell chamber are provided in Figure 4.11. As expected, as the number of gA ion channels in the membrane increases, the associated conductance of the membrane increases. As seen from Figure 4.11, the membrane conductance G_m depends on both the concentration of gA present, and on the voltage bias V_b. As the concentration of gA increases, the associated membrane conductance G_m increases. This is expected as the more gA ion channels are present the higher the permeability of the membrane. An immediate result from Figure 4.11 is that the gA ion channel is cation selective. For a negative bias $V_b = -100$ mV, a surplus of cations is present at the electrode surface. Since gA is a cation-selective ion channel, when a surplus of cations is present, this will increase the associated conductance of the membrane. However, for $V_b = +100$ mV this causes a surplus of anions to be present at the electrode surface. Since gA is cation selective, this causes a decrease in the membrane conductance. How many gA ion channels are present in the membrane compared to lipid molecules? The surface area of the membrane is 2.1 mm^2, and the area per lipid is approximately 75 Å^2. Therefore, there are approximately 2.8×10^{12} lipids present in each membrane. The conductance of each gA ion channel is approximately 22.5 pS; and the membrane conductance is in the range 1–8 μS across the flow chambers 2–6, then there are approximately 1×10^6 to 8×10^6 gA ion channels present within each of the flow cell chambers 2–6. For every gA ion channel, there are roughly 1×10^7 lipid molecules present in the engineered membrane. Even though the number of lipid molecules is several magnitudes larger than the number of gA ion channels, the presence of these ion channels non-negligibly impacts the conductance response of the engineered tethered membrane as seen from the results in Figure 4.11.

The results presented in this section provide an introduction to constructing engineered membranes and how to perform experimental measurements using engineered membranes. The construction of all engineered tethered membrane devices (ICS biosensor, pore formation measurement platform [PFMP], electroporation measurement platform [EMP], and ERP) follows similar laboratory procedures as presented in this section. Additionally, though impedance measurements can be used to determine several important features of gA in engineered tethered membranes (e.g., population of conducting gA dimers as a function of concentration), there are several other experimental techniques that can be used to determine the structure and dynamics of gA in tethered membranes. In Chapter 7 we discuss how electric, spectroscopic, and imaging techniques can be used to determine the structure and dynamics of gA in engineered tethered membranes.

4.5 Complements and Sources

In this chapter we discussed methods for building precisely engineered tethered membranes and how to insert peptides and proteins into such membranes. Over the past century, work by many groups has led to the current Singer–Nicolson [387] physical model for a biological membrane. Since its publication in 1972 new insights have challenged significant aspects of the model. In particular, the view that an extensive patch of unperturbed lipid is an appropriate foundation for a biological membrane is now questioned. It has been found that high concentrations of protein exceeding 25 percent in the membrane area are commonplace in all biological membranes. These protein inclusions break up the patches of free lipid into smaller zones of order 30–100 lipids and stabilize cell membranes against both electrical and mechanical stress. A further property of biological membranes not included in the Singer–Nicolson model is the presence of an adjacent hydrated stabilizing surface that possesses a lowered water activity. The role of this layer remains conjectural but it may play an important function in membrane stability and in ionic storage and release during periods of membrane activity. The engineered tethered membrane provides a biomimetic platform that allows the dynamics of peptides and proteins to be studied in a controlled environment. Three methods were discussed for inserting peptides and proteins into engineered tethered membranes, namely, the spontaneous insertion method, the electrochemical insertion method, and the proteoliposomal insertion method.

The spontaneous insertion method can be used for proteins and peptides that will spontaneously insert into the engineered tethered membrane. Note that this spontaneous insertion process depends on the tethering density: as the size of the peptide and protein increases, the associated tether density must decrease to allow for sufficient space for the molecule to fit into the lipid bilayer. Examples of the spontaneous insertion method are provided in [84, 336, 434, 442, 474] for the insertion of alamethicin, valinomycin, α-hemolysin, gA, and CLIC1. The electrochemical insertion method can be used to promote the insertion of soluble proteins and peptides into tethered membranes. Examples of the electrochemical insertion method are provided in [88] for promoting the insertion of PGLa, and in [474] for the insertion of alamethicin. The proteoliposomal insertion process also allows insertion of insoluble peptides and proteins into tethered membranes directly from a cell membrane. For the insertion of the larger channels, incorporation from a proteoliposome method is favored [335]. Examples of the proteoliposomal insertion method include [101, 212] for the formation of an engineered tethered membrane containing the voltage-dependent anion channels VDAC and hVDAC1, and [409] for the insertion of photosynthetic antenna proteins.

In addition to these three insertion methods, a fourth method is the detergent cosolvation method. The detergent cosolvation method is similar to the proteoliposomal insertion method; however, in the detergent cosolvation method the detergent-lipid-protein complex, illustrated in Figure 4.9 on page 75, binds directly with the tethered lipid monolayer to form the lipid bilayer with the inserted protein. Details on the detergent cosolvation method are provided in [86] for the bacterial sodium channel NaChBac, and in [269, 323] for the bacterial mechanosensitive channels MscS and MscL. Examples of commonly used detergent molecules in the detergent cosolvation method are Brij 58, CYMAL-5, TWEEN-20, Triton X-100, and DDM, all of which have a high aggregation number.

4.6 Closing Remarks

This chapter has presented methods for constructing engineered tethered membranes with different tether densities and lipid compositions. Additionally, three techniques (spontaneous insertion, electrochemical insertion, and incorporation from a proteoliposome) were discussed for inserting various peptides and proteins into an engineered tethered membrane. These techniques are used to construct the four engineered tethered membrane sensing devices: the ICS biosensor, the PFMP, the EMP, and the ERP. Given the engineered tethered membrane devices, how can we estimate important biological parameters using either impedance measurements or current response measurements? This requires the development of dynamic models of these tethered membrane devices. The construction of these dynamic models is detailed in Part III of the book.

5 Ion-Channel Switch (ICS) Biosensor

5.1 Introduction

In this chapter we describe the ion-channel switch (ICS) biosensor. By a biosensor, we mean a sensor built out of synthetic biological material – in our case, built out of an artificial membrane, ion channels, spacers, and tethers. The purpose of the biosensor is to detect the presence of analyte molecules.

The ICS biosensor provides an example of how devices built out of tethered engineered membranes can be used for high-resolution detection of specific target (analyte) molecules. When it senses the presence of analyte molecules, the ICS biosensor responds by changing its electrical impedance significantly. The amount the impedance changes by depends on the concentration of the analyte molecules. Therefore, by examining the current flowing through the biosensor, one can detect both the presence of the analyte and its concentration. A key feature of the ICS biosensor is that the detection is performed by measuring the time-dependent conduction of the engineered membrane, which is dependent on the ensemble of aqueous pores and a single pair of conducting gramicidin (gA) dimers present. A gramicidin dimer [83, 84, 466] is an example of a conducting ion channel which allows ions to pass through the tethered membrane. Using specific molecular components and excitation potentials, the ICS can be designed to detect specific analyte molecules of interest.

The ICS biosensor provides an interesting example of engineering at the nanoscale. It is remarkable that the functionality of the device depends on approximately 100 lipids, and a pair of ion channels modulating the flow of billions of ions in a typical sensing event of approximately 5 min. Since the gramicidin monomers (each with a conducting pore of diameter 0.4 nm and length 2.8 nm) move randomly in the outer lipid leaflet of the membrane (1.4 nm thick), we can view the biosensor as a fully functioning nanomachine with moving parts. Indeed, each individual gramicidin monomer diffuses randomly over an area of order 1 μm^2. Furthermore, the 4-nm-thick lipid bilayer is tethered 2–3 nm away from the gold surface by hydrophilic spacers, thereby allowing ions to diffuse between the membrane and gold. This permits a flux in excess of 10^6 ions per second to traverse each gramicidin channel comprised of two gA monomers.

Devices for detecting biomolecules include the nanogap biosensor [202, 286] and the nanoneedle biosensor [114]. The nanogap biosensor relies on detecting impedance changes of the electrode surface which is proportional to the concentration of target molecules. An issue with the nanogap biosensor and similar sensors is that spurious electrochemical reactions resulting from proteins and ions binding to the electrode surface can interfere with measurements. As the nanogap biosensor utilizes a redox active electrode (e.g., Ag/AgCl) [286], the electrode ablates and releases metal ions into solution which can conformationally change the biomolecules being detected, resulting in measurement errors. The nanoneedle biosensor utilizes the change in conduction between two poly-silica phosphorous-doped electrodes sandwiched between silicon dioxide layers. The electrodes are placed 30 nm apart allowing for the detection of biomolecules. As silicon is poorly soluble it does not release harmful ions into solution; however, since silicon dioxide has an isoelectronic point of 3 (i.e., the pH at which silicon dioxide carries no net electrical charge), the adsorption of certain proteins and peptides on the surface is a possibility [114]. Given the electrode size and detection mechanism of the nanoneedle biosensor, it is difficult to perform concentration estimates in cluttered electrolyte environments.

To overcome the above limitations, the ICS biosensor employs an inert bioelectronic interface and an engineered membrane for detection and measurement. The electrical instrumentation of the ICS is connected to the electrolyte solution via gold electrodes. Using gold electrodes as the bioelectronic interface provides a superior interface as compared with using redox active electrodes for two reasons [88]. First, if redox active electrodes are used, the metal will ablate, causing the tethers to dissociate from the electrode surface, destroying the membrane. Second, redox active electrodes release metal ions into solution which can interfere with the electrophysiological response of proteins and peptides. The inert gold electrode capacitively couples the electronic domain to the physiological domain without the issues associated with redox electrodes; however, the capacitive effects of the electrode must be accounted for when modeling the ICS biosensor.

A schematic of the ICS biosensor is given in Figure 3.6. From a practical point of view, the ICS biosensor can be designed whereby specific binding sites with the membrane have a lifetime of several months [84, 88, 152, 273, 331, 336]. The engineered membrane is composed of a self-assembled monolayer of mobile lipids and gA monomers, and a self-assembled monolayer of mobile lipids and gA monomers. The tethered components are anchored to the gold electrode via polyethylene glycol chains. Spacer molecules are used to ensure the tethers are evenly spread over the gold electrode. The intrinsic spacing between tethers and spacers is maintained by the benzyl disulfide moieties which bond the spacers and tethers to the electrode surface. A time-dependent voltage potential is applied between the electrodes to induce a transmembrane potential of electrophysiological interest; this results in a current $I(t)$ related to the charging of the electrical double layers and the conductance of the engineered membrane.

The ICS biosensor is capable of detecting femtomolar concentrations of target species including proteins, hormones, polypeptides, microorganisms, oligonucleotides, DNA segments, and polymers in cluttered electrolyte environments [220, 221, 283]. This remarkable detection ability is achieved using engineered receptor sites connected to mobile gA monomers and biotinylated lipids in the tethered membrane (refer to Figure 3.6). By measuring the dynamics of the membrane conductance, the concentration of a specific analyte can be estimated. In reference to Figure 3.6, the mobile gA monomer is tethered to a biological receptor such as a nucleotide or antibody which binds to specific target species. In the neighborhood of the mobile gA, a tethered monolayer lipid is present with the tethered receptor in contact with the analyte solution. When the receptor binds to the analyte, the mobile gA monomer diffuses to the tethered lipid, causing the conducting gA dimer to break. As an ensemble of gA dimers dissociates, the conductance of the membrane decreases. Measurement of the conductance change allows both the detection of the analyte species and an estimate of the concentration of the analyte species in cluttered environments because the receptor is designed to bind to a specific target species. To relate the analyte concentration to changes in membrane conductance requires the use of an electrodiffusive model for the analyte coupled with the surface reactions present at the tethered membrane surface. Such electrodiffusive models are presented in Part III.

5.2 ICS Biosensor: Construction and Formation

This section describes how the ICS biosensor is constructed and how the quality of the ICS can be measured prior to performing experimental measurements.

The engineered membrane of the ICS is supported by a $25 \times 75 \times 1$ mm^3 polycarbonate slide. Six 100-nm-thick sputtered gold electrodes, each with dimensions 0.7×3 mm^2, rest on the polycarbonate slide. Each electrode is in an isolated flow cell with a common gold return electrode. The formation of the tethered membrane on the gold electrode is performed in two stages using a solvent-exchange technique [84, 307, 464].

Stage 1: The first stage of formation involves anchoring of the tethers and spacers to the gold surface. The tethers provide structural integrity to the membrane and mimic the physiological response of the cytoskeletal supports of real cell membranes. The spacers laterally separate the tethers, allowing patches of mobile lipids to diffuse in the membrane. The tethers and spacers both contain benzyl disulfide components (i.e., MSLOH, DLP, ether-DLP, tether-gA, MSLB, and spacer molecule (SP) in Figure 3.6). The benzyl disulfide bonds to the gold surface with the disulfide bond maintained [84, 220]. This bonding structure has been detected experimentally from X-ray photoelectron spectra. The use of the disulfide has the advantage that the thiols do not oxidize upon storage, allowing the membrane to have a lifetime of several months. From experimental measurements, the electrodesorption of the thiol-to-gold bond is negligible for electrode potentials below 800 mV [439].

To form the anchoring layer, an ethanolic solution containing 370 μM of engineered ratios of benzyl disulfide components is prepared. The ratio of benzyl disulphide components defines the tethering density of the membrane. For example, for a 10 percent tethered membrane, for every nine spacer molecules there is one tether molecule. This solution is exposed to the gold surface for 30 min, then the surface is flushed with ethanol and air dried for approximately 2 min. Note that in the special case of 100 percent tethering, the engineered membrane is composed of a tethered archaebacterial-based monolayer with no spacer molecules. As experimentally illustrated in [152], it is not possible to construct a 0 percent tethered membrane because any formed membrane binds to the gold surface. As the electrolyte reservoir separating the membrane and electrode surface is required for the normal physiological function of the membrane, and noting that all prokaryotic and eukaryotic cell membranes contain cytoskeletal supports with a 1–10 percent tether density, the inability to construct a 0 percent tethered membrane is of little importance.

Stage 2: The second stage involves the formation of the tethered membrane. A solution containing a mixture of mobile lipids is brought into contact with the gold bonded components from stage 1. Several lipid solvents can be used [84, 273]; however, in most cases the lipids selected to form the bilayer are soluble in ethanol. As an example, let us consider the formation of a 70 percent DphPC and 30% GDPE mixed tethered membrane. 8 μL of 3 mM of the 70 percent DphPC and 30 percent GDPE ethanolic solution is added to the flow chamber. The ethanol solution also contains biotinylated gA monomers (tether-gA in Figure 3.6) with the biotin linked to the gA monomer via a 5-aminocaproyl linker. The solution is incubated for 2 min at 20°C in which the tethered membrane forms. After the 2-min incubation period, 300 μL of phosphate-buffered saline is flushed through each flow chamber. The tethered membrane is equilibrated for 30 min prior to performing any experimental measurements. For the detection of streptavidin a biotin receptor is used. The associated antibody fragments used to detect ferritin, thyroid-stimulating hormone, and hCG are the antiferritin F'_{ab}, thyrotropin binding inhibitory immunoglobulin (anti-TSH F'_{ab}), and immunoglobulin G (anti-hCG F'_{ab}), respectively. Details on how the antibodies are connected to the MSLB and mobile gA monomers, in Figure 3.6, is provided in [84].

Remark: The membrane stability is primarily enhanced by tethering the inner membrane leaflet to the gold surface. However, additional stability is achieved by substituting a major fraction of the tethered lipids with archaebacterial lipids. These are lipids modeled on constituents found in bacteria capable of surviving extremes of temperature and hostile chemical environments. Characteristics of these lipids are that the hydrocarbon chains span the entire membrane and that all ester linkages are replaced with ethers [93, 226]. Bilayer lipid membrane (BLM) films have previously been formed from archaebacterial lipids and resulted in membranes that are stable to temperatures in excess of 90°C [139]. A stable membrane incorporating ion channels can be self-assembled on a clean, smooth gold surface using a combination of sulfur-gold chemistry and physisorption [291]. Most studies of the ICS biosensor have used antibody F'_{ab} fragments as the receptor; however, the approach has also been demonstrated to operate using oligonucleotide probes, heavy-metal chelates, and cell surface receptors.

Quality control and measurement: The quality of the tethered membrane is measured continuously using electrical impedance spectroscopy measurements. Typically these impedance measurements are performed at frequencies of 1000, 500, 200, 100, 40, 20, 10, 5, 2, 1, 0.5, and 0.1 Hz and an excitation potential with magnitude not exceeding 50 mV. The defect density in the formed membrane can be estimated from the impedance measurements using the protocol presented in [435]. For all experimental measurements, the membrane contained negligible defects. To detect the presence of electrodesorption and release of portions of tethered membrane into solution, the capacitance of the gold electrode and membrane is monitored via the impedance. Using the experimentally measured impedance and lumped circuit model of the biosensor, the integrity of the sulphur-gold bond and membrane is ensured by comparing the associated capacitances before and after all experimental measurements. Note that alternative measures of the membranes impedance characteristics may be obtained using a potentiostat where typically a ramp or step excitation potentials is applied and the current response measured.

5.3 Operation of the ICS Biosensor

The ICS biosensor provides a rapid detection mechanism for both low-molecular-weight drugs and high-molecular-weight proteins. This section discusses how the ICS biosensor operates to detect analyte molecules. Also experimental results are given for the detection of a biotinylated antibody fragment (b-F_{ab}).

5.3.1 Large and Small Analyte Detection

Large analyte molecules include proteins, hormones, polypeptides, microorganisms, oligonucleotides, DNA segments, and polymers. In the same manner that an enzyme-linked immunosorbent assay (ELISA) may be fabricated based on a complementary antibody pair, the ICS biosensor may be adapted to the detection of any antigenic target for which a suitable antibody pair is available. The bacterial ion channel gramicidin A is assembled into a tethered lipid membrane and coupled to an antibody targeting a compound of diagnostic interest. The binding of the target molecule causes the conformation of the gramicidin channels to switch from predominantly a conducting dimer to predominantly nonconducting monomers as shown in Figure 5.1. For target analytes with low molecular weights such as therapeutic drugs where the target is too small to use a two-site sandwich assay, a competitive adaptation of the ICS is available. This is shown in Figure 5.2.

5.3.2 Impedance Response of ICS Biosensor for Digoxin and b-F_{ab}

Let us consider the interaction of analyte molecules with the ICS biosensor illustrated in Figure 5.2. The ICS biosensor contains a modified gramicidin channel in which an analog of the target analyte is covalently attached to the gramicidin ethanolamine group, forming gA-dig (gA ion channels with a water-soluble hapten digoxigenin) and tethered-membrane-spanning lipids. The analyte molecule of interest is the small-molecular-weight digoxin (781 Da). Digoxin is known to bond with the molecule b-F_{ab}, a biotinylated antibody fragment. Therefore, the ICS biosensor is designed to include membrane-spanning lipids with tethered b-F_{ab} molecules that bind with digoxin. With

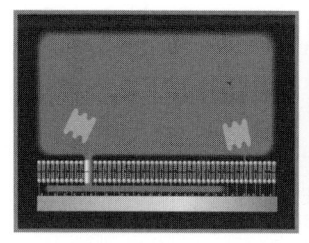

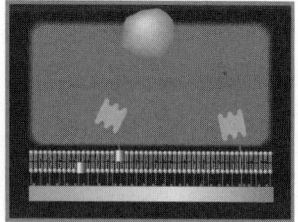

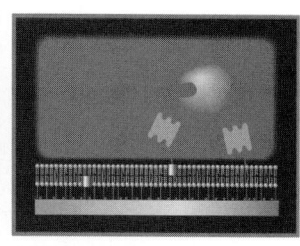

(a) Conducting gA dimer.   (b) Noncoducting monomers.   (c) Nonconducting gA monomers.

Figure 5.1 A color version of this figure can be found online at www.cambridge.org/engineered-artificial-membranes. Large-analyte transduction mechanism. The binding of analyte (green) to the antibody fragments (F_{ab}) (red) causes the conformation of gramicidin A (gA) to shift from conductive dimers to nonconductive monomers. This causes a loss of conduction of ions across the membrane. The scale can be visualized by the fact that the tethered lipid bilayer is 4 nm thick. The dynamics of the analyte binding in the ICS are illustrated in the YouTube video www.youtube.com/watch?v=6Ti83oO2ml4.

no digoxin in the analyte solution the gA-dig and b-F_{ab} tethered to the membrane-spanning lipid cross-link to reduce the population of gA dimers in the membrane, as illustrated in Figure 5.2. If digoxin is added to the analyte solution, then it will diffuse to the membrane surface and reverse the cross-link between the gA-dig and the b-F_{ab}. This allows an increase in the population of mobile gA-dig in the membrane which results in an increased formation of gA dimers. This increase in the population of gA dimers causes an increase in the overall conductance of the membrane. Using impedance measurements, it is possible to detect the change in conductance of the membrane that results from digoxin coming into contact with the molecules gA-dig and b-F_{ab} on the membrane surface. The specific chemical reactions that are present at the ICS biosensor surface for this example are provided in Figure 5.3.

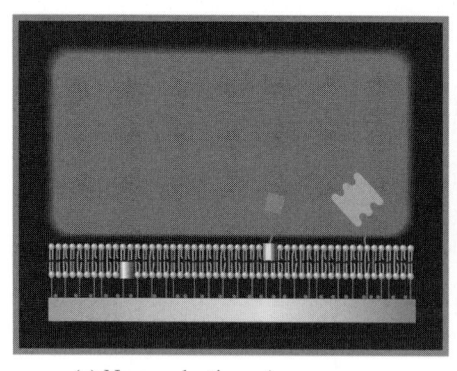

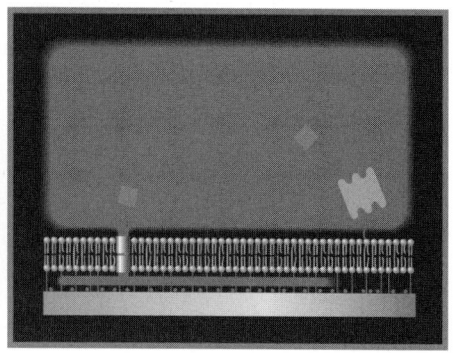

(a) Nonconducting gA monomers.   (b) Conducting gA dimer.

Figure 5.2 Small-analyte transduction mechanism. In the absence of analyte, the mobile channels cross-link to antitarget b-F_{ab}s anchored at the tether sites. Dimer formation is prevented and the biosensor conductance decreases. The introduction of analyte competes off the hapten (target analog) and increases the biosensor conductance.

5.3 Operation of the ICS Biosensor

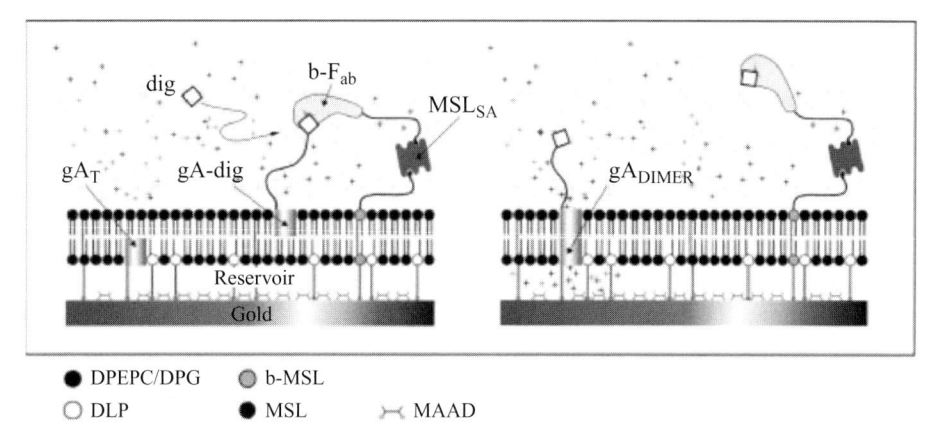

Figure 5.3 Schematic of the chemical reactions present in the ICS biosensor for the detection of b-F_{ab} and dissociation via the analyte molecule digoxin. gA_T is tethered gramicidin, gA-dig is gramicidin digoxin, dig is digoxin, b-F_{ab} is biotinylated antibody fragment, MSL_{SA} is membrane spanning lipids with streptavidin.

How can changes in the ICS membrane conductance be detected using the results of impedance measurements? A detailed analysis of impedance plots and models is provided in Part III; here we provide a brief explanation. Figure 5.4(a) provides the computed impedance for different values of membrane conductance in the frequency range of 0.1 Hz to 1 kHz. The results illustrate that, as the membrane conductance increases, the local minima in the phase $\angle Z(f)$ increases in frequency. Additionally, the associated magnitude $|Z(f)|$ of the impedance decreases for increasing membrane conductance. Therefore, the phase and magnitude from the impedance measurements provide a qualitative measure of changes in conductance of the tethered membrane.

Using the above guide for how the impedance (phase and magnitude) change as a result of conductance changes in the membrane, we can now use impedance measurements to detect the presence of digoxin. The experimentally measured impedance after the addition of digoxin to the ICS biosensor is illustrated in Figure 5.4(b). When digoxin comes into contact with the membrane surface, the local minima in the phase angle increase and the associated magnitude decreases. This suggests that the addition of digoxin causes a decrease in the membrane conductance of the ICS biosensor. Therefore, using the ICS biosensor and measuring the impedance allows us to detect the presence of the analyte digoxin.

The next question is this: Can an accurate estimate of the concentration of digoxin be made using impedance measurements from the ICS biosensor? The answer is yes; however, this requires a mathematical model of the ICS biosensor that accounts for the chemical reactions present and the electrodiffusive dynamics of the analyte molecules. In Chapters 9 and 10 we construct such models of the ICS biosensor that can be used to estimate the concentration of specific analyte molecules present using the experimentally measured impedance response.

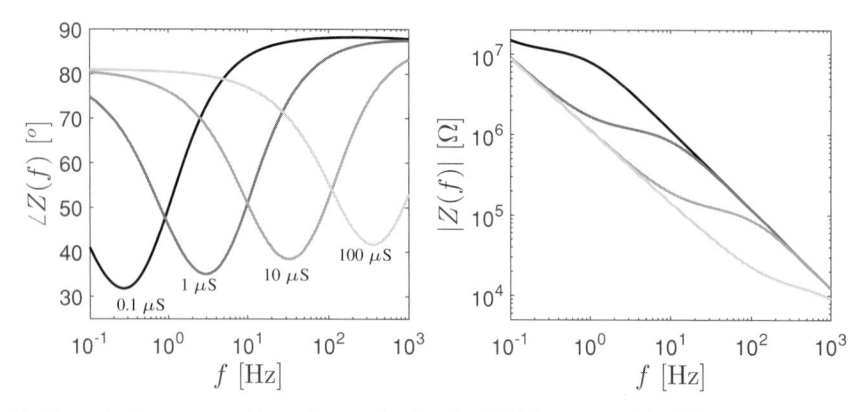

(a) Numerically computed impedance plot for the ICS biosensor with different membrane conductances. The phase $\angle Z(f)$ (in degrees) and magnitude $|Z(f)|$ (log scale in ohms) are displayed.

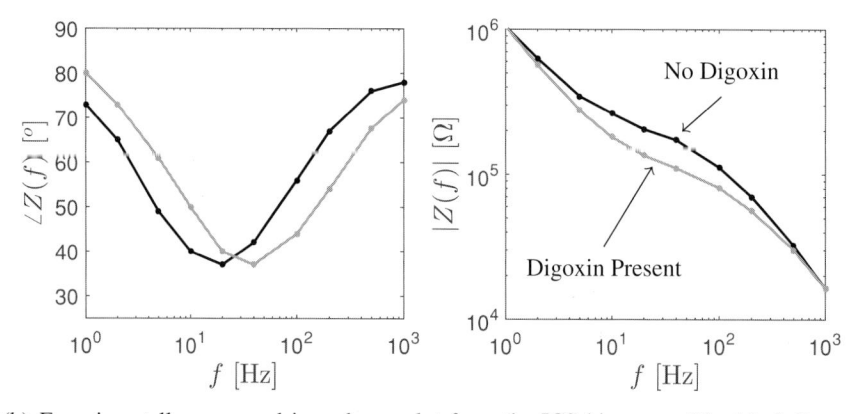

(b) Experimentally measured impedance plot from the ICS biosensor. The black lines indicate the impedance of the ICS biosensor when no digoxin is present in the analyte solution. The light gray lines indicate the change in impedance that results from the addition of digoxin to the analyte solution. The conductance of the membrane increases as a result of the digoxin reversing the bond between the gramicidin digoxin and the b-F_{ab} biotinylated antibody fragment that is connected to the tethered-membrane-spanning lipids as illustrated in Figure 5.3. The time duration between the two impedance measurements is 10 min, and the amount of digoxin added to the analyte was 0.38 ng [463].

Figure 5.4 Impedance response of the ICS biosensor to b-F_{ab} and digoxin.

5.4 ICS Biosensor: Flow Velocity, Binding-Site Density, and Specificity

This section describes how the impedance response of the ICS biosensor depends on the flow velocity of the electrolyte in the flow chamber and the binding-site density on the membrane surface. A schematic of the flow chamber of the ICS biosensor is provided in Figure 5.5. The analyte is transported to the reacting surface (biomimetic surface of the ICS) by diffusion and advection, where it reacts with the immobilized receptors (illustrated in Figure 5.3). As the analyte binds to the receptors and other mobile gA monomers, this reduces the number of mobile gA monomers on the membrane surface

5.4 ICS Biosensor: Flow Velocity, Binding-Site Density, and Specificity

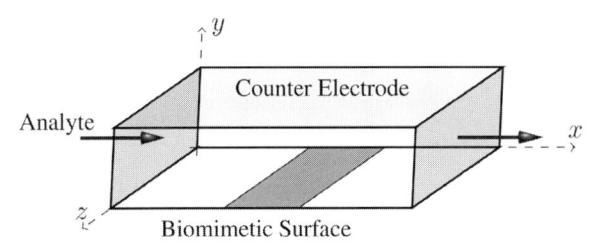

Figure 5.5 A color version of this figure can be found online at www.cambridge.org/engineered-artificial-membranes. Schematic of the flow chamber of the ICS biosensor. The biomimetic surface (dark gray region) indicates the location of the engineered membrane and the gold electrode. Recall that Figure 5.3 provides a detailed schematic of the engineered membrane of the ICS biosensor. The solution containing the analyte enters the flow chamber at the left-hand side, $x = 0$, and flows along the x axis. Impedance measurements are performed between the gold electrode (dark gray region) and gold counter electrode (yellow region). Typically the flow chamber of the ICS has dimensions $L_x = 6$ mm, $L_y = 0.1$ mm, and $L_z = 3$ mm in the x, y, z coordinate axes.

that are required to form conducting gA dimers. As such, as the number of mobile gA monomers decreases, this causes a change in the measured impedance of the ICS biosensor. Mathematical models of the ICS biosensor are constructed in Chapters 9 and 10. Here we illustrate how flow rate and binding-site density impact the response of the ICS biosensor. Additionally, we illustrate the performance of the ICS biosensor for detecting specific analyte species in cluttered environments such as blood.

5.4.1 Flow Velocity and Binding-Site Density

The values of the flow velocity and binding-site density in the ICS biosensor are important as they determine if the ICS biosensor is operating in the reaction-rate-limited regime or the mass-transport or diffusion-limited regime. Using the results from an advection-diffusion model (presented in Part III), Figure 5.6 shows quantitative predictions of the change in the biosensor resistance R_m per unit time for various binding-site densities and sample flow rates, for high analyte concentrations (100 pM) and low analyte concentrations (10 fM). Note that the conductance G_m (in siemens) is $1/R_m$, where R_m is the membrane resistance (in ohms). As shown in Figure 5.6, at high analyte concentrations the biosensor response is insensitive to flow rate. This corresponds to the reaction-rate-limited kinetics in which the diffusion dynamics of the analyte solution negligibly contribute to the biosensor resistance dynamics. However, for low analyte concentration and high binding-site density, a high flow rate is required to achieve a measurable response. This corresponds to mass-transport-influenced kinetics [441], in which the diffusion dynamics of the analyte solution must be accounted for. It is also apparent from Figure 5.6 that a high binding-site density is essential for high sensitivity. With high binding-site density, target molecules collide more frequently with receptors and are thus captured more quickly. The greater the ratio of binding-site density to analyte concentration, the faster the response of the biosensor.

Are the numerical results in Figure 5.6 in agreement with the experimentally measured response of the ICS biosensor for different flow rates and concentrations of

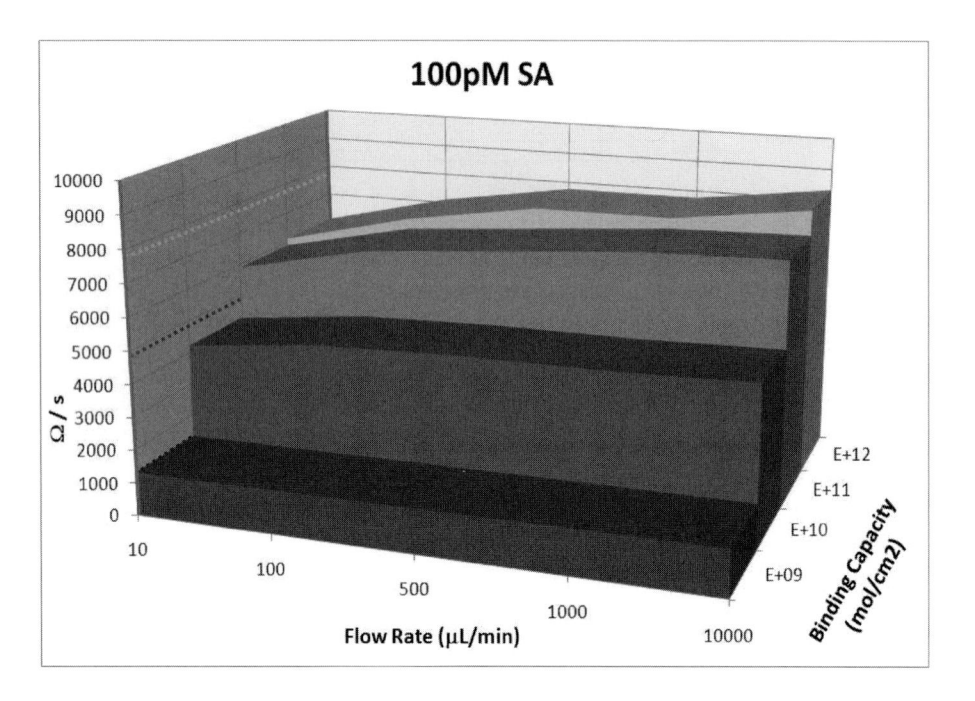

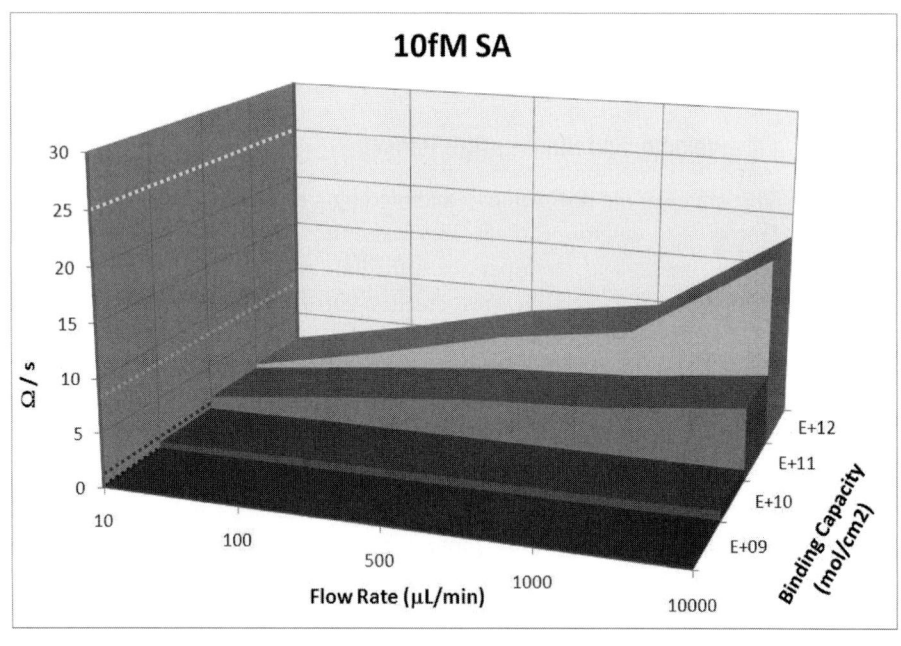

Figure 5.6 Biosensor response to 100 pM and 10 fM concentrations of streptavidin (SA) for various binding-site densities and flow rates. At 100 pM and low binding-site densities, increasing flow rate (left to right) has a negligible effect on the response. At 10 fM and high binding-site densities, the flow rate has a significant effect on the response. High flow limits are shown as dashed lines at the left.

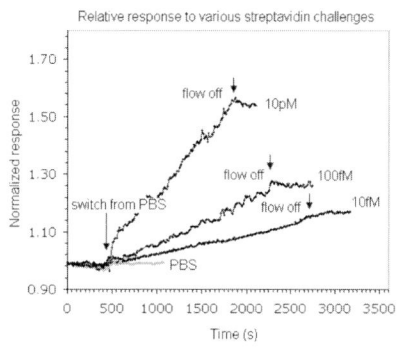

5.4 ICS Biosensor: Flow Velocity, Binding-Site Density, and Specificity

(a) Normalized resistance R_m of ICS biosenso with 150 μL/min flow rate. PBS denotes the phosphate-buffered saline that contains no streptavidin. The time point where streptavidin is added is indicated by the "switch from PBS" arrow. The time point where no further streptavidin is added is indicated by the "flow off" arrow.

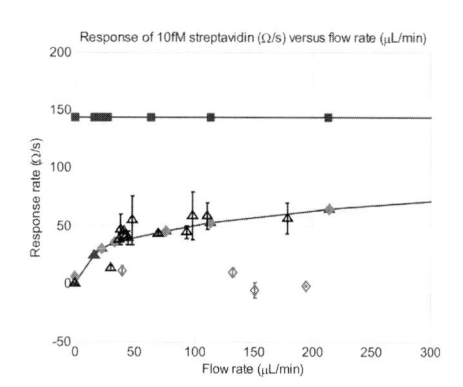

(b) Response of ICS sensor with 10 fM streptavidin where black indicates the experimentally measured response, and blue is the numerically estimated response.

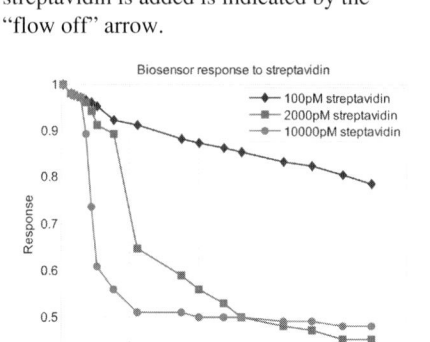

(c) Normalized conductance G_m of ICS biosensor with zero flow rate.

Figure 5.7 A color version of this figure can be found online at www.cambridge.org/engineered-artificial-membranes. Experimentally measured response of the ICS biosensor to different concentrations of streptavidin and flow rates.

analyte? Figure 5.7 provides the response of the ICS biosensor to different concentrations of streptavidin and flow rates and illustrates the highly sensitive response of the biosensor and how this response depends on the analyte flow rate. Figure 5.7(a) shows the experimentally observed increase in the tethered-membrane resistance for increasing analyte concentrations at a flow rate of $\upsilon = 150\,\mu\text{L/min}$. Remarkably the biosensor has a measurable change in resistance in the presence of only 10 fM of streptavidin. Interestingly, the initial slope of the resistance R_m is proportional to the concentration of streptavidin present. Given a data set of the initial slopes of R_m and the associated concentrations, a linear model can be constructed to estimate the concentration of analyte present [231]. Note, however, that such a model cannot be used to design the ICS biosensor as it only provides the relationship between initial slopes of resistance and analyte concentration for the specific ICS biosensor and analyte used to construct the linear model. The ICS biosensor response in Figure 5.7(b) illustrates how the increase in flow rate results in an increase in the associated change in membrane resistance in the reaction-rate-limited regime. For these flow rates and streptavidin concentration, the increase in membrane resistance as a function of time is linear; therefore, only the change in membrane resistance is reported. For zero flow rate, the concentration of streptavidin must be sufficiently high to ensure diffusion to the membrane surface to elicit a measurable response – this is an example of the diffusion-limited regime of the ICS biosensor. Figure 5.7(c) provides the experimentally measured response of the ICS biosensor to different concentrations of streptavidin. As shown, for low concentrations of streptavidin the response is approximately linear. This is an example of the diffusion-limited regime as the analyte concentration at the surface of the membrane cannot be assumed to be constant. However, for very large streptavidin concentrations, the analyte concentration at the membrane surface is approximately constant; this provides an example of the reaction-rate-limited regime and provides an approximately exponential response. The results in Figure 5.7(c) provide a clear demonstration of the different kinetic regimes of the ICS biosensor, namely the reaction-rate-limited and diffusion-limited regimes [83, 465].

5.4.2 Specificity in Complex Environments

Below we illustrate how the ICS biosensor can accurately detect a specific analyte molecule in whole blood.[1] Note that whole blood contains several different types of cells (erythrocytes, leukocytes, and thrombocytes), all of which are contained in plasma composed of sugars, lipids, vitamins, minerals, hormones, enzymes, antibodies, and other proteins. Therefore, whole blood provides an excellent environment to test if the ICS biosensor can detect a specific protein.

For this analysis we use the ICS biosensor to detect the analyte molecule ferritin in whole blood. Ferritin is the principal iron-transporting protein in human blood and is a large biomolecule with a molecular weight of approximately 450 kDa. Of clinical interest is the measurement of ferittin at concentrations between 10 and 1000 pM. Figure 5.8(a) provides ICS biosensor response (tethered-membrane conductance) as a function of time in response to whole blood containing no ferritin, and whole blood containing 50 pM of ferritin. As seen, the ICS biosensor only provides a dynamic change in conductance with the whole blood containing 50 pM of ferritin. To test the clinical accuracy of the ICS biosensor we compared the results of the ICS biosensor with a Diagnostics Products Corporation (Los Angeles, CA) IMMULITE analyzer for measuring ferritin concentration in whole blood from 100 patients chosen to provide a wide spread of clinical values. The results are displayed in Figure 5.8(b). As shown, the ICS

---

[1] Whole blood refers to human blood from a standard blood donation.

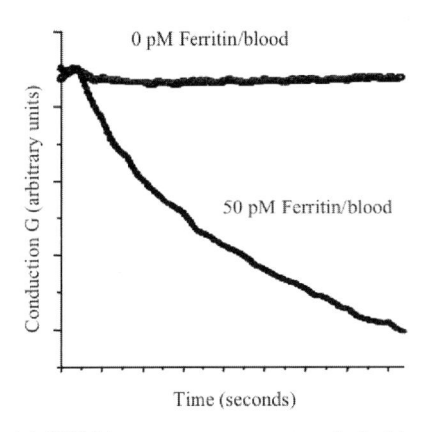

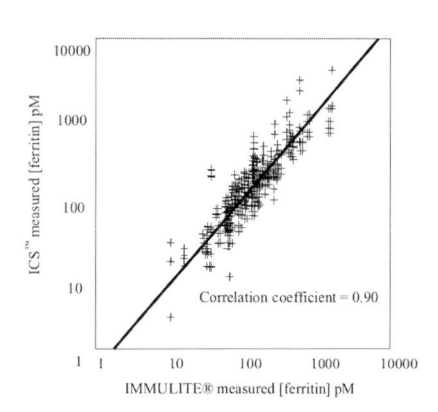

(a) ICS biosensor response to whole blood containing 0 pM of ferritin and 50 pM of ferritin over a time period of 300 s.

(b) A comparison of estimated ferritin concentrations from 100 patients measured by the ICS biosensor and the IMMULITE analyzer.

Figure 5.8 ICS biosensor response to varying concentrations of ferritin in whole blood [374].

biosensor has performs comparably to this state-of-the-art clinical method over a wide range of analyte concentrations under clinical conditions.

5.5 Detection of Influenza A in Clinical Samples

The ICS biosensor offers a rapid method for the detection of microorganisms within 10 min at room temperature, without the time-consuming steps of specimen extraction, specimen washing, and incubation that are currently necessary in an ELISA or polymerase chain reaction test for virus detection [206, 404]. To illustrate the remarkable sensitivity of the ICS biosensor, this section illustrates how the ICS biosensor can be used to detect the presence of the influenza A virus. Influenza is a highly contagious respiratory infection that is spread by aerosol transmission or close personal contact. Rapid detection of the virus is crucial for prompt patient management and implementation of public health alert and control measures.

5.5.1 ICS Biosensor Preparation and Clinical Trials for Rapid Influenza A Diagnosis

To detect the influenza A virus, we must select the appropriate biotinylated antibody fragments (b-F_{ab}) in the ICS biosensor (Figure 5.1). A complementary pair of monoclonal antibodies were selected for reactivity to a specific strain of influenza A subtype. F_{ab} fragments were prepared from two commercially available influenza A nucleocapsid-specific monoclonal antibodies and fragmented using papain digestion (www.piercenet.com) and biotinylated with iodoacetyl-LC-biotin according to a proprietary method that was developed in house. An equimolar mix of the two monoclonal antibodies was used.

To evaluate the performance of the ICS biosensor, we compared the results with the commercially available detection kit obtained from Medix Biochemica (www.medixbiochemica.com) and the culture or antigen ELISA [206, 366] tests. The Medix test kit consists of a chromatographic strip paper impregnated with influenza A antibodies. An internal control was included in the test strip. After treating the specimen in the extraction solution for 5 min at room temperature, the immunochromatographic strip was immersed into the treated sample mix for an additional 10 min and read against a color standard. The culturing or antigen ELISA techniques were as previously reported in [206, 366].

5.5.2 Influenza A Clinical Samples

Two groups of respiratory samples were collected across the state of South Australia and tested to assess the effects of interference in the untreated clinical samples and the ICS biosensor response to the targeted influenza A virus.

> **Group 1:** This group consisted of 74 samples drawn from nasopharyngeal aspirates, sputum, bronchial or tracheal aspirates, and nose and throat swabs during the period July–August 2006. The samples were stored at 4°C, and tested within 2 days by ICS, Medix, as well as by culture. However, during the period July–August 2006, when group 1 samples were collected, no cases of influenza A occurred in South Australia. So the samples in group 1 serve as a useful test for false positives.
>
> **Group 2:** This group consisted of 34 randomly selected samples that had been collected during an outbreak of influenza A in July–September 2005. These samples had been stored at −70°C. (The number of clinical samples in this group was limited by the sample populations the authors were able to negotiate from the South Australian government.) These specimens had previously been submitted for routine virus culture.

5.5.3 Results of Influenza A Clinical Trial

The test results for groups 1 and 2 (defined in §5.5.2) were analyzed in terms of sensitivity, specificity, positive predictive value (PPV), and negative predictive value (NPV). These performance metrics are defined as follows:

$$\text{Sensitivity} = \frac{\text{number of true positives}}{\text{number of true positives} + \text{number of false negatives}}$$

$$\text{Specificity} = \frac{\text{number of true negatives}}{\text{number of true negatives} + \text{number of false positives}}$$

$$\text{PPV} = \frac{\text{number of true positives}}{\text{number of true positives} + \text{number of false positives}}$$

$$\text{NPV} = \frac{\text{number of true negatives}}{\text{number of true negatives} + \text{number of false negatives}} \quad (5.1)$$

Table 5.1 Comparison of ICS and Medix rapid tests with culture for detection of influenza A virus in 34 respiratory specimens.

	ICS %	Medix %
Sensitivity	52	57
Specificity	82	100
Positive Predictive Value	86	100
Negative Predictive Value	45	52

The associated detection rates of influenza A for groups 1 and 2 (defined in §5.5.2) from the ICS biosensor, Medix Kit, and antigen ELISA are provided below.

Group 1: Of 74 samples in group 1, no influenza A virus was detected using the ICS biosensor, the Medix kit, or by culture. Thus the ICS did not yield any false positives. Also, using culture or antigen ELISA [206, 366], it was found that 14 of 74 samples yielded a positive result for influenza B, adenovirus, respiratory syncytial virus, or parainfluenza 3 virus. This means that the ICS showed no cross-reactivity with unrelated viral antigens or interference by heterogeneous respiratory specimens. These results illustrate the specificity of the ICS.

Group 2: The 34 samples in group 2 were tested using the ICS and Medix tests. The specimens were diluted with an equal volume of phosphate-buffered saline prior to addition to the ICS and Medix tests. When compared to the culture as a reference, the ICS and Medix tests showed very similar sensitivities, specificities, positive predictive values, and negative predictive values in detecting influenza A virus. The results are shown in Table 5.1.

These results suggest that the ICS biosensor can effectively detect the influenza A virus from fresh or frozen clinical samples without the need for detergent disruption or sample preparation. Remarkably the ICS biosensor provides very similar results to existing commercial immunochromatographic test strips. After this trial, the importance of sample flow in detection sensitivity of the ICS biosensor has been explored and it was found that a substantial improvement resulted from sample flow rates of 10–50 μL/min. Recall that the design of the ICS biosensor can be performed given a predictive model of the biosensor. Such a predictive model is provided in Part III. Further studies on the detection of Influenza A using the ICS sensor can be found in [231].

5.6 ICS for Multianalyte Detection

In this section we describe how the ICS biosensor can be constructed to detect multiple analytes. The principle of operation is the same as in monoanalyte detection; however, now several tethered membranes are used with each having a different antibody receptor. Note that this allows the ICS biosensor to be used to simultaneously detect for both large- and small-analyte molecules using the methods illustrated in Figures 5.1 and 5.2.

Ion-Channel Switch (ICS) Biosensor

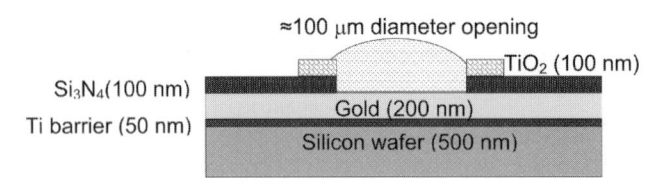

Figure 5.9 Cross section of one element in the biosensor array. The design incorporates five layers: (i) an underlying silicon wafer, (ii) a 50-nm titanium (Ti) barrier, (iii) a 200-nm gold layer, (iv) a 100-nm silicon nitride (Si_3N_4) layer, and (v) a patterned ring of titanium oxide (TiO_2). The titanium oxide ring is designed to provide a hydrophilic surface at the membrane edge.

5.6.1 Biosensor Arrays

ICS biosensor arrays can be fabricated using silicon nitride, silicon carbide, and glass substrates. We now show that by using an ICS array it is possible to detect multiple analytes simultaneously in a single sample. Multianalyte detection is useful as it permits onboard calibration to correct for systematic variations which can occur across an electrode array and to correct for electrode-to-electrode variation between different biosensor electrodes. A novel element in the design of these arrays is the use of a titanium oxide ring at the perimeter of the electrode opening. The titanium ring is designed to provide a mechanical seal for the outer leaflet, preventing it from diffusing beyond the area of the tethered inner leaflet lipids. In addition, the titanium seal retains water during the patterning of antibodies, and during the drydown process for storage. A schematic of the design of an element in the biosensor array is shown in Figure 5.9.

Examples of biosensor electrode arrays recorded by optical microscopy are shown in Figure 5.10. Figure 5.10(a) illustrates a 16-element array of 150 μm-diameter electrodes. Figure 5.10(b) depicts a test array of four electrodes ranging from 150 to 20 μm

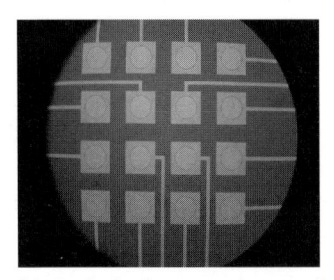

(a) Optical microscopy image of a 16-element biosensor array with 150-μm diameter electrodes.

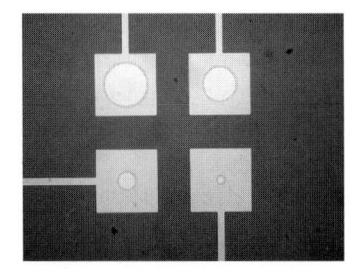

(b) Optical microscopy image of a test array with four electrode elements of 150-, 100-, 50-, and 20-μm diameter.

Figure 5.10 The apparently square geometry of the sensor elements arises from the gold being patterned as rectangles and the silicon nitride openings being round. The thin (100 nm) and transparent silicon nitride allows the gold to be viewed through the nitride layer. Also visible is the light grey 2-μm-wide titanium oxide ring.

in diameter. A consequence of reducing the electrode diameter from 150 to 20 μm is a reduction of the membrane capacitance and an increase in the membrane resistance. Both measures scale with membrane area: the capacitance linearly and the resistance inversely.

Although the impedance of these electrodes is dependent on the electrode area, the time constant of the response is independent of the electrode area. A dependence is expected, however, when the spacing of the ion channels or antibodies is comparable in size to the electrode dimensions. For example, if the electrode dimensions have a diameter of 0.1–1 μm then the response is expected to be dependent on the electrode area. To reduce such effects, the ICS biosensor can be designed with a 20-μm-diameter or larger electrode. Using an electrode array to estimate the concentration of target species is useful as the variance in the estimated response rate can be reduced compared with using measurements from a single electrode of similar size. In fact, an exponential fit to the admittance decay curve measured across 16 electrodes yielded coefficients of variation (standard deviation/mean) of well below 10 percent. This indicates that the biosensor array silicon chip fabrication procedure can provide a highly reproducible electrode geometry and structure. The distributed sensing array possesses a statistical improvement of 16 independent measurements rather than one. A further benefit of miniaturization is the detection of single-channel noise, which permits stochastic analysis of single-molecule detection.

5.6.2 Multi-Analyte Detection

Biosensor arrays also have the ability to measure multiple target concentrations from a single sample addition to the sensor. A key problem when fabricating an array capable of multi-analyte detection is the site-specific decoration of the chip with different antibodies. The approach discussed here is shown in Figure 5.11. A fluid handling spotter (sciFLEXARRAYER leased from Scienion AG Berlin) was loaded with the appropriate antibody solution and directed to a biosensor chip surface that had been partially dried via glycerol, trehalose, polyvinylpyrrolidone, or their combination.

The lower size limit of the electrodes was determined by the resolution achievable by the spotter-surface characteristics and not the constraints of the electrode-membrane characteristics. The limiting dimension set by the spotter-surface combination was 80 μm diameter whereas the limit set by the electrode-membrane combination was 10 μm. Figure 5.11 shows an example of a biosensor chip array with four arbitrary antibody receptors. That is, one quadrant contains four electrodes, with the pregnancy hormone hCG, two quadrants have antibody receptors for influenza A, and a further quadrant uses the reference receptor to a target not in the test sample. Figure 5.12(a) shows the response to four samples containing 150 mIU/mL hCG, 100 ng/L influenza A virus, or neither. The reference electrode cluster yielded a null result for all three challenges; the hCG cluster yielded a positive response (reduction in admittance read as a negative slope) to the 150 mIU/mL hCG sample but zero to the influenza A challenge, whereas the influenza A clusters yielded a positive response to the challenge

(a) Site-specific coating. (b) 20 pL volume drop. (c) Multi-analyte biosensor chip.

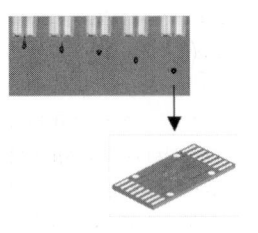

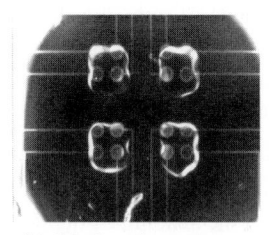

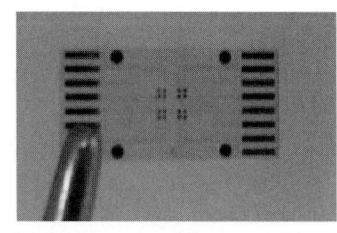

Figure 5.11 (a) Site-specific coating was achieved by directing a metered volume of biotinylated antibody F_{ab} onto an electrode element (230 μm diameter) in the array. The electrode surface was pretreated with 2–5 percent trehalose solution and dried prior to spotting, the surface. Prior to spotting, the membranes had been assembled to a common structure across all electrodes, including a streptavidin linker to the ion channels and membrane-spanning lipids. (b) The sciFLEXARRAYER (Scienion AG Berlin) provided a stream of 20 pL volume drop. Typically 15 drops were applied to each electrode resulting in 1.2 nL per quadrant of four electrodes. Each quadrant received a different biotinylated antibody fragment. (c) The biosensor chip used here was a cluster of four 2 × 2 electrode arrays, each 230 μm diameter on either a glass or silicon substrate.

with influenza A but zero to hCG. Figure 5.12(b) shows the layout of the four quadrants. These results illustrate the ability of a biosensor array to detect multiple target species in one sample. In this case the sample volumes used were 100 μL but these can be reduced to 10–20 μL using a coplanar return electrode.

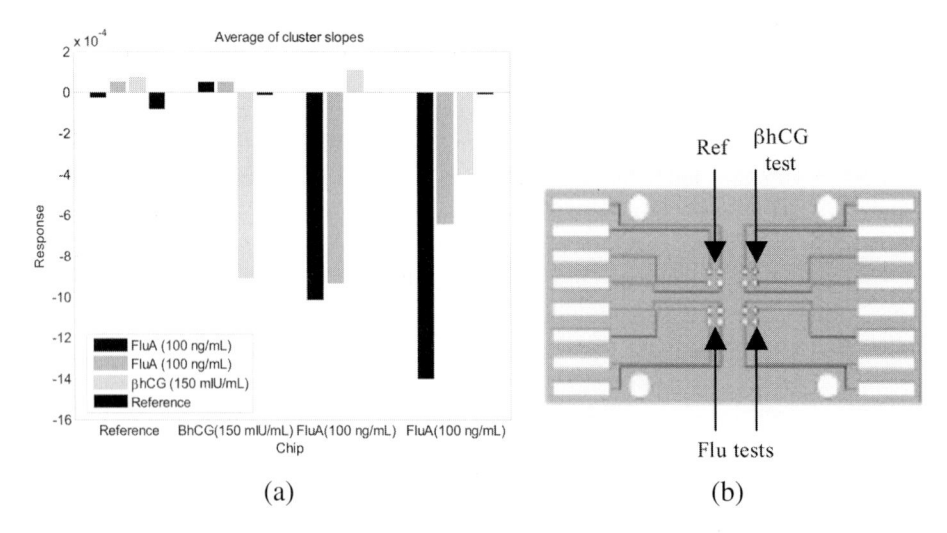

(a)   (b)

Figure 5.12 (a) From left to right, the first set of bars corresponds to electrodes with no receptors, the second set corresponds to electrodes with hCG receptors, and the third and fourth sets contain influenza A receptors. Chip challenged with single sample possessing 150 mIU/mL βhCG and 100 ng/mL influenza A virus. (b) Array geometry and the distribution of antibodies on each of the four clusters.

5.7 Complements and Sources

The ICS biosensor that incorporates gramicidin A ion channels into a tethered engineered membrane was developed in [84]. During the past decade many novel functionalities have been added to successive generations of the biosensor [83, 85, 219–221, 283, 374, 465]. These include the covalent linkage of F_{ab} (fragment antigen-binding portion of antibody) to the gramicidin channels, the use of flow cells and the miniaturization of electrode dimension from 1 mm to 20 μm. Krishnamurthy et al. [219] constructed a biosensor by incorporating bis-gA ion channels into the lipid bilayer membrane of giant unilamellar liposomes and then excising small patches (1 μm in diameter) of the lipid membrane using a patch-clamp micropipette.

Neher provides an overview of the interface between ion channels and microelectronics [295]. Several companies and research groups have developed biosensors based on synthetic lipid monolayers and bilayers. For example, OhmX Corporation developed a reagentless biosensor system using self-assembled monolayers tethered to a gold surface for the electronic detection of biomarkers in clinical samples [131]. Stochastic signal analysis has been employed by Bayley's group at Oxford and has made substantial contributions in the advancement of ion-channel biosensors [168, 322]. The detection of single-gramicidin-channel currents in a tethered membrane is described in [131].

The first attempt at developing a practical membrane-based biosensor device was reported in [241]. The poor stability of the receptor-membrane complex limited the range of applications of the device. One of the first examples of a functionally active biomimetic surface was reported in [292], in which an active cytochrome C was incorporated into a tethered membrane.

The stabilization of the BLM has been a central theme in the development of ion-channel biosensors [240]. Many strategies have been developed. The primary focus has been on physisorbing or chemically attaching a layer of hydrocarbon to a silicon [155], hydrogel [257], polymer [290], or metal surface [403]. Subsequently a second layer of mobile lipids is fused onto the tethered monolayer to form a tethered bilayer lipid membrane. Earlier work on BLM stabilization is reviewed in [354] and [327]. In [285] peptide nanotubes were fabricated within a supported self-assembled monolayer. The ICS biosensor employs an alkane disulfide bond to stabilize the bilayer lipid membrane at the electrode surface.

A key requirement of an ion-channel biosensor is to engineer a switching mechanism that modulates the flow of ions when an analyte is detected [374]. Switching mechanisms range from antichannel antibodies that disrupt ion transport [55] to molecular plugs that block the channel entrance [252]. OmpF porin channels from *E. coli*, were incorporated into a tethered BLM and their conduction modulated using the channel blocker colicin [405].

Miniaturization and patterning are two further opportunities for tethered-membrane technologies [116, 389]. The functionalities that may be brought to tethered bilayers are becoming extensive. Topographical templates for chemiselective ligation of antigenic peptides to self-assembled monolayers were fabricated in [364]. Also, advances in microfluidics and transduction technologies have led to the development of biosensors capable of bimolecular detection at very low (e.g., femtomolar) concentrations of analyte without the need to label the participating molecules [28, 29]; see also [220, 221, 283].

5.8 Closing Remarks

In this chapter we have introduced the ICS biosensor. It is a remarkably high-resolution sensor which can detect as low as femtomolar concentrations of target species including proteins, hormones, polypeptides, microorganisms, oligonucleotides, DNA segments, and polymers in cluttered electrolyte environments. The ICS biosensor employs engineered receptor sites connected to mobile gA monomers and biotinylated lipids in a tethered membrane. As the analyte binds to these receptor sites the associated conductance of the tethered membrane decreases. Note that more than 50 different antibodies as well as cell-surface receptors for growth factor detection, oligonucleotide probes for DNA strand detection, lectins for glucose detection, and metal chelates for heavy-metal detection have been utilized in the ICS biosensor. Further receptors can be designed by exploiting the receptor association dynamics discussed in [229]. The major requirement is an attachment chemistry that will not inactivate the receptor for a particular analyte. Additionally, this chapter illustrated how biosensor arrays can be constructed to improve the detection capability of an analyte, but also to simultaneously detect multiple-analyte species. To use the experimentally measured response from the ICS biosensor to perform concentration estimation or to optimally design an ICS biosensor for the detection of a specific analyte requires a dynamic model of the response.

Part III of this book presents several methods for mathematically modeling the dynamics of the ICS biosensor from the atomistic level for gramicidin dimer dissociation, the continuum for the reaction-diffusion processes present in the analyte and membrane surface, and reaction-rate models for the experimentally measurable conductance from the ICS biosensor. All these models can be used to design novel ICS biosensors and to interpret the experimentally measurable conductance from the biosensor.

6 Physiochemical Membrane Platforms

6.1 Introduction

This chapter describes the construction and operation of three important tethered-membrane-based sensor platforms, namely, the pore formation measurement platform (PFMP), the electroporation measurement platform (EMP), and the electrophysiological response platform (ERP). These three platforms are constructed to mimic real cell membranes and allow experimental measurements of the dynamics and biochemical reaction pathways present in biological membranes (e.g., the physiochemical properties of membranes). As such, the results from the PFMP, EMP, and ERP elucidate the structural and dynamic properties of real biological membranes. An understanding of the structural and dynamic properties of biological membranes' interaction with macromolecules is also useful for rational drug design and therapeutic protocols as discussed in §3.7. A schematic diagram of these platforms is provided in Figure 6.1; recall that these platform devices were introduced in Chapter 3.

The PFMP, EMP, and ERP are designed to perform specific tasks. The PFMP is designed to estimate the reaction dynamics of pore-forming peptides and proteins. The dynamics of pore-forming peptides and proteins are crucial to the attack and defense mechanisms of biological organisms. Understanding the chemical kinetics of pore-forming peptides and proteins provides vital information of use to pharmacologists to target specific classes of peptides and proteins for in-depth pharmaceutical screening of novel drugs. The principal operation of the PFMP is that changes in the membrane conductance can be related to the number of formed pores in the membrane. These changes in membrane conductance can be measured experimentally using impedance measurements. The EMP is designed to study the dynamics of electroporation in tethered membranes. Electroporation is a process in which a transmembrane potential induces the formation of aqueous pores in the tethered membrane. The formation of aqueous pores is important for the transport of molecules through the membrane. The ERP is designed to measure the conductance dynamics of embedded ion channels[1] and to measure the electrophysiological response of cells grown on the surface of the tethered membrane. These results are important for drug screening and diagnosis of channelopathic diseases. Since the conductance dynamics of the EMP and ERP are on the scale of microseconds, and unique excitation potentials are desirable to induce electroporation and study the

---

[1] Embedded ion channels are ion channels that have been inserted into the tethered membrane.

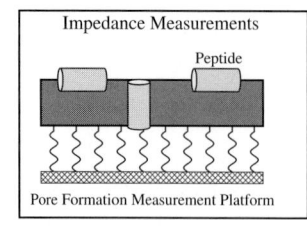

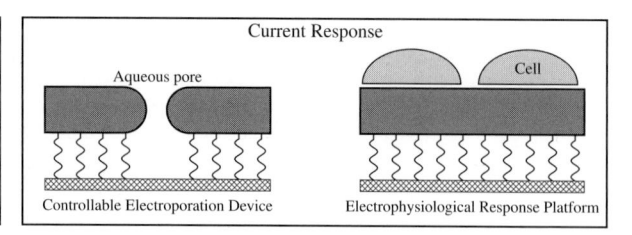

Figure 6.1 A schematic diagram of the three devices described in this chapter, namely, the PFMP, EMP, and ERP tethered-membrane devices. The tethered membrane is depicted in gray, and the gold bioelectronic interface by the crosshatch pattern.

gating dynamics of ion channels, measurements from the EMP and ERP are gathered using the time-dependent current response and not the impedance response. For both the impedance measurements and the current response measurements the results are dependent on the charging dynamics of the bioelectronic interface and membrane, and the conductance dynamics of the membrane(s) present. Note that by using dynamic models with experimental measurements it is possible to estimate the membrane conductance from the current response of these platforms. Part III provides a detailed description of such models.

6.2 Device 1: Pore Formation Measurement Platform (PFMP)

Here we introduce the pore formation measurement platform (PFMP) and show how it can be used to study the pore formation dynamics of proteins and peptides in an engineered membrane.

6.2.1 Pore Formation Measurement Platform: Introduction

Pore-forming peptides and proteins are crucial to the attack and defense mechanisms of biological organisms. Understanding the chemical kinetics of pore-forming peptides and proteins provides vital information of use to pharmacologists to target specific classes of peptides and proteins for in-depth pharmaceutical screening of novel drugs. The PFMP can be utilized to gain key insights into the specificity and dynamics of pore-forming peptides and proteins.

Here we consider using the PFMP to study the pore formation dynamics of α-hemolysin from methicillin-resistant *Staphylococcus aureus* (MRSA) [397], and the antimicrobial peptide PGLa (peptidyl-glycine-leucine-carboxyamide) originally isolated from frogs. α-Hemolysin is secreted from *Staphylococcus aureus* and binds with human platelets, erythrocytes, monocytes, lymphocytes, and endothelial cells, ultimately causing cell death. The pore formation dynamics of α-hemolysin are well known (refer to [106, 470] for details). PGLa is a membrane-active antimicrobial peptide produced in specialized neuroepithelial cells in the African frog *Xenopus laevis* [477]. In addition to the antimicrobial activity of PGLa, it also contains anticancer [25, 457], antiviral [71], and antifungal properties [347]. The remarkable feature of PGLa is that it provides a potential source of new antibiotics that can be used against increasingly common multiresistant pathogens (i.e., "superbugs") such as MRSA [82].

Using molecular dynamics and nuclear magnetic resonance (NMR) techniques several key insights have been provided for the pore formation dynamics of PGLa. For example, using ^{2}H-, ^{15}N-, and ^{19}F-NMR spectroscopy the topology of PGLa in membranes has been observed, which provides insight into the intermediate states of PGLa leading to pore formation [7, 137, 357, 406]. PGLa is in an α-helical conformation when membrane bound. The orientation of the membrane-bound PGLa is dependent on the lipid composition, peptide-to-lipid ratio, hydration, temperature, and pH [357]. From molecular dynamics [431] and ^{19}F-NMR [137] the long axis of PGLa is shown to be aligned parallel to the membrane surface in a monomeric state for peptide-to-lipid concentrations below 1:200. However, for high concentrations ($\geq$ 1:50) a tilted dimerization state is observed. The transmembrane state of PGLa has only been observed at temperatures below 15°C when the membrane is in the gel phase [7]. When PGLa is in the transmembrane state, an oligomerization process is suggested which leads to the formation of conducting toroidal pores. Using the observed NMR topologies and orientation, the reaction mechanism for PGLa pore formation is suggested to involve PGLa binding to the membrane, translocation via thermal fluctuations to the transmembrane state, and finally oligomerization to form conducting pores. The results provided by the PFMP can be used to complement these molecular-level studies and validate their results.

6.2.2 Pore Formation Measurement Platform: Construction

The PFMP is constructed in a similar way to the ICS biosensor presented in Chapter 5. The tethered membrane is supported by a $25 \times 75 \times 1$ mm^3 polycarbonate slide. Six 100-nm-thick sputtered-gold electrodes, each with dimensions 0.7×3 mm^2, rest on the polycarbonate slide. Each electrode is in an isolated flow cell with a common gold return electrode. The formation of the tethered membrane on the gold electrode is performed in two stages using the solvent-exchange technique presented in [84]. The first stage of formation involves anchoring of the tethers and spacers to the gold surface. The tethers provide structural integrity to the membrane and mimic the physiological response of the cytoskeletal supports of biological cell membranes. The spacers laterally separate the tethers, allowing patches of mobile lipids to diffuse in the membrane. The spacer is composed of a benzyl disulfide connected to a four-oxygen ethylene glycol group terminated by an OH; the tethers are composed of a benzyl disulfide connected to an eight-oxygen ethylene glycol group terminated by a C20 hydrophobic phytanyl chain. To form the anchoring layer, an ethanolic solution containing 370 μM of engineered ratios of benzyl disulfide components is prepared. This solution is exposed to the gold surface for 30 min; then the surface is flushed with ethanol and air dried for approximately 2 min. Note that in the special case of 100 percent tethering, the engineered membrane is composed of a tethered archaebacterial-based monolayer with no spacer molecules. Stage 2 involves the formation of the tethered membrane. The tethered membrane mobile lipids are composed of engineered mixtures of zwittrionic C20 diphytanyl-ether-glycero-phosphatidylcholine (DphPC) lipids, and C20 diphytanyl-diglyceride ether (GDPE) lipids. The solution containing the engineered mixture of mobile lipids is brought into contact with the gold-bonded components from stage 1. Several lipid solvents can be used [84, 273]; however, in most cases the lipids selected to form the bilayer are soluble in ethanol. As an example, let us consider the formation of a 70 percent DphPC and 30 percent GDPE mixed tethered membrane. 8 μL of 3 mM of the 70 percent DphPC and 30 percent GDPE ethanolic solution is added to the flow chamber. The solution is incubated for 2 min at 20°C in which the tethered membrane forms. Following the 2-min incubation, 300 μL of phosphate-buffered saline (PBS) is flushed through each flow chamber. The tethered membrane is equilibrated for 30 min prior to performing any experimental measurements. Note that several different types of lipids and other molecules can be constructed. For example, constructing a PFMP to mimic the response of *Escherichia coli* and *Saccharomyces cerevisiae* membranes requires that the associated lipids from these cells be used in the tethered membrane. Once the desired lipids are in the solution, a procedure similar to that described above is used to form the PFMP; as such, it is omitted here.

6.2.3 Pore Formation Measurement Platform: Operation and Experimental Measurements

We now consider the application of the PFMP to study the pore formation dynamics of the protein toxin α-hemolysin and the antimicrobial peptide PGLa.

Figure 6.2 provides the membrane conductance computed after the addition of α-hemolysin and PGLa. As Figure 6.2(a) illustrates, α-hemolysin rapidly forms pores in uncharged membranes[2] which remain active for the duration of the measurement from the PFMP (approximately 2 h). This is an expected result from the pore formation dynamics suggested in [106, 470]; however, without the use of a dynamic model we cannot validate the results presented in [106, 470] using the measured conductance from the PFMP. Figure 6.2(b) provides the membrane conductance as a function of time resulting from the positively charged PGLa interacting with varying negatively charged membrane surfaces. As expected, Figure 6.2(b) illustrates that the negatively charged palmitoyloleoyl-phosphatidylglycerol (POPG) lipids promote the binding of positively charged PGLa peptides: as the percent of POPG lipids increase, the membrane conductance increases. However, an interesting result is that as time progresses (past 2000 s) it appears that the conductance of the membrane decreases, suggesting that the PGLa pores close. Using the results in Figure 6.2(b) we claim that as the negative charge of the membrane increases there is an increase in the number of PGLa pores and pore lifetime. This makes PGLa especially effective for killing biological membranes containing a net negative charge. To gain further insight into the formation dynamics of PGLa would require a dynamic model that accounts for the diffusion of

---

[2] Uncharged membranes have an equal number of negative and positive charges. Such membranes are typically comprised of lipids that have a net zero charge.

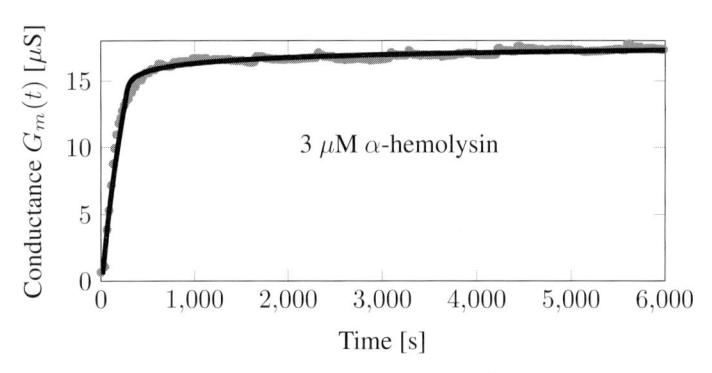

(a) PFMP membrane conductance as a result of the addition of α-hemolysin.

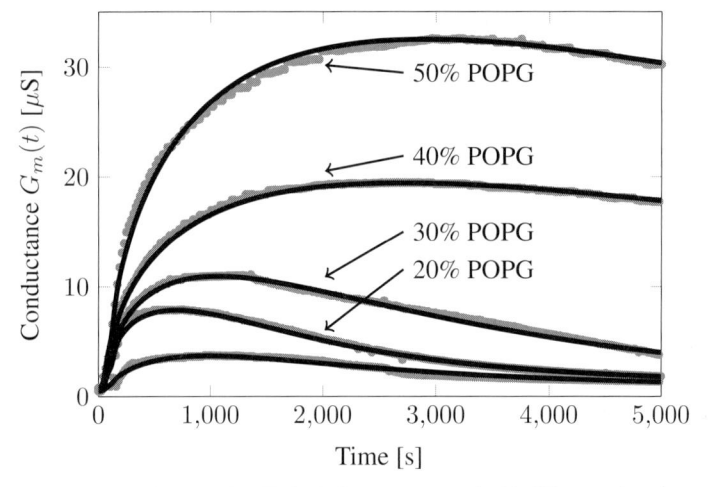

(b) Membrane conductance for PFMP platforms constructed with different ratios of negatively charged POPG lipids as a result of the addition of 30 μM PGLa.

Figure 6.2 PFMP membrane conductance as a result of the addition of α-hemolysin and PGLa.

the peptide to the surface, and the chemical reactions leading to pore formation and closure.

6.3 Device 2: Electroporation Measurement Platform (EMP)

This section discusses the construction and operation of the EMP, which provides a precisely controllable platform for electroporation studies.

6.3.1 Electroporation Measurement Platform: Introduction

The EMP consists of membranes engineered from synthetic archaebacterial lipids, and lipids extracted from prokaryotic and eukaryotic cells. The tethers mimic cytoskeletal supports in biological membranes, thereby facilitating in vivo measurements of electroporation. Electroporation is the phenomenon where aqueous pores form spontaneously in a cell membrane when a high transmembrane potential is applied. Electroporation facilitates the passage of otherwise impermeable molecules across the membrane into a cell and is used to catalyze the uptake of chemotherapeutic agents, DNA molecules, and neuron-specific proteins in drug delivery applications. It is reported [111] that electroporation can circumvent poor delivery of medications to the central nervous system and hence is a potential tool in treating Alzheimer's disease, Parkinson's disease, and some brain cancers. Though widely used, the electroporation process remains poorly understood, which is hindering the development of novel electrochemotherapy protocols. Key to the design of successful electrochemotherapy protocols is the construction of a robust and controllable platform for the study of electroporation, and an atomistic-to-observable model of the electroporation process.

Two classes of platforms exist for the study of electroporation. The first is in vivo cells [95, 96, 233, 452], which provide a physiological system for validation. A complication with using cells for model validation is that it is impossible to fully define the physiological environment which affects properties associated with electroporation. The second class is synthetic bilayer lipid membranes [162, 317, 318], which model the physiological response of biological cell membranes. Synthetic membranes benefit as the physiological environment is controllable; however, components of biological cell membranes such as the cytoskeletal network, and membrane species including proteins and peptides, are not included. This motivates the need for a well-defined model to study electroporation that incorporates the components of natural cell membranes in a highly controllable physiological environment.

Models for the complex dynamics of electroporation at the atomistic level are typically constructed using molecular dynamics. Recent results using molecular dynamics have shown that the mechanism of pore formation in symmetric and asymmetric membranes is dependent on the lipid composition [145, 324, 329]. That is, pure DphPC membranes have a higher resistance to electroporation compared to that of a 1,2-dipalmitoyl-sn-glycero-3-phosphocholine (DPPC) membrane as a result of the mobility of the hydrophobic tails of the DphPC lipid [329]. Using molecular dynamics it is shown that the *Escherichia coli* membrane has a higher resistance to electroporation compared to that of *S. aureus* membranes as a result of the presence of lipopolysaccharides [324]. This result suggests that membranes containing lipopolysaccharides would have a higher resistance to electroporation compared to prokaryotic membranes that do not contain lipopolysaccharides. To our knowledge no experimental evidence has been provided to support these results.

6.3.2 Electroporation Measurement Platform: Formation

The EMP is constructed using the rapid solvent-exchange technique discussed in Chapter 4. A schematic of the constructed EMP is provided in Figure 3.7. The density of tethers, physiological environment, and membrane composition are all adjustable in the EMP. Using the fundamentally different physical formation process of the EMP compared to that of black lipid membranes[3] allows the EMP to have a lifetime of several months and the ability to withstand excitation potentials of up to 800 mV. The EMP has a significantly longer lifetime compared with typical BLMs, which have a lifetime of tens of minutes. The EMP is composed of engineered archaebacterial lipids, tethers, spacers, and lipids from biological cell membranes such as *E. coli* and *S. cerevisiae*. We denote the tethering density as the ratio of spacer molecules to tethering molecules. For example, for a 10 percent tethered membrane, for every nine spacer molecules there is one tethering molecule. The long life of the EMP is a result of how the tethers and spacers are bonded to the inert gold surface. The tethers and spacers both contain a benzyl disulfide attachment chemistry to the gold. The sulfur bonded to the benzyl bonds to the gold surface with the disulfide bond maintained. This bonding structure has been detected experimentally from X-ray photoelectron spectra. The use of the disulfide has two advantages: the thiols do not oxidize on storage, allowing the membrane components to have a lifetime of several months, and the disulfide maintains the spacing between the spacers and tethers [84, 88, 273, 331, 336].

6.3.3 Electroporation Measurement Platform: Operation and Experimental Measurements

Measurements with the EMP are performed by applying a drive voltage $V_s(t)$ between the gold electrodes of the EMP and measuring the current response $I(t)$. For large excitation potentials it is expected that the conductance of the membrane will increase due to the formation of conducting aqueous pores – the process of electroporation. Figure 6.3 provides an example of the current response $I(t)$ from the EMP resulting from the drive voltage $V_s(t)$. As seen by comparing the maxima of the measured current response, shorter duration ramps with higher slope cause an increase in the population of conducting aqueous pores compared to longer duration ramps with lower slope. The current response is a function of the charging dynamics of the bioelectronic interface and membrane, and the formation of aqueous pores in the membrane. The reason that shorter duration ramps with higher slope cause an increase in the measured current response is that a significant portion of the initial charge is stored by the membrane at not the bioelectronic interface. However, for longer duration ramps, the bioelectronic interface retains more charge, causing a reduction in the associated formation of aqueous pores. The shorter duration ramps cause an increase in the current response because the charge accumulation at the membrane surface is larger compared to the charge accumulation dynamics of the longer duration ramps. Initially, this larger charge accumulation causes the formation of aqueous pores that allow larger amounts of charge to reach the gold bioelectronic interface compared to the longer duration ramps. Once the applied potential decreases, the depolarization of the bioelectronic interface occurs, which causes the current to decrease. The magnitude of this decrease depends on the population of

---

[3] The black lipid membrane was the first synthetic engineered membrane platform constructed in 1962 [287] and contains a lipid bilayer supported between two Teflon partitions in a similar structure to the FLB discussed in §3.2.

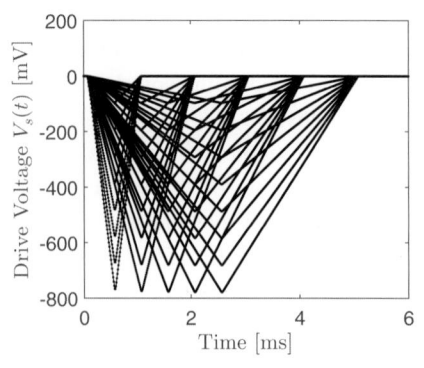

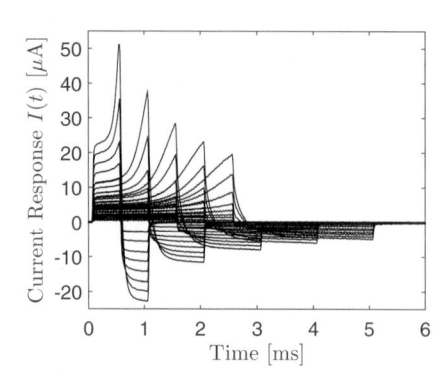

(a) Drive voltage of a 10 percent tethered DphPC membrane.

(b) Current response of a 10 percent tethered DphPC membrane.

Figure 6.3 EMP drive voltage and associated current response of a 10 percent tethered DphPC membrane.

aqueous pores present and the charge accumulation on both the membrane and the gold bioelectronic interface. At the time point where the current begins to decrease, the charge accumulation at the bioelectronic interface is larger for the longer ramps compared to shorter ramps. However, the associated population of aqueous pores is larger for the shorter ramps compared to the longer ramps. As such, the depolarization of the bioelectronic interface for the shorter ramps occurs on a faster time scale and with a higher-magnitude current compared to the depolarization that results from the longer ramps, as can be seen in the current responses illustrated in Figure 6.3. To gain further details on these important charging dynamics requires the use of a dynamic model. These types of dynamic models are constructed in Part III.

6.4 Device 3: Electrophysiological Response Platform (ERP)

This section discusses the construction and operation of the ERP for measuring the response of ion channels and the electrophysiological response of cells grown on the tethered-membrane surface of the ERP. Mathematical models for the ERP are constructed in Part III at various levels of abstraction.

6.4.1 Electrophysiological Response Platform: Overview

The ERP is an engineered membrane platform for measuring the response of embedded ion channels and cells grown on the tethered-membrane surface. Equivalently, the ERP is designed to measure the electrophysiological response of cells and embedded ion channels in controllable membrane and electrolyte environments, which allows the platform to be used for drug screening and diagnosing diseases in which ion-channel functionality is disrupted (i.e., channelopathies). Two applications of the ERP are presented in this section. The first application is the measurement of the electrophysiological response of the voltage-gated NaChBac ion channel. The second is the electrophysiological measurement of skeletal myoblasts grown on the surface of the ERP.

Electrophysiological response of the NaChBac ion channel: Voltage-gated ion channels are specialized proteins that only allow the passage of specific ions at a rate determined by the electric potential gradient across the membrane (i.e., the transmembrane potential). Voltage-gated channels initiate action potentials in nerve, muscle, and other excitable cells and are vital for transcellular communication. In this chapter the ERP is applied to measure the electrophysiological response of the prokaryotic sodium channel NaChBac, from *Bacillus halodurans*. The NaChBac channel was first reported in 2001 [341] and is likely an evolutionary ancestor of the larger four domain sodium channels in eukaryotes [65, 68]. The gating dynamics of NaChBac have been studied using patch clamping [227]; however, given the slow kinetics of NaChBac it was not possible to detect the gating dynamics at room temperature. In [227] the patch-clamping measurements were therefore performed at an elevated temperature of $28°C$. The results in [227] suggest that NaChBac has several closed states with voltage-dependent transitions. Using the ERP it is possible to measure the electrophysiological response of NaChBac at room temperature in an engineered membrane that mimics the response of biological membranes. The advantage of using the tethered membrane compared to that of black lipid membranes for the study of NaChBac gating dynamics is its stability and robustness. This allows the effects of electroporation to be accounted for while measuring the electrophysiological response of NaChBac in the tethered membrane.

Electrophysiological response of cells: The electrophysiological measurement of cells can lead to detection methods for the diagnosis of channelopathic diseases such as cystic fibrosis and Bartter syndrome. Consider that in cystic fibrosis, the *cystic fibrosis transmembrane conducting regulator* protein blocks the flow of chloride and thiocyanate ions, which cause a decrease in the cell membrane conductance. The ERP utilizes an engineered membrane for the noninvasive electrophysiological measurement of cells which can be used to detect such a change in conductance. Typically an adhesion protein, such as Glycocalyx or Fibronectin, is used ensure there is sufficient adhesion of the cell to the sensing surface [400]. Using a suitably designed tethered membrane containing molecular groups that adhere to the cell surface, cells can reach a sufficient coverage and adhesion to allow for their electrophysiological measurement. It was determined that a 100 percent tethered archaebacterial-based monolayer provides a suitable membrane to promote cell growth for electrophysiological measurement. Remarkably this allows the response to be measured using a noninvasive technique in the proximity of a synthetic membrane that mimics the electrophysiological response of biological membranes.

Though we only provide results for the measurement of the NaChBac ion channel, and of skeletal myoblasts, the procedures in Chapter 4 in combination with the methods

112   Physiochemical Membrane Platforms

presented here can be used to construct ERPs for measuring the electrical dynamics of several types of ion channels and cells.

6.4.2 Electrophysiological Response Platform: Formation

The ERP can be used to measure the electrophysiological response[4] of

(i) ion channels embedded in the tethered membrane of the ERP and
(ii) cells grown on the tethered membrane surface of the ERP.

Remarkably, the ERP does not use adhesion proteins; however, cells can still reach a sufficient coverage and adhesion to allow for their electrophysiological measurement.

The formation of the ERP is outlined in Chapter 4. For ion channels that spontaneously insert, the formation of the ERP follows an identical procedure to the EMP. However, for ion channels that do not spontaneously insert, the channels must be inserted when the tethered membrane is being formed. For the analysis of the voltage-gated NaChBac ion channel the ERP contains a palmitoleic phytanyl phosphatidylcholine tethered membrane with a 10 percent tether density. The following protocol is used to acquire and insert the NaChBac channels into the ERP for measurement.

- The first step is to begin with cells that contain NaChbac with a polyhistidine tag. In this analysis the tagged NaChBac is obtained from *B. halodurans*, a Gram-positive bacterium found in soil.
- The second step is to fragment the cells to extract the polyhistidine-tagged NaChBac proteins. This can be done using either a French press, or a mortar and pestle fragmentation protocol.
- The third step involves anchoring the cell fragments in a minicolumn. The anchored cells are then rinsed with imidazole to release the tagged NaChBac proteins.
- The fourth step is to rinse the released NaChBac proteins with a detergent. Several detergents can be used; however, it is recommended to use a detergent with a high aggregation number such as Brij 58, CYMAL-5, TWEEN-20, Triton X-100, or DDM. For this analysis the CYMAL-5 detergent micelles are used to produce a solution containing 100 nM of NaChBac.
- Add the solution containing 100 nM of NaChBac to the mobile lipid solution and proceed with the formation of the membrane, provided in Chapter 4.

The formation of the ERP for electrophysiological measurement of cells is similar to that for a standard engineered membrane as presented in Chapter 4; however, the membrane is required to have a high tethering density to support the cells and contain bis-gramicidin A to increase the permeability of the engineered membrane.

---

[4] When an electric field is applied to a membrane this can cause mechanical deformations, charge accumulation at the surface, and the formation of aqueous pores in the membrane, which results in a change in permeability of the membrane. The electrophysiological response of ion channels is associated with the change in permeability of ions through the ion channels that result from an electric field. These are all examples of an *electrophysiological response*.

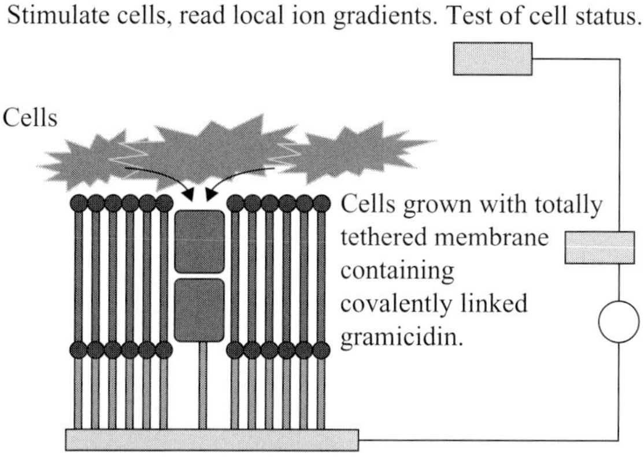

6.4 Device 3: Electrophysiological Response Platform (ERP)

Stimulate cells, read local ion gradients. Test of cell status.

Cells

Cells grown with totally tethered membrane containing covalently linked gramicidin.

Figure 6.4 A color version of this figure can be found online at www.cambridge.org/engineered-artificial-membranes. Architecture of the engineered membrane for the electrophysiological measurement of cells. The bis-gramicidin A is illustrated in red, the electrode and counter electrode in yellow, and the electrical measurement equipment in pink.

Additionally, there is an equilibration step which involves the cells in solution binding to the surface and growing. For the electrophysiological measurement of cells, the engineered membrane must have a 100 percent tethering density to ensure membrane stability during cell growth and measurement. The method to form the 100 percent tether density membrane is provided in Chapter 4. Given the engineered membrane containing bis-gramicidin A, cells are then placed on the membrane surface. An incubation period is required that promotes cell differentiation and growth on the membrane surface. The experimental techniques to perform cell placement and growth are cell specific – different cell types (e.g., myocardial cells, neuronal cells) will respond differently to the same proteins, peptides, pH, and temperature. Once sufficient cell coverage is achieved, the final result is the ERP illustrated in Figure 6.4. Below we describe a procedure for skeletal myoblast cell placement and growth on the membrane surface.

Given an engineered membrane with 100 percent tethering density with inserted bis-gramicidin A (refer to Figure 6.4), how can we place skeletal myoblast cells on the surface of the membrane to perform electrophysiological measurements? First, progenitor cells are used to produce skeletal myoblast cells. To induce differentiation, we exposed skeletal myoblasts to a range of growth factors (proteins and peptides) currently commercial-in-confidence to Victor Chang Institute.[5] After thawing the skeletal myoblasts, they were resuspended in Genea skeletal muscle thawing solution and plated onto the membrane surface. Immediately after plating, the passaging buffer was replaced with 450 μL of PBS. The engineered membrane and skeletal myoblasts were then transferred to a 37°C, 75 percent relative humidity incubation cabinet. During the incubation process, cell coverage on the engineered membrane was monitored using impedance measurements with an excitation potential of 30 mV. To perform electrophysiological measurements of the cells requires that the cells completely cover

---

[5] www.victorchang.edu.au/

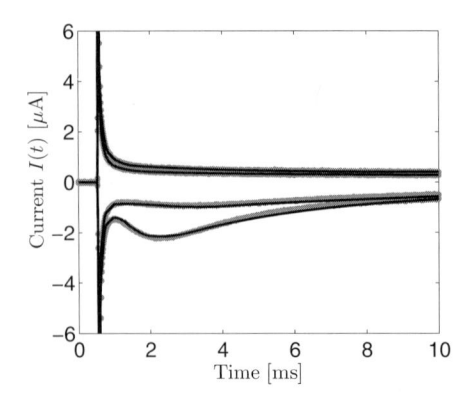

Figure 6.5 ERP current response for inserted NaChBac ion channels for different square waveform voltage excitations.

the surface of the tethered membrane. If this is not the case, the measured current response of the engineered membrane and cells would be dominated by the current flowing through the leakage paths between the cells. In such cases, the current response could not be used to estimate the electrophysiological response of the cells.

6.4.3 Electrophysiological Response Platform: Operation and Experimental Measurements

Measuring the electrophysiological response of embedded ion channels and cells requires applying an excitation potential to the ERP. In this section we examine the current response of the ERP with NaChBac and of the ERP with skeletal myoblasts grown on the surface. The measured current response results from the gating dynamics of ion channels, membrane electroporation dynamics, and the charging dynamics of the bioelectronic interface.

Let us first consider the electrophysiological response of the NaChBac ion channel. Figure 6.5 provides the measured current response of the NaChBac ion channel as a result of a square voltage excitation. As seen, the current response is dramatically different for the positive and negative excitations. This is expected as the NaChBac ion channel is a voltage-gated ion channel designed for transporting positively charged sodium ions. Therefore, if negatively charged ions are promoted to cross the membrane interface then this can only be performed efficiently via the process of electroporation. From Figure 6.5 we see that the current response for the case in which sodium ions cross the membrane is much larger as compared with the current response when chloride ions must cross the membrane.

Given the engineered membrane with skeletal myoblasts cells (formed using the process in §6.4.2), how can we measure the electrophysiological response of the skeletal myoblasts? The first task is to detect if sufficient cell coverage of the membrane has been achieved. Recall that if the leakage current between cells is dominant, then the measured current response $I(t)$ (in Figure 6.6) is dominated by the current flowing

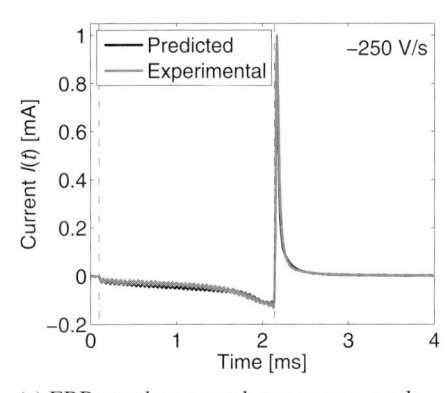

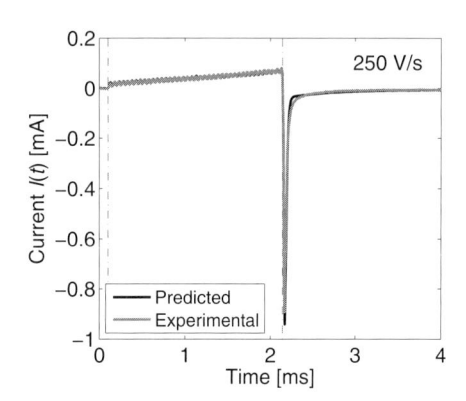

(a) ERP membrane conductance as a result of a positive excitation potential.

(b) ERP membrane conductance as a result of a negative excitation potential.

Figure 6.6 ERP current response for skeletal myoblasts grown on the membrane surface.

through the leakage paths between the cells. Additionally, if the cells have asymmetric shapes and the leakage current between cells is negligible, then the current response of the engineered membrane will be dependent on the polarity of the excitation potential. Therefore, the current response can be used to detect for sufficient cell coverage if the cells have asymmetric shapes. To measure the electrophysiological response of the skeletal myoblasts, a sawtooth voltage excitation waveform was utilized with a slope of 250 V/s for 2 ms, and another with a slope of -250 V/s for a 2 ms. The measured current response $I(t)$ for these excitation potentials is illustrated in Figure 6.6. Comparing the results in Figures 6.6(a) and 6.6(b), the current response is asymmetric. Therefore, we conclude that the leakage paths between the cells are sufficiently low and that the cell shapes are asymmetric. To estimate the electrophysiological response of the cells requires that we use experimental measurements in combination with a dynamic model of the engineered membrane. In Chapter 11 we construct a dynamic model of the engineered membrane to estimate the electrophysiological response of the cells given the experimentally measured current response.

Though the measured current response can be used to gain insight into the electrophysiological response of ion channel and cells, a dynamics model is required to estimate the contribution each has on the measured current response. Such dynamic models are constructed in Part III.

6.5 Complements and Sources

In this chapter we discussed the formation and operation of the PFMP, EMP, and ERP devices. Experimental measurements from the PFMP, EMP, and ERP in combination with dynamic models of these devices allow important biological parameters to be estimated. Also, the experimental measurements and dynamic models can be used to design therapeutic protocols and for rational drug design. In Part III we construct the dynamic models of these important engineered tethered-membrane devices.

The PFMP can be used in combination with a dynamic model to determine the pore formation dynamics of peptides and proteins on the surface of an engineered tethered membrane. The PFMP has been used to model the pore formation dynamics of the antimicrobial peptide PGLa [160], gramicidin, large oligomeric bacterial toxins such as α-hemolysin [442] from methicillin-resistant *S. aureus* superbugs, the anthrax ion channel (*Bacillus anthracis* PA63) [384], influenza virus matrix protein 2 (M2) [367], as well as a host of redox proteins which are involved in the reduction-oxidation processes in biological cells [182]. Additionally, the results from the PFMP can be used in combination with other experimental techniques to elucidate the pore formation dynamics, such as lytic experiments, gel electrophoresis, site-directed mutagenesis, and cryoelectron microscopy [23, 110, 259, 309]. Knowledge of how these ion channels form can be used in the fight against human disease – the rational design of drugs that can be used to disrupt the pore formation process of bacterial, viral, and superbug pore-forming peptides and proteins.

Experimental results from the EMP in combination with a dynamic model can be used to determine important biological parameters of membranes that affect the process of electroporation. Recall that electroporation is a microbiology technique focused on methods for engineering the excitation potential to increase the permeability of the cell membrane to allow the passage of macromolecules, drugs, DNA, and antimicrobial peptides into the cell or to cause cell death. Therefore, the EMP and dynamic model provide a method for designing electroporation-based therapeutic protocols that are targeted at specific cellular membranes. Targeting specific cellular membranes is important because increasing the permeability of all cellular membranes in the region of the excitation potential is not desirable, for example, when causing cancer cell death. For a general introduction to these therapeutic protocols, refer to books such as [239, 314, 410].

The ERP is used to perform experimental measurements that can be used to elucidate the electrophysiological response of membrane bound ion channels and cells. This allows the ERP to be used for drug screening and diagnosing diseases in which ion-channel functionality is disrupted (i.e., channelopathies). Classical techniques for measuring the electrophysiological response of membrane ion channels and cells are labor intensive and require complex experimental setup. For example, patch clamping [156, 281, 356] provides information-rich data of use for validating ion-channel gating models. Additionally, for the electrophysiological measurement of cells, substrate-integrated microelectrode arrays [46, 400], and sharp or patch microelectrodes that puncture the cell membrane [400], are used. The ERP is noninvasive and does not require complex experimental setup to measure the electrophysiological response of membrane ion channels and cells.

Though we have focused on using the ERP for studying the gating dynamics of ion channels and the response of cells, the ERP can also be used to study the dynamics of charge-carrier ionophores[6] in engineered tethered membranes. An example of an ionophore is the antibiotic valinomycin, which binds with a single potassium ion and

---

[6] An ionophore is a lipid-soluble macromolecule that can reversibly bind with ions and transport these ions across the membrane.

transports it across the membrane. Notice that valinomycin does not transport the potassium ion across the membrane via an ion channel. Ionophores are used extensively in the farming industry for killing bacteria and to reduce methane emission from livestock.

6.6 Closing Remarks

This chapter discussed the formation and operation of three important tethered-membrane platforms (PFMP, EMP, and ERP) for studying the pore formation dynamics of peptides and proteins, studying the process of electroporation, and measuring the electrophysiological dynamics of membrane ion channels and cells grown on the surface of the tethered membrane. Using the tools in Chapter 4 for constructing tethered membranes, several types of PFMP, EMP, and ERP can be constructed with different membranes, ion channels, and cells. This allows a host of studies to be performed related to pore formation dynamics, electroporation, and in-depth screening of novel drugs and diagnosis of various channelopathies.

In Part III we construct mathematical models that allow the experimentally measured response from the PFMP, EMP, and ERP to be used to estimate important biological parameters such as reaction rates of pore formation, the size and population of aqueous pores, and the deconvolution of the conductance resulting from the tethered membrane, ion channels, and cellular membranes from cells grown on the surface of the tethered membrane.

7 Experimental Measurement Methods for Engineered Membranes

7.1 Introduction

In previous chapters we described how to construct engineered membranes and devices including the ion-channel switch biosensor, electroporation measurement platform, and electrophysiological response platform. An important question is: How can one verify that the artificial membranes constructed in previous chapters have the precisely engineered structure that we have claimed? This chapter focuses on how experimental measurements can be performed on engineered membranes to estimate important biological parameters such as the membrane conductance and capacitance, and also how such measurements can be used to verify the structure of the membrane. We describe three important measurement methods to determine the properties of engineered membranes:

 (i) impedance measurements,
 (ii) time-dependent electrical measurements, and
(iii) spectroscopy and imaging techniques.

Impedance measurement and time-dependent electrical measurement techniques require minimal experimental setup and low-cost laboratory equipment compared with optical and imaging techniques. For example, a standard computer sound card can be used to perform impedance measurements; in comparison, spectroscopic and imaging techniques require both labor-intensive experimental setup and access to equipment that can perform fluorescence correlation spectroscopy, X-ray and neutron reflectometry, and electron microscopy. In practice this is why either impedance measurements or time-dependent electrical measurements are used for estimating properties of engineered membranes. The measurement methods in this chapter also play an important role in Part III, where we construct dynamic models of engineered membranes which can be used to relate the experimental measurements from tethered membranes to biological parameters of interest (e.g., reaction pathway of pore-forming peptides, the concentration estimation of analyte species, and the population and dynamics of aqueous pores in the membrane).

7.2 Electrical Response of Engineered Membranes

A very useful yet straightforward procedure for performing measurements of an engineered membrane is to apply an excitation potential across the membrane and then

7.2 Electrical Response of Engineered Membranes

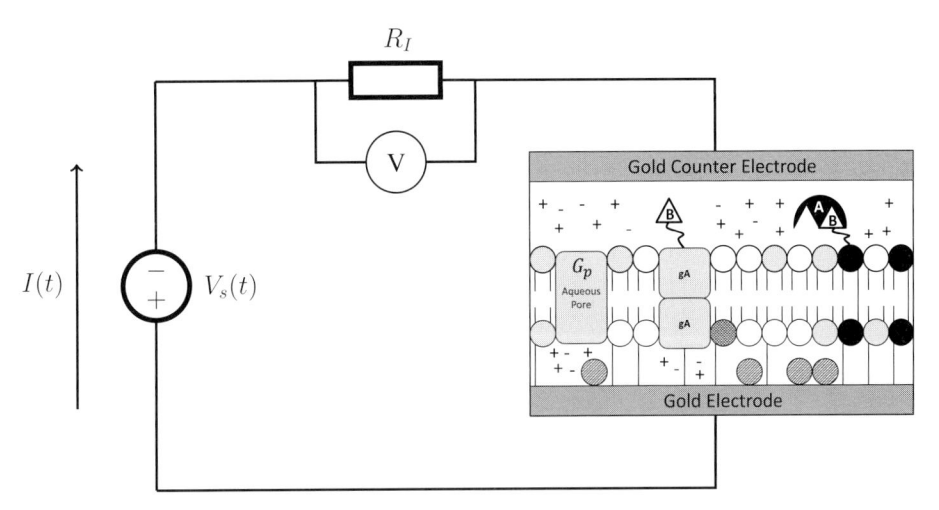

Figure 7.1 Schematic of the electrical measurement equipment. $V_s(t)$ is the voltage (an electric potential difference) produced by a function generator, R_I is a small resistance, V is a voltmeter, $I(t)$ is the current response, and the content between the gold counter electrode and gold electrode is the engineered membrane.

record the current response. Indeed, in Part III, we construct mathematical models for the electrical response of engineered membranes at several levels of abstraction (from atoms to device).

When the engineered tethered membrane is excited by a low-level (below 50 mV) external alternating voltage, the ions in the electrolyte solution migrate to and from the gold electrode surface and insulating barrier of the membrane. Additionally, ions can also pass through aqueous pores and ion channels in the membrane. However, unlike in an electrochemical system, there is no exchange of electrons – there is no redox reaction present at the surface of the electrode.[1]

A schematic of the experimental setup for performing electrical measurements is illustrated in Figure 7.1. A function generator inputs a time-varying voltage $V_s(t)$ across the engineered membrane which is in series with a small resistance R_I. The experimentally measured response is obtained using the voltage drop across the resistor R_I which provides information about the current response from the membrane. The voltage drop across the resistor R_I is measured using a voltmeter (denoted as V in Figure 7.1). Using Ohm's law, the current through the resistor R_I, and the engineered membrane, is given by

$$I(t) = \frac{V_I(t)}{R_I}. \tag{7.1}$$

---

[1] An issue with using any of the redox reactive systems is the toxicity of the solute redox reagents necessary to cause electron transfer at the electrode surface. Silver both ablates and is toxic to biological samples, which prevents Ag from being used with in vivo or ex vivo measurements. A further problem when working with redox model systems is the propensity of metal ions such as Ag^+ to block ion channels and perturb the phase behavior of lipids. The work described here focuses exclusively on capacitive coupled electrodes as it is these that dominate the implantable medical device industry. This carries the limitation that only changing potentials will result in a current flow.

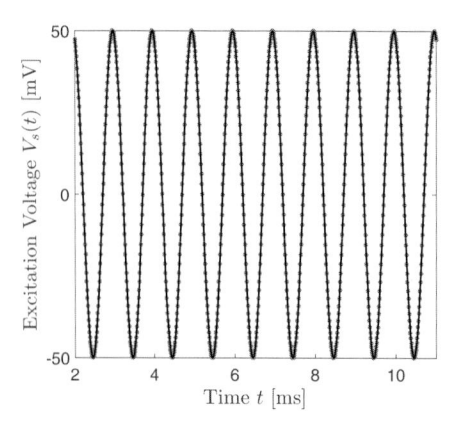

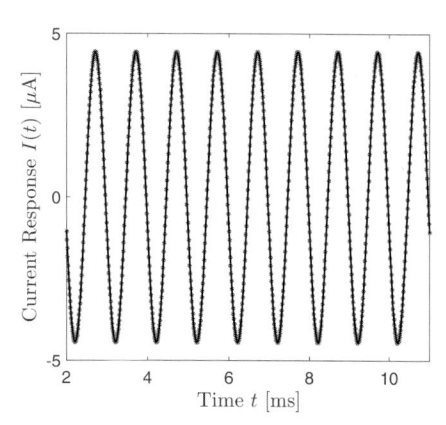

(a) Function generator output for the excitation voltage $V_s(t)$.

(b) Measured current response $I(t)$.

Figure 7.2 Experimentally measured current response $I(t)$ of an engineered membrane to the sinusoidal excitation voltage $V_s(t)$ with a frequency of $f = 1$ kHz.

There are two types of electrical measurements that can be obtained. The first is impedance measurements in which the function generator produces a periodic sinusoidal signal $V_s(t) = V_0 \sin(2\pi f t)$ at a frequency f and with magnitude V_o. The second is time-dependent electrical measurements in which the function generator produces a general time-dependent signal $V_s(t)$. In the following sections we give a detailed discussion of both of these electrical measurement techniques for engineered membranes.

7.2.1 Electrical Impedance Measurements

Electrical impedance measurements of engineered membranes provide useful information on both the quality and the conductance of the tethered membrane. Such impedance measurements are performed by applying a sinusoidal excitation voltage $V_s(t) = V_0 \sin(2\pi f t)$, where f denotes frequency in hertz, t denotes continuous time, and V_o denotes the magnitude of the sinusoidal signal. Note that the period of this signal is given by $T = 1/(2\pi f)$ s. A typical impedance measurement response for an engineered membrane is provided in Figure 7.2. Notice that the measured current $I(t)$ has a phase shift compared to the excitation voltage $V_s(t)$, and additionally the current $I(t)$ follows a sinusoidal waveform. This phase shift results from the complex interaction of the electrolyte solution with the bioelectronic interface, and the dynamics of the engineered membrane. The applied voltage $V_s(t)$ and measured current $I(t)$ in Figure 7.2 can be modeled using the following equations:

$$V_s(t) = V_0 \sin(2\pi f t),$$
$$I(t) = I_0 \sin(2\pi f t + \phi), \qquad (7.2)$$

where ϕ is the phase difference between the two sinusoidal signals in Figure 7.2. The impedance Z (which is composed of real and imaginary numbers, e.g., complex valued)

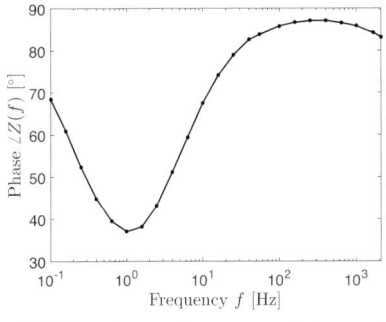

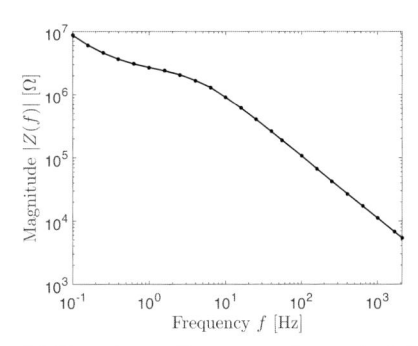

7.2 Electrical Response of Engineered Membranes

(a) Experimentally measured phase $\angle Z$.

(b) Experimentally measured magnitude $|Z|$.

Figure 7.3 Experimentally measured impedance (magnitude $|Z|$ and phase $\angle Z$) for a sinusoidal excitation voltage $V_s(t) = 0.05 \sin(2\pi f t)$ from frequencies $f = 0.1$ Hz to $f = 10$ kHz.

that relates the two sinusoidal signals $V_s(t)$ and $I(t)$ is

$$Z = \frac{V_0}{I_0} e^{j\phi}, \tag{7.3}$$

where $j = \sqrt{-1}$ and $e^{j\phi}$ is the complex exponential[2] with the phase ϕ expressed in radians. For the applied voltage $V_s(t)$ and measured current $I(t)$ in Figure 7.2 the associated impedance is given by $Z = 11250 e^{j1.49}$, which has magnitude $|Z| = 11.25$ kΩ and a phase of $\angle Z = 85.82°$.

To estimate important biological parameters (such as membrane conductance and capacitance) of the engineered membrane, we require the impedance Z to be measured at several frequencies f ranging from 0.1 Hz to several kilohertz. Using the same method as above, we apply a low-magnitude excitation voltage, record the associated current response, and compute the impedance $Z(f)$ at different frequencies. The results are illustrated in Figure 7.3 for a 1 percent tether density zwittrionic C20 diphytanyl-ether-glycero-phosphatidylcholine lipid (DphPC) engineered membrane. The experimentally measured impedance of the engineered membrane can be used to determine properties of the membrane. Specifically, the high-frequency content of the impedance is related to the properties of the electrolyte solution, and the low-frequency content is related to the dynamics of the electrode-electrolyte interface (i.e., the bioelectronic interface). The impedance in between the low and high frequencies provides information related to the membrane conductance and capacitance, as seen in Figure 5.4 for digoxin. Using electrical impedance spectroscopy it is also possible to monitor the formation dynamics of the engineered membrane. As the membrane forms, the minima of the phase $\angle Z$ decrease as the conductance of the membrane decreases. Further details on how the impedance depends on the membrane properties requires a dynamic model of the engineered membrane.

Note that to perform impedance measurements it is important that the magnitude of the applied voltage $V_s(t)$ be below 50 mV. If this is not the case then electroporation and

---

[2] Recall Euler's formula: $e^{j\phi} = \cos(\phi) + j\cos(\phi)$, where $j = \sqrt{-1}$.

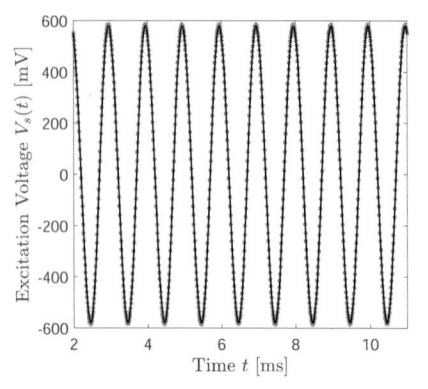

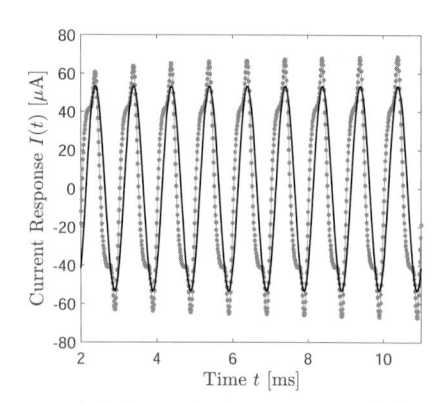

(a) Function generator output for the excitation voltage $V_s(t)$.

(b) Measured current response $I(t)$.

Figure 7.4 Experimentally measured current response $I(t)$ to an excitation voltage $V_s(t)$ for a 1 percent tether density DphPC engineered membrane. The black lines indicate the perfectly sinusoidal response, and the grey lines indicate the experimentally measured response.

polarization effects can cause the associated current response $I(t)$ to not be sinusoidal. An example of this is illustrated in Figure 7.4. As seen for this large drive voltage, the measured current is not sinusoidal as a result of the pore formation dynamics in the engineered membrane, which causes the conductance of the membrane to change. Additionally, notice that the peak of the current response increases with time. This results because not all pores have sufficient time to close prior to the increase in the membrane voltage (transmembrane voltage). If this excitation was to continue for a long period of time then the membrane would eventually rupture as a result of irreversible electroporation. To prevent this effect the applied voltage $V_s(t)$ must be sufficiently low to prevent electroporation and polarization effects, which will cause the current response to fail to be sinusoidal.

7.2.2 Time-Dependent Electrical Measurements

We now discuss the second method for obtaining the electrical response of an engineered membrane. We are interested in studying the gating dynamics of ion channels and the electrophysiological response of the engineered membrane to nonsinusoidal excitations. This requires the application of a time-dependent excitation voltage $V_s(t)$. The measured current response $I(t)$ of such an excitation can provide information related to the population dynamics of aqueous pores and the gating dynamics of inserted ion channels, which cannot be measured using electrical impedance measurements. The reason that impedance measurements (discussed above) cannot be used is because the dynamics of aqueous pores and ion channels are directly dependent on the dynamics of $V_s(t)$; that is, the dynamics of aqueous pores and ion channels have memory that is dependent on the history of the applied voltage $V_s(t)$. Therefore, time-dependent excitation voltages $V_s(t)$ are required to measure these memory dependent dynamics.

7.2 Electrical Response of Engineered Membranes

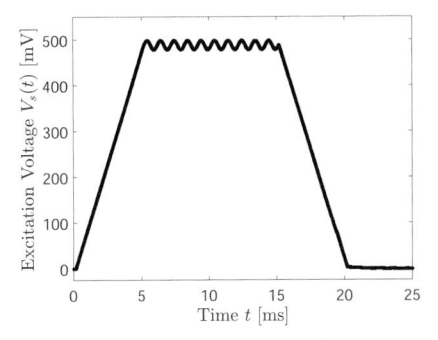

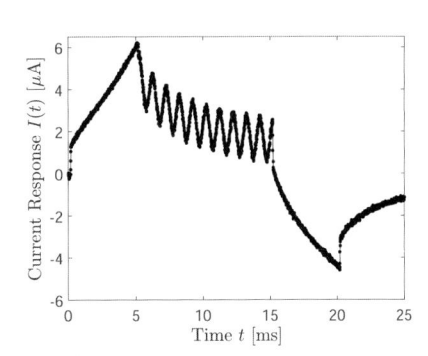

(a) Function generator output for the excitation voltage $V_s(t)$.

(b) Measured current response $I(t)$.

Figure 7.5 Time-dependent excitation voltage $V_s(t)$ and experimentally measured current response $I(t)$ of a 1 percent tether density DphPC engineered membrane.

Example 1: DphPC membrane. Let us consider the application of a time-dependent excitation to a 1 percent tether density DphPC membrane formed using the procedure presented in Chapter 4. The excitation voltage and resulting current response from the engineered membrane are displayed in Figure 7.5. The linear increase in the current response results as the membrane polarizes; that is, ions migrate to the membrane surface. As the drive voltage decreases in slope the current decreases as the ions migrate in the opposite direction. At the average excitation potential of 495 mV with a small sinusoidal voltage, once a sufficient number of ions have reached the membrane surface, aqueous pores form in the membrane, causing a nonlinear decrease in the current response. When the excitation voltage begins to linearly decrease, the associated current response decreases as these aqueous pores close and the membrane depolarizes. To gain further insight from the current response requires a dynamic model.

Example 2: DphPC membrane with alamethicin. Alamethicin is a pore-forming antimicrobial peptide. Experimental measurements have indicated that the alamethicin peptides form voltage-gated ion channels in biological membranes that can comprise up to 12 alamethicin monomers [60, 360, 467]. Using the spontaneous insertion method discussed in Chapter 4, we construct a 10 percent tether density DphPC membrane that contains alamethicin ion channels. Then, we apply voltage steps in the range of −300 to 300 mV and measure the current response. The measured current response is illustrated in Figure 7.6. Alamethicin ion channels are cation selective. Therefore, a negative excitation potential will cause positive cations (Na^+) to be attracted to the electrode surface. Therefore, for negative excitation potentials we see an increase in the measured current response of the engineered tethered membrane. However, for positive excitation potentials, since now negative anions (Cl^-) are attracted to the electrode surface, a decrease in the current response is observed as the alamethicin ion channels are cation selective. Hence, the resultant current responses in Figure 7.6 follow the expected potential and polarity selectivity of alamethicin. Given a dynamic model of the engineered membrane and these experimental measurements, it would be possible to validate gating dynamics models of the alamethicin ion channels.

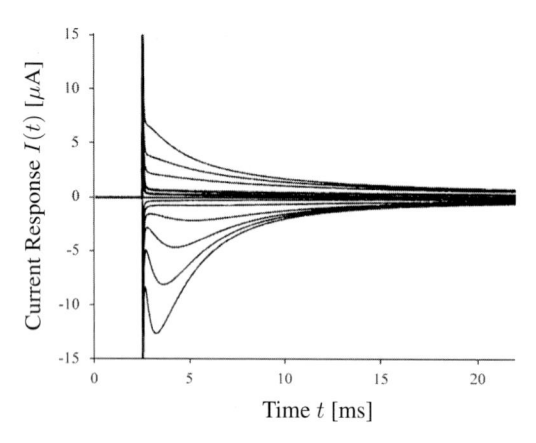

Figure 7.6 Measured current response $I(t)$ of a 10 percent tether density DphPC membrane containing alamethicin that results from voltage excitation steps in the range of -300 to 300 mV. The voltage steps are applied at $t = 2.5$ ms.

Example 3: Voltage ramps. When a time-varying excitation potential is applied to the engineered tethered membrane, both capacitive and conductive currents are observed. However, if the membrane conductance G_m is negligible or the excitation potential varies rapidly, then the current response will be governed solely by the charge accumulation dynamics at the electrode and membrane surface. Rapid excitations will only cause changes in the charge accumulation at the electrode and membrane surface because there is insufficient time for ions to migrate through the membrane, which in essence causes the membrane conductance to be negligible. The relationship between current I, charge Q, and capacitance C is

$$I = \frac{dQ}{dt} = C\frac{dV}{dt}. \tag{7.4}$$

Thus, if a potential ramp be applied with a slope of dV/dt, and the membrane conductance is negligible, then the magnitude of the current response I will be constant and proportional to the capacitance C. Figure 7.7 illustrates the current response I of an engineered tethered membrane to a ramp potential up to 500 mV. It should be recognized that action potentials, the electrical signaling in the body, seldom exceed -100 to $+200$ mV. Thus 500 mV is many times the potential gradient normally experienced in biology. One of the applications of engineered tethered membranes is the ability to study the properties of lipid bilayer membranes under high potential gradients.

The results in Figure 7.7 suggest that at extremely high transmembrane potentials, when applied rapidly compared to the characteristic time of any discharge pathways within the membrane, express potential gradients across the lipid bilayer cause morphological changes in the membrane structure, inducing ion pathways across the membrane. This morphological change of the membrane as a result of the transmembrane potential is known as electroporation and is discussed in detail in Chapters 11 and 12. This change in the membrane conductance is found to be essentially reversible for drive potentials below 800 mV. The main result here is that the engineered membrane does not merely act as an ideal capacitor, but contains complex morphological dynamics that

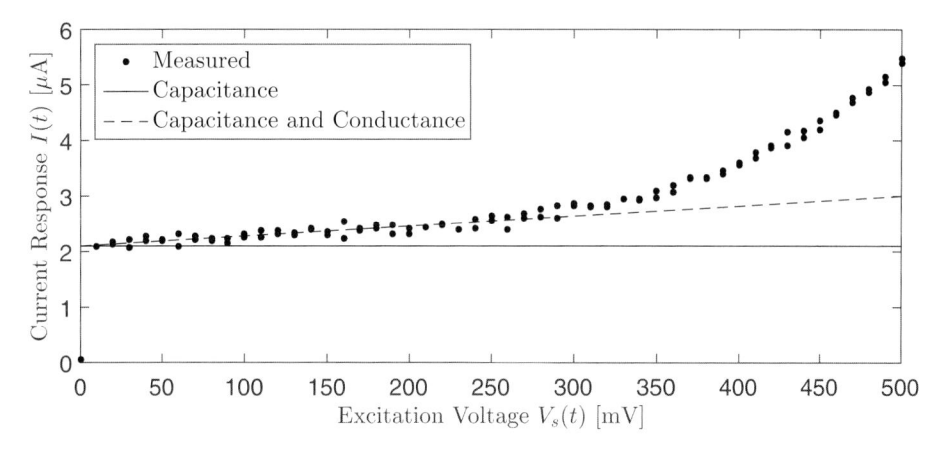

Figure 7.7 Measured current response of a 10 percent tether density DphPC membrane resulting from a linear potential ramp from 0 mV to 500 in 5 ms. The rate of potential increase is 100 V/s resulting in a current increase of 2.1 μA for a 21-nF membrane capacitance. The constant capacitance model (black line) only fits the initial rise in current that results from the applied linear potential ramp. The constant capacitance and conductance model (dashed black line) provides an accurate fit to the measured current response until approximately 250 mV, at which point membrane morphological changes cause the membrane conductance to vary.

change the permeability of the membrane. Hence, the engineered membrane is composed of both a capacitive element to account for charge accumulation and a conductive element to account for the dynamics of the permeability of the membrane.

7.2.3 Interpretation of Measured Current Response

The interpretation of the measured current response from frequency- and time-dependent measurements is generally performed by comparison with the well-established libraries of electrical equivalent circuits, composed largely of capacitors and resistors as illustrated in Figure 7.8. In Part III a detailed description of equivalent circuit models of engineered tethered membranes is discussed; here we merely provide a preliminary discussion to illustrate the use of such dynamic models. The most important message from the discussion in this section is that, unless the excitation potential is time varying and rapid relative to the inherent time constants of the electrode and membrane charge accumulation, then the differential relationship between the current and the potential must be included in any analysis. This is a feature that complicates all implanted medical devices as a consequence of the need to use capacitively coupled electrodes.

In Figure 7.8, the capacitors C_{tdl}, C_m, and C_{bdl} model the ion accumulation at the bioelectronic interface (gold electrode surface) and at the membrane surface, whereas the resistor G_m models the conduction of ions through the membrane, and R_e accounts for the electrolyte resistance. For an excitation voltage step of 500 mV, the first observation from Figure 7.8 is the inverse relationship between the potential across each capacitor (counter electrode capacitance C_{tdl}, membrane capacitance C_m, and electrode capacitance C_{bdl}) and its capacitance. A less apparent result is that, because of the membrane

126 Experimental Measurement Methods for Engineered Membranes

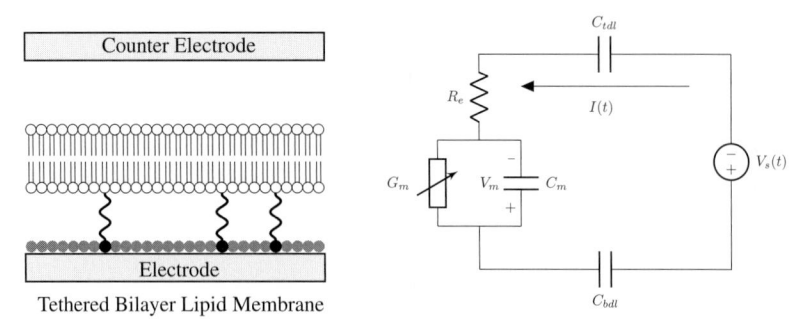

(a) Equivalent circuit model of engineered membrane.

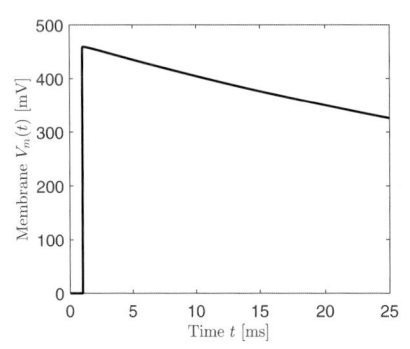

(b) Model-predicted membrane potential $V_m(t)$.

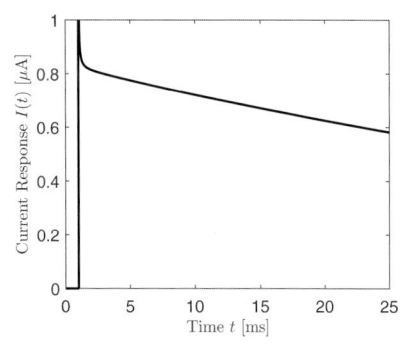

(c) Model-predicted current response $I(t)$.

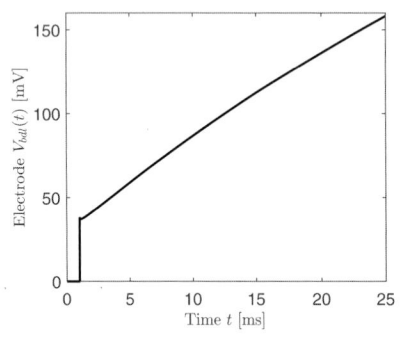

(d) Model-predicted electrode potential $V_{bdl}(t)$.

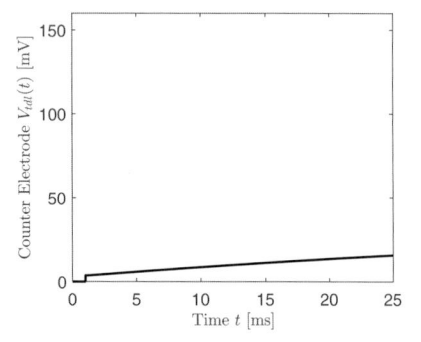

(e) Model-predicted counter electrode potential $V_{tdl}(t)$.

Figure 7.8 In (a) the schematic of the equivalent circuit model of the engineered tethered membrane composed of capacitors and resistors is illustrated. C_{tdl} is the the counter electrode capacitance, C_m is the membrane capacitance, C_{bdl} is the electrode capacitance, G_m is the membrane conductivity, and $V_s(t)$ indicates the excitation potential. The numerically predicted (b) membrane potential, (c) current response, (d) electrode potential, and (e) counter electrode potential that result from a 500-mV voltage step excitation potential $V_s(t)$ with circuit parameters $C_m = 15$ nF, $C_{bdl} = 150$ nF, $C_{tdl} = 1500$ nF, $G_m = 2.0$ and μS, $R_e = 200$ Ω.

conductance G_m bypasses the membrane capacitance C_m, the potential across the membrane decays to zero and the potential across the electrode grows to essentially the full excitation potential $V_s(t)$. A very small residual potential remains across the counter electrode but, because of its large capacitance, this potential is negligible.

A subtlety that arises when modeling the behavior of ions rather than electrons in the electric circuit in Figure 7.8 is that the bulkiness of ions and their slow mobility in solution are not accounted for by ideal capacitors. Interesting phenomena are experienced when measuring currents arising from ions moving in solution. Ions at a surface can repel further ions from accumulating at the surface, resulting in a compact layer of ions at an electrode surface (called the Helmholtz layer) and a diffuse layer of ions that propagate back into the solution away from the electrode. The extent of this diffuse layer is dependent on the ionic concentration and properties of the ions (e.g., charge and hydration size). At high ionic strengths of approximately 100 mM, as found in biological solutions, the range of the diffuse layer is only a few ionic radii thick. At lower concentrations the diffuse layer can extend to many ionic radii. As a consequence of the slow mobility and self-interaction of ions in solution, capacitors do not instantaneously charge and discharge as the electronically used symbol normally suggests. To account for the electrolyte charging dynamics at the electrode and membrane, the ideal capacitors are replaced with constant-phase elements (CPEs). The CPE is a useful simplifying model for the complex behavior of ions in solution. The CPE is defined in relation to the impedance Z of the electrode or membrane and is described by

$$Z = \frac{1}{C(j\omega)^p}, \tag{7.5}$$

where C is the capacitance and $0 \leq p \leq 1$ is a fractional order parameter. If $p = 1$, then (7.5) defines an ideal capacitor with capacitance C, and if $p = 0$ then (7.5) defines an ideal resistor with resistance $1/C$. The reason for the use of the CPE is to account for the slow charge accumulation at the electrode and membrane that results from the limited mobility of the ions in solution [87, 436]. This is particularly true of the space between the membrane and the tethering gold electrode. This space is only 2 to 4 nm wide and is crowded with ethylene glycol groups. For engineered tethered membranes, typically the electrode charging dynamics are modeled using a CPE, and the membrane charging dynamics with an ideal capacitor.

7.3 Spectroscopy and Imaging Techniques for Engineered Tethered Membranes

In this section we briefly outline how optical and imaging techniques can be used to verify the structure and dynamics of engineered membranes. Specifically, we describe the following methods:

(i) X-radiation (X-ray) reflectometry to measure the area per lipid,
(ii) neutron reflectometry (NR) for measuring membrane thickness and reservoir thickness,

(iii) fluorescence recovery after photobleaching (FRAP) for measuring lipid diffusion coefficients, and

(iv) nuclear magnetic resonance (NMR) for measuring the conformation and orientation of gramicidin A.

Note that these four experimental techniques require expensive laboratory equipment. Hence they are mainly used to obtain parameters of membranes that cannot be obtained via electrical measurement techniques.

7.3.1 X-Ray Reflectometry for Measuring Area per Lipid

X-ray reflectometry is used to measure the structural properties of engineered tethered membranes, including the membrane thickness and area per lipid, and the atomic structure of proteins (X-ray crystallography).

The main idea of X-ray reflectometry is to transmit an incident beam of electromagnetic X-rays to an object and then measure the intensity and direction of the reflected X-rays. The wavelength of X-rays is on the scale of angstroms, and so they are ideally suited for determining the atomic structure of molecules. X-ray photons interact with electrons in atoms, which results in the X-ray photons being either elastically or inelastically scattered.[3] An elastic interaction occurs if the scattered photon from the incident beam does not lose any energy. An inelastic interaction occurs if some of the incident photon energy is transferred to the electron, resulting in the scattered photon having a different wavelength than the incident photon. In X-ray reflectometry, only the elastically scattered photons are measured. These elastically scattered photons contain information about the electron distribution in molecules, which can be used to determine important structural information about the atoms. For example, if the molecules are arranged in a periodic fashion, as in lipid bilayers, the ensemble-average measured intensity of elastically scattered photons will have distinct peaks with the same symmetry as the spatial distribution of the molecules. Therefore, measuring the intensity of scattered photons can be used to infer the structural distribution of atoms and molecules in a membrane.

For engineered membranes, X-ray reflectometry is used to determine important membrane properties such as membrane thickness and area per lipid [13, 223, 273, 274, 433, 458, 459, 468]. X-ray reflectometry has a resolution of approximately 1 Å (1×10^{-10} m) and requires lengthy exposures which are longer than the survival lifetime of most model membrane systems. However the technique becomes practical given the stability and long lifetimes of those found in the engineered membrane. Otherwise the intensity that results from the scattered photons is too small to detect. In [273, 274] the authors utilized X-ray scattering to measure the thickness and surface roughness of the gold electrode support which had a root-mean-square surface roughness of ±5 Å. In [468], X-ray scattering was used to measure the DphPC area per lipid, which was found to be approximately $A_L = 75$ Å^2.

---

[3] If the scattering of the photon is elastic, this is also known as Thompson scattering. Additionally, if the photon is scattered inelastically, this is known as Compton scattering.

7.3.2 Nuclear Magnetic Resonance Measurements of the Conformation and Orientation of Gramicidin A

NMR spectroscopy can be used to measure the conformation and orientation of peptides and proteins in engineered membranes.

The main idea of NMR is to adjust the quantum spin of nuclei and then measure the absorption or desorption of energy that results from changing that quantum spin. Only nuclei that possess a nonzero quantum spin number[4] (intrinsic angular momentum of the nuclei) can be detected by NMR. Qualitatively, NMR can only be used to measure properties of nuclei that have electron orbitals that are nonspherical. The quantum spin of a nucleus depends on the number of protons and neutrons in the nucleus. Commonly used nuclei for NMR include hydrogen-1 (^{1}H), carbon-13 (^{13}C), fluorine-19 (^{19}F), and phosphorus-31 (^{31}P), all of which have a quantum spin of $1/2$. Each nucleus has a magnetic moment that is related to the quantum spin number of the nucleus. When these nuclei are inserted into a magnetic field, the nuclei align either with or against the magnetic field – this is known as the Zeeman effect. If, additionally, an electromagnetic wave is applied the nuclei that align with the magnetic field can absorb the electromagnetic energy and transition to align against the magnetic field. After the electromagnetic wave is removed, the nuclei will transition back to being aligned with the magnetic field. It is possible to measure either the adsorption of the electromagnetic energy or the release of energy as the nucleus transitions back to its equilibrium alignment in the magnetic field. The characteristics of the release or adsorption of electromagnetic energy in NMR are a result of both the applied magnetic field and the magnetic field of other nuclei with nonzero quantum spin. As such, the measured energy contains information about the coordinates of atoms in a molecule, or the alignment of molecules to the external magnetic field. Therefore, NMR provides a method for measuring the conformation and orientation of peptides and proteins in membranes.

As discussed in Chapter 5, gramicidin A (gA) plays an essential role in the switching mechanism for detecting specific analyte molecules. Using NMR, it is possible to measure the orientation of gA in engineered tethered membranes. For example, in [16, 200] NMR was used to measure the tilt angle of the α-helix, which was found to be $5° \pm 9°$ to the membrane normal. Recall that the conformational structure of gA is given in Figure 2.7 on page 20 which was computed from X-ray crystallography. Using the measurements from NMR and X-ray crystallography provides a complete atomistic description of gA in engineered tethered membranes.

7.3.3 Fluorescence Recovery after Photobleaching for Measuring Lipid Diffusion

FRAP can be used to measure the diffusion of lipids or the binding of proteins in engineered membranes.

The main idea of FRAP for measuring the diffusion of lipids is to insert fluorophores (either fluorescent probes or fluorescent proteins) into the engineered membrane. Then a high-power laser is used to irreversibly photobleached a small region of the membrane

---

[4] Also known as the orbital angular momentum quantum number, which describes the shape of the electron orbitals around a nuclei. The quantum spin number is different from the electron spin.

(a) Photobleaching region with intact fluorophore probes.

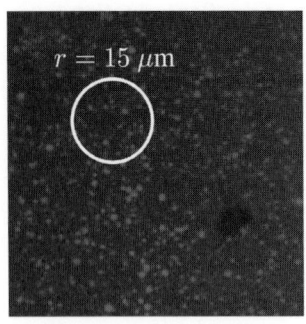

(b) After photobleaching with 15-mW pulse laser.

Figure 7.9 Confocal microscope images of the DphPC lipids in a 10 percent tether density membrane with ALEX-488 fluorophore probe present before and after photobleaching. The photobleaching region has a radius of $r = 15$ μm.

that contains fluorophores. The diffusion coefficient can then be inferred by measuring the recovery in fluorescence that results from irreversibly photobleached fluorophores diffusing out of this small region and being replaced with intact fluorophores. The principal operation of FRAP is straightforward; however, the design of the fluorophores, experimental setup, and relating the fluorescence recovery to the diffusion of the fluorophores is nontrivial. Since FRAP is a noninvasive technique, it can be used in both in vitro and in vivo to study the molecular movement of both fluorescent probes and fluorescent proteins.

Our aim is to measure the diffusion of DphPC lipids in a 10 percent tether density engineered membrane using FRAP. To perform the FRAP measurement, we introduce the ALEX-488 fluorophore probe into the engineered tethered membrane. A confocal microscope is used to measure the small region that will be photobleached by the high-power (15-mW) laser. A disk-shaped bleaching geometry of radius $r = 15$ μm is used for this experiment. Examples of the intact membrane pre- and post-photobleaching from the confocal microscope are illustrated in Figure 7.9. As seen, due to photobleaching, the fluorescence dramatically decreases in the disk-shaped bleaching region. The intensity of this region is measured while the process of fluorescence recovery occurs. To relate this information to the diffusion coefficients of the DphPC lipids requires a model between the measured intensity and the diffusion coefficient. For a disk-shaped bleaching geometry, the normalized fluorescence intensity is [398]

$$\bar{I}(t) = e^{-2\tau/t} \left(I_o\left(\frac{2\tau}{t}\right) + I_1\left(\frac{2\tau}{t}\right) \right) \tag{7.6}$$

$$\tau = \frac{r^2}{16D}$$

$$I_i\left(\frac{2\tau}{t}\right) = \frac{1}{\pi} \int_0^\pi \exp\left(\frac{2\tau}{t}\cos(\xi)\right) \cos(i\xi)\, d\xi \quad \text{for } i \in \{1, 2\}$$

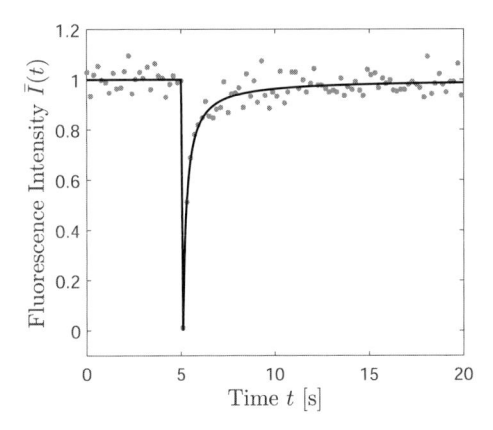

Figure 7.10 Experimentally measured (gray dots) and model-predicted (black line) normalized fluorescence intensity $\bar{I}(t)$ for a DphPC 10 percent tether density membrane. The photobleaching was applied to the disk-shaped region at $t = 5$ s. The model-predicted intensity is computed using (7.6) with $r = 15$ μm and $D = 172$ nm²/μs.

where r is the diameter of the disk-shaped bleaching region, τ is the characteristic time scale of fluorescence recovery, D is the lipid diffusion coefficient, and $I_i(\cdot)$ are modified Bessel functions.

Figure 7.10 provides the experimentally measured and model-predicted fluorescence intensity. The resulting diffusion coefficient of the DphPC lipids in the 10 percent tether density is $D = 172$ nm²/μs. The diffusion coefficient of DphPC lipids in 0 percent tether density membranes are in the range of 256 to 290 nm²/μs, and for 25 percent tether density membranes are in the range of 87 to 128 nm²/μs as computed from coarse-grained molecular dynamic simulations. As expected, since the tethers impede the diffusion of lipids, the diffusion coefficient of the lipids in a 10 percent tether density membrane lies between the diffusion coefficient for the 0 percent tether density and that of 25 percent tether density engineered membranes.

7.3.4 Neutron Reflectometry for Measuring Membrane Thickness and Reservoir Thickness

NR provides a method to measure the thickness, density, and roughness of the engineered membrane [54, 152, 190, 273]. Additionally, NR can also be used to gain insight into the geometries of proteins inserted into the engineered membrane [151, 274].

The main idea of neutron reflectometry is to measure the scattering of neutrons from planar surfaces. The scale of neutron scattering changes dramatically between hydrogenated and deuterated materials. The inclusion of a neutron in the hydrogen nucleus to transmute it to a deuterium increases the coherent scattering cross section by a factor of approximately 3. By exchanging the bathing media of an engineered tethered membrane sample from H_2O (water) to heavy water which comprises HDO (hydrogen deuterium oxide) and D_2O (deuterium oxide), the progressively altered neutron-scattering pattern permits a reconstruction of the structure of the engineered tethered membrane.

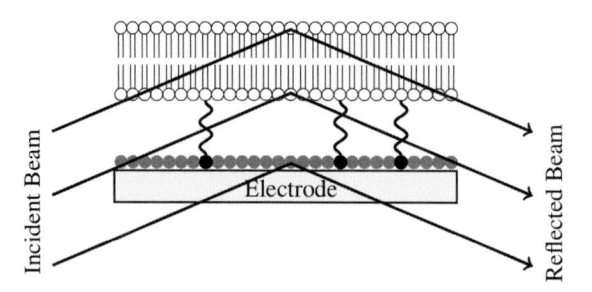

Figure 7.11 Schematic of the neutron reflectometry measurement. An incident beam is brought into contact with the engineered artificial membrane and the associated reflected beam is measured. The reflected beam contains information about the mechanical structure of the engineered artificial membrane because the reflected light waves undergo constructive and destructive interference.

To measure the thickness of a membrane, the double bonds of natural phospholipids are hydrogenated, and the electrolyte solution is composed of heavy water instead of standard H_2O. Figure 7.11 provides a schematic of the neutron beam for measuring the properties of the engineered membrane. Neutrons are subatomic elementary particles and can be produced efficiently from nuclear reactors via the fission of uranium nuclei, or by spallation where protons are accelerated and hit a heavy metal such as tungsten or lead. Note that neutrons produced from fission and spallation have very high energies of the order of MeV and are typically thermalized through successive collisions with deuterium atoms in heavy water at room temperature such that energy is reduced to approximately 25 meV. Several elaborate methods have been employed to filter the spectrum and increase in the neutron flux; the reader is referred to [123] for details. Given the complexity and cost of operation of such devices, samples are generally sent to NR facilities for analysis. A list of all the major NR facilities are provided in [108, 420]. A benefit of NR for biological applications is that any typical horizontal or vertical cold neutron reflectometer in a standard configuration can be utilized for the measurement of biological membranes. However, for the measurement of inserted proteins, a high neutron flux is required with precise measurement of the neutron reflectivity. This results because the concentration of inserted protein is low compared with the lipid and substrate molecules – as such most of the neutron reflectivity results from the lipid bilayer interface and supported substrate. Even though imaging of structures at a 5Å resolution is possible, further development is required for accurately measuring proteins inserted in the membrane [263].

Another major challenge with NR is that it is difficult to obtain deuterated materials – specialized techniques and facilities are required to construct deuterated molecules. Note that deuteration changes the coherent neutron-scattering-length density of the deuterated molecules. Selection of which molecules to deuterate is a key step in the experimental setup for NR measurements. If the molecules of interest have a coherent neutron-scattering-length density similar to that of the surrounding material then it is impossible to use the results of NR measurements to estimate the structural properties of the membrane and/or inserted protein. Typically for NR measurements both the lipid

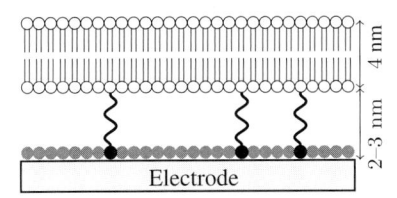

7.3 Spectroscopy and Imaging Techniques for Engineered Tethered Membranes

Figure 7.12 Building blocks of a engineered tethered membrane. The distance between the gold bioelectronic interface and the membrane surface is approximately 2–3 nm (measured from NR experiments) depending on the tethering chemistry (tether density, lipid type, and embedded macromolecules), and the membrane has an approximate thickness of 4 nm.

hydrocarbon tails and the protein need to be deuterated. Another challenge with NR is that a flat reflective surface on the order of several square centimeters is required for performing measurements [151]. Conceptually this is simply done by observing the angular spread of neutrons scattered from an engineered membrane of area 10–50 cm^2. These experiments are technically challenging and require hours of data acquisition at each D/H ratio[5]. It is only because of the stability of engineered tethered membranes that the NR approach can be used. The reconstruction of the deuteron-scattering profile is performed by iteratively comparing the predicted neutron-scattering profile from a multilayered sandwich model with the experimental result. The sandwich model is one where each layer is assigned a coherent scattering value. The initial layer in this multilayered model is, for example, a base of gold, then the next layer is assigned to be 0.2–0.4 nm of sulfur chemistries, the next to 2–3 nm of reservoir forming polyethylene glycols, and then further sequences of layers describing the lipid bilayer. The fit to the experimental data is compared for different layer thicknesses. When a fit is obtained the model profile is taken as the result.

The NR method has successfully been applied for measuring structural properties of engineered membranes with no excitation voltage [152], containing proteins [151, 274], and with an excitation voltage [191]. Depending on the experimental data quality, up to eight to ten slices may be meaningfully modeled. As is common to all such diffraction or scattering reconstruction techniques, the sign and phase contributed by each layer to the Fourier synthesized scattering profile are determined from a family of D/H ratios. Each ratio will provide a different profile as the deuteration level changes. More subtle detail may be obtained by site-specific deuteration of the membrane in either the initial self-assembled monolayer or in the subsequently formed engineered tethered membranes. The structure that has been confirmed by many NR studies is shown in Figure 7.12.

7.3.5 Summary

Many experimental techniques are available to determine the structure of freestanding nontethered lipid bilayers. Spanning more than 50 years, these include X-ray diffraction,

---

[5] The D/H ratio between deuterium ^{2}H and hydrogen ^{1}H in the electrolyte.

X-ray scattering, NMR spectroscopy, electron spin resonance spectroscopy, quartz crystal microbalance, and calorimetry, which have in combination provided a solid foundation on which to base an understanding of the atomic structure and dynamics of lipid bilayers. In this section we have discussed how X-ray diffraction can be used to measure the area per lipid, NMR can be used to measure the orientation of gA, FRAP can be used to measure the diffusion of lipids, and NR can be used to measure the membrane thickness and reservoir thickness of engineered membranes.

Despite the impressive power of these spectroscopic and imaging techniques, to determine the residual morphology and structural coherence of highly disordered systems (lipid membranes, lipid-peptide interaction, etc.), the information provided is inevitably that of a homogeneous bulk and does not elucidate the rare, low-probability events upon which most biological functions depend. The major benefit of the engineered tethered membrane is that it provides a means for measuring rare but vital events that drive biology.

The biological function and relevance of lipid bilayers depend on their ability to block the passage of ions. However, it is rare events that give rise to the modulation of ionic fluxes that traverse membranes that both sense the world around us via our senses of sight, sound, taste, touch, and smell, and permit our every response to those sensory signals. Signaling in the brain and the transmission of instructions to our muscles all depend on these transmembrane ion fluxes. Adding to these core functions in biology the metabolic processes of the respiratory cycle in which oxygen and nutrition result in the establishment of ion gradients which are essentially the batteries that drive these ion fluxes, it is evident that modeling and understanding the properties of transmembrane ion fluxes are central to the understanding of biology. The unique role in experimental biophysics brought by engineered tethered membranes is the ease and convenience with which these rare but essential ionic fluxes may be observed and manipulated.

7.4 Complements and Sources

In this chapter we have described experimental measurement techniques for engineered membranes that included electrical impedance, X-ray reflectometry, NMR, FRAP, and NR. There are, however, several other spectroscopic and imaging techniques that can be used to determine the structure and function of bilayer membranes and embedded macromolecules. For a general introduction to these techniques, refer to books such as [61, 66, 90, 113, 230, 302]. The use of confocal fluorescence, atomic force, electron, and magnetic resonance microscopy techniques can be used to determine the structure of molecules and bilayers with a resolution on the scale of nanometers to micrometers. Optical tweezers and fluorescence correlation spectroscopy can be used to manipulate and determine the forces exerted on individual molecules. Spectroscopic techniques (FRAP, NMR, X-ray reflectometry, etc.) provide a wealth of information that can be used to determine the structure, function, and dynamics of macromolecules and lipids contained in biological membranes. All these methods are, however, labor intensive and require costly experimental equipment. For example, to measure the structure of a macromolecule using X-ray crystallography requires the construction of a protein crystal. There is no general optimal protocol for constructing protein crystals; therefore, thousands of different protocols are generally required in the hopes of forming a protein crystal that can be used for X-ray crystallography. As such, electrical measurement techniques are the preferred method for measuring the properties of engineered tethered membranes.

7.5 Closing Remarks

In this chapter we have described experimental measurement techniques for engineered membranes. The two most widely used methods are impedance measurements and time-dependent electrical measurements. However, to measure properties such as lipid diffusion, membrane thickness, and other structural properties we also briefly discussed optical and imaging techniques of use for engineered membranes, including X-ray reflectometry, NMR, FRAP, and NR. In Part III we construct dynamic models of engineered membranes which can be used to relate the experimental measurements from tethered membranes to biological parameters of interest (e.g., the reaction pathway of pore-forming peptides, the concentration estimation of analyte species, and the population and dynamics of aqueous pores in the membrane). The design of novel protocols for measuring important biological parameters is critical to validate any dynamic models of engineered membranes, from the atomistic details to the reaction-rate theory.

Part III

Dynamic Models for Artificial Membranes: From Atoms to Device

Levels of Abstraction				
Ab Initio Molecular Dynamics	Classical Molecular Dynamics	Coarse-Grained Molecular Dynamics	Continuum Theory	Macroscopic Theory
nm fs	nm ns	nm μs	μm μs	m s

Part III deals with models for going from structure to function for the four engineered membrane devices (ion-channel switch, pore formation measurement platform, electroporation measurement platform, and electrophysiological response platform) discussed in Part II. These engineered membranes (hardware) are used together with mathematical models (software) to go from structure to function. The hardware discussed in Part II allows us to consider various components in their natural state (as they interact with other components), whereas the software discussed in Part III then zooms into specific subparts in isolation. This combination of hardware (ex vivo and in vitro) and software (in silico) allows us to engineer the membrane at various levels of abstraction: from atoms to the macroscopic device.

In this part of the book, we focus on the microscopic-to-macroscopic levels of abstraction of engineered membranes. The combination of microscopic (molecular dynamics and coarse-grained molecular dynamics), mesoscopic (continuum theory), and macroscopic (reaction-rate theory, which constitutes the highest level of abstraction) provides a complete description of the membrane. The macroscopic models of the engineered membranes can be used in combination with experimental measurements to estimate important biological parameters such as membrane capacitance and conductance, and also to both estimate the quality of the tethered membrane and to detect if subdiffusion or anomalous diffusion processes are present at the gold bioelectronic interface. The macroscopic models are represented by a set of ordinary differential equations (ODEs), or a set of fractional-order differential equations. The mesoscopic models can be used to estimate the dynamics of electroporation, aqueous pore conductance, and the dynamics of pore formation, and to study the electrical double-layer charging dynamics at the bioelectronic interface. The mesoscopic models are represented by a set of partial differential equations (PDEs) which include elliptic and parabolic PDEs with Neumann, Cauchy, and Dirichlet boundary conditions. The microscopic models (namely, coarse-grained molecular dynamics and classical molecular dynamics) provide atomistic details of the dynamics of the engineered membrane. These molecular dynamics–based microscopic models are simulation models. They can be used to estimate important biological parameters such as lipid diffusion, aqueous pore density, dynamics of pore formation, and the reaction dynamics of chemical species in the engineered membrane. The microscopic model is composed of a set of ODEs, one for each atom, that follow Newton's laws of motion. A key assumption when using results from the microscopic simulation model is that the system must satisfy the *ergodicity property*: the time-averaged trajectories obtained from the simulation equal the ensemble average. If the microscopic simulations are ergodic and long enough then meaningful information can be derived from the time-dependent atom trajectories such as the diffusion of lipids.

It is important to note that, conceptually, Part III can be viewed as a two-step approximation to real-life biological membranes. The first approximation step involves constructing an artificial membrane to mimic a real membrane (as detailed in Part II). The second approximation is to model the dynamics of this artificial membrane. This two-level modeling procedure provides useful insight into the properties of living membranes.

8 Reaction-Rate-Constrained Models for Engineered Membranes

8.1 Introduction

Part II gave a detailed description of fabricating synthetic biological devices using engineered membranes and also discussed methods to perform experimental measurements. Constructing mathematical models that accurately predict the behavior of engineered membranes is important, both for the design of devices built out of the membrane and for estimating parameters such as analyte concentration and membrane conductance. This chapter focuses on constructing *reaction-rate models* for engineered membranes. Such models comprise a system of nonlinear ordinary differential equations and can be viewed as a black-box phenomenological model[1] for the underlying synthetic biological device. Reaction-rate models account for the dynamics of tethered membranes that are excited by voltages less than 50 mV; therefore, the process of electroporation is not present. Specifically these models in combination with experimental measurements can be used to estimate the structural integrity of the tethered membrane and the disulfide bonds that anchor the tethers to the gold bioelectronic interface.

When an electrical potential is established across a tethered membrane, the macroscopic reaction-rate model for the current flowing through the membrane is equivalent to a lumped electrical circuit composed of resistors and capacitors as shown in Figure 8.1. In other words, the simple resistor–capacitor (RC) circuit of Figure 8.1 models the electrical response of the engineered membrane at a macroscopic level. The resulting current voltage response is a second-order linear system of ordinary differential equations. Though widely used to study the dynamics of tethered membranes, this reaction-rate model cannot account for the restricted diffusion dynamics that occur at the electrode surface. Modeling these effects requires introducing fractional-order operators. The fractional-order macroscopic model of the tethered membrane accounts for both the charging dynamics of the membrane and the restricted diffusion dynamics at the electrode surface. This chapter also presents a fractional-order macroscopic model for tethered membranes that contain sterol molecules (e.g., cholesterol). All the reaction-rate models in this chapter are validated using real-world experimental data

---

[1] A phenomenological equation is a mathematical model constructed to match experimental results but not derived from first-principles physical laws. An example is a black-box model where the system is viewed in terms of its input and outputs, without using physics for the internal workings.

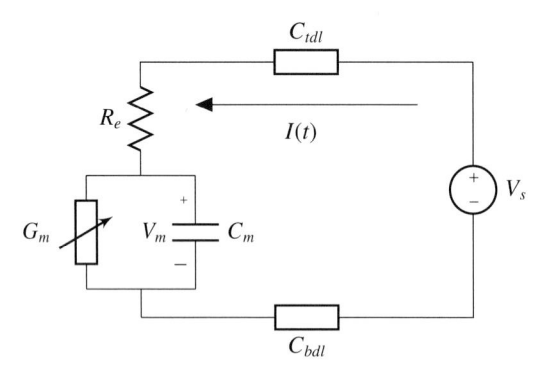

Figure 8.1 Lumped circuit macroscopic model of an artificial tethered membrane. The parameter R_e is the electrolyte resistance, C_m is the membrane capacitance, G_m is the membrane conductance, C_{bdl} is the electrode capacitance, C_{tdl} is the counter electrode capacitance, and V_m is the tethered transmembrane electric potential difference. The total electrode capacitance is given by the fractional-order capacitance C_{dl} which is equal to the fractional-order capacitance of C_{bdl} in series with the fractional-order capacitance C_{tdl} (see (8.2)). Typical values of these parameters are specified in Table C.6 on page 418.

from engineered membranes with different tether densities, lipid compositions, and analyte solutions.

Remark: Digital data acquisition, i.e., how the measured current is time sampled and digitized for processing on a computer, is not described in this book. Excellent resources for the design of analog-to-digital converters, time discretization, and antialiasing filters are [52, 102].

Before proceeding, we briefly discuss typical values of the parameters of the tethered-membrane model displayed in Figure 8.1; Appendix C also gives a more detailed list. A high-quality membrane with negligible defects will have a low conductance G_m in the range of 0.04 to 1.00 μS/mm^2, and an associated membrane capacitance C_m in the range of 4 to 10 nF/mm^2. The gold electrodes typically have a double-layer capacitance C_{dl} between 47 and 85 nF/mm^2. The fractional-order parameter p for the tethered membrane is in the range of 0.85 to 0.95, which illustrates that fractional-order dynamics are present at the electrode surface of the tethered membrane. Finally the electrolyte resistance R_e is in the range of 47 to 380 Ω/mm^2. Note that both C_m and G_m are dependent on the tethering density as seen from the experimentally measured values in Table C.6 on page 418. If the membrane contains significant defects then the values of G_m and C_m will be outside of these ranges of values. Therefore, to test the integrity of the membrane requires that we ensure these parameters (G_m, C_m, C_{dl}, and p) are contained in the above ranges.

8.2 Fractional-Order Macroscopic Model

This section describes a fractional-order macroscopic model for the electrical response of an engineered membrane. Using this model, the current $I(t)$ at continuous

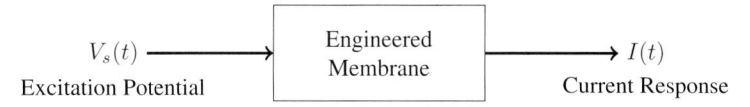

Figure 8.2 Black-box model of the engineered membrane. $V_s(t)$ denotes the input excitation voltage and $I(t)$ is the associated current response at time t. The engineered membrane is modeled by Figure 8.1.

time[2] t through the membrane can be computed given the excitation voltage $V_s(t)$, as illustrated in Figure 8.2. The model uses fractional-order derivatives to account for long-range memory effects that are associated with charge accumulation at the bioelectronic interface. Details on the evaluation of these fractional-order operators are also described below.

Current response of the membrane. The current response $I(t)$ contains important information about the charging dynamics of the bioelectronic interface and the tethered membrane, as well as the conductance of the tethered membrane. The fractional-order macroscopic model of the tethered membrane is given by (see Figure 8.1)

$$\frac{dV_m}{dt} = -\left(\frac{1}{C_m R_e} + \frac{G_m}{C_m}\right)V_m - \frac{1}{C_m R_e}V_{dl} + \frac{1}{C_m R_e}V_s,$$

$$\frac{d^p V_{dl}}{dt^p} = -\frac{1}{C_{dl} R_e}V_m - \frac{1}{C_{dl} R_e}V_{dl} + \frac{1}{C_{dl} R_e}V_s \qquad p \leq 1,$$

$$I(t) = \frac{1}{R_e}(V_s - V_m - V_{dl}), \tag{8.1}$$

where the double-layer capacitance C_{dl} is

$$C_{dl} = \left[\frac{1}{C_{tdl}} + \frac{1}{C_{bdl}}\right]^{-1}. \tag{8.2}$$

In (8.1), d/dt denotes the time derivative. On the other hand, d^p/dt^p denotes the pth-order fractional derivative which is discussed in (8.5) below. Choosing $p < 1$ models C_{dl} as a nonstandard capacitor to account for long-range memory due to charge accumulation at the bioelectronic interface. Also, V_m is the transmembrane potential, and V_{dl} is the double-layer potential. For a membrane with $V_s(t) = 0$ for $t \leq 0$, the initial conditions of (8.1) are $V_m = 0$ and $V_{dl} = 0$. That is, the transmembrane potential and double-layer potential are both at equilibrium for a tethered membrane that was not previously excited by an external voltage potential.

The parameters C_m (membrane capacitance), G_m (membrane conductance), R_e (electrolyte resistance), C_{dl} (double-layer capacitance), and p (fractional order) completely specify our macroscopic model (8.1) of the tethered membrane. Choosing $p = 1$ yields the classical model of Figure 8.1. More specifically, the tethered-membrane platform is composed of three distinct regions: the bulk electrolyte solution (R_e), the tethered

---

[2] Throughout this book continuous time t belongs to the set of non-negative real numbers. To perform digital processing, the continuous-time signals are sampled and digitized. Such digital data acquisition is classical and is not described in this book.

membrane (C_m, R_m), and the bioelectronic interface at the gold electrodes (C_{dl}, p). We now discuss these parameters in detail.

Electrolyte parameter R_e. The electrolyte dynamics depend on the concentration, hydration size, Coulomb correlations, polarization and screening effects, charge accumulation, diffusion-limited charge transfer, reaction-limited charge transfer, and adsorption dynamics of the electrolyte and analyte molecules, as discussed in Section 3.6 on page 76. Detailed models for these physical phenomena are covered in Chapter 11. Here we make the following assumptions:

(i) The bulk electrolyte is electroneutral.[3]
(ii) The concentrations of ions and analyte molecules in the bulk electrolyte are fixed in space and time.

With these assumptions, the bulk electrolyte dynamics can be characterized by the conductance of the electrolyte, σ_e, and the electrical permittivity of the electrolyte, ε_e. The electrical permittivity gives a measure of the ability of the molecules to reorient to reduce the effects of an applied electric field. In water, when an electric field is applied, the water molecules will rotate to reduce the effects of the applied electric field. This process is known as the polarization effect. The time scale of this reorientation to the applied electric field indicates if polarization effects can be neglected.

The characteristic relaxation time of the electrolyte, $\tau_e = \sigma_e/\varepsilon_e$, yields a measure of the time scale needed for the water molecules to rotate in response to an applied electric field. If $\tau_e \ll 1$, then polarization effects can be neglected and only the conductance of the electrolyte, σ_e, will contribute to the bulk electrolyte dynamics. The typical value of the bulk electrolyte solution in engineered tethered membranes is 150 M NaCl. For this concentration, the bulk electrolyte conductance $\sigma_e \approx 1.6$ S/m, and electric permittivity $\varepsilon_w = 2\varepsilon_0$ where ε_0 is the vacuum permittivity. The characteristic relaxation time of the electrolyte $\tau_e = 11$ ps. Since the excitations typically used for engineered membranes are on the microsecond time scale, we can neglect the polarization effects of the bulk electrolyte solution. Therefore, the bulk electrolyte solution can be modeled by the electrolyte resistance R_e.

Engineered membrane parameters C_m and G_m. The membrane capacitance C_m and membrane conductance G_m account for the charge accumulation and permeability dynamics of the tethered membrane. The membrane is assumed to be polarizable and to also contain aqueous pores as a result of random thermal fluctuations; that is, random thermal fluctuations allow the energy barrier to be crossed for the conformational change of lipids, allowing the formation of transient aqueous pores. This allows the tethered membrane to be modeled by an effective permittivity with capacitance C_m in parallel with the tethered membrane conductance G_m. The membrane conductance depends on the population of aqueous pores, embedded ion channels, and the conductance dynamics of these aqueous pores and channels. For example, in the ion-channel switch biosensor the conductance of the membrane, G_m, is dependent on the population of aqueous pores and conducting gramicidin dimers present. The population and size of aqueous pores in the membrane are governed by the process of

---

[3] Electroneutrality means that the total concentration of anions (negative ions) and cations (positive ions) in any region of the electrolyte result in a net zero charge.

electroporation (discussed in Chapter 12) which results from nonzero transmembrane potentials. As such, the membrane conductance G_m also depends on the transmembrane potential. Note that negligible changes in the population and size of aqueous pores result if the transmembrane potential is below 50 mV. In such cases it is reasonable to assume that the membrane conductance remains at its equilibrium value.

Engineered membrane parameters C_{dl} and p. The double-layer capacitance C_{dl} (8.2) and fractional-order p are used to model the dynamics of the bioelectronic interface.[4] Several physical phenomena may be present at the bioelectronic interface including charge accumulation, diffusion-limited charge transfer, reaction-limited charge transfer, and ionic adsorption dynamics. These double-layer charging effects can be modeled using fractional-order operators. Notice that if only charge accumulation is present, then the bioelectronic interface can be modeled using a standard capacitor with capacitance C_{dl} in farads with $p = 1$. However, if a combination of these double-layer charging effects is present then the bioelectronic interface can be modeled using a constant-phase element composed of a capacitance C_{dl} and the fractional-order operator $p < 1$.

The electrode capacitance adjacent to the tethered membrane is denoted by C_{bdl}, and the counter electrode capacitance by C_{tdl}. Note that C_{bdl} and C_{tdl} provide the connection from the ionic solution to the electrical domain. One face of the capacitor is charged by the presence of ions and the other is charged with electrons, producing the current $I(t)$ in the circuit.

Let us now discuss the fractional-order derivatives that appear in (8.1). The model of the electrical double-layer capacitance is given by

$$C_{dl} \frac{d^p V_{dl}}{dt^p} = I(t) \tag{8.3}$$

where $I(t)$ is the current flowing through the double-layer capacitance C_{dl}. The units of C_{dl} depend on the value of the fractional-order operator p to ensure the left-hand side of (8.3) is in amperes. The SI units of C_{dl} are $s^{(p+3)}A^2/m^2kg$. For $p = 1$, (8.3) is the classical current-voltage relation for a capacitor with C_{dl} having the units of farads. For the case that $p < 1$, the current-voltage relation is defined by a fractional-order differential equation. The fractional derivative can be evaluated using either the Riemann–Liouville, the Caputo, or the Grünwald–Letnikov definitions. The Riemann–Liouville definition of the fractional-order operator in (8.3) is[5]

$$\frac{d^p V_{dl}}{dt^p} = \frac{1}{\Gamma(1-p)} \frac{d}{dt} \int_0^t V_{dl}(\xi)(t-\xi)^{-p} d\xi, \quad 0 < p \leq 1, \tag{8.5}$$

---

[4] Recall from §2.6 that the bioelectronic interface comprises a gold electrode that interfaces ions in the electrolyte with electrons in the gold electrode.

[5] For engineered tethered membranes the fractional-order parameter p is the range of 0.5 to 1. However, the evaluation of the fractional derivative can be performed for any real positive number p. The Riemann–Liouville definition of the fractional-order derivative in the general case is

$$\frac{d^p V_{dl}}{dt^p} = \frac{1}{\Gamma(\eta-p)} \frac{d^\eta}{dt^\eta} \int_0^t V_{dl}(\xi)(t-\xi)^{\eta-p-1} d\xi, \quad p > 0, \tag{8.4}$$

where η is an integer satisfying $p \leq \eta < p + 1$.

where $\Gamma(\cdot)$ is the gamma function.[6] For engineered tethered membranes the typical value of p is in the range of 0.5 to 1. As seen in (8.5), the fractional-order operator is not local. This means that the value of $d^p V_{dl}/dt^p$ depends on all the values of V_{dl} in the entire time interval from 0 to t. This dependency on time allows the fractional-order operator to model path-dependent phenomena such as subdiffusion processes. In the engineered membrane the electrode surface is composed of tethers and spacers and may contain irregularities that impede the flow of ions in the electrolyte. This impedance of flow causes the electrolyte dynamics to violate Brownian motion (e.g., the mean-squared displacement of particles is not proportional to the diffusion times time, or the relative change in position of an ion between time increments are not independent). This type of long-range memory, or path-dependent phenomenon, can be modeled using the fractional-order operator.

Note that for high double-layer potentials V_{dl} above 500 mV the double-layer capacitance C_{dl} also becomes voltage-dependent as a result of Coulomb correlations between atoms at the bioelectronic interface. §12.9 discusses how these Coulomb correlations can be modeled to estimate the voltage dependence of C_{dl}.

8.2.1 Fractional-Order Derivatives: Double-Layer Capacitance and Charging Dynamics

In this section we obtain expressions for the transfer function and impedance, and we analyze the charging dynamics of the electrical double-layer capacitance C_{dl} (8.3). We show that C_{dl} in Figure 8.1 on page 140 is equivalent to the constant-phase element used in electrochemistry to account for the charge accumulation at rough and fractal electrode surfaces [11, 92, 181, 418]. Additionally, we show that the electrical double-layer capacitance (8.3) is equivalent to the generalized Warburg impedance (also known as a constant-phase element) classically used to account for charge accumulation, diffusion-limited charge transfer, reaction-limited charge transfer, and ionic adsorption dynamics that may be present at electrode-to-electrolyte interfaces. To illustrate key differences between the charging dynamics of the double-layer capacitance C_{dl} compared with an ideal capacitor, we compute analytical expressions for the charge $Q_{dl}(t)$ in the double-layer region for an electrode-to-electrolyte interface for current-step and voltage-step excitations.

Double-layer transfer function $Z_{dl}(s)$ (Laplace domain). The Laplace transform converts differential equations in time $t \in [0, \infty)$ into algebraic equations as a function of complex frequency $s = \sigma + j\omega$, where σ and ω are real variables and $j = \sqrt{-1}$. The Laplace transform of a function $V_{dl}(t)$ is

$$V_{dl}(s) = \mathcal{L}[V_{dl}(t)] = \int_0^\infty V_{dl}(t) e^{-st} \, dt. \tag{8.6}$$

The aim here is to construct the transfer function $Z_{dl}(s) = V_{dl}(s)/I_{dl}(s)$ of the double-layer capacitance C_{dl} defined in (8.3). To construct $Z_{dl}(s)$ we first derive $I_{dl}(s)$ by

---

[6] The gamma function $\Gamma(z)$ for any real z is

$$\Gamma(z) = \int_0^\infty \xi^{z-1} e^{-\xi} \, d\xi.$$

evaluating the Laplace transform of (8.3):

$$I_{dl}(s) = \mathcal{L}[I_{dl}(t)] = C_{dl}\mathcal{L}\left[\frac{d^p V_{dl}}{dt^p}\right]$$

$$= C_{dl}\int_0^\infty \frac{d^p V_{dl}}{dt^p} e^{-st}\, dt$$

$$= C_{dl}\int_0^\infty \left[\frac{1}{\Gamma(1-p)}\int_0^t \frac{dV_{dl}}{dt}\bigg|_{t=\xi}(\xi - t)^{-p}\right] e^{-st}\, d\xi\, dt$$

$$= C_{dl} s^p \mathcal{L}[V_{dl}] - C_{dl} s^{p-1} V_{dl}(0). \tag{8.7}$$

Here we have used the Caputo definition[7] of the fractional-order derivative to construct the relation between $I_{dl}(s)$ and $V_{dl}(s)$. If initially the double layer contains no charge ($V_{dl}(0) = 0$), then the transfer function of the double-layer capacitance is

$$Z_{dl}(s) = \frac{\mathcal{L}[V_{dl}(t)]}{\mathcal{L}[I_{dl}(t)]} = \frac{1}{C_{dl} s^p}. \tag{8.9}$$

Note that the transfer function (8.9) in the Laplace domain for the double-layer capacitance is only applicable for time-domain functions defined in $t \in [0, \infty)$.

Double-layer impedance $Z_{dl}(\omega)$ (Fourier domain). To estimate the impedance $Z_{dl}(\omega)$ of the double-layer capacitance C_{dl} defined in (8.3) requires that we consider time-domain functions defined in $t \in (-\infty, \infty)$. The Fourier transform of $V_{dl}(t)$ is

$$V_{dl}(\omega) = \mathcal{F}[V_{dl}(t)] = \int_{-\infty}^\infty V_{dl}(t) e^{-j\omega t}\, dt, \tag{8.10}$$

where $\omega = 2\pi f$ is the angular frequency. To construct the impedance $Z_{dl}(\omega) = V_{dl}(\omega)/I_{dl}(\omega)$, we first derive $I_{dl}(\omega)$ by taking the Fourier transform of (8.3):

$$\mathcal{F}[I_{dl}(t)] = C_{dl}\mathcal{F}\left[\frac{d^p V_{dl}}{dt^p}\right]$$

$$= C_{dl}\int_{-\infty}^\infty \frac{d^p V_{dl}}{dt^p} e^{-j\omega t}\, dt$$

$$= C_{dl}\int_{-\infty}^\infty \left[\frac{1}{\Gamma(1-p)}\int_{-\infty}^t \frac{dV_{dl}}{dt}\bigg|_{t=\xi}(\xi - t)^{-p}\right] e^{-j\omega t}\, d\xi\, dt$$

$$= C_{dl}(j\omega)^p \mathcal{F}[V_{dl}]. \tag{8.11}$$

---

[7] The Caputo definition of the fractional derivative in the general case is

$$\frac{d^p V_{dl}}{dt^p} = \frac{1}{\Gamma(\eta - p)}\int_0^t \left[\frac{d^\eta V_{dl}}{dt^\eta}\bigg|_{t=\xi}\right](t - \xi)^{\eta - p - 1}\, d\xi, \quad p > 0, \tag{8.8}$$

where η is an integer satisfying $p \leq \eta < p + 1$. Note that for the evaluation of the Caputo fractional derivative, the double-layer voltage $V_{dl}(t)$ must be defined for the time interval $t \in [0, \infty)$.

Here we have used the Caputo–Weyl definition[8] of the fractional-order derivative to construct the relation between $I_{dl}(\omega)$ and $V_{dl}(\omega)$. Given (8.11), the double-layer impedance $Z_{dl}(\omega)$ is

$$Z_{dl}(\omega) = \frac{\mathcal{F}[V_{dl}(t)]}{\mathcal{F}[I_{dl}(t)]} = \frac{1}{C_{dl}(j\omega)^p}. \tag{8.13}$$

The key result here is that if $p = 1$, then (8.13) is equal to the impedance of an ideal capacitor. Additionally, if $p = 0.5$, then (8.13) is equal to the Warburg impedance [11, 181, 418]. For engineered tethered membranes, a generalized Warburg impedance with $0 < p \leq 1$ is used to account for charge accumulation, diffusion-limited charge transfer, reaction-limited charge transfer, and ionic adsorption dynamics at the surface of the bioelectronic interface.

Charging dynamics of the double layer. The double-layer capacitance C_{dl} in (8.3) has units of $s^{(p+3)}A^2/m^2kg$. Therefore, if we were to naively compute the double-layer charge using the ideal capacitor charge-voltage relation $Q_{dl} = C_{dl}V_{dl}$, the units of Q_{dl} would be $s^p A$, which do not equal the unit of Coulombs unless $p = 1$. For a general double-layer current $I_{dl}(t)$, the total charge accumulation in the double layer is

$$Q_{dl}(t) = \int_0^t I_{dl}(\xi)\,d\xi, \tag{8.14}$$

where $Q_{dl}(t)$ has units of farads. In some cases, it is also possible to compute $Q_{dl}(t)$ in closed form using the Laplace transform. Below we illustrate how $Q_{dl}(t)$ can be computed in closed form for an electrode-to-electrolyte interface.

Consider an electrode-to-electrolyte interface that we model using the electrolyte resistance R_e in series with the double-layer capacitance C_{dl}. Note that this is equivalent to the lumped circuit model in Figure 8.1 with no membrane present. The aim is to compute the effective time-dependent capacitance $\bar{C}_{dl}$ and double-layer charge Q_{dl} for a constant current-step excitation and constant voltage-step excitation.

Constant current step. The response of the electrode-to-electrolyte interface for a constant current step $I(t) = I_o$ for $t > 0$ can be determined using the Laplace transforms method. The transfer function $Z(s)$ and current input $I(s)$ are related to the Laplace domain voltage $V(s)$ by

$$V(s) = Z(s)I(s) = \left[R_e + \frac{1}{C_{dl}s^p}\right]\frac{I_o}{s}, \tag{8.15}$$

---

[8] The Caputo–Weyl definition of the fractional derivative operator (fractional-order derivative) in the general case is

$$\frac{d^p V_{dl}}{dt^p} = \frac{1}{\Gamma(\eta - p)} \int_{-\infty}^{t} \left[\frac{d^\eta V_{dl}}{dt^\eta}\bigg|_{t=\xi}\right](t - \xi)^{\eta - p - 1}\,d\xi, \quad p > 0, \tag{8.12}$$

where η is an integer satisfying $p \leq \eta < p + 1$. Note that for the evaluation of the Caputo fractional-order derivative, the double-layer voltage $V_{dl}(t)$ must be defined for the time interval $t \in (-\infty, \infty)$.

where p is the fractional-order operator. The inverse Laplace transform of (8.15) is

$$V(t) = I_o R_e + \frac{I_o t^p}{C_{dl}\Gamma(1+p)} = V_e(t) + \frac{I_o t}{\bar{C}_{dl}(t)} = V_e(t) + V_{dl}(t), \quad (8.16)$$

where $V_e(t)$ is the voltage across the electrolyte resistance R_e, $V_{dl}(t)$ is the voltage across the double-layer capacitance, and $\bar{C}_{dl}(t) = C_{dl}\Gamma(1+p)t^{1-p}$ is the effective time-dependent capacitance with units of farads. The charge accumulation at the double layer can be computed using an effective time-dependent capacitance and is equal to $Q_{dl} = \bar{C}_{dl}(t)V_{dl}(t) = I_o t$. This is expected because, for an input current step $I(t) = I_o$ for $t > 0$, (8.14) requires that $Q_{dl} = I_o t$.

Constant voltage step. The response of the electrode-to-electrolyte interface for a constant voltage step $V(t) = V_o$ for $t > 0$ can be constructed using the Laplace-transform method. The transfer function $Z(s)$ and voltage input $V(s)$ are related to the Laplace domain current $I(s)$ by

$$I(s) = \frac{V(s)}{Z(s)} = \left[R_e + \frac{1}{C_{dl}s^p}\right]^{-1}\frac{V_o}{s}, \quad (8.17)$$

where p is the fractional-order operator. The inverse Laplace transform of (8.17) does not have an analytical solution. However, for negligible electrolyte resistance $R_e \ll 1$, the inverse Laplace transform of (8.17) is

$$I(t) = \frac{V_o}{C_{dl}}t^p\Gamma(1-p) = \frac{(1-p)\bar{C}_{dl}V_o}{t}, \quad (8.18)$$

where $\bar{C}_{dl}$ is the effective time-dependent capacitance with units of farads. The charge accumulation at the double layer can be computed by substituting (8.18) into (8.14) and evaluating the integral to obtain

$$Q_{dl}(t) = \left[\frac{C_{dl}t^{1-p}}{\Gamma(2-p)}\right]V_o = \bar{C}_{dl}V_o. \quad (8.19)$$

The key insight here is that for a constant voltage step $Q_{dl}(t) \propto t^{1-p}$. For $p = 1$ the value of $Q_{dl} = C_{dl}V_o$ is a constant; however, for $0 < p < 1$ the accumulated charge $Q_{dl}(t)$ increases via a power law at a rate of t^{1-p}. The smaller the value of p, the larger the rate of charge accumulation. This is counterintuitive because it is expected that the result of diffusion-limited charge transfer, reaction-limited charge transfer, and ionic adsorption dynamics would reduce the rate of charge accumulation. However, as we show here, the opposite occurs and the smaller the value of the fractional-order parameter p, the larger the rate of charge accumulation in the double layer.

8.2.2 Fractional-Order Macroscopic Model: Sinusoidal and Time-Varying Excitation Potential

Given the excitation potential $V_s(t)$ and lumped circuit parameters C_m, G_m, C_{dl}, p, and R_e, the fractional-order macroscopic model (8.1) can be used to compute the current response $I(t)$ of the engineered membrane. However, in the special case when $V_s(t)$

is small (below 50 mV) and sinusoidal, then $I(t)$ can be evaluated in steady state as an algebraic expression (rather than a differential equation). Below we discuss both methods for computing $I(t)$ using (8.1).

Using a sinusoidal excitation voltage $V_s(t) = V_o \sin(2\pi f t)$ with frequency f (in hertz) and magnitude V_o below 50 mV, the current response of the tethered membrane can be computed using a set of algebraic equations. Converting (8.1) into the complex domain with $V_s(t) = V_o \sin(2\pi f t)$, the current response at frequency f is given by the complex-valued phasor[9]

$$I(f) = V_o \left[R_e + \frac{1}{G_m + j2\pi f C_m} + \frac{1}{(j2\pi f)^p C_{dl}} \right]^{-1}. \quad (8.20)$$

In (8.20), j denotes the complex number $\sqrt{-1}$. Note that the impedance $Z(f)$ of the tethered membrane is given by the expression in $[\cdot]$ of (8.20). Assuming $G_m(t)$ is static during the measurement of $I(f)$ (which occurs if the magnitude of $V_s(t)$ is below 50 mV), the membrane conductance $G_m(t)$ is constant and can be computed using a least-squares estimator with a cost function given by the difference between the measured current and the computed current from (8.20). Note that the parameters R_e, C_m, p, and C_{dl} are constant over the frequency range of measurement (i.e., 0.1 Hz to 10 kHz).

If the excitation voltage $V_s(t)$ is not periodic, then we need to solve the system of fractional-order differential equations in (8.1). The evaluation of the fractional-order derivative can be performed using the Adomian decomposition method [179], which is a popular semianalytical method for solving ordinary and nonlinear partial differential equations. Another method is the variational iteration method [282] typically used for numerically solving nonlinear partial differential equations. Linear multistep methods, such as the Adams–Bashforth–Moulton method [104], can also be used to evaluate the fractional-order derivative.

8.2.3 Determining the Quality of an Engineered Membrane Using the Fractional-Order Macroscopic Model

Since the fractional-order macroscopic model (8.20) can accurately predict the response of the membrane (as illustrated in §8.3), it stands to reason that this model can also be used to detect if the engineered membrane has defects. Such defects would be manifest in the circuit parameters C_m, C_{dl}, R_e, and G_m in (8.20). So the natural question is this: Can variations in the circuit parameters C_m, C_{dl}, R_e, and G_m in (8.20) be detected given experimental measurements of the current voltage response of the membrane? Figure 8.3 provides the numerically computed impedance for variations in C_m, C_{dl}, and R_e. As seen from Figure 8.3(a)–(c), using an excitation frequency from 0.1 Hz to 1 kHz allows the estimation of the parameters C_m, C_{dl}, and R_e. In Figure 8.3(a) we see that the major effect of C_m on the impedance occurs at frequencies above 1 Hz, and from Figure 8.3(b) the effects of C_{dl} are pronounced for frequencies below 10 Hz. This results because C_m is about a factor of 10 smaller then C_{dl} for the tethered-membrane platform.

---

[9] Equivalently, the steady-state current is a sinusoid with frequency f, magnitude $|I(f)|$, and phase $\angle I(f)$.

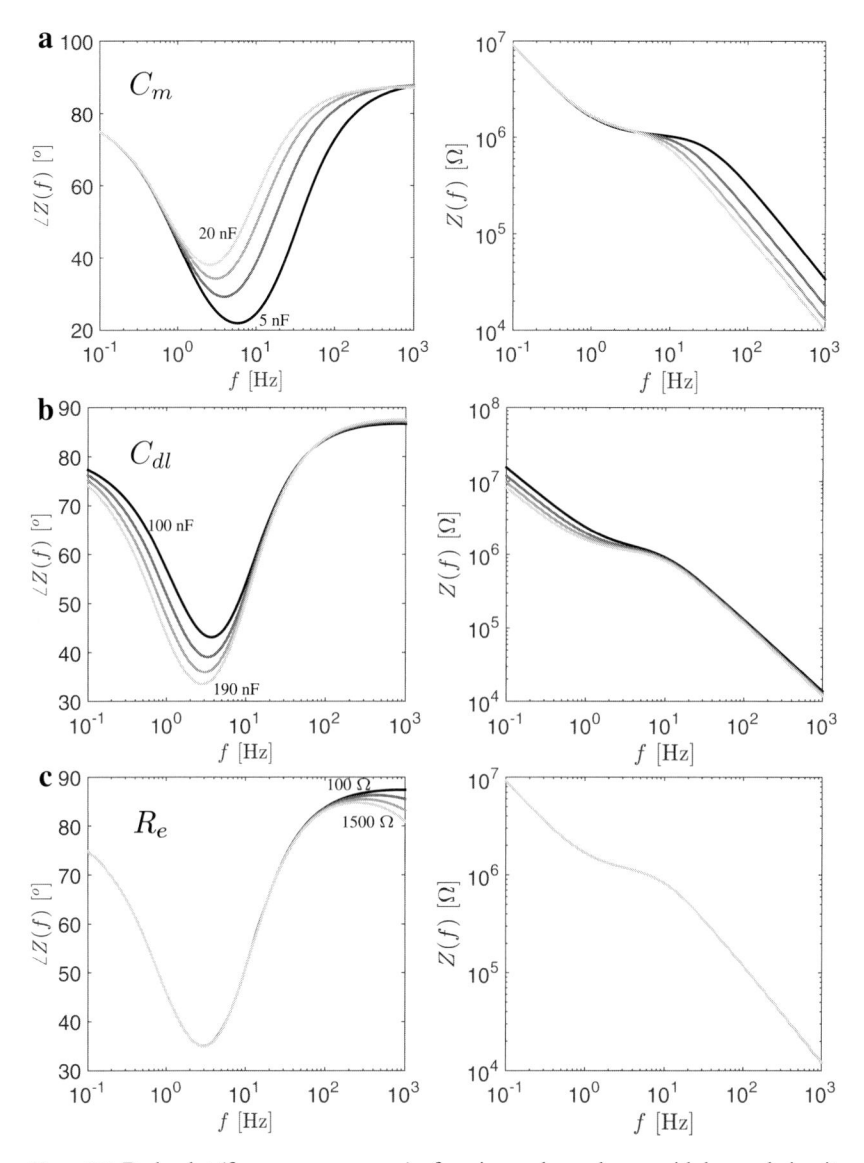

Figure 8.3 Bode plot (frequency response) of engineered membrane with lumped circuit model in Figure 8.1. (a) Illustration of how the impedance (phase and magnitude) changes for capacitance C_m values of 5, 10, 15, and 20 nF with all other parameters fixed. (b) Illustration of how the impedance changes with C_{dl} values of 100, 130, 160, and 190 nF with all other parameters fixed. (c) Illustration of how the impedance changes for the R_e values of 100, 500, 1000, and 1500 Ω with all other parameters fixed. The predicted impedance (phase is represented by $\angle Z(f)$ in degrees and magnitude by $Z(f)$) is computed using (8.20).

Figure 8.3(c) shows that R_e only causes measurable effects on the phase at frequencies above 110 Hz. Note that once the parameters C_m, C_{dl}, and R_e have been estimated they do not change during the experimental measurements; as such, changes would indicate a catastrophic change in the tethered-membrane stability. The only parameter

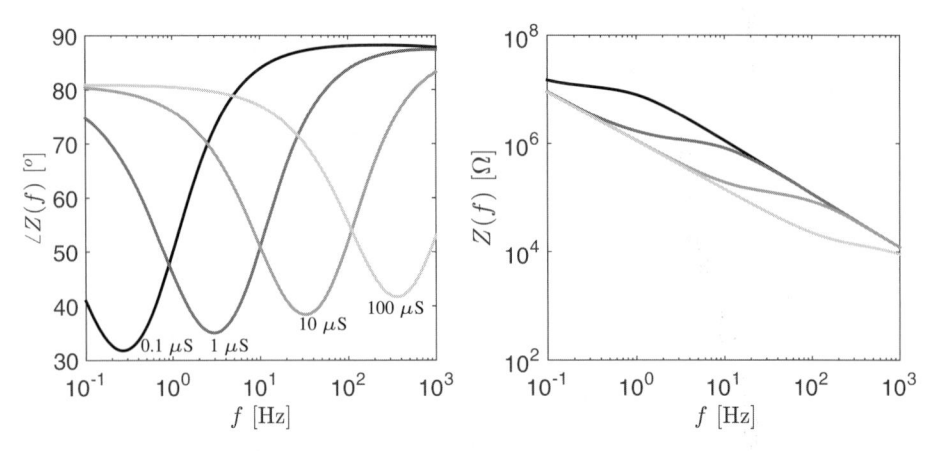

Figure 8.4 Bode plot of engineered membrane with the lumped circuit model in Figure 8.1 (phase is represented by $\angle Z(f)$ in degrees, and magnitude by $Z(f)$) for various membrane conductances G_m with C_m, C_{dl}, and R_e fixed. The impedances are computed using (8.20).

that varies throughout the experimental measurements is the membrane conductance G_m. Figure 8.4 illustrates how changes in G_m can be estimated from the experimentally measured impedance response. As seen, using an excitation voltage of 0.1 Hz to 1 kHz allows the measurement of G_m in the range of 0.1 to 100 μS.

How can we experimentally verify that the formed membrane does not contain significant defects? Possible membrane defects include patches where the gold electrode is directly exposed to the bulk electrolyte, or where portions of the bilayer are sandwiched together. Using the measured current response resulting from an excitation voltage below 50 mV, we can compute the mean-squared error (MSE) between the predicted current from (8.1) and the experimentally measured current. If a significant MSE is obtained then the model of a homogeneous membrane (8.1) is not suitable and the membrane is concluded to contain inhomogeneities (i.e., defects). A major concern when performing electroporation experiments is the detection of the catastrophic voltage breakdown of the membrane causing separated areas of membrane to degrade. This effect can be detected by a high MSE and a significant increase in the estimated membrane conductance C_m resulting from the electrode surface capacitance coming into contact with the bulk electrolyte. Typical values for membrane capacitance and conductance are in the ranges 0.5–1.3 μF/cm^2 and 0.5–2.0 μS for an intact 1–100 percent tethered membrane with surface area 2.1 mm^2.

8.3 Experimental Measurements: Fractional-Order Macroscopic Model

In this section we apply the fractional-order macroscopic model (8.20), illustrated schematically in Figure 8.2, to study diffusion-limited processes at the electrode surface. We also discuss how the model accounts for different electrolyte ion concentrations, tether densities, and lipid compositions.

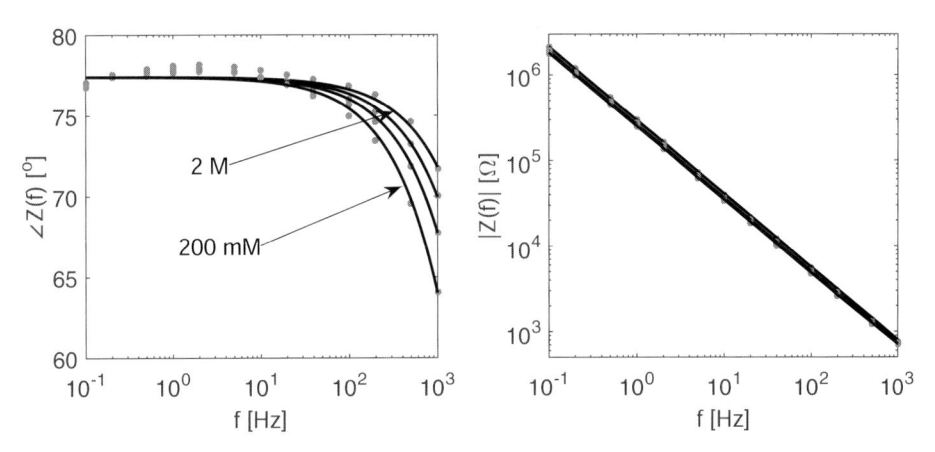

Figure 8.5 The model-predicted (black line) and experimentally measured (gray dots) impedance (phase is represented by $\angle Z(f)$ in degrees, and magnitude by $|Z(f)|$) of the spacer surface for various electrolyte molar concentrations of NaCl. The results are for NaCl solutions of concentrations 2 M, 1 M, 500 mM, and 200 mM. All predictions are computed using the model (8.20).

8.3.1 Spacer Surface and Electrolyte Concentration

Recall that the fractional-order macroscopic model (8.20) has parameters that model three distinct regions in the engineered membrane: the tethered membrane (C_m, R_m), the bulk electrolyte solution (R_e), and the bioelectronic interface at the gold electrodes (C_{dl}, p). If only the bulk electrolyte solution and the bioelectronic interface are present, then we can directly study the effect of the electrolyte NaCl concentration on the parameters R_e, C_{dl}, and p.

In this section we perform impedance measurements of a spacer surface[10] (only the bulk electrolyte solution and the bioelectronic interface are present) in contact with an electrolyte solution containing, respectively, 2 M, 1 M, 500 mM, and 200 mM NaCl. Using the macroscopic model (8.20) and experimental measurements of the spacer surface allows one to accurately estimate the parameters R_e, C_{dl}, and p. The experimentally measured and model-predicted impedance from the fractional-order macroscopic model are provided in Figure 8.5. In the absence of a diffusion-limited process, $p = 1$, and $\angle Z(f) \approx 90°$ at low frequencies (approximately $f < 1$ Hz), where the capacitive effects of the bioelectronic interface are dominant. As seen from Figure 8.5, for frequencies below 1 Hz the phase angle $\angle Z(f) \approx 76°$; therefore, for all concentrations measured there is a diffusion-limited process present with a fractional-order parameter $p = 0.86$. The fractional-order behavior at the electrode surface may be caused by diffusion-limited charge transfer and adsorption on the electrode [388]. Though the source of the behavior is unknown, the dynamics of the interface can be modeled using the fractional-order macroscopic model (8.20) as illustrated in Figure 8.5.

---

[10] Recall from §2.6 that spacers are an integral component of engineered membranes.

8.3.2 Variation in Membrane Types and Tether Density

Here we use the fractional-order macroscopic model (8.20) to estimate the following circuit parameters of the tethered-membrane model using experimental measurements: membrane capacitance C_m, membrane conductance G_m, the bulk electrolyte solution resistance R_e the bioelectronic interface capacitance C_{dl}, and fractional-order parameter p (please see Figure 8.1 on page 140 for the equivalent circuit model of the tethered membrane). We consider several different engineered membranes with different lipid compositions and tether densities. The membrane compositions considered are zwitterionic C20 diphytanyl-ether-glycero-phosphatidylcholine (DphPC) and C20 diphytanyl-diglyceride ether (GDPE) lipids from archaea, lipids from the eukaryotic species *Saccharomyces cerevisiae* (a type of yeast), and lipids from the prokaryote species *Escherichia coli*.[11]

Figure 8.6 provides the model-predicted and experimentally measured impedance of the tethered membranes. The predicted impedances are computed using the fractional-order macroscopic model (8.20) illustrated in Figure 8.2. As seen, excellent agreement is obtained between the predicted and measured impedance. Comparing the impedance of the 1 and 10 percent tethered DphPC, *E. coli*, and *S. cerevisiae* membranes, we see that the resistance to membrane defects from highest to lowest is DphPC, *E. coli*, and *S. cerevisiae*. Recall from §8.2.3 that the frequency where the minimum in phase occurs gives an indication of the membrane conductance G_m. Therefore, comparing the minima between two phase plots allows a comparison in the membrane conductance between the two tethered membranes. As expected the resistance to membrane defects increases as the tether density increases. The difference in the resistance to membrane defects of DphPC and GDPE compared to that of *E. coli* and *S. cerevisiae* is a result of the phytanyl chain packing properties and the ether linking the phytanyl to the lipid head group. This structure promotes the hydrogen bonding in DphPC and GDPE lipids compared to the lipids from the *E. coli* and *S. cerevisiae* organisms. That is, the structural properties of the lipid tails of DphPC allow these lipids to form a dense packing structure which is more resistant to the flow of ions and water compared with the lipid tails from the *E. coli* and *S. cerevisiae* organisms. A surprising observation is that the *E. coli* membrane is more resistant to membrane defects compared to that of *S. cerevisiae* even though the membrane thickness of *S. cerevisiae* is larger than that of *E. coli*. A possible mechanism for this difference is that defects in the *E. coli* membrane are primarily formed by the flip-flop of specific groups of phospholipids [324].

8.3.3 Estimating the Dielectric Constant of the Membrane

The fractional-order macroscopic model (8.20) can be used to estimate important biological parameters such as the dielectric permittivity of the tethered membrane, ε_m. To compute these estimates we use the results from the impedance fit in Figure 8.6. The

---

[11] DphPC and GDEP are the primary components of most engineered membranes, as discussed in Chapter 4. Note that *E. coli* is one of the most widely studied prokaryotic organisms and is used widely in the field of biotechnology.

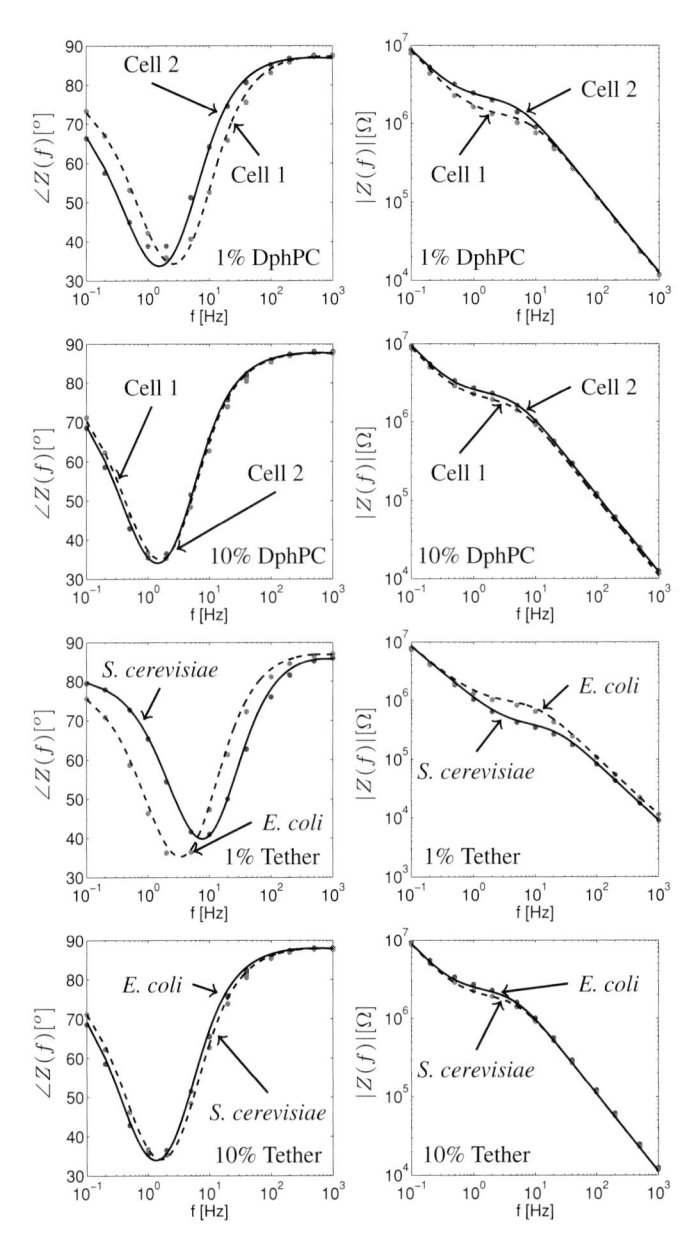

Figure 8.6 Model predicted and actual frequency response impedance (phase $\angle Z(f)$ and magnitude $|Z(f)|$) for 1% and 10% tether density engineered membranes comprised of DphPC lipids, lipids extracted from S. cerevisiae, and lipids extracted from E. coli. The solid and dashed lines indicate the model predicted impedances while the dots indicate the measured frequency response. Cell 1 and Cell 2 denote each respective flow-cell chambers that the DphPC membranes were prepared in. All predicted impedances are computed using the fractional-order macroscopic model (8.20) with the parameter values defined in Table C.6.

capacitance of the membrane typically satisfies $C_m = \varepsilon_m A_m / h_m$, where A_m is the area of the membrane, and h_m is the thickness of the membrane. For the 10 percent tether density DphPC membrane $h_m = 3.5$ nm, $A_m = 2.1$ mm^2, and C_m is the range of 12.5 to 15.5 nF. Therefore, the relative permittivity[12] ε_m is in the range 2.35 to 2.92. For the 1 percent tether density DphPC C_m is in the range of 16 to 17.5 nF, and, assuming the thickness of the tethered DphPC is $h_m = 3.4$ nm (this quantity can be computed using molecular dynamics models), the relative permittivity ε_m is in the range 2.92 to 3.20. Note that the relative permittivity of the membrane, ε_m, must be between the values of 2 for pure hydrocarbon and 80 for an electrolyte at physiological concentrations. This shows that the dielectric permittivity is dependent on the tether density. To estimate the permittivity of the tethered *E. coli* and *S. cerevisiae* membranes we chose $h_m = 3.29$ nm for the *E. coli*, and $h_m = 4.30$ nm for the *S. cerevisiae* as justified by the molecular dynamics results [185, 324]. The associated relative permittivity of *E. coli* is in the range 2.5 to 3.0, and the relative permittivity of *S. cerevisiae* is in the range 3.2 to 4.2. These are in excellent agreement with the experimentally measured relative permittivity results of *E. coli* and *S. cerevisiae* cell membranes documented in [337].

8.4 Modeling Membranes with Sterol Components

The purpose of this section is to construct a macroscopic mathematical model for how cholesterol[13] affects the dynamics of an artificial membrane. Since the artificial membrane mimics a biological membrane, we can then understand how cholesterol affects a biological membrane. Indeed, cholesterol is a major sterol component in most eukaryotic membranes and is known to regulate important membrane properties such as lipid diffusion and membrane stability. See §2.2 for further details on cholesterol.

For engineered membranes that contain large concentrations of sterols such as cholesterol, the fractional-order macroscopic model in Figure 8.1 is not suitable for modeling the current response. As the concentration of sterols in the membrane increases, it brings about the formation of lipid rafts,[14] which significantly contribute to the current flowing through the engineered membrane. For example, consider the fluorescence micrographs for a cholesterol-enriched 1,2-dioleoyl-sn-glycero-3-phosphocholine membrane provided in Figure 8.7. As seen, as the concentration of cholesterol increases, the cholesterol aggregates, forming lipid rafts in the engineered membrane.

8.4.1 Fractional-Order Model for Cholesterol in Engineered Membranes

The fractional-order macroscopic model that we construct for a tethered membrane containing cholesterol is illustrated schematically in Figure 8.8. The model in Figure 8.8 is

---

[12] The relative permittivity is given by $\varepsilon_r = \varepsilon/\varepsilon_0$, where ϵ is the absolute permittivity and $\varepsilon_0 = 8.854187817 \times 10^{-12}$ F/m is the permittivity of vacuum.

[13] How cholesterol affects the dynamics of an artificial membrane is of significant importance. We will study this topic at several levels of abstraction. In §14.7 we construct coarse-grained molecular dynamics simulation models to determine how cholesterol affects an artificial membrane at the molecular level.

[14] Lipids rafts are discussed in Chapter 2. Briefly, they are large patches of lipids with a high concentration of cholesterol.

8.4 Modeling Membranes with Sterol Components

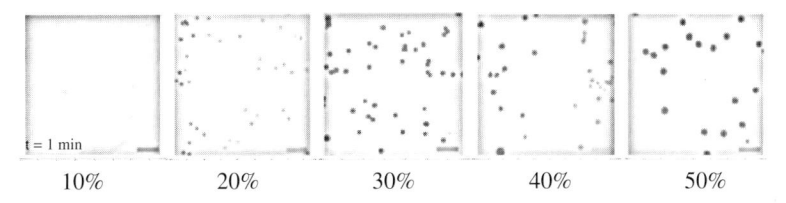

Figure 8.7 Fluorescence micrographs of engineered artificial cell membrane containing 10 to 50 percent cholesterol supported by a glass substrate as described in [412]. All fluorescence measurements are recorded after $t = 1$ min. The dark patches indicate the locations of lipid rafts containing high concentrations of cholesterol.

similar to that of the tethered membrane in Figure 8.1; however, it models two additional important functionalities in the membrane. The first is the *lipid domain*, which contains low concentrations of cholesterol. The second is the *lipid raft domain*, which contains lipid rafts comprised of high concentrations of cholesterol. Note that for low concentrations of cholesterol only the lipid domain is present as lipid rafts do not form in the engineered membrane. The construction of the fractional-order macroscopic model is similar to that presented in §8.2; however, now we construct an equivalent circuit model for each respective domain. Just as before, both the lipid domain and the lipid raft domain in the tethered membrane are composed of three distinct regions: the bioelectronic interface at the gold electrodes, the lipid domain or lipid raft domain, and the bulk electrolyte solution. The lipid and lipid raft domain are assumed to be polarizable and to also contain aqueous pores as a result of random thermal fluctuations; that is, random thermal fluctuations allow the energy barrier to be crossed for the conformational change of lipids or cholesterol, allowing the formation of transient aqueous

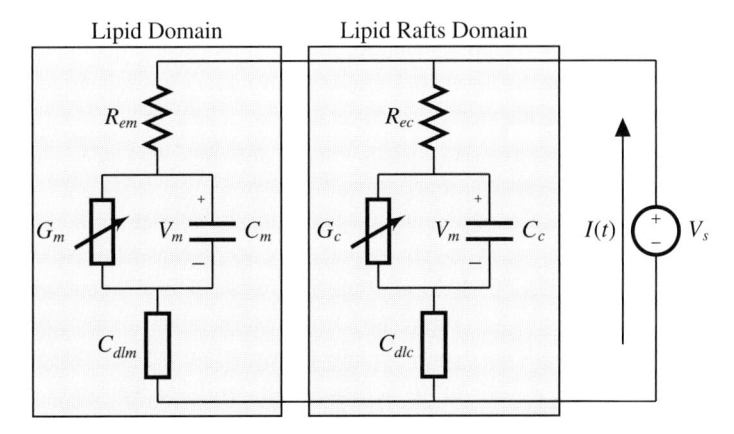

Figure 8.8 Fractional-order macroscopic model of an engineered membrane containing high concentrations of cholesterol. The electrical response of the engineered membrane can be modeled by the circuit shown in the figure. The lipid domain contains patches of tethered membrane that have low concentrations of cholesterol, and the lipid raft domain contains patches of tethered membrane with high concentrations of cholesterol. The circuit parameters are defined in §8.4.

pores. This allows the lipid domain to be modeled by the capacitance C_m in parallel with the conductance $G_m(t)$, and the lipid raft domain to be modeled using the capacitance C_c in parallel with the conductance $G_c(t)$. G_m and G_c are both dependent on the population of aqueous pores in each respective domain. The bulk electrolyte solution is assumed to be purely ohmic, which allows the electrolyte resistance for the lipid domain and lipid raft domain to be modeled by R_{em} and R_{ec}, respectively. The electrical double layer at the bioelectronic interface of the lipid domain and lipid raft domain is modeled using a capacitor if diffusion-limited charge transfer, reaction-limited charge transfer, and ionic adsorption dynamics are not present. If these double-layer charging effects are present then the bioelectronic interface can be modeled using a constant-phase element composed of a capacitance C_{dl} and the fractional-order operator p. The associated double-layer capacitance for the lipid domain and lipid raft domain are given by C_{dlm} and C_{dlc}, respectively, with the same order parameter p. The value of p must be equivalent for both the lipid domain and the lipid raft domain as the dynamics of the lipid membrane do not significantly contribute to the diffusion properties of ions adjacent to the bioelectronic interface. An excitation voltage $V_s(t)$ is applied across the two electrodes of the tethered membrane and the current response $I(t)$ is measured. Using the same techniques presented in §8.2, the associated impedance of the tethered membrane containing sterol components can be constructed based on the fractional-order macroscopic model in Figure 8.8.

8.4.2 Impedance Analysis of Engineered Membranes Containing Sterol Molecules

In this section we use the fractional-order macroscopic model for tethered membranes containing high concentrations of cholesterol, given by (8.8), to study how cholesterol impacts the dynamics of the tethered membrane. The inclusion of cholesterol creates two domains in the membrane consisting of: a cholesterol-rich domain (the lipid raft domain) and a cholesterol-poor domain (the lipid domain). The formation of these domains impacts both the dynamics and adsorption proteins and peptides in the membrane which is important for cellular signaling. This includes the production of bile acids, vitamin D, steroid hormones, and metabolism. Therefore, having a method to estimate the amount of cholesterol in a membrane provides an important tool for the detection of cholesterol homeostasis complications.

Figure 8.9 provides the experimentally measured and model-predicted impedance (from the fractional-order reaction-rate model) for a 10 percent tethered DphPC membrane containing 0, 10, 20, 30, 40, and 50 percent cholesterol concentrations, respectively. As seen from Figure 8.9, the results from the fractional-order model (8.8) are in excellent agreement with the experimentally measured impedance response from the tethered membrane. The results at 40 percent cholesterol concentration deserve special mention. Comparing the phase plot of the 40 percent cholesterol with the phase plot of the 0, 10, 20, 30, and 50 percent cholesterol membranes, we see that the local phase minimum of the 40 percent cholesterol occurs at an approximate frequency of $f = 0.8$ Hz, while the others have a local phase minimum at an approximate frequency of $f = 0.4$ Hz. Using the results in §8.4 that relate the frequency of the local phase

(a) Phase represented by $\angle Z(f)$ in degrees.

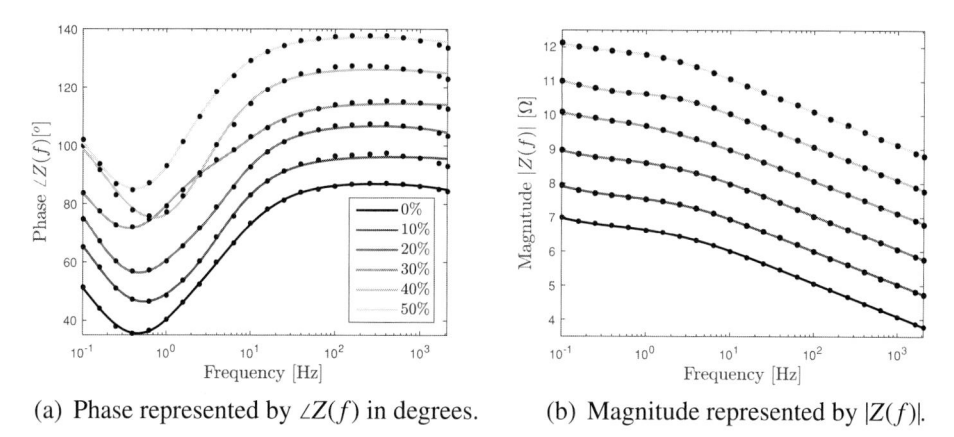

(b) Magnitude represented by $|Z(f)|$.

Figure 8.9 Experimentally measured (black dots) and model-predicted (solid line) phase and impedance for a 10 percent tethered DphPC membrane containing 0, 10, 20, 30, 40, and 50 percent cholesterol concentrations. The black line indicates the 0 percent cholesterol concentration, and the top light gray line indicates the 50 percent cholesterol concentration.

minimum to membrane conductance, this change in the frequency of the phase minima results because the membrane conductance of the 40 percent cholesterol is larger than the membrane conductance of the other membranes. The increase in conductance of the 40 percent cholesterol membrane suggests that this membrane has a lower stability than the 0, 10, 20, 30, and 50 percent cholesterol membranes. These results suggest that large concentrations of cholesterol in the archaebacterial membrane cause the disentanglement of the phytanyl chains in the DphPC and GDPE lipids. The disentanglement of the phytanyl chains results in the formation of membrane defects which cause an increase in the membrane conductance.

8.5 Complements and Sources

The electrical circuit model for the lipid membrane, interfacial capacitance, and electrolyte resistance in Figure 8.1 is similar to that used in electrophysiological models of cell membranes; see [156] for a textbook treatment. The conceptual idea behind electrophysiological models originates from the work of Cole, who pioneered the notion that cell membranes could be likened to an electronic circuit [81]. The key difference between the electric circuit models in [81, 156, 161] is that the fractional-order macroscopic model includes a constant-phase element, composed of a capacitance C_{dl} and fractional-order parameter p, to account for charge accumulation, diffusion-limited charge transfer, reaction-limited charge transfer, and ionic adsorption dynamics that may be present at electrode-to-electrolyte interfaces [164]. In electrochemistry, if $p = 0.5$ this constant-phase element is known as a Warburg impedance [11, 92, 181, 418], and for $0 < p \leq 1$ the constant-phase element is known as a generalized Warburg impedance. For engineered tethered membranes, p is in the range of 0.5 to 1.0.

The dynamics of the current response of the engineered tethered membrane is described by the fractional-order macroscopic model, which contains fractional-order derivatives. Fractional-order differential operators are used in several domains, including chemistry, physics, and bioengineering, to account for memory phenomena which are neglected using integer-order differential operators. The physical meaning of the fractional-order derivative is still an open problem; however, it has been proposed that the order of the fractional-order operator provides a measure of the memory of a process. For example, if $0 < p \leq 1$ then the fractional-order derivative is nonlocal and possesses infinite memory because it takes into account the entire history of input argument. Specifically for engineered tethered membranes, if $p = 1$ then no memory is present and the bioelectronic interface can be modeled using an ideal capacitor which neglects electrodiffusive dynamics. If these electrodiffusive dynamics are accounted for then $0 < p \leq 1$ as the process of ionic migration to the electrode surface contains memory. In Chapter 10 we provide a direct link between the continuum electrodiffusive dynamics and the reaction-rate fractional-order operator for modeling charge accumulation at the electrode surface. For an introduction to fractional-order operators and their applications, refer to the books [62, 92, 328].

8.6 Closing Remarks

This chapter has discussed reaction-rate models for engineered membranes. These models are useful for modeling the dynamics of tethered membranes that are excited at drive voltages less than about 50 mV where the process of electroporation is not present, and when a sinusoidal drive voltage is used. The models are applicable to mimicking the response of an engineered membrane on the time scale of several minutes. These models can be used to measure the structural integrity of the tethered membrane and the disulfide bonds of the tethers which anchor the membrane. The models account for the restricted diffusion dynamics that occur at the electrode surface and can model the dynamics of membranes containing sterol molecules. The models were validated using experimental data from engineered membranes with different tether densities, lipid compositions, and analyte solutions.

The fractional-order macroscopic models presented in this chapter provide the basis to study several interesting phenomena that occur in tethered membranes. These include measuring the dynamics of embedded ion channels, estimating the properties of cells grown the surface of the tethered membrane, and studying the dynamics of electroporation. To model these dynamics with the fractional-order macroscopic model requires the use of continuum and molecular dynamics models, which are the subjects of the subsequent chapters.

9 Reaction-Rate-Constrained Models for the ICS Biosensor

9.1 Introduction

Recall that the ion-channel switch (ICS) biosensor is a synthetic biological device built out of an artificial membrane, ion channels, spacers, and tethers. Its purpose is to detect the presence of analyte molecules in fluid. When it senses the presence of analyte molecules, the biosensor responds by changing its electrical impedance significantly. The amount the impedance changes by depends on the concentration of the analyte molecules in the fluid. Therefore, by examining the current flowing through the biosensor, one can detect both the presence of analyte and its concentration.

This chapter studies how the current flowing through the ICS biosensor can be modeled using a two–time scale nonlinear system of ordinary differential equations (for chemical reactions) coupled with the fractional-order macroscopic model of Chapter 8 (for electrical dynamics). The fractional-order macroscopic model of the tethered membrane accounts for both the electrical dynamics of the membrane and the restricted diffusion dynamics at the electrode surface (as detailed in Chapter 8), while the two–time scale nonlinear system of differential equations models the chemical reaction dynamics that occur on the membrane surface of the ICS bisensor. We call the resulting system a *reaction-rate-constrained model*. Such reaction-rate models are useful for modeling the dynamics of tethered membranes that are excited by potentials less than 50 mV (so that electroporation does not occur).

After developing reaction-rate-constrained models for the ICS biosensor (with μm^2 electrode size), we consider biosensors with very small electrodes (area on the order of 1×10^{-12} m$^2 = \mu$m^2). We construct a stochastic model for the current response of such microelectrode ICS (mICS) biosensors as a noisy finite-state Markov chain observed in noise, i.e., a hidden Markov model (HMM), where each state is associated with the number of conducting gramicidin A (gA) dimers. When the electrode size is very small, individual dimer events are recorded – which is a finite-state process – as opposed to the ensemble average of millions of such events in a large electrode, which gives an ordinary differential equation.

The typical current response of the ICS biosensor and mICS biosensor is illustrated schematically in Figure 9.1. All the models in this chapter are validated using real-world experimental data from engineered membranes with different tether densities, lipid compositions, and analyte solutions.

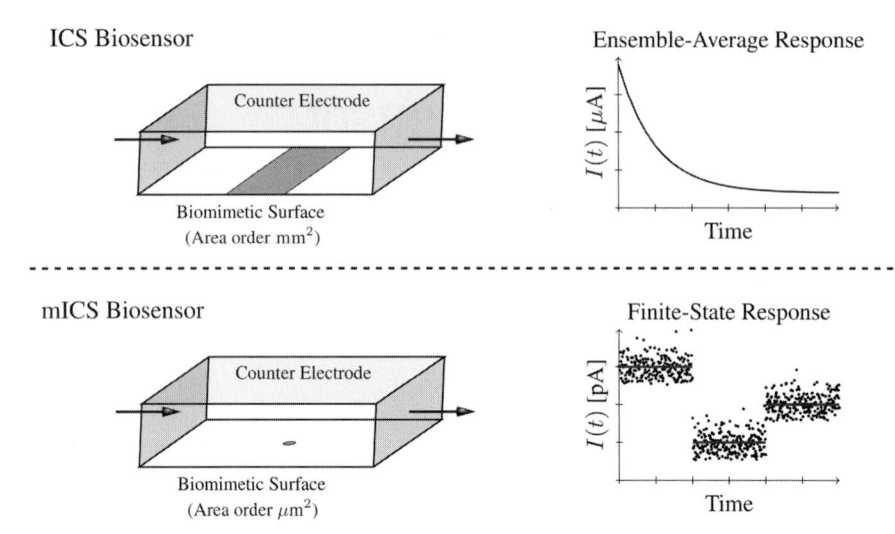

Figure 9.1 Schematic of the ICS biosensor and the microelectrode ICS (mICS) biosensor flow chambers. The biomimetic surface (dark gray region at the bottom) indicates the location of engineered membrane and gold electrode of the two biosensors. The ICS biosensor has a biomimetic surface area on the order of $mm^2 = 1 \times 10^{-6}$ m^2 while the mICS biosensor has a biomimetic surface area on the order of $\mu m^2 = 1 \times 10^{-12}$ m^2. In the ICS biosensor, the current response of the engineered membrane is a result of the dynamics of millions of gA dimers. In such cases the current response $I(t)$ is the result of the ensemble-average conductance of all the gA dimers. However, in the mICS biosensor the current response of the engineered membrane is controlled by the dynamics of a few gA dimers. In such cases the current response (illustrated by dots) can be represented by a finite-state response (lines) that depends on the number of conducting gA dimers in the membrane. Note that, as a result of Johnson–Nyquist noise, ionic concentration noise, and ion-channel shot noise, the exact number of conducting gA dimers cannot be directly measured from the current response.

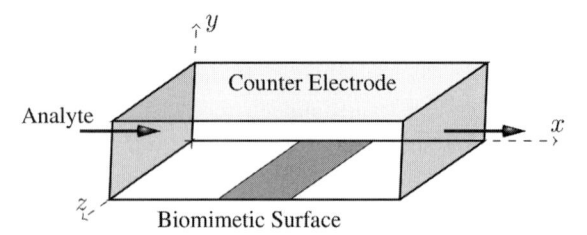

Figure 9.2 A color version of this figure can be found online at www.cambridge.org/engineered-artificial-membranes. Enlarged schematic of the flow chamber of the ICS biosensor. The biomimetic surface (dark gray region at the bottom) indicates the location of engineered membrane and gold electrode. Figure 5.3 on page 89 provides a detailed schematic of the engineered membrane of the ICS biosensor. The solution containing the analyte (target molecules) enters the flow chamber at $x = 0$ and flows along the x axis. Impedance measurements are performed between the gold electrode (below the dark gray region) and counter gold electrode (yellow region). The main assumption of this chapter is that the fluid flow has attained equilibrium and so the model of the biosensor does not involve the fluid flow dynamics. In comparison, Chapter 10 studies the nonequilibrium case where the fluid flow dynamics affect the sensing.

Before proceeding, it is important to emphasize that the ICS biosensor operates in a flow chamber as illustrated in Figure 9.2 (recall Chapter 5). The solution containing analyte (target molecules) enters the flow chamber at $x = 0$ with flow directed along the x axis. In this chapter we assume that the analyte concentration and/or fluid flow velocity is sufficiently large, so that the fluid flow (advection-diffusion) dynamics of the analyte can be neglected.[1] Thus we only need to consider the chemical reaction at the biomimetic surface of the ICS biosensor to estimate the impedance response as the concentration in the analyte flow chamber has reached its equilibrium value. So the surface chemical reactions are represented by a system of ordinary differential equations. In comparison, Chapter 10 considers the more complex case where P_e or D_a are large (fluid velocity is small or the analyte concentration is small). That is, in Chapter 10, the ICS biosensor operates in the transport-limited regime where we explicitly model dynamics of the fluid flow in the chamber as an advection diffusion partial differential equation.

9.2 Detection of Analyte Species in the Reaction-Rate Regime

This section constructs a reaction-rate-constrained model for the ICS biosensor to estimate the analyte concentration given the measured current flowing through the biosensor. The model combines the fractional-order macroscopic model for the electrical response with a system of nonlinear differential equations for the surface chemical reaction dynamics in the ICS biosensor. By exploiting the two–time scale property of the model, singular perturbation theory is then used to approximate the conductance of the tethered membrane as a function of the conducting dimer concentration and population of aqueous pores.

The main result of this section is that, in the reaction-rate regime, the electrical (and chemical) response of the ICS biosensor can be modeled as a system of nonlinear ordinary differential equations. We show that such models yield excellent predictive capabilities for the response of the ICS biosensor. In particular, the accuracy of the reaction-rate model is evaluated by comparing the model-predicted conductance response to the experimentally measured conductance for different analyte concentrations of the human chorionic gonadotropin (hCG) pregnancy hormone.

9.2.1 Aside: From Chemical Equations to Reaction-Rate Differential Equations

To model the chemical reactions in the ICS biosensor, we first review how to go from a chemical equation to the corresponding reaction-rate system of differential equations. Given two reactants (chemicals) A and B, consider a generic chemical reaction given by

$$nA + mB \underset{r}{\overset{f}{\rightleftharpoons}} C. \tag{9.1}$$

---

[1] This occurs if the Péclet number P_e is small, or the Damköhler number D_a is small. The Péclet number defines the ratio of diffusive flux to convective flux. The Péclet and Damköhler numbers are dimensionless constants that are widely used to characterize flows in chemical engineering and fluid dynamics. The Damköhler number defines the ratio of reaction flux to diffusive flux.

Here n is the number of molecules of chemical A and m is the number of molecules of chemical B that need to react to form the chemical product C. The constants f and r (above and below the arrows) denote the forward and reverse reaction rates, respectively.

How can we evaluate the rate of change of A, B, and C given the forward reaction rate f and the reverse reaction rate r? The first thing we need to compute is the rate R of the chemical reaction (9.1). The key point is that R satisfies a rate equation.[2] The rate equation gives the concentration change of the chemical product C in (9.1) as a function of time for the chemical reaction (9.1). Additionally, using the stoichiometric coefficients n and m in (9.1), we can relate the change in concentration of C to the change in concentrations of A and B. The complete system of nonlinear differential equations that characterize (9.1) are given by

$$\frac{dC}{dt} = R = fA^n B^m - rC,$$

$$\frac{dA}{dt} = -nR = -nfA^n B^m + nrC,$$

$$\frac{dB}{dt} = -mR = -mfA^n B^m + mrC. \qquad (9.2)$$

9.2.2 Reaction-Rate Model of the ICS Biosensor

We are now ready to construct a model for the electrochemical response of the ICS biosensor. The model consists of ordinary differential equations for the chemical reactions coupled with ordinary differential equations for the electrical response. The final summary of the key equations of the model are given at the end of this subsection.

The electrical response of the ICS biosensor consist of charge accumulation (refer to §3.6) at the bioelectronic interface and membrane surface, and the conductance of the membrane and electrolyte. These electrolyte dynamics can be accounted for using the fractional-order macroscopic model. Recall from Chapter 8 that the fractional-order macroscopic model consists of a system of fractional-order nonlinear differential equations, as illustrated in Figure 9.3. This is repeated below for convenience:

$$\frac{dV_m}{dt} = -\left(\frac{1}{C_m R_e} + \frac{G_m(t)}{C_m}\right) V_m - \frac{1}{C_m R_e} V_{dl} + \frac{1}{C_m R_e} V_s,$$

$$\frac{d^p V_{dl}}{dt^p} = -\frac{1}{C_{dl} R_e} V_m - \frac{1}{C_{dl} R_e} V_{dl} + \frac{1}{C_{dl} R_e} V_s, \quad p \leq 1,$$

$$I(t) = \frac{1}{R_e}(V_s - V_m - V_{dl}). \qquad (9.3)$$

Recall that a membrane that was not previously excited ($V_s(t) = 0$ for $t \leq 0$) will have initial conditions of zero transmembrane potential $V_m(0) = 0$ and zero double-layer potential $V_{dl}(0) = 0$. The key parameter that links the above electrical response with the chemical reactions is the time-varying conductance $G_m(t)$ which depends on the

---

[2] The rate equation links the forward and reverse reaction rates with the concentrations of the reactants. The units of the rate are mol/s (moles per second).

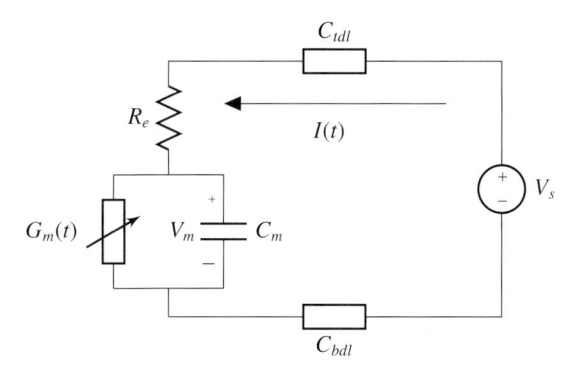

9.2 Detection of Analyte Species in the Reaction-Rate Regime

Figure 9.3 Lumped circuit macroscopic model of an engineered tethered membrane. The parameter R_e is the electrolyte resistance, C_m is the membrane capacitance, $G_m(t)$ is the membrane conductance, C_{bdl} is the electrode fractional-order capacitance, C_{tdl} is the counter electrode fractional-order capacitance, and V_m is the tethered-transmembrane electric potential difference. The total electrode fractional-order capacitance is given by C_{dl}, which is equal to the capacitance of C_{bdl} in series with C_{tdl} (see (8.2)). In the fractional model of order p, C_{dl} is a fractional-order capacitor with frequency response $1/(2j\pi f)^p$ at frequency f. Typical values of these parameters are specified in Table C.6 on page 418.

concentration of gA dimers present in the biosensor. The dimer concentration in turn depends on the target analyte concentration (which we wish to detect using the biosensor). To relate the membrane conductance $G_m(t)$ to the target analyte concentration requires a model that accounts for the electrodiffusive effects of the analyte in the electrolyte, and the surface reactions present on the tethered-membrane surface of the ICS biosensor.

The chemical reactions occurring at the ICS biosensor involve analyte molecules binding to the tethered antibody sites followed by a cross-linking of the mobile gA monomers to the captured analytes. The primary species involved in this process include the analytes a, binding sites b, mobile gA monomers c, tethered gA monomers s, and the dimers d, with respective concentrations $\{A, B, C, S, D\}$. Other intermediate chemical complexes that arise in the chemical reactions are denoted by $w, x, y,$ and z with concentrations $\{W, X, Y, Z\}$. The chemical reactions that relate these chemical species are described by the following set of reactions:

$$a+b \underset{r_1}{\overset{f_1}{\rightleftharpoons}} w, \quad a+c \underset{r_2}{\overset{f_2}{\rightleftharpoons}} x, \quad w+c \underset{r_3}{\overset{f_3}{\rightleftharpoons}} y, \quad x+b \underset{r_4}{\overset{f_4}{\rightleftharpoons}} y,$$

$$c+s \underset{r_5}{\overset{f_5}{\rightleftharpoons}} d, \quad a+d \underset{r_6}{\overset{f_6}{\rightleftharpoons}} z, \quad x+s \underset{r_7}{\overset{f_7}{\rightleftharpoons}} z. \tag{9.4}$$

In (9.4), r_i and f_i, for $i \in \{1, 2, 3, 4, 5, 6, 7\}$, denote the reverse and forward reaction rates for the chemical species $\{a, b, c, d, s, w, x, y, z\}$.

The chemical reactions in (9.4) give a complete symbolic description of the operation of the ICS biosensor that was qualitatively described in Chapter 5. Let us now parse the various components of (9.4). The forward part of the first equation reports on an analyte molecule a being captured by a binding site b and the resulting complex is denoted by w. The third equation in (9.4) says that a free-moving gramicidin monomer c in the outer leaflet of the bilayer lipid membrane (BLM) can bind to the complex w, thus producing another complex, denoted by y. An analyte molecule can also be captured by the binding site linked to the freely diffusing monomer, c. The second equation in (9.4) says that this results in the production of the complex x. The complex x can still diffuse on the outer leaflet of the BLM, and so can move toward a tethered binding site, b, and bind to it, resulting in the complex y (fourth equation in (9.4)). However, the complex x can diffuse on top of the tethered ion-channel monomer s, which results in the production of complex z (seventh equation in (9.4)). The event that determines the biosensor conductance (and thus the current flowing through the biosensor) is the binding of the free-moving ion-channel monomer c, and the tethered ion-channel monomer s. This results in the formation of a dimer d (fifth equation in (9.4)). Indeed, the biosensor conductance is proportional to the dimer concentration, i.e., $G(t) = \text{constant} \times D(t)$. Finally, an analyte molecule can also bind to an already formed dimer, which again produces the complex z (sixth equation in (9.4)).

Applying the procedure in §9.2.1, we can formulate the chemical reactions in (9.4) as a system of nonlinear ordinary differential equations with reaction rates given by

$$R_1 = f_1 AB - r_1 W, \qquad R_2 = f_2 AC - r_2 X,$$
$$R_3 = f_3 WC - r_3 Y, \qquad R_4 = f_4 XB - r_4 Y,$$
$$R_5 = f_4 CS - r_5 D, \qquad R_6 = f_6 AD - r_6 Z,$$
$$R_7 = f_7 XS - r_7 Z. \qquad (9.5)$$

Using the reaction rates (9.5), the time evolution of the chemical species in the ICS biosensor is specified by the system of nonlinear ordinary differential equations

$$\frac{du}{dt} = Mr(u(t)), \qquad (9.6)$$

where $M = \begin{pmatrix} -1 & 0 & 0 & -1 & 0 & 0 & 0 \\ 0 & -1 & -1 & 0 & -1 & 0 & 0 \\ 0 & 0 & 0 & 0 & 1 & -1 & 0 \\ 0 & 0 & 0 & 0 & -1 & 0 & -1 \\ 1 & 0 & -1 & 0 & 0 & 0 & 0 \\ 0 & 1 & 0 & -1 & 0 & 0 & -1 \\ 0 & 0 & 1 & 1 & 0 & 0 & 0 \\ 0 & 0 & 0 & 0 & 0 & 1 & 1 \end{pmatrix}$,

where $u(t) = [B, C, D, S, W, X, Y, Z]'$ denotes the concentration of the chemical species, and $r(u) = [R_1, R_2, \ldots, R_7]'$ is the vector of reaction rates. The matrix M in (9.6) is the stoichiometry matrix relating u and $r(u)$. The initial conditions of the ordinary differential equation (9.6) are given by

$$u(0) = [B(0), C(0), D(0), S(0), 0, 0, 0, 0]'. \qquad (9.7)$$

Note that (9.4) and (9.6) do not model the dynamics of the analyte concentration A. That is, we assume A is a constant. Accounting for the analyte dynamics, which do not occur in the reaction-rate-limited regime, requires continuum theories for electrodiffusive flow which are addressed in Chapter 10.

The membrane conductance $G_m(t)$ is linearly proportional to the gA ion-channel concentration $D(t)$. Therefore, the time evolution of $G_m(t)$ is computed from the solution of the fractional-order macroscopic model, defined in Chapter 8, and (9.6) given the initial concentration of the chemical species, reaction rates, diffusivity constant, and flow velocity, and assuming we are in the reaction-rate-limited regime. The reaction-rate-limited regime of the ICS biosensor is satisfied if large analyte flow rates of mL/min, micromolar analyte concentration, or low binding-site densities less than $10^8/\text{cm}^2$, are present.

Summary. The reaction-rate-constrained dynamics of the ICS biosensor are given by the fractional-order model (9.3), (9.6), and (9.7). Equation (9.3) accounts for the double-layer charging dynamics of the membrane and bioelectronic interface, as well as the electrolyte resistance. The membrane conductance $G_m(t)$ is governed by the surface reaction dynamics defined in (9.6) with the initial conditions given in (9.7). Given the input analyte concentration, the ICS biosensor model can be used to compute the time-evolving current response. Note that the reaction-rate model (9.3), (9.6), and (9.7) is useful when the ICS biosensor operates in the reaction-rate-limited regime and when the effects of electroporation are negligible. Methods for modeling the diffusion-limited regime are provided in Chapter 10, and those that account for electroporation are provided in Chapter 11.

9.2.3 Singular Perturbation Analysis of Dimer Concentration

For large analyte flow rates of mL/min, micromolar analyte concentration, or low binding-site densities less than $10^8/\text{cm}^2$, the assumption that the analyte concentration is approximately constant over space and time is reasonable. Then, singular perturbation theory (see [207]) can be used to approximate the time evolution of the dimer concentration $D(t)$ which determines the ICS biosensor conductance $G_m(t)$ (see also [221]).

The intuition behind singular perturbation theory is as follows. Consider the system of differential equations (f and g below are generic functions)

$$\frac{dx}{dt} = f(x, y), \quad \epsilon \frac{dy}{dt} = g(x, y), \tag{9.8}$$

where ϵ is a small positive real number. As ϵ approaches zero, $y(t)$ evolves much faster than $x(t)$, resulting in a two–time scale system. Indeed, setting $\epsilon = 0$, $y(t)$ satisfies the algebraic equation $g(x(t), y(t)) = 0$. Suppose the solution of this algebraic equation is $y(t) = h(x(t))$. Then the differential equation for x becomes $dx/dt = f(x, h(x))$, which is called the slow or quasisteady-state model.

To summarize: singular perturbation analysis says that for sufficiently small ϵ, the two–time scale system (9.8) can be approximated by the simplified system

$$\frac{dx}{dt} = f(x, h(x)), \quad y(t) = h(x(t)).$$

For a textbook treatment of singular perturbation in nonlinear systems see [201].

In order to apply singular perturbation analysis to the ICS biosensor dynamics, we first need to identify which variables evolve quickly and which variables evolve slowly. To do so we nondimensionalize the differential equations as described in Example 2 in Appendix A.3 on page 397. Then for the parameter values in Table C.1 on page 413, it can be seen that the decay rate of species Y and Z is much faster than the decay rate of the other species by analysis of the eigenvalues of the linearized version of (9.6). Let us denote the parameters $\gamma = \{Y, Z\}$ for the fast species, and $\beta = \{B, C, D, S, W, X\}$ for the slow species. We can then represent (9.5) and (9.6) as a two–time scale system,

$$\frac{d\beta}{dt} = f(\beta, \gamma), \quad \epsilon \frac{d\gamma}{dt} = g(\beta, \gamma), \qquad (9.9)$$

where f and g denote vector fields of the fast variables and the slow variables, and $\epsilon \approx \frac{1}{|\lambda_{max}|}$ with λ_{max} the largest eigenvalue of the linearized version of (9.6). Tikhonov's theorem combined with the approximation $S \approx S(0)$ allows for the simplification of the above two–time scale system of chemical dynamics equations (9.9) [201, §11.1]. The following theorem based on singular perturbation analysis provides an equation to evaluate the evolution of the biosensor conductance versus analyte concentration.

Theorem 1 ([221]). *Consider the two–time-scale system (9.9) for the dynamics of the chemical species. As $\epsilon \to 0$, the dimer concentration $D(t)$ converges to $\overline{D}(t)$ given by the following differential equation:*

$$\frac{d}{dt}\overline{D}(t) = -\overline{D}(t)(r_5 + f_6 A^*) + \left(f_5 C + \frac{r_6 f_7 X}{r_6 + r_7}\right) S(0) \qquad (9.10)$$

with constants r_5, r_6, r_7, f_5, f_6, and f_7 defined in (10.12), A^* the analyte concentration, and $S(0)$ the initial number of tethered gA monomers. Specifically, if the initial dimer concentration $D(0)$ at time $t = 0$ is within an $O(\epsilon)$ neighborhood of $\gamma = h(\beta)$, where $h(\beta)$ is the solution of the algebraic equation $g(\beta, \gamma) = 0$ for g given in (9.9), then for all time $t \in [0, T]$, $|D(t) - \overline{D}(t)| = O(\epsilon)$, where $T > 0$ denotes a finite time horizon. □

The proof of Theorem 1 is given in [201]. It can be shown that as $\epsilon \to 0$, the fast dynamics approach the quasisteady state, $h(\beta)$ defined in the theorem. This quasisteady state $h(\beta)$ of the fast variables β is then substituted in the slow dynamics in (9.9), which results in the following approximate dynamics for the slow species: $d\bar{\beta}/dt = f(\bar{\beta}, h(\bar{\beta}))$. We are interested in a specific component of $\bar{\alpha}$, namely, the approximate dimer concentration $\bar{D}$, which can be shown to evolve according to (9.10).

Using Theorem 1, the conductance of the ICS biosensor for different analyte concentrations A^* can be computed using (9.10) when the operation of the biosensor is in the reaction-rate-limited regime.

9.2.4 Detection of Human Chorionic Gonadotropin (hCG)

To illustrate the application of the ICS biosensor and predictive model (9.10) with (8.1) for estimating the analyte concentration, we compare the model-predicted membrane

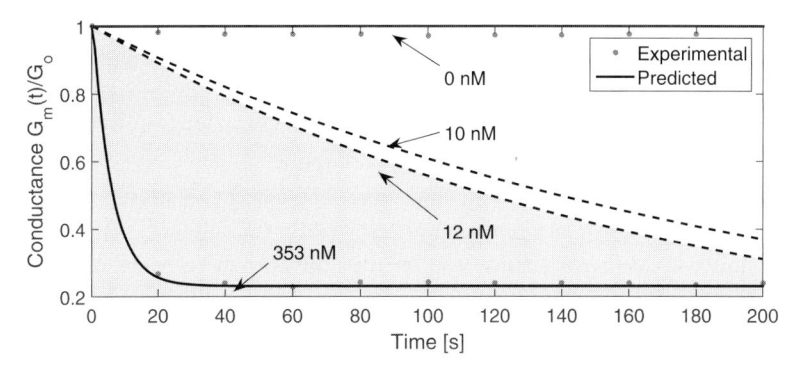

Figure 9.4 Experimentally measured (gray dots) and model-predicted concentration (solid black lines) of hCG present in an analyte solution using the ICS biosensor. The ground truths for the hCG concentrations were 0 and 353 nM. The dashed black lines for the 10 and 12 nM concentrations are associated with the a blastocyst or mammalian embryogenesis being present. Above 10 nM a mammalian embryogenesis is present; below 12 nM a mammalian embryogenesis is not present. Intermediate values can be associated with either a mammalian embryogenesis being present or not. All predictions are performed using the reaction-rate-limited model (9.10) and (8.1) with the parameters defined in Table C.1 on page 413.

conductance with the experimentally measured membrane conductance for the analyte hCG.[3]

Figure 9.4 presents the measured and model-predicted conductance of the ICS biosensor for the concentration estimation of hCG. The ground truths for hCG concentration were 0 and 353 nM, respectively. As seen from Figure 9.4, these two concentrations produce very different membrane conductance dynamics. From Figure 9.4, at $t = 40$ s, the change in the number of conducting gA dimers is negligible. Therefore, the membrane conductance is at its equilibrium value and results because only aqueous pores are present for $t > 40$ s. As seen from Figure 9.4, using the reaction-rate-limited dynamic model (9.10) in combination with the fractional-order macroscopic model (8.1) it is possible to estimate the concentration of hCG present.

9.3 Microelectrode ICS (mICS) Biosensor and Hidden Markov Model (HMM)

This section discusses a *micrometer*-sized electrode version of the ICS biosensor; we call this the mICS bisensor. Since the electrode size is μm^2 (micrometer × micrometer 1×10^{-6} m^2), individual dimers forming and dissociating need to be modeled. Our aim

---

[3] hCG is commonly called the pregnancy hormone. More technically, the concentration of hCG is an excellent indicator of the presence of blastocyst or mammalian embryogenesis (pregnancy). hCG concentrations above 10 nM in blood or urine indicate that a blastocyst or mammalian embryogenesis is present.

is to construct a black-box stochastic model[4] for the electrical response of the mICS biosensor. This is in comparison to previous sections, where the current response of the classical (millimeter-sized electrode) ICS biosensor was formulated as a deterministic reaction-rate ordinary differential equation model obtained as the ensemble average of the formation of millions of gA dimers; see Figure 9.1.

But there is an additional issue to resolve. The current response of the mICS biosensor can be interpreted as a random finite-state "digital" output signal corrupted by thermal noise. If no thermal noise was present, the current response of the mICS would be proportional to the number of conducting gA channels. However, in electrical and biological systems Johnson–Nyquist noise[5] (thermal noise) is present. Additionally in the mICS, the gA channels also introduce noise from ionic conductance noise[6] and ion-channel shot noise.[7] Therefore, a useful stochastic model and associated statistical signal processing algorithm is required to estimate the state (number of gA dimers) given the measured (noisy) current response.

In this section we construct a stochastic model called a hidden Markov model (HMM) for the mICS biosensor and then describe associated signal processing algorithms to estimate the number of gA dimers given the measured current.

9.3.1 Hidden Markov Model for mICS Biosensor

We consider discrete-time models (unlike the rest of the book, which uses continuous-time models) since we are eventually interested in formulating discrete-time statistical signal processing algorithms for numerical implementation.

Let $k = 0, 1, 2, \ldots$ denote discrete time. The discrete-time HMM[8] for the mICS biosensor has two ingredients:

1. Finite-state Markov chain. The formation and disassociation of gA dimers is modeled explicitly as a Markov chain. Assume that each individual lipid monolayer is composed of $(N-1)$ gA monomers. When a monomer in the bottom lipid layer couples with a monomer in the top layer a gA dimer is formed, which forms a conducting

---

[4] Recall that in §8.2 we constructed the fractional-order deterministic black-box model of the membrane. Here we construct a different, stochastic type of black-box model. The model is stochastic since in mICS we measure only a few dimer events at each time instant. In the standard ICS we measure several thousands of dimer events at each time and so the stochasticity of individual dimer events averages out.

[5] Johnson–Nyquist noise results from the random movement of ions and other charged molecules that result from thermal fluctuations. The variance of Johnson–Nyquist noise is proportional to temperature. A first-principles derivation of the Johnson–Nyquist noise distribution (Gaussian distribution) uses statistical physics and the fluctuation-dissipation theorem.

[6] Ionic conductance noise results from the random configuration (or shape) changes of the gA channels that result from thermal fluctuations.

[7] Ion-channel shot noise results from the discontinuous movement of ions through an ion channel. Ion-channel shot noise is present in the gA channel because only a discrete number of ions can pass through the ion channel at any given time. The number of discrete ions that pass through the gA channel is random.

[8] For a quick intuitive visualization of an HMM, please see Figure 9.5 on page 171. An HMM consists of a random finite-state (digital) signal depicted in black corrupted by random noise. Given the noisy measurements (irregular grey line) how can the finite-state signal be estimated? How can the parameters that determine the evolution of the finite-state signal be estimated?

channel through the membrane. Assume each such channel has a conductance G_p. The monomers couple and decouple typically at a millisecond time scale. Therefore, at each time instant k, the total current resulting from the conducting gA channels (dimers) can take on one of N possible levels $\{\mu_1, \ldots, \mu_N\}$ with $\mu_i = (i-1)V_m G_p$. Here, V_m is the transmembrane potential. The transition dynamics between these N possible current levels can be modeled using an N-state discrete-time Markov chain X_k with state space $i \in \{1, \ldots, N\}$ associated with the current levels $\{\mu_1, \ldots, \mu_N\}$. Denote the transition probabilities of this Markov chain as

$$a_{ij} = P(X_k = \mu_j | X_{k-1} = \mu_i), \quad i, j \in \{1, \ldots, N\}. \tag{9.11}$$

The transition matrix of the Markov chain is defined by A with elements given by a_{ij}. The initial distribution of the Markov chain is denoted by

$$\pi = [\pi_1, \pi_2, \ldots, \pi_N]' = [P(X_1 = \mu_1), P(X_1 = \mu_2), \ldots, P(X_1 = \mu_N)]', \tag{9.12}$$

where π is an $N \times 1$ column vector. (Recall that the prime denotes transpose.)

2. Noisy observations. The measured current Y_k is a noisy observation of the total current X_k at each discrete time instant k. The noise results from thermal noise, the antialiasing effect from sampling, and an open-channel noise (also known as $1/f$ noise). To model this correlated noise stochastic process we use an autoregressive Gaussian process that comprises a white Gaussian noise process[9] W_k filtered by an all-pole filter. The measured current Y_k is given by

$$Y_k = \sum_{i=1}^{M} h_i Y_{k-i} + X_k + \sum_{i=1}^{M} h_i X_{k-i} + W_k, \tag{9.13}$$

where h_i for $i \in \{1, \ldots, M\}$ are the parameters of the filter. Equivalently, denoting the transfer function of the all-pole filter as $H(q^{-1}) = 1 + h_1 q^{-1} + \cdots + h_M q^{-M}$, where q^{-1} is the delay operator, the observed current is

$$Y_k = X_k + \frac{W_k}{H(q^{-1})} \quad \text{or equivalently } H(q^{-1})Y_k = H(q^{-1})X_k + W_k. \tag{9.14}$$

The model (9.13) or equivalently (9.14) constitutes an HMM; it consists of a Markov chain observed in noise.

However, from a signal processing algorithm point of view the HMM (9.13) has a dimensionality problem. $1/f$ noise is a long memory process which requires choosing the filter order M to be large. The process $H(q^{-1})X_k$ in (9.14) is equivalent to a Markov chain with N^M states, and HMM signal processing algorithms will require the order of N^{2M} computations at each time instant. So at first sight the HMM (9.13) is not practically useful since the computational cost of the associated signal processing algorithms increases exponentially with the long memory dimension M.

The dimensionality problem can be resolved by the following approximation which exploits time-scale separation: the formation and splitting of gA dimers is a much slower

---

[9] By zero mean, we mean an independent and identically distributed sequence of Gaussian random variables $\{W_k\}, k = 0, 1, \ldots$.

process relative to the sampling rate and most of the power of the stochastic process resides at direct current (zero frequency). Then one can closely approximate (9.14) with the following HMM:

$$Y_k = \frac{X_k + W_k}{H(q^{-1})} \text{ or equivalently } Y_k = \sum_{i=1}^{M} h_i Y_{k-i} + X_k + W_k. \quad (9.15)$$

Summary. We now have a tractable HMM (9.15) for the mICS biosensor; an N-state Markov chain X_k observed in noise via the measurement process Y_k. It is convenient to model the observation noise W_k as state-dependent noise; that is, the noise variance is σ_i^2 when the Markov chain is in state $i \in \{1, 2, \ldots, N\}$.

Writing $h = [1, h_1, \ldots, h_M]'$, $\sigma^2 = [\sigma_1^2, \ldots, \sigma_N^2]$, and $\mu = [\mu_1, \ldots, \mu_N]$, the mICS biosensor is completely specified by the HMM with parameters $\theta = (A, \pi, \mu, \sigma^2, h)$.

9.3.2 Hidden Markov Model Statistical Signal Processing

Given the HMM stochastic model of the mICS biosensor, we are interested in solving the following two classical statistical signal processing problems:

Maximum likelihood parameter estimation. Given the sequence of observations $Y_1, \ldots, Y_t$ from the mICS biosensor, compute the maximum likelihood estimate (MLE)

$$\theta^* = \arg\max_{\theta} P(Y_1, \ldots, Y_T | \theta). \quad (9.16)$$

That is, compute the parameters θ^* that best fit the data $Y_1, \ldots, Y_T$ in terms of the likelihood function. The maximum likelihood criterion (9.16) is widely used for estimating the parameters of an HMM. The optimization problem (9.16) is nonconvex in general. A local maximum θ^* can be obtained by using a general-purpose numerical optimization algorithm, or the well-known expectation maximization algorithm. Once the MLE is computed, we have useful information about the gA dimer association and disassociation rate from the transition probabilities.

State estimation. Suppose that the HMM parameters θ are known or the maximum likelihood estimate θ^* has been computed. The state estimation problem involves estimating the unobserved current response X_k at each time instant $k = 1, 2, \ldots$, given the HMM observations $Y_1, \ldots, Y_k$ from the mICS biosensor. The state estimate is computed using the hidden Markov model filtering algorithm which yields the conditional mean state estimate[10] $\hat{X}_k = E[X_k | Y_1, Y_2, \ldots, Y_k]$. Numerical implementation of the HMM filter requires the order of N^2 multiplications at

---

[10] Here $E[X_k | Y_1, Y_2, \ldots Y_k]$ denotes the conditional expectation (conditional mean) of X_k given observations $Y_1, \ldots, Y_k$. The conditional mean state estimate is optimal in the minimum mean-square (minimum variance) sense and is widely used in signal processing. In simple terms, given the data $Y_1, \ldots, Y_k$, we seek an estimator $g(Y_1, \ldots, Y_k)$ of the state X_k to minimize the mean-square error $E[(X_k - g(Y_1, \ldots, Y_k))^2]$. It is easily shown that the choice of estimator $g(Y_1, \ldots, Y_k)$ that minimizes this mean-square error is the conditional mean estimate $E[X_k | Y_1, Y_2, \ldots, Y_k]$.

(a) 0 M of MTOP present. **(b) 0.5 μM of MTOP present.**

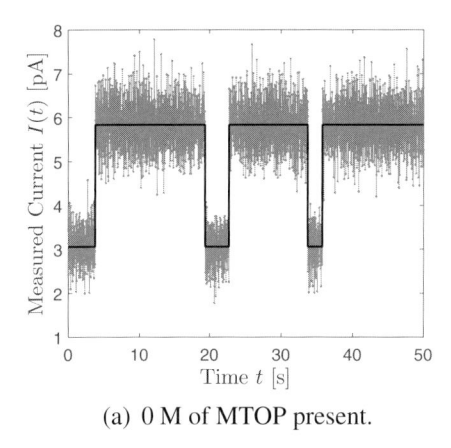

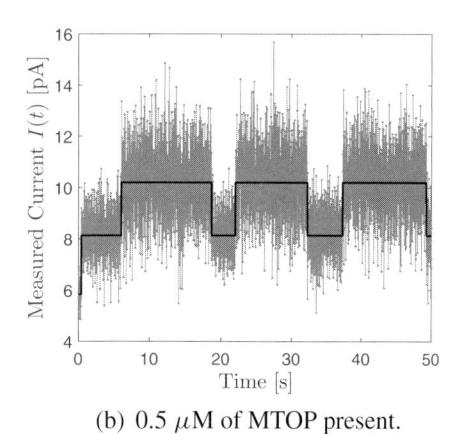

Figure 9.5 Experimentally measured (gray dots) and model-predicted (solid line) current response of the mICS biosensor to different concentrations of MTOP. All numerical predictions are computed using the HMM filter algorithm based on the stochastic model of the mICS biosensor.

each time instant to compute the state estimate. Alternatively, the Viterbi algorithm can be used which yields the maximum likelihood state sequence estimate. The state estimate $\hat{X}_k$ directly yields the estimated number of dimers formed in the mICS biosensor at each time k. It allows detecting the presence of analytes.

Since state and parameter estimation for HMMs are well-studied classical problems in signal processing, we will not provide the algorithms here; please refer to [216, 334] for the precise equations and implementation details.

9.3.3 Detection of Monoterpene Oxidation Product (MTOP)

In this section we compute the MLE θ^* of the HMM parameters and then use the HMM filter (conditional mean) estimate $\hat{X}_k$ to detect for the presence of the analyte monoterpene oxidation product (MTOP); for further discussion including detailed evaluation of how well the HMM fits the mICS data, see [219]. The detection method will be based on the measured current response of the mICS biosensor between the case with zero MTOP and with 50 μM of MTOP.

The mICS biosensor response is recorded using an applied voltage $V_s = 50$ mV. The electrolyte solution is composed of 0.5 M NaCl. The experimentally measured and model-predicted current response for MTOP is illustrated in Figure 9.5. The solid black line in Figure 9.5 indicates the most likely current level as a function of time. The lowest current is measured at a value of 3.057 pA. Is this current associated with the conductance of a gA channel and/or the current resulting from aqueous pores in the membrane? The conductance of the membrane at a current $I = 3.057$ pA can be computed using $G_m = I/A_m V_m$ and has a value of $G_m \approx 19.5$ S/m^2; a typical membrane with no pores has a conductance in the range of 0.04 S/m^2 to 1.00 S/m^2. Therefore, this current is primarily a result of a conducting gA channel. Comparing the current response from the mICS biosensor with no MTOP present and with MTOP present in Figure 9.5, we see that more gA channels are conducting with MTOP present. Therefore, we can construct a detection method based on a threshold of the current response of the mICS biosensor that only depends on having a few gA channels open simultaneously. The use of the stochastic model is crucial for estimating the value of this detection threshold and for estimating the associated current response from noisy measurements from the mICS biosensor.

9.4 Complements and Sources

In this chapter we constructed reaction-rate-limited models of the ICS biosensor. The chemical kinetics in §9.2 result in a system of nonlinear ordinary differential equations. The book [229] is an excellent example of such chemical kinetics and binding. Similar models have been adopted in a lateral flow bioreactor in [332]. Though the system of nonlinear ordinary differential equations can be solved numerically, it is possible to use singular perturbation methods to approximate the ICS biosensor system of nonlinear ordinary differential equations into a first-order linear ordinary differential which has an algebraic solution. This approximation is accurate for large analyte flow rates of mL/min, micromolar analyte concentration, or low binding-site densities less than $10^8/\text{cm}^2$. The singular perturbation methods used in §9.2 are well known in nonlinear systems theory; see [201] for a textbook treatment. More sophisticated stochastic singular perturbation methods are studied in [225].

For the ICS biosensor the electrode size is on the order of mm^2 (1×10^{-6} m^2) which means that the membrane conductance is a result of an ensemble of thousands of gA ion channels. However, for the mICS biosensor the electrode size is on the order of μm^2 (1×10^{-12} m^2), in which case the formation of individual conducting gA ion channels cause a measureable change in the membrane conductance. In such cases we cannot use deterministic models for the membrane conductance; we must use statistical signal processing techniques. The statistical signal processing of ion-channel currents in §9.3 has been an active area of research with several papers since the 1990s. HMMs have been widely used for extracting ion-channel currents given noisy measurements; see [73, 76, 219] and references therein. One of the major difficulties in the construction of HMMs is the reliable identification of the conductance states and the transitions between them [380]. Recently, hierarchical HMMs have been proposed that account for both the transition between different conducting states of the ion channels, and the stochastic opening and closing within each state [381]. For detailed descriptions of HMMs see [63, 334].

9.5 Closing Remarks

This chapter studied reaction-rate-limited models of the ICS biosensor. Despite the complexities in the dynamics of the ICS biosensing device (moving gramicidin channels in an engineered membrane containing tethers and spacers and a sophisticated bioelectronic interface), the reaction-rate model is surprisingly accurate at the macroscopic level. The model accounts for the charging dynamics of the electrical double layers, conductance dynamics of the tethered membrane, and the surface reactions that occur in the ICS biosensor. Reaction-rate-limited models are applicable to the ICS biosensor dynamics when the magnitude of the excitation potential $V_s(t)$ is less than about 50 mV where the process of electroporation is not present, and when a sinusoidal excitation potential is used. Additionally, it is assumed that large analyte flow rates of mL/min, micromolar analyte concentration, or low binding-site densities less than $10^8/cm^2$ are present. We also presented future research directions on using 1-μm radius electrodes with tethered membranes for "digital" analyte detection and illustrated how noisy finite-state Markov chains can be used to detect if the analyte is present or not. All the models in this chapter are validated using real-world experimental data from ICS biosensors.

The reaction-rate-limited models of the ICS biosensor presented in this chapter provide a basis to study several interesting phenomena that occur in tethered membranes. These include the reaction dynamics of pore-forming chemical species. However, in the case that a large excitation voltage, or one of the assumptions related to analyte flow rate, analyte concentration, or binding-site densities is violated, it is necessary to use continuum and/or molecular dynamics models, which are the subject of the subsequent chapters.

10 Diffusion-Constrained Continuum Models of Engineered Membranes

10.1 Introduction

In this chapter (and the following three chapters) we study mesoscopic mathematical models for the dynamics of engineered membranes. This chapter constructs mesoscopic models for the following two synthetic biological devices built out of artificial membranes:

(i) the ion-channel switch (ICS) biosensor and
(ii) the pore formation measurement platform (PFMP).

Mesoscopic models deal with physical phenomena at the micrometer length scale and millisecond time scale – their level of abstraction lies between the macroscopic reaction-rate models and the microscopic atomistic models. To put this chapter into perspective, recall that Chapter 4 dealt with the construction of engineered membrane devices. Also, Chapters 8 and 9 constructed *macroscopic* reaction-rate models of these devices which comprised ordinary and fractional-order differential equations. At the end of this chapter we illustrate how approximations of the continuum models can be used to construct the parameters of the macroscopic models in Chapters 8 and 9.

As discussed in Chapter 9 the ICS biosensor operates in either the *mass-transport-influenced kinetics* or the *reaction-rate-limited kinetics* regime. A schematic of the ICS biosensor and PFMP in the flow chamber is provided in Figure 10.1. If the analyte concentration and/or fluid flow velocity is sufficiently large, the advection-diffusion dynamics of the analyte can be neglected and only the macroscopic surface reaction dynamics govern the impedance response of the ICS and PFMP. These reaction-rate-limited kinetics models were the focus of Chapter 9. In this chapter we consider the mass-transport-influenced kinetics regime, where the advection-diffusion dynamics cannot be neglected in the fluid flow chamber. Therefore, the dynamic mesoscopic models involve diffusion-type partial differential equations (PDEs) with boundary conditions determined by reaction-rate-type ordinary differential equations (ODEs).

Figure 10.1 illustrates the mesoscopic model which couples the electrolyte dynamics with the surface reactions. The electrolyte dynamics involve the advection-diffusion of molecules and ions, and may also include electrodiffusion effects if the molecules and ions are charged. The study of the electrodiffusion of charged particles such as molecules and ions in the presence of an applied external electric field is of great

Flow Chamber

[Diagram: 3D flow chamber with Counter Electrode on top, dimensions labeled L, W, h, axes x, y, z, with Analyte entering from the left.]

Advection-Diffusion Equation

$$\frac{\partial c}{\partial t} = \nabla \cdot \left[D\nabla c - v\nabla c \right]$$

Analyte c_o, Q

$$n \cdot D\nabla c = \frac{ds}{dt}$$

$$\frac{ds}{dt} = R_s = Mr(s, c)$$

Surface Reaction Dynamics

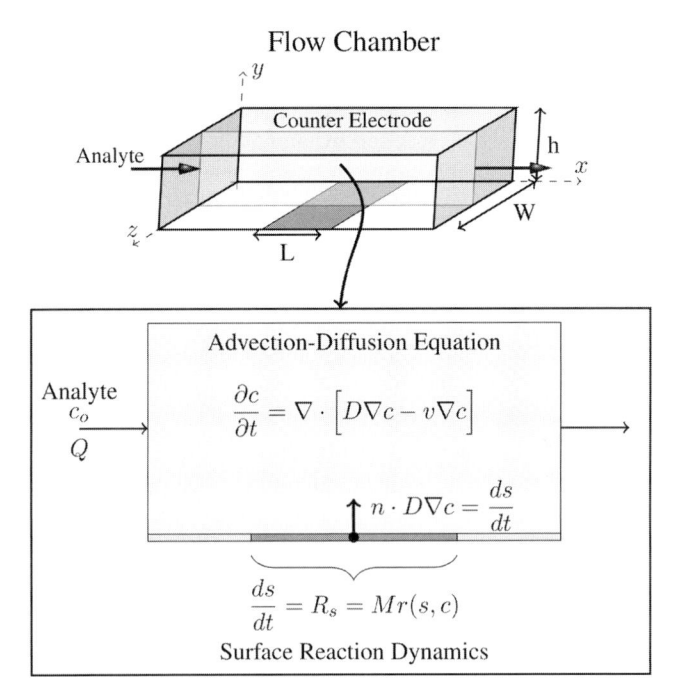

Figure 10.1 A color version of this figure can be found online at www.cambridge.org/engineered-artificial-membranes. Schematic of the flow chamber of the ICS biosensor and the PFMP. The biomimetic surface (dark gray region at the bottom of flow chamber with length L) indicates the location of the engineered membrane and gold electrode. The electrolyte solution containing the analyte (target molecules) enters the flow chamber at $x = 0$ and flows along the x axis. Impedance measurements are performed between the gold electrode (dark gray region) and counter gold electrode (yellow region). Typically the flow chamber has dimensions $L = 0.7$ mm, $h = 0.1$ mm, and $W = 3$ mm. The total length of the flow chamber in the x coordinate is 6 mm. The light gray region indicates a slice of the flow chamber that will be used for modeling the dynamics of the ICS and PFMP.

Unlike Chapter 9, in this chapter we explicitly consider the fluid flow dynamics of the analyte molecules. The advection-diffusion partial differential equation models the fluid flow dynamics of the analyte concentration c over space and time. The parameters are D, diffusion coefficient of the analyte; v, velocity of the analyte in the flow chamber; Q, input flow rate; and c_o, input analyte concentration to the flow chamber.

The surface reaction dynamics of surface species s are modeled by an ordinary differential equation with stoichiometric matrix M, and reaction-rate equation $r(s, c)$ which is dependent on the surface concentration and concentration of analyte c adjacent to the surface. The equations are coupled via the nonlinear boundary condition with n the normal vector pointing into the electrolyte domain.

importance in a number of disciplines. For example, in semiconductors these mesoscopic models describe the dynamics of electrons and holes and are used for the design of modern electronic components such as transistors, diodes, and infrared lasers [80]. In the biological applications, the process of interest is the flow of inorganic ions (K^+, Na^+, Ca^{2+}, Cl^-, etc.) through aqueous pores in lipid bilayer membranes. Additionally, the study of organic molecules and pore-forming peptides and proteins is also of importance. Biochemical signal communication in living organisms heavily relies on the transport and concentration of both inorganic and organic species.

The main idea behind constructing continuum models for the ICS biosensor and PFMP is to ensure that the surface reaction dynamic equations adequately account for the chemical reactions that occur. For the construction of both the ICS biosensor and the PFMP the electrolyte dynamics are coupled to the surface dynamics via a "Langmuir–Hinshelwood"-like equation[1] which is classically used to model surface adsorption dynamics. Additionally, from Chapter 9, we know the parametric structure of the chemical reactions at the surface of the ICS biosensor. However, for the PFMP we do not know the chemical reactions present that lead to pore formation. In this chapter we illustrate how a least-squares estimator can be used to estimate the chemical reactions present.

Finally, the continuum models of the ICS biosensor and PFMP proposed in this chapter are validated using real-world experimental data for the detection of streptavidin, thyroid-stimulating hormone (TSH), ferritin, and human chorionic gonadotropin (hCG), and estimating the pore formation dynamics of peptidyl-glycine-leucine-carboxyamide (PGLa).

10.2 Mass Transport versus Reaction-Rate-Limited Kinetics

The ICS biosensor, illustrated schematically in Figure 10.1, operates in one of two possible regimes: mass-transport-influenced kinetics and reaction-rate limited kinetics. The behavior of the biosensor is very different in these regimes. The operating regime depends on the dynamics of the surface reactions $Mr(s, c)$, diffusion of the analyte D, and the velocity field v of the analyte in the flow chamber. Determining the operating regime is crucial for accurately modeling the dynamics of the ICS biosensor.

10.2.1 Damköhler and Péclet Numbers

We now discuss the conditions for each operating regime in terms of two important dimensionless constants, the Damköhler and Péclet numbers.

Consider the boundary of the membrane surface illustrated in Figure 10.1 which links the analyte concentration c to the surface-bound analyte concentration s via the flux boundary condition $n \cdot D\nabla c = ds/dt$. The flux boundary condition states that the rate of arrival of analyte molecules c on the membrane surface equals the rate of the surface-binding reaction $R_s = Mr(s, c)$. Then the surface reaction satisfies a Langmuir binding mechanism[2] if

$$\frac{ds}{dt} = R_s = k_f c(s_{\max} - s) - k_r s, \qquad (10.1)$$

---

[1] The Langmuir–Hinshelwood equation describes the absorption of molecules on a surface via the following chemical reactions: $c^i + c^i_m \rightleftharpoons c^i_s$, $c^i_s + c^j_s \to c^k_s$, where c^i is the concentration of the ith electrolyte species $i \in \{1, \ldots, n\}$, c^i_m is the concentration of surface species that i can bind with, and c^k_s represents the surface concentration of the chemical product of the reaction of two surface-bound species.

[2] The Langmuir binding mechanism is described by the chemical reaction $c \underset{r_1}{\overset{f_1}{\rightleftharpoons}} s$, where c is the analyte concentration and s is the surface-bound analyte concentration.

Table 10.1 Possible ICS biosensor operating regimes, namely, mass-transport-influenced regime and reaction-rate-limited regime.

	$D_a \ll 1$	$D_a \gg 1$
$P_e \ll 1$	Mass Transport	Reaction Rate
$P_e \gg 1$	Mass Transport	Mass Transport

where k_f and k_r are the forward and reverse binding reaction rates, s is the surface concentration of bound analyte, and c is the analyte concentration.

The Damköhler and Péclet numbers are important dimensionless constants that characterize the fluid dynamics of the analyte. For the advection-diffusion equation in Figure 10.1 and the boundary condition with reaction rate R_s (10.1), the Damköhler number D_a and the Péclet number P_e (refer to Appendix A for details) are, respectively,

$$D_a = \frac{\tau_D}{\tau_R} = \frac{k_f c_o h^2}{D}, \quad P_e = \frac{\tau_D}{\tau_C} = \frac{Q}{DW}, \qquad (10.2)$$

where c_o, Q, D, h, and W are defined in Figure 10.1 and k_f is defined below (10.1), τ_D is the characteristic mixing time, τ_R is the characteristic surface reaction time, and τ_C is the characteristic time of convection. Here the Damköhler number D_a defines the ratio of mixing time to surface reaction time, and the Péclet number P_e defines the ratio of mixing time to advection time to travel across the channel width h.

10.2.2 Characterization of Operating Regime

The Damköhler number provides information about the surface-binding kinetic regime of the biosensor, and the Péclet number provides information about the analyte transport regime in the flow chamber of the biosensor. The Damköhler and Péclet numbers determine the operating regime of the ICS biosensor as described in Table 10.1.

We now discuss the four operating regimes of Table 10.1.

(i) $D_a \ll 1$ indicates that the chemical reactions on the ICS biosensor surface are in the reaction-limited surface-binding kinetic regime. This occurs if the characteristic time for an analyte molecule to reach the ICS biosensor surface is significantly faster than the characteristic time of the binding reactions of the analyte to the biosensor surface. Hence, the surface-binding reaction is the rate-limiting step.

(ii) $D_a \gg 1$ indicates that the chemical reactions on the ICS biosensor surface are in the mass-transport-limited kinetic regime. In this case, the diffusion of molecules to the sensor surface is the rate-limiting step.

(iii) $P_e \ll 1$ indicates that the analyte dynamics in the flow chamber are in the advection-limited analyte-transport regime. This occurs if the characteristic time for the analyte molecules to reach the sensor surface is significantly less than the characteristic advection time for the molecules to travel the same distance parallel to the sensor surface. Hence, in this case the transport of analyte molecules in the flow chamber is advection limited.

(iv) $P_e \gg 1$ indicates that the analyte dynamics in the flow chamber are in the mass-transport-limited analyte-transport regime. In this case the time for analyte molecules to traverse the flow chamber (parallel to the biosensor surface) is significantly faster than the characteristic diffusion time of the molecules to reach the sensor surface. In such cases, the bulk analyte concentration in the flow chamber remains approximately constant.

Though each of the above operating regimes plays an important role in the dynamics of the ICS biosensor, our interest is in whether the operation of the ICS biosensor is in the reaction-rate-limited regime or in the mass-transport-limited regime. For the ICS biosensor to operate in the reaction-rate-limited regime, the surface-binding kinetics must be reaction limited and the bulk analyte concentration must remain approximately constant. This occurs if $D_a \ll 1$ and $P_e \gg 1$, where large input flow rates Q (milliliters per minute), large analyte concentrations c_o (micromolar concentrations), and low binding-site densities (less than 10^8) are present. In the reaction-limited regime, the dynamics of the ICS biosensor can be described by a system of fractional-order differential equations; refer to §9.2 for details. If either $D_a \gg 1$ or $P_e \ll 1$ occurs, then the ICS biosensor operates in the mass-transport-limited regime.

Remark: The Damköhler number provides a rough approximation for the binding kinetic regime (reaction limited or mass transport limited) of the biosensor, and the Péclet number provides a rough approximation for the analyte transport regime (advection limited or mass-transport limited) in the flow chamber. Different expressions can be formulated for the Damköhler and Péclet numbers depending on the approximations used to construct the characteristic times (or fluxes), surface chemical reactions, geometry, and boundary conditions in which the advection-diffusion equation is evaluated. For example, the Damköhler and Péclet numbers in (10.2) were constructed using nondimensionalization constants of the advection-diffusion equation with a Langmuir surface-binding mechanism (refer to Appendix A for details). Equivalently, we could have constructed these constants using the characteristic time scales of diffusion, advection, and surface-binding reactions. In §10.4 we formulate the Damköhler number in terms of the ratio of initial reactive flux to the diffusive flux based on the analyte depletion region above the sensor surface. Though different expressions can be constructed for the Damköhler and Péclet numbers, the key idea is that the Damköhler number provides a measure of the analyte-binding kinetic regime on the biosensor surface, and the Péclet number provides a measure of the analyte-transport regime in the flow chamber. Both Damköhler and Péclet numbers are required to evaluate if the biosensor operates in the reaction-limited regime.

10.3 Mass-Transport-Limited Model of the ICS Biosensor Dynamics

Consider the ICS biosensor in a fluid flow chamber as illustrated in Figure 10.1. This section constructs a mass-transport-limited model of the ICS biosensor for estimating the concentration of analyte species given the current-voltage response. Recall from

10.3 Mass-Transport-Limited Model of the ICS Biosensor Dynamics

```
Q, c_o^a, c_s^b, c_s^c,        ┌──────────┐
c_s^s, c_s^d        ─────────▶│   ICS    │─────────▶ Ĝ_m(t)
                                │ Biosensor│
Input Parameters                └──────────┘        Membrane Conductance
```

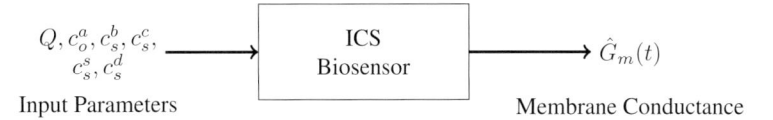

Figure 10.2 Black-box model of the ICS biosensor which is used estimate the membrane conductance $\hat{G}_m(t)$ as a function of the parameters Q, c_o^a, c_s^b, c_s^c, c_s^s, and c_s^d. Q is the input flow rate, c_o^a is the input analyte concentration, c_s^b is the binding-site density, c_s^c is the concentration of mobile gA monomers, c_s^s is the concentration of tethered gA monomers, and c_s^d is the concentration of gA dimers.

§9.2 that the detection of analyte molecules by the ICS biosensor is performed by detecting changes in the engineered membrane conductance $G_m(t)$ resulting from the molecules of interest binding to the engineered binding sites. As molecules bind, the population of conducting gramicidin A (gA) dimers decreases, which decreases the total conductance of the tethered membrane. The main aim of the mass-transport-limited-model is to specify $G_m(t)$ as a function of the input analyte concentrations, flow rate, and chemical reactions present on the surface of the membrane. The mass-transport-limited model is depicted schematically in the black-box model of Figure 10.2 with $\hat{G}_m(t)$ representing the estimated conductance.

The movement of analyte molecules in the electrolyte solution of the ICS biosensor flow chamber is governed by advection-diffusion and electromigration processes. Additionally, the analyte molecules bind through surface reactions on the surface of the ICS biosensor. Below we illustrate how these physical and chemical processes can be accounted for using partial differential equations. Then, using experimental measurements with the constructed mass-transport-limited model, we illustrate how the analyte concentration can be accurately estimated.

The reader who wishes to skip the background material should consult the schematic in Figure 10.3 and then jump to (10.10) on page 183 and subsequent material for the main ICS model. The final model is (10.14).

```
                    n · J^a = 0
           ┌─────────────────────────────┐
           │                             │
Q ⇒∂Ω_in   │  ∂c^a/∂t = −∇·(J^a) − v(z)∇c^a  │ h
           │  v(z) = (6Q/Lwh)(z/h)(1 − z/h)  │ n·J^a = 0
  z axis, n│                             │
     ∂Ω_b  └─────────────────────────────┘ ∂Ω_b
                      ∂Ω_surf
                        L
```

Figure 10.3 Schematic of the ICS biosensor flow chamber model. The analyte fluid enters the flow chamber at surface $\partial\Omega_{\text{in}}$ with a flow rate of Q. $\partial\Omega_{\text{surf}}$ denotes the sensing surface comprised of the tethered membrane, and $\partial\Omega_b$ denotes the boundary points of the membrane. n denotes the inward normal vector from the surface. Typical numerical values for the dimension of the flow chamber and parameters are given in Table C.1 on page 413.

10.3.1 Poisson's Equation: Electrostatics

In engineered membranes, the analyte solution is composed of ions and analyte molecules that may have nonzero net charge (due to the presence of charged ions or molecules). For example, the Na$^+$ and Cl$^-$ ions have an elementary charge[3] of $+1$ and -1, respectively. All the charged molecules in the engineered membrane exert a force on each other, which contributes to their dynamics. In this section we illustrate how the force between charged molecules can be estimated using Poisson's equation.

In general, ions and charged molecules interact through an electric field and a magnetic field – this is known as an electrodynamic interaction. A general description of the electrodynamic interaction of charged molecules is given by Maxwell's equations[4] in combination with the Lorentz force[5] (the force exerted by the electric and magnetic fields on the molecules). However, for engineered membranes, only the electric field contributes to the electrodynamics of ions and charged molecules in the engineered membrane. This allows us to approximate the electrodynamic interaction of ions and charged molecules using Poisson's equation in combination with Coulomb's force $F = qE$ (discussed in Chapter 2).

Poisson's equation determines the electric potential $\phi(x, t)$ resulting from the charge density $\rho(x, t)$ and is given by

$$\nabla \cdot (\varepsilon \nabla \phi(x, t)) = -\nabla \cdot (\varepsilon E(x, t)) = -\rho(x, t), \tag{10.3}$$

where ε is the dielectric permittivity, $E(x, t)$ is the electric field, and $\nabla \cdot$ denotes the divergence of a vector field.[6] The results of Poisson's equation are used to account for the electrostatic forces (Coulomb interactions) between ions and molecules.

If the concentration of ions and charged molecules in the analyte solution follows an advection-diffusion processes such that $\rho(x, t)$ is time dependent, then why is Poisson's equation used to estimate the "electrostatic" potential? A complete description of the electric field in the ICS biosensor would require solving Maxwell's equations. However, electromagnetic wave phenomena occur on a time scale of nanoseconds, while the time scale of interest for the electrodiffusion processes in the ICS biosensor occurs at the millisecond time scale. Therefore, assuming the electrostatic approximation of Maxwell's equations[7] holds, we use the fundamental theorem of vector calculus to relate the electric field to the potential field via $E = -\nabla \phi$.

---

[3] The elementary charge q is the charge of a single proton and is equal to $1.60217662 \times 10^{-19}$ C.
[4] Maxwell's equations are composed of a system of PDEs. They consist of Gauss's law for electricity and magnetism, Faraday's law for induction, and Ampere's law which relates current and magnetism.
[5] The Lorentz force is the force F exerted on a molecule with net charge q and is given by $F = qE + qv \times B$, where E is the electric field, v is the velocity of the molecule, and B is the magnetic field.
[6] For a vector field $F = F_x \hat{i} + F_y \hat{j} + F_z \hat{k}$ defined in three-dimensional Euclidean space, the divergence of F is defined as $\nabla \cdot F = \partial F_x/\partial x + \partial F_y/\partial y + \partial F_z/\partial z$.
[7] $\nabla \times E = 0$ where $\nabla \times$ denotes the curl operator of the vector field E.

10.3.2 Nernst–Planck Equation: Advection and Diffusion

Here we present the Nernst–Planck equation, which is used to the model the advection and diffusion of molecules and ions in the flow chamber of the ICS biosensor; recall Figure 10.1. Our eventual goal is to construct a model for the ICS biosensor comprising a fluid flow advection-diffusion PDE coupled with a chemical reaction ODE.

The derivation of the Nernst–Planck equation begins by assuming the ion interactions and continuum descriptions of concentration and electric potential are valid; i.e., the mean-field approximation holds. In the mean-field approximation, time and spatial averages of the atom positions are used to compute the concentration of atoms. Also in this continuum model, only the concentration of atoms is modeled, not the positions of individual atoms. Consider an electrolyte composed of ions or molecules of species $i \in \{1, 2, \ldots, M\}$. For example, $i = 1$ could denote Na^+ ions and $i = 2$ could denote Cl^- ions. Let $c^i(x, y, z, t)$ denote the spatially dependent concentration of species i at time t at spatial coordinate x, y, z in a three-dimensional Euclidean space. We impose *mass conservation*, which states that the time change of the concentration c^i, for an ion or molecule of species i, must be equal to the divergence of the total flux J^i. The Nernst–Planck equation is given by

$$\frac{\partial c^i}{\partial t} = -\nabla \cdot J^i, \quad i \in \{1, 2\}, \tag{10.4}$$

where t is time and $\nabla \cdot$ is the divergence operator. The total flux J^i of species i is dependent on Fick's law of diffusion, electrostatic forces (which are dependent on Poisson's equation), and advection. Below we use the Nernst–Planck equation to evaluate the total flux J^i which accounts for all these effects.

The total flux J^i in (10.4) of species i results from diffusive flux J_d^i, an electrical-migration flux J_e^i, and a velocity field flux denoted by J_v^i. Each of these flux terms is discussed below.

Diffusive flux. The diffusive flux J_d^i produced by concentration gradients is given by

$$J_d^i = -D^i \nabla c^i, \tag{10.5}$$

where D^i is the diffusion coefficient. Equation (10.5) is known as Fick's first law of diffusion. Qualitatively, (10.5) states that the ionic species i will diffuse from higher-concentration spatial regions to lower-concentration regions at a rate proportional to the difference in concentration.

Electrical-migration flux. The electrical-migration flux J_e^i of the ions is the number of moles of ions passing through a unit area per second and is given by

$$J_e^i = \mu_i c^i q^i F E, \tag{10.6}$$

where μ^i is the mobility,[8] c^i is the concentration, q^i is the charge, F is Faraday's constant, and E denotes the electric field.

Velocity flux. The velocity flux J_v^i is given by

$$J_v^i = vc^i, \tag{10.7}$$

where v is the velocity field. The velocity field can be estimated from the Navier–Stokes equation[9]:

$$\varrho\left(\frac{\partial v}{\partial t} + v \cdot \nabla v\right) = -\nabla p + \eta\nabla^2 v + (\varepsilon\nabla^2\phi)\nabla\phi. \tag{10.8}$$

Here ϱ is the density of the electrolyte, v is the velocity field, p is the pressure, η is the viscosity of the electrolyte, and $(\varepsilon\nabla^2\phi)\nabla\phi$ is the electrostatic force on the electrolyte from Poisson's equation (10.3).

Using (10.5), (10.6), and (10.7) the total flux of species i is given by the Nernst–Planck equation:

$$\frac{\partial c^i}{\partial t} = -\nabla \cdot J^i,$$

$$J^i = J_e^i + J_d^i + J_v^i$$

$$= -D^i\nabla c^i + \mu_i c^i q^i F E + vc^i$$

$$= -D^i(\nabla c^i - \frac{q^i}{k_B T}c^i E) + vc^i, \tag{10.9}$$

where the last relation is obtained by substitution of the Einstein relation $\mu^i = D^i/k_B T$, where k_B is the Boltzmann constant and T the temperature of the solution. The electrical field E, in (10.6), is obtained from Poisson's equation (10.3).

To summarize, the Nernst–Planck equation (10.9) accounts for the change in concentration of species i that results from diffusive flux J_d^i, electrical-migration flux J_e^i, and velocity field flux J_v^i. The Nernst–Planck equation depends on the velocity field v and the electric field E. The velocity field v can be evaluated using the Navier–Stokes equation (10.8), and the electric field E can be evaluated using Poisson's equation (10.3). Below we discuss the boundary conditions and initial conditions of (10.9) used to model the concentration dynamics of ions in the engineered tethered membrane.

10.3.3 Poisson–Nernst–Planck Equation

The model for electrolyte dynamics in the ICS biosensor combines Poisson's equation (10.3) for electrostatics and the Nernst–Planck equation (10.4) for the advection-diffusion fluid flow of the electrolyte species. This combination of PDEs is known as the Poisson–Nernst–Planck (PNP) model, which constitutes a classical continuum theory model for ion transport.

---

[8] Mobility μ^i is the ratio of the ith species drift velocity to an applied force. The Einstein relation $D^i = \mu^i k_B T$ relates the diffusion coefficient of species i to the mobility μ^i, where k_B is Boltzmann's constant and T is the temperature.

[9] The Navier–Stokes equation is a PDE for modeling the conservation of linear momentum in fluids.

10.3 Mass-Transport-Limited Model of the ICS Biosensor Dynamics

The ICS biosensor is placed in a flow chamber as illustrated in Figure 10.1. Let $c^i(x, y, z, t)$ denote the spatially dependent concentration of species i (with spatial coordinates x, y, z) at time t. The PNP model for the analyte concentration dynamics in the flow chamber is the following partial differential equation:

$$\frac{\partial c^i}{\partial t} = -\nabla \cdot J^i = \nabla \cdot \left[D^i \left(\nabla c^i + \frac{q^i}{k_B T} c^i \nabla \phi \right) - v c^i \right],$$

$$\nabla \cdot (\varepsilon \nabla \phi) = -\rho = -F \sum_i q^i c^i \quad \text{in } \Omega_e,$$

for ion species $i = \{1, 2, \ldots, M\}$, (10.10)

where Ω_e defines the electrolyte domain in the flow chamber. In (10.10) $i \in \{1, 2, \ldots, M\}$ denotes the ionic species, c^i is the concentration of species i, D^i is the diffusivity of species i, q^i is the charge of species i, F is Faraday's constant, k_B is Boltzmann's constant, T is the temperature, ε is the dielectric permittivity, J^i is the ionic flux, and ϕ is the potential.

We now specify the boundary and initial conditions of the PNP model (10.10).

The boundary conditions provide constraints on the concentration of c^i. These boundary conditions account for the surface chemical reactions in the ICS biosensor (refer to Chapter 9 for details). A schematic of the electrolyte domain Ω_e of the flow chamber is provided in Figure 10.3. The parameters L and h are the length and height of the simulation cell, Q is the input analyte flow rate, $\partial\Omega_{\text{surf}}$ defines the surface boundary where the chemical reactions take place, and $\partial\Omega_{\text{in}}$ defines the input boundary to the ICS biosensor.

The boundary and initial conditions of (10.10) for uncharged analyte species $z^i = 0$ are given by

$$n \cdot D^i \nabla c^i = R_s^i \text{ in } \partial\Omega_{\text{surf}}, \quad n \cdot D^a \nabla c^i = 0 \text{ otherwise},$$

$$n \cdot (\varepsilon_m \nabla \phi_m - \varepsilon_w \nabla \phi_w) = -F \sum_j q^j c_s^j \text{ in } \partial\Omega_{\text{surf}},$$

$$c^i(0) = c_o^i \text{ in } \partial\Omega_{\text{in}}, \quad n \cdot D_s^j \nabla_s c_s^j = 0 \text{ in } \partial\Omega_b. \quad (10.11)$$

Here $\partial\Omega_{\text{surf}}$, $\partial\Omega_{\text{in}}$, and $\partial\Omega_b$ are defined in Figure 10.3, n is the normal vector to the membrane surface $\partial\Omega_{\text{surf}}$, ε_m and ε_w are the dielectric permittivities of the membrane and analyte solution, respectively, $\nabla \phi_m$ denotes the potential drop in the interior of the membrane adjacent to $\partial\Omega_{\text{surf}}$, and $\nabla \phi_w$ denotes the potential drop in the exterior of the membrane adjacent to $\partial\Omega_{\text{surf}}$. The parameter $\nabla \phi_m$ depends on the dynamics of ions and charged molecule dynamics in the tethering reservoir and the electrode potential. The initial conditions are given by $c^i(0) = c_o^i$ and $c_s^j(0)$ defined in Table C.1 on page 413 for analyte molecules $i \in \{1, \ldots, M\}$ and surface molecules $j \in \{1, \ldots, m\}$. In (10.11), R_s^i is the surface reaction rate which couples[10] the change in analyte concentration (units

---

[10] Physically R_s^i represents the surface binding of all analyte molecules in a small volume (several nanometers) adjacent to the membrane surface. The small volume means that all molecules in this region either bind to the surface or not, and the volume is assumed to be sufficiently small so that diffusion does not impact the dynamics of surface binding in this region.

of mol/m^3) to the change in surface-bound species concentration (units of mol/m^2), and c_o^i are the input analyte concentrations.

Computing the surface reaction rate R_s^i requires knowledge of the chemical reactions on $\partial\Omega_{\text{surf}}$. From Chapter 9, the chemical species in $\partial\Omega_{\text{surf}}$ for the ICS biosensor include the analytes a, binding sites b, mobile gA monomers c, tethered gA monomers s, and dimers d, with respective concentrations $\{c^a, c_s^b, c_s^c, c_s^s, c_s^d\}$. Other chemical complexes present include $w, x, y,$ and z with concentrations $\{c_s^w, c_s^x, c_s^y, c_s^z\}$. The chemical reactions that relate these chemical species are described by the following set of reactions (discussed in Chapter 9):

$$a+b \underset{r_1}{\overset{f_1}{\rightleftharpoons}} w, \quad a+c \underset{r_2}{\overset{f_2}{\rightleftharpoons}} x, \quad w+c \underset{r_3}{\overset{f_3}{\rightleftharpoons}} y, \quad x+b \underset{r_4}{\overset{f_4}{\rightleftharpoons}} y,$$

$$c+s \underset{r_5}{\overset{f_5}{\rightleftharpoons}} d, \quad a+d \underset{r_6}{\overset{f_6}{\rightleftharpoons}} z, \quad x+s \underset{r_7}{\overset{f_7}{\rightleftharpoons}} z. \quad (10.12)$$

In (10.12), r_j and f_j, for $j \in \{1, 2, 3, 4, 5, 6, 7\}$, denote the reverse and forward reaction rates for the chemical species $\{a, b, c, d, s, w, x, y, z\}$. The dynamics of the surface species are described by the system of ordinary differential equations

$$\frac{dc_s^j}{dt} = R_s^j, \quad j \in \{1, \ldots, 7\} \quad \text{in } \partial\Omega_{\text{surf}}. \quad (10.13)$$

The expression for R_s^j in (10.13) is specified by the chemical reactions (10.12) using the methods presented in Chapter 9.

Summary. The complete mass-transport-limited model of concentration $c^i(x, y, z, t)$, as a function of space (x, y, z) and time t, for each ion species $i = 1, 2, \ldots, M$ in the ICS biosensor is given by the following coupled system of the advection-diffusion PDE, together with the reaction-rate ODEs that constitute the boundary condition:

$$\frac{\partial c^i}{\partial t} = -\nabla \cdot J^i = \nabla \cdot \left[D^i(\nabla c^i + \frac{q^i}{k_B T} c^i \nabla \phi) - v c^i \right], \quad (10.14\text{a})$$

$$\nabla \cdot (\varepsilon \nabla \phi) = -\rho = -F \sum_i q^i c^i \quad \text{in } \Omega_e, \quad (10.14\text{b})$$

$$\frac{dc_s^j}{dt} = R_s^j \quad \text{in } \partial\Omega_{\text{surf}}, \quad (10.14\text{c})$$

$$n \cdot D^i \nabla c^i = R_s^i \text{ in } \partial\Omega_{\text{surf}}, \quad n \cdot D^a \nabla c^i = 0 \text{ otherwise}, \quad (10.14\text{d})$$

$$n \cdot (\varepsilon_m \nabla \phi_m - \varepsilon_w \nabla \phi_w) = -F \sum_j q^j c_s^j \text{ in } \partial\Omega_{\text{surf}}, \quad (10.14\text{e})$$

$$c^i(0) = c_o^i \text{ in } \partial\Omega_{\text{in}}, \quad n \cdot D_s^j \nabla_s c_s^j = 0 \text{ in } \partial\Omega_b, \quad (10.14\text{f})$$

for $i = \{1, 2, \ldots, M\}$ and $j \in \{1, \ldots, 7\}$.

Note that the ordinary differential equations (10.13) and PNP partial differential equation (10.10) are coupled through the boundary condition $n \cdot D^i \nabla c^i = R_s^i$ in (10.14c); see also (10.11).

Remark 1: Notice that if the charge $q^i \neq 0$ of species i is nonzero, then an expression for $\nabla \phi_m$ (this potential drop in the interior of the membrane is defined below (10.11)) and the initial condition for ϕ are required to solve the ordinary differential equations (10.13) and PNP partial differential equation (10.10). Therefore, if $q^i \neq 0$, then the full system of nonlinear partial differential equations and ordinary differential equations in (10.10) must be solved with additional boundary conditions for ϕ. The electrodynamics of ions and charged molecules can be neglected if the excitation potential $V_s(t)$ is sufficiently small and the surface charge density, which results from charged molecules on the surface, is negligible.

Remark 2: In (10.10), solving for the velocity field v requires solving the Navier–Stokes equation (10.8). However, given the time scale of the conductance measurements is seconds, we assume that the velocity field v in (10.10) is a fully developed laminar flow with a parabolic velocity profile given by

$$v(z) = \frac{6Qz}{Lh^2}\left(1 - \frac{z}{h}\right), \tag{10.15}$$

where Q, L, and h are illustrated in Figure 10.3 and typical values are specified in Table C.1 on page 413. Q is the input flow rate, L is the length of the surface of the tethered membrane, and h is the height of the electrolyte region of the surface of the tethered membrane with respect to the counter gold electrode.

10.3.4 Estimating the Reaction Rates in the ICS Biosensor

We now present a least-squares method to estimate the surface reaction rates in (10.13) using results from the continuum model (10.10) in combination with the experimentally measured conductance $G_m(t)$. Recall that these surface reactions ultimately determine the concentration of conducting dimers and therefore the conductance (electrical response) of the ICS biosensor.

The continuum model (10.10) provides a model of the concentration of conducting gA dimers in the ICS biosensor. Given (10.10) with boundary conditions (10.11), we can estimate the membrane conductance $\hat{G}_m(t)$ given the parameters $\theta = \{f_1, \ldots, f_6, r_1, \ldots, r_6, D^a, c_o^a, D_s^1, \ldots, D_s^6\}$ with the flow rate Q and initial conditions known. The concentration of gA dimers is given by c_s^d and is related to the estimated membrane conductance via

$$\hat{G}_m(t; \theta) = \kappa \int_{\partial \Omega_{\text{surf}}} c_s^d(t) dS + G_o. \tag{10.16}$$

Here κ is the proportionality constant relating the conductance of the gA dimers to the number of gA dimers, G_o is the equilibrium membrane conductance (the conductance when the transmembrane potential $V_m = 0$), and $\partial \Omega_{\text{surf}}$ is defined in Figure 10.3.

The experimental measurements of the electrical response of the ICS yield a time series of conductance measurements $G_m = \{G_m(T_1), G_m(T_2), \ldots, G_m(T_K)\}$ at time instants $T_1, \ldots, T_K$. Estimating the ICS model parameters in the least-squares sense involves solving the following constrained optimization problem (constrained since $\theta \in \mathbb{R}_+^n$):

$$\theta^* \in \underset{\theta \in \mathbb{R}^+}{\arg\min} \left\{ \sum_{i=1}^{K} (G_m(T_i) - \hat{G}_m(T_i))^2 \right\}. \quad (10.17)$$

In (10.17), the parameter θ^* denotes the solution to the constrained optimization problem, and $\hat{G}_m(T_i)$ is evaluated using (10.16). Note that typically only the diffusion D^a, and initial concentration c_o^a are not known when fitting (10.17) to experimental data. The surface concentration of gA dimers $c_s^d(i; \theta)$ is computed using (10.10), (10.11), and the parameters in θ. To estimate θ^* we utilize the "Levenberg–Marquardt" algorithm.[11] If the conductance measurements of the ICS are corrupted by independent stationary white Gaussian noise, this nonlinear least-squares method provides the *maximum likelihood estimate* of the model parameters θ.

10.3.5 Experimental Results: Streptavidin, TSH, Ferritin, and hCG

In this section we compare the experimentally measured conductance response $G_m(t)$ of the ICS biosensor to the model-predicted conductance $\hat{G}_m(t)$ using the continuum model (10.10) for the following four analyte species: streptavidin, TSH, ferritin, and hCG. By model-predicted conductance, we mean that the ICS biosensor model developed in §10.3.3 is used to numerically compute the membrane conductance given the analyte concentration.

Figure 10.4 displays the experimentally measured and model-predicted conductance of the ICS biosensor for the concentration estimation of streptavidin, TSH, and ferritin. As seen, the experimentally measured conductance is in excellent agreement with the model-predicted conductance from the reaction-diffusion model coupled with fluid flow dynamics. For low (i.e., picomolar) analyte concentrations, the diffusive dynamics of the analyte significantly influences the population of gA dimers present in the tethered membrane. As seen in Figure 10.4, the dynamic model presented in §10.3.3 can account for the diffusive dynamics. This allows the dynamic model to be used for not only concentration estimation, but also the design of the ICS biosensor. If a specific concentration of analyte is to be measured using the ICS biosensor, the dynamic model presented in §10.3.3 can be used to select the number of binding sites and flow rate necessary for measurement. This procedure was applied for the design of the number of binding sites for the detection of TSH from 100 fM to 350 pM, and for the detection of ferritin from 100 fM to 100 pM. As seen in Figure 10.4(c), using the selected number of binding sites, mobile gA monomers, and flow rate, the concentration of TSH in the range of 100 fM to 350 pM can be estimated using the ICS. Figure 10.4(d) presents the measurement of ferritin in whole blood. Note that, as mentioned, the ICS is designed to

---

[11] The Levenberg–Marquardt algorithm is a classical damped nonlinear least-squares solution method. Several other constrained optimization solvers for nonconvex problems can be used such as augmented Lagrangian methods [44] which under reasonable conditions converge to a local stationary point.

(a) Streptividin concentrations c^A: 10 pM, 100 pM, 1000 pM.

(b) Ferritin in PBS, concentrations c^A: 200 pM, 400 pM, 600 pM.

(c) TSH concentrations c^A: 312 pM, 2 pM, 100 fM.

(d) Ferritin in whole blood, concentrations c^A: 0 pM, 50 pM.

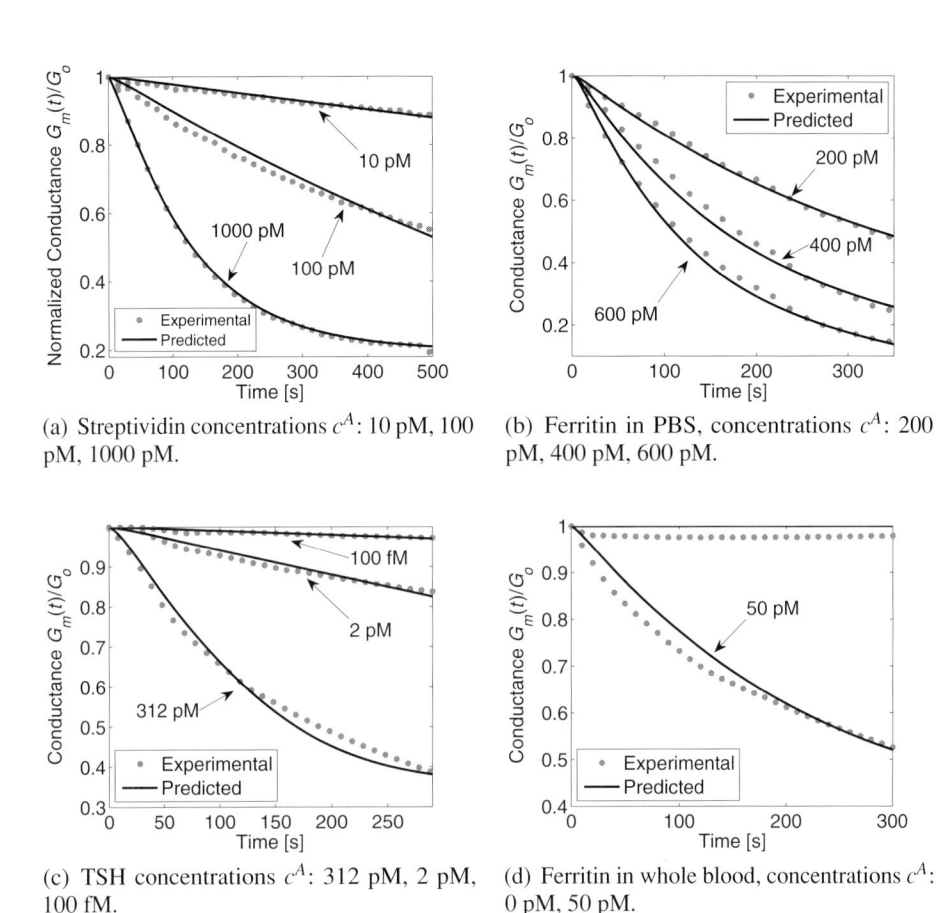

Figure 10.4 Experimentally measured and model-predicted normalized conductance $G_m(t)/G_o$ for streptavidin, TSH, and ferritin. Recall that G_o is the equilibrium membrane conductance with no analyte present in the flow chamber. The numerical predictions are computed using the ICS biosensor model in §10.3.3 with the parameters defined in Table C.1 on page 413.

only detect specific target species. As thousands of molecular compounds are present in human blood, a remarkably good estimate of ferritin can be obtained using the measured impedance of the ICS and the predictive model presented in §10.3.3.

10.4 Biosensor Arrays: Numerical Case Study

This section considers the ICS biosensor operating in the mass-transport-influenced regime. First, based on the concentration of analyte, flow rate of analyte, and size of the sensing surface, we show that the mass-transport-influenced regime itself has four subregimes. Second, we explore via numerical modeling how an array of smaller sensing surfaces that are spatially spread out can perform better than a single larger sensing surface.

Table 10.2 Parameter values for streptavidin-biotin interaction on ICS biosensor.

Chamber and biosensor dimensions (Figure 10.6)	
Sensor length L (mm)	2
Chamber width W (mm)	3
Chamber height h (mm)	0.1
Primary species concentration	
Binding site B^* (molecules/cm^2)	1×10^{11}
Free monomers C^* (molecules/cm^2)	1×10^9
Tethered monomers S^* (molecules/cm^2)	1×10^{10}
Reaction rate constants	
$f_1 = f_2 = f_6$ (M^{-1}s^{-1})	8×10^6
$f_3 = f_4$ (cm^{2}molecule^{-1}s^{-1})	5×10^{-9}
$f_5 = f_7$ (cm^{2}molecule^{-1}s^{-1})	1×10^{-10}
$r_1 = r_2 = r_3 = r_4 = r_6$ (s^{-1})	1×10^{-6}
$r_5 = r_7$ (s^{-1})	1.5×10^{-2}
Diffusion constant D (cm^{2}s^{-1})	1.5×10^{-6}
Flow rate Q (μL/min)	10
Dimensionless parameters	
Péclet number P_e	370.37
Dimensionless sensor length $\lambda = \frac{L}{h}$	20
Damköhler number D_a	2.29
Ratio of the collection rate to the input flux, k	0.21

The ICS biosensor (and other surface-based sensors) have receptors on the sensing surface that grab analyte molecules and react with them. Therefore, the sensor affects the concentration of the analyte.[12] Suppose the capture rate of analyte molecules at the sensing surface is faster than passive diffusion can replenish. Then most of the analyte molecules are captured close to the entry port of the flow chamber. If a single large sensing surface is used, depletion of analyte occurs downstream within the flow chamber. At these downstream areas, capture of analyte at the sensor surface is delayed until the analyte depletion near the entry port is saturated. This results in a sensor response rate that is slower than that determined by the analyte-receptor reaction rates at the sensor surface. Figure 10.5(a) illustrates this depletion effect for a flow cell of capillary height h and analyte concentration c_o (the sensor and its parameters are specified in Table 10.2).

Instead of using a single sensing surface, it is advantageous to use an array of sensors that are sufficiently spaced apart so that the depletion caused by one sensor does not affect the measurements made by subsequent sensors. That is, a single sensor of area A

---

[12] Put differently, the measurement affects the state of the system (analyte concentration) that we wish to estimate. This is unusual from a signal processing/estimation point of view.

10.4 Biosensor Arrays: Numerical Case Study

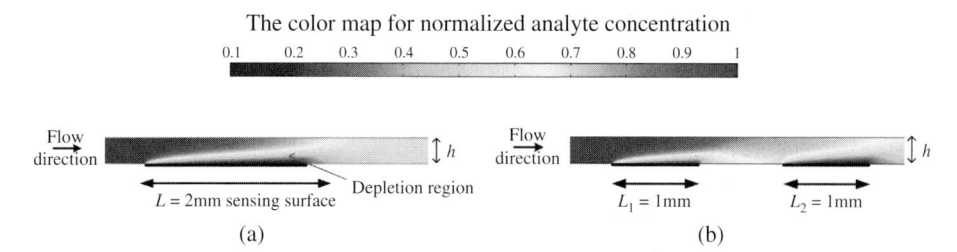

Figure 10.5 A color version of this figure can be found online at www.cambridge.org/engineered-artificial-membranes. Depletion in analyte concentration along a sensor. (a) The analytes are bound close to the entry to the flow chamber, causing a depletion in analyte concentration for subsequent binding sites and an overall slowing of the sensor response. (b) Replacing the sensor with two spaced-apart half-sized sensors allows analyte to be replenished between the individual sensors.

is replaced by N individual elements of area $\frac{A}{N}$ separated by a distance within the flow chamber that permits analyte to be replenished between the individual elements in the array (as shown in Figure 10.5(b)). When optimized, the analyte collection rate on the array is substantially improved (up to 10 percent) compared to the rate for a single large sensing element (as shown in Table 10.3 on page 195).

10.4.1 Biosensor Array Model

Consider a flow chamber with a rectangular cross section where analyte molecules move in a fluid flow past N sensors. The sensors form a linear array along the flow direction on the floor of the flow chamber as shown in Figure 10.6. The width of each sensor is assumed to be equal to the width of the flow chamber. Assuming that the ratio of the height h to the width W of the flow chamber is selected to be less than $1/20$, the system is symmetric about the width of the flow chamber. Two-dimensional Cartesian coordinates are introduced with the x axis along the flow direction and the y axis along the height of the flow chamber and perpendicular to the sensors. The dynamics of analyte concentration $c(t, x, z)$ are described by the two-dimensional advection-diffusion PDE

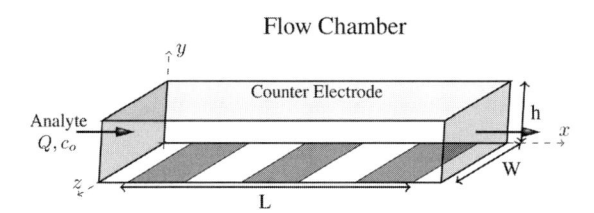

Figure 10.6 A linear array of $N = 3$ sensors in a rectangular flow chamber of length L, height h, and width W. The fluid containing analytes enters from the left side of the chamber. The analyte concentration at the inlet is c_o as expressed in the boundary condition (10.19), and the flow velocity of the inlet is Q.

(10.18), subject to the initial and boundary conditions in (10.19):

$$\frac{\partial c}{\partial t} = D\left(\frac{\partial^2 c}{\partial x^2} + \frac{\partial^2 c}{\partial y^2}\right) - v(y)\frac{\partial c}{\partial x}, \qquad (10.18)$$

$$c(t=0,x,y) = 0, \quad c(t,x=0,y) = c_o, \qquad (10.19)$$

$$\frac{\partial c}{\partial x}(t,x=L,y) = 0, \quad \frac{\partial c}{\partial y}(t,x,z=h) = 0,$$

$$D\frac{\partial c}{\partial y}(t,x \in I_i, y=0) = R(c, \mathbf{u}_i), \text{ for } i \in \{1,2,\ldots,N\},$$

$$\frac{\partial c}{\partial y}(t, x \notin \cup_{i=1}^{N} I_i, y=0) = 0.$$

In (10.18) and (10.19), t denotes time, D is the analyte diffusion constant, and $v(y)$ describes the fully developed velocity profile in the flow chamber; c_o denotes the analyte injection concentration, L and h denote the length and height of the flow chamber, respectively, and I_i denotes the range of x along the length of the flow chamber where sensor i is located. The number of sensors is denoted by N. The boundary condition on each sensor is determined by the diffusive flux of analytes on that surface, which is equal to the rate by which analytes are grabbed by immobilized species at the surface. This rate is determined by the reactions on the sensor and is denoted by $R(c, \mathbf{u}_i)$ for sensor i in (10.19). The adsorption rate $R(c, \mathbf{u}_i)$ is a function of analyte concentration c on the sensor surface and the surface concentration $\mathbf{u}_i$ of chemical species on sensor i.

The dynamics of the chemical species on sensor i are described by a system of ODEs as

$$\frac{d\mathbf{u}_i(t,x)}{dt} = G(\mathbf{u}_i(t,x), c), \quad \mathbf{u}_i(0,x) = u_0. \qquad (10.20)$$

Here, $\mathbf{u}_i$ is a vector that contains the concentration values of the chemical species. The function $G(\cdot)$ is determined by the rate law of reactions on the sensor; see (10.12) on page 184 for the specific reactions. The constant u_0 is the initial concentration of the chemical species.

10.4.2 Mass-Transport Phase Diagram

In the mass-transport-limited regime, the ICS biosensor can operate in one of four subregimes depending on the analyte concentration, flow rates, and binding kinetics. These four subregimes are illustrated in a phase diagram in Figure 10.7. In order to explain this, we first define the analyte collection rate and also Damköhler and Péclet numbers.[13]

1. Analyte collection rate. The analyte collection rate in molecules per unit time at the sensing surface is $J = -D\nabla c \times$ area, where approximations will be made to the gradient term.

---

[13] §10.2 defined the Damköhler and Péclet numbers in terms of characteristic times. Here, we define them in terms of fluxes.

2. Damköhler number. The dimensionless Damköhler number D_a determines the relative influence of mass-transport effects and the surface chemical reactions in the binding kinetics. Here, we define the Damköhler number as the ratio of the initial reactive flux J_R to the diffusive flux J_D. The maximum initial reactive flux at the surface of the biosensor is

$$J_R = k_f s_{\max} c_o L W, \qquad (10.21)$$

where k_f is the first-order forward reaction rate, $s_{\max}$ is the total binding-site concentration on the sensor, c_o is the input analyte concentration, L is the length of the sensor, and W is the width of the sensor. Note that in constructing (10.21) we are assuming that the analyte species binds to the biosensor surface via the Langmuir binding mechanism described by (10.1). The initial diffusive flux through the depletion region of the sensor surface is

$$\begin{aligned} J_D &= D \nabla c W L \\ &\approx \frac{D(c_o - c_s) W L}{\delta_s} \\ &\approx \frac{D c_o W L}{\delta_s}, \end{aligned} \qquad (10.22)$$

where δ_s is the length of the depletion region of analyte above the sensor surface, and c_s is the analyte concentration at the sensor surface. To construct an expression for the Damköhler number D_a using (10.21) and (10.22) requires an expression for the depletion region δ_s. The depletion region δ_s can be estimated by considering the characteristic time τ_L it takes for the analyte to cross the sensor surface and the characteristic time it takes the analyte to diffuse to the sensor surface starting from the boundary of the depletion region, $\tau_{DS} = \delta_s^2 / D$. If the depletion region satisfies $\delta_s \ll h$, then the analyte velocity in this region is approximately[14]

$$v(y) \approx \frac{6Q}{W h^2} y. \qquad (10.23)$$

Therefore, the characteristic time it takes an analyte molecule to cross the sensor surface in the depletion region is

$$\tau_L = \frac{W h^2 L}{6 Q \delta_s}, \qquad (10.24)$$

where $6Q\delta_s/Wh^2$ is the average velocity of the analyte in the depletion region. The boundary of the depletion region results where $\tau_{DS} = \tau_L$, where

$$\delta_s = \left[\frac{W h^2 L D}{6 Q} \right]^{1/3}. \qquad (10.25)$$

---

[14] Setting $y \ll h$ in (10.15), then the approximate velocity adjacent to the sensor surface (10.23) results.

Using (10.21)–(10.25), the Damköhler number is[15]

$$D_a = \frac{J_R}{J_D} = k_f s_{\max} \left[\frac{W h^2 L}{6 Q D^2}\right]^{1/3}. \qquad (10.26)$$

When $D_a \ll 1$, the reactive flux J_R is much smaller than the diffusive flux J_D and the binding kinetics are reaction limited. When $D_a \gg 1$, the kinetics are mass-transport influenced. In order to benefit from arrayed-sensor measurement in terms of total analyte collection rate, the kinetics conditions should be mass-transport influenced; i.e., the Damköhler number D_a should be sufficiently large.

3. Péclet number. While the Damköhler number determines the binding kinetics regime, the mass-transport condition is determined by the Péclet number, defined as

$$P_e = \frac{Q}{DW}, \qquad (10.27)$$

where Q is the volumetric flow rate, W is the width of the flow chamber, and D is the analyte diffusion constant as described in (10.18). The flow profile and the mass-transport limit on analyte collection rate are determined by the Péclet number and the dimensionless length of the sensor is defined as $\lambda = \frac{L}{h}$ which is the ratio of the sensor length to the height of the flow chamber.

Phase diagram. Figure 10.7 shows a phase diagram (see [402] for a detailed description) where the mass-transport conditions are categorized into four regions based on the values of the Péclet number P_e and the sensor dimensionless length λ. In each region, the thickness of the depletion zone and the mass-transport limit on analyte collection rate have different formulations in terms of λ and P_e.

- Region (i) in Figure 10.7 corresponds to full collection regimes where the analyte collection rate on the sensor is equal to the input convective flux. So the depletion zone extends far upstream along the flow chamber. In this case multiple sensing surfaces are of no use since the depletion zone is very long.
- In region (ii) the depletion zone that forms is substantially thinner than the sensor length and flow chamber height. This is typically where the ICS biosensors operates when it is in the mass-transport-influenced regime.
- In region (iii) the depletion zone is thinner than the channel but thicker than the sensor. The ICS biosensor parameters are such that it does not operate in region (iii).
- The ICS biosensor parameters are such that it does not operate in region (iv).

---

[15] In general, the Damköhler number is

$$D_a = \frac{k_f s_{\max} L}{DF},$$

where F is a dimensionless flux which is computed in different ways under different mass-transport conditions. Refer to [402] for details of different methods to evaluate F.

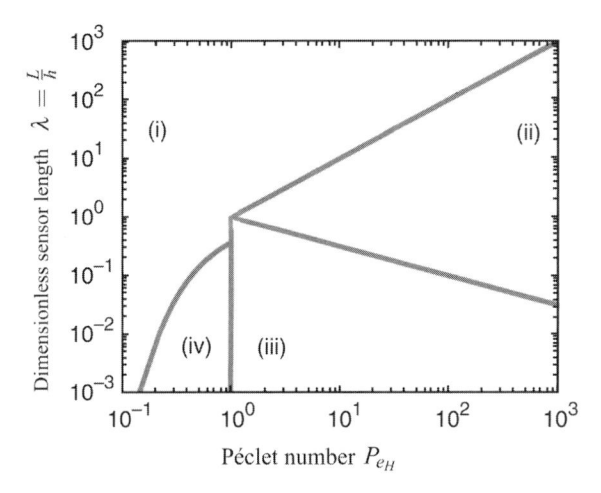

Figure 10.7 A typical mass-transport phase diagram for the binding kinetics on a surface-based sensor such as ICS. In each of the four regions, the analyte collection rate has a different formulation in terms of λ and P_e. Region (i) corresponds to full collection regimes where the analyte collection rate on the sensor is equal to the input convective flux. In region (ii), the depletion zone on the sensor surface is thin compared to the height of the flow chamber and the sensor length. In this region, the steady-state analyte collection rate on the sensor is obtained by (10.30). The analyte collection rate in region (iii) is studied in [4].

10.4.3 Sensor Array Can Mitigate Mass-Transport Limits

The focus here is on region (ii) in the phase diagram (Figure 10.7), where arrayed-sensor measurements can enhance total analyte collection rate on the sensors. The concentration c_s at the surface of the biosensor can be evaluated by solving c_s when the diffusive flux J_D (10.22) through the depletion region is equal to the initial reactive flux $J_R = k_f s_{\max} c_s LW$ to the sensor. The result is

$$\frac{c_s}{c_o} = \frac{1}{1 + D_a}, \qquad (10.28)$$

where the Damköhler number D_a is defined in (10.26). The steady-state collection rate J in the depletion region on the surface of a single sensor is

$$\begin{aligned} J &\approx \frac{D(c_o - c_s)LW}{\delta_s} \\ &\approx Dc_o W \left(1 - \frac{c_s}{c_o}\right)\left(\frac{L}{\delta_s}\right) \\ &\approx Dc_o W \left(\frac{D_a}{1 + D_a}\right)\left(\frac{L}{\delta_s}\right) \\ &\approx 6^{1/3} Dc_o W \left(\frac{D_a}{1 + D_a}\right) \lambda^{2/3} P_e^{1/3}, \end{aligned} \qquad (10.29)$$

where the Péclet number P_e is defined in (10.27), and $\lambda = L/h$ is the dimensionless length. In the general case the steady-state collection rate J is

$$J = Dc_oW \left(\frac{D_a}{1+D_a}\right) F, \quad (10.30)$$

where F is the dimensionless flux [402]. If we set $F = 6^{1/3}\lambda^{2/3}P_e^{1/3}$ in (10.30), then the collection rate in (10.29) results.

Next consider replacing a single sensor with the array of N sensors described above while the total sensing surface area remains fixed. Each of the N sensors has the same width W equal to the width of the flow chamber. The total dimensionless length of the sensors in the array can thus be written as $\sum_{i=1}^{N} \lambda_i = \lambda$, which is equal to the length of the initial single sensor. The following result can be obtained:

Result. The steady-state analyte collection rate J_i for sensor $i \in \{1, 2, \ldots, N\}$ is

$$J_i \approx WDc_iF_i\frac{D_{a,i}}{1+D_{a,i}}, \quad c_i = \left(1 - \frac{F_{i-1}D_{a,i-1}}{P_e(1+D_{a,i-1})}\right)c_{i-1}, \quad (10.31)$$

where W is the width of the flow chamber, D is the diffusion constant, and P_e is the Péclet number obtained by (10.27). The dimensionless flux F_i on sensor i is obtained by replacing λ with λ_i in (10.30). Similarly, the Damkohler number $D_{a,i}$ of sensor i is obtained by (10.26) for the corresponding length l_i and dimensionless flux F_i. Here, we define $F_0 = 0$ and $c_0 = c_o$, which is the analyte injection concentration defined in (10.19) and illustrated in Figure 10.6.

Define the improvement in analyte collection rate, obtained by replacing a single sensor with an array of N spaced sensors, as

$$C(\lambda_1, \ldots, \lambda_N) = J^{-1} \sum_{i=1}^{N} J_i - 1,$$

where J is the analyte collection rate on the initial single sensor obtained by (10.30). Then the maximum value of the defined improvement with respect to the values of the dimensionless length of the array of sensors is obtained as

$$\max C(\lambda_1, \ldots, \lambda_N) \text{ wrt } \lambda_1, \ldots, \lambda_N \text{ subject to } \sum_{i=1}^{N} \lambda_i = \lambda. \quad (10.32)$$

Derivation of (10.31). Using (10.30), the validity of (10.31) for the first sensor ($i = 1$) is straightforward. The steady-state analyte concentration c_2 above the second sensor is evaluated by deducting the analyte collection rate J_1 of sensor 1 from the input convective flux $J_C = c_oQ$. It is assumed that the sensors are sufficiently spaced apart such that analyte can be replenished between the individual sensors. Therefore, the analyte concentration c_2 above the second sensor is $c_2 \approx \frac{1}{Q}(c_oQ - J_1)$. Using (10.27) and (10.31) to compute J_1, c_2 can be obtained as $c_2 \approx c_o\left(1 - \frac{F_1 D_{a,1}}{P_e(1+D_{a,1})}\right)$. Similarly, (10.31) for the steady-state analyte collection rate on any sensor in the array is derived by deducting the total collection rate on the previous sensors from the input convective flux.

Table 10.3 Maximum attainable improvement in analyte collection rate, defined in (10.32). The improvement is obtained by replacing a single sensor with a uniform array of N smaller sensors of similar type. The analyte injection concentration is $c_o = 1$ pM. The total length of the sensors in each array is $L = 2$ mm, which is equal to the length of the initial single sensor. The sensor spacing is considered to be sufficiently large in each case to allow the analyte concentration to reach its maximum value in the gaps between the sensors.

	Sensor length (mm)	Damkohler number $D_{a,i}$ (for sensor i)	Collection rate improvement $\left(J^{-1} \sum_{i=1}^{N} J_i - 1 \right)$ (%) compared to one sensor
$N = 2$	1	1.82	11.86
$N = 4$	0.5	1.44	24.20
$N = 5$	0.4	1.34	27.94

Summary. In the mass-transport-influenced regime, the sensing surface of the ICS biosensor grabs analyte molecules, resulting in a depletion of analyte concentration downstream; in other words, sensing affects the underlying system state (concentration). In such cases, for a fixed sensing surface area, using an array of sensors instead of a single sensor can substantially improve the analyte collection rate, which results in a higher response rate and detection sensitivity. Table 10.3 illustrates this improvement.

10.5 Pore Formation Dynamics: Models for PGLa Antimicrobial Peptides

We now move from diffusion models for the ICS biosensor to the second part of this chapter, namely, diffusion models for the pore formation measurement platform (PFMP). Recall that the PFMP is a synthetic biological device whose construction using engineered membranes was discussed in Chapter 6. In this section we study how the PFMP can be used to understand the dynamics of antimicrobial pore-forming peptides. Pore-forming peptides are crucial in the attack and defense mechanisms of biological organisms. Understanding the chemical kinetics of pore-forming peptides provides vital information for pharmaceutical screening of potential drugs. In this section a dynamic model of the PFMP is presented. This can be used to gain insight into the pore formation reaction mechanism and potency of pore-forming antimicrobial peptides. Conceptually, this section can be viewed as a two-level modeling approximation to pore formation in real-life biological membranes. The pore formation measurement platform is a biomimetic approximation of the real-life membrane. Next, the mathematical models capture the dynamics of the pore formation measurement platform.

We use the constructed dynamic model and experimental measurements from the PFMP to study the pore formation dynamics of PGLa, a membrane-active antimicrobial peptide produced in specialized neuroepithelial cells in the African frog *Xenopus laevis* [477]. In addition to the antimicrobial activity, PGLa also has anticancer [25, 457], antiviral [71], and antifungal properties [347]. The remarkable feature of PGLa is that it provides a potential source for new antibiotics against increasingly common

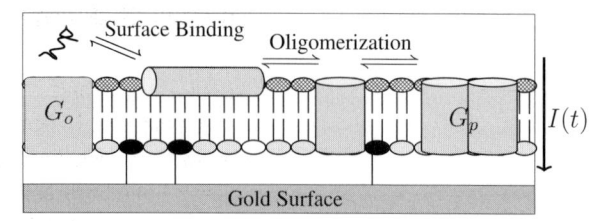

Figure 10.8 Schematic of the pore formation measurement platform (PFMP). An excitation voltage is applied between the gold electrode and counter electrode (not shown) and the current response $I(t)$ is measured. The PGLa peptide binds to the membrane surface, then undergoes oligomerization steps to create a PGLa pore in the membrane with conductance G_p. The current response $I(t)$ of the PFMP is dependent on the number of conducing PGLa pores and the equilibrium number of aqueous pores with conductance G_o present in the tethered membrane. The dashed beads represent mobile lipids on the membrane surface, the gray beads are mobile lipids adjacent to the gold electrode, and the black beads are tethered lipids. The construction and formation of the PFMP is provided in Chapter 4.

multiresistant pathogens (i.e., "superbugs") such as methicillin-resistant *Staphylococcus aureus* [82].

The interaction of pore-forming peptides with the PFMP is illustrated in Figure 10.8. As seen, the action mechanism of antimicrobial peptides involves the molecule binding to the membrane surface, then undergoing orientational, conformational, and oligomerization processes to create a conducting pore. Recall from Chapter 2 that this illustrates the formation of a peptide ion-channel. The PFMP can be used to study the dynamics of peptide ion-channel formation in a controllable environment because the tethered-membrane surface can be engineered to mimic prokaryotic, eukaryotic, and archaebacterial membranes. The benefit of using the PFMP to study pore formation dynamics compared to methods such as lytic experiments, gel electrophoresis, site-directed mutagenesis, and cryoelectron microscopy is that the tether density, electrolyte composition, membrane composition, and applied transmembrane potential can be controlled by the experimentalist.

The main idea behind the detection mechanism of the PFMP is that, by measuring the current response of the PFMP, it is possible to estimate the time-dependent conductance of the tethered membrane using a dynamic model. The membrane conductance in turn depends on the number of conducting PGLa pores (peptide ion channels) and aqueous pores present in the tethered membrane. Additionally, a dynamic model could be constructed to elucidate the reaction pathway leading to PGLa pore formation. Such a model must account for diffusion of the peptides in the electrolyte, and the reaction-diffusion processes present on the membrane surface, as illustrated in Figure 10.8. In this section we construct the dynamic model of the PFMP for computing the current response $I(t)$ of the PFMP. The computed current response $I(t)$ can be compared with the experimentally measured current response from the PFMP to estimate both the membrane conductance and possible reaction pathways that lead to PGLa pore formation. That is, we adjust the parameters of the dynamic model to fit the experimentally

10.5 Pore Formation Dynamics: Models for PGLa Antimicrobial Peptides

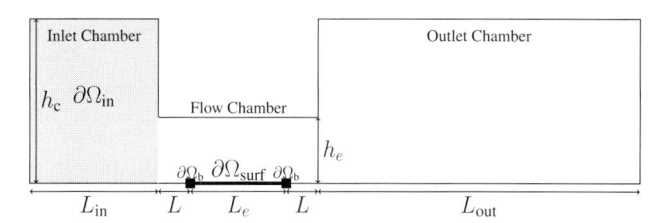

Figure 10.9 PFMP and a schematic of the computational domain for the generalized reaction-diffusion continuum model. The parameters are defined in Table C.4 on page 415. $\partial\Omega_b$ denotes the boundary of the tethered membrane illustrated by the black boxes, and $\partial\Omega_{in}$ is the analyte input flow chamber indicted in gray. Note that the PFMP computational domain explicitly includes the inlet chamber (height h_c and length L_{in}), flow chamber (height h_e and length $2L + L_e$), and outlet chamber h_c and length L_{out}). The width of the inlet, flow, and outlet chambers are all W.

measured data and interpret the fitted parameters in terms of the underlying physical phenomena present.

10.5.1 Generalized Reaction-Diffusion Equation

This section presents an advection-diffusion equation is coupled to a surface reaction-diffusion equation to model the dynamics of the membrane conductance for a given pore formation reaction mechanism.[16] The reaction mechanism is validated when agreement between the experimentally measured conductance $G_m(t)$ from the fractional-order model and the predicted membrane conductance $\hat{G}_m(t)$ is realized.

The computational domain of the continuum model is specified in Figure 10.9. Recall that the flow chamber of the PFMP, illustrated in Figure 10.1 on page 175, is a three-dimensional structure. However, a two-dimensional spatial domain is considered for the PFMP as the flow chamber width and height of the PFMP are $W = 3$ mm and $h_e = 0.1$ mm, respectively. As shown in [51], for $h_e/W < 0.1$ the variation in concentration along the width of the chamber is negligible. For the PFMP, the aspect ratio is $h_e/W = 0.03$; therefore, a two-dimensional domain that neglects the width can be used to model the reaction-diffusion dynamics in the PFMP.

10.5.2 Analyte and Surface Reaction Mechanism of PGLa

The chemical reactions leading to pore formation occur at the surface of the tethered membrane; this surface is denoted by $\partial\Omega_{\text{surf}}$ in Figure 10.9. The time-evolving conductance $G_m(t)$ of the membrane is dependent on the concentration of conducting pores in the membrane. To compute the concentration of conducting pores we consider the following reaction mechanism:

$$a \underset{k_d}{\overset{k_a^1}{\rightleftharpoons}} p_m \overset{k_p}{\rightarrow} p_1, \quad np_1 \overset{k_1}{\rightarrow} p_n, \quad mp_n \overset{k_c}{\rightarrow} c, \quad (10.33)$$

---

[16] Please see Appendix A for a short overview of such partial differential equations.

where p_m is the membrane-bound monomer, p_1 is the protomer,[17] p_n is the conducting pore containing n protomers, and c is a closed pore. The first reaction in (10.33) accounts for the binding of peptides to the membrane surface, and then these bound peptides become protomers. The second reaction in (10.33) models the oligomerization of protomers to form a conducting pore. The third reaction in (10.33) models the rate of closure of the conducting pores. Note that, after a peptide binds to the surface, it may undergo conformational and/or orientational changes prior to forming the pore. Once these changes have completed, the monomer is denoted as a protomer. In (10.33), the parameters below and above the arrows denote reaction rates: k_d is the dissociation constant, k_p is the rate of protomer formation, $\mathbf{k}_a^1$ the association rate constant, k_1 the rate of protomer binding which includes the translocation of the peptide from the surface to the transmembrane orientation, and k_c the rate of pore closing with n and m denoting stoichiometric numbers. We define the association rate constant $\mathbf{k}_a^1$ as decreasing as the number of membrane-bound peptides increases. If we denote $p_{\max}$ as the maximum concentration of bound peptides, then the association constant is defined as

$$\mathbf{k}_a^1 = k_a(p_{\max} - p_m - p_1 - np_n - nmc). \qquad (10.34)$$

Note that if the analyte concentration a, adjacent to the membrane surface $\partial\Omega_{\text{surf}}$, was a constant, then the surface binding mechanism in (10.33) resembles the Langmuir–Hinshelwood equation that is classically used to describe the dynamics of adsorption processes at surfaces.

Remark 3: PGLa binds to the membrane surface (lipids) via a noncovalent molecular bonding process. This interaction of PGLa with lipids is an example of a van der Waals interaction discussed in Chapter 2. Given that PGLa has a charge of +4 and is an amphiphilic molecule,[18] the van der Waals interaction results from a combination of "nonspecific hydrophobic association" (hydrophobic regions of PGLa and the lipid are both resistant to water), and "peptide-lipid electrostatic interactions" for negatively charged lipids.

Remark 4: The reaction mechanism (10.33) is a Hill-type approximation[19] of a sequential reaction mechanism for pore formation given by

$$n_1 p_1 \to p_2, \quad p_1 + p_2 \to p_3, \quad \ldots, \quad p_1 + p_{n-1} \to p_n \quad \text{for } n_1 < n.$$

---

[17] A protomer is composed of one or several polypeptide chains that is used to construct a protein. Note that a monomer is just a single polypeptide chain.

[18] Refer to Chapter 2: an amphiphilic molecule contains both water-resistance and water-seeking sections. The PGLa molecule contains four positively charged hydrophilic amino acids on one side, and hydrophobic amino acids on the other side.

[19] The Hill equation (also known as the Hill–Langmuir equation) is given by the reaction mechanism $c^i + nl^j \rightleftharpoons c$, where c^i is an electrolyte species, l^j is a membrane-bound species, n is the number of binding sites of the species l^j, and c represents the chemical product of the reaction. This reaction mechanism is typically used in biochemistry to describe the binding of ligands (l^j) to proteins (c^i). The process of ligand binding to a protein is a sequential process; however, the Hill equation assumes that all ligands bind to the protein simultaneously.

10.5 Pore Formation Dynamics: Models for PGLa Antimicrobial Peptides

The PFMP model cannot be used to differentiate whether the reaction mechanism (10.33) or the sequential reaction mechanism results in the formation of pores. That is, the model is not identifiable.[20] In Chapter 14 we illustrate how the tools of coarse-grained molecular dynamics can be used to estimate the reaction mechanism of PGLa. When used in combination with the PFMP model this alleviates the problem of model identifiability.

10.5.3 Dynamic Model of Electrolyte and Surface Diffusion of PGLa

The tethered-membrane conductance of the PFMP is dependent on the chemical reactions and diffusion dynamics of PGLa. To model the diffusion dynamics of PGLa in solution we utilize the PNP system of equations, (10.10). However, in the PFMP the flow velocity $v = 0$, and we assume the electrodiffusive effects of PGLa are negligible. This assumption holds when $Fq_a u_a a \nabla \phi \ll 1$, where F is Faraday's constant, q_a is the charge of PGLa, and u_a is the electric mobility[21] of PGLa. Therefore, the dynamics of PGLa in the electrolyte solution satisfy[22]

$$\frac{\partial a}{\partial t} = \nabla \cdot (D_a \nabla a), \tag{10.35}$$

where a is the concentration of PGLa in solution, and D_a is the diffusion coefficient of PGLa.

The membrane has surface-bound PGLa molecules with concentration m, protomer[23] monomers, protomer dimers, higher-order protomer complexes, and conducting pores, with concentrations given by $a, p_m, p_1, p_2, \ldots, p_n$. The dynamics of the PGLa peptide complexes in the membrane are governed by the following surface reaction-diffusion partial differential equations:

$$\begin{aligned}\frac{\partial p_m}{\partial t} &= \nabla_s \cdot (-D_m \nabla_s p_m), \\ \frac{\partial p_i}{\partial t} &= \nabla_s \cdot (-D_i \nabla_s p_i) + R_i, \\ &\text{for } i \in \{1, 2, \ldots, n\}.\end{aligned} \tag{10.36}$$

In (10.36) ∇_s is the surface gradient, D is the surface diffusion coefficient of the respective species, R_m denotes the change in concentration as a result of PGLa binding to the membrane surface, and R_i are the subsequent chemical reactions leading to pore

---

[20] The PFMP model is nonidentifiable as several possible reaction mechanisms could be used to obtain equivalent results.
[21] The electric mobility u_a relates the average velocity of the molecule to the applied electric field.
[22] Recall $\nabla \cdot$ denotes the divergence operator of a vector field. Refer to the footnote below (10.3) for notational details.
[23] Recall, from Chapter 2, that a protomer is a chemical species composed of one or several polypeptide chains that is used to construct an ion channel.

formation. The boundary conditions of (10.35) and (10.36) are given by

$$n \cdot D_a \nabla a = R_a \text{ in } \partial \Omega_{\text{surf}}, \qquad n \cdot D_a \nabla a = 0 \text{ otherwise}$$
$$n \cdot D_m \nabla_s p_m = 0 \text{ in } \partial \Omega_b, \qquad n \cdot D_i \nabla_s p_i = 0 \text{ in } \partial \Omega_b, \qquad (10.37)$$

where $\partial \Omega_{\text{surf}}$ denotes the membrane surface, $\partial \Omega_b$ denotes the boundary of the membrane surface, and n is the unit normal vector, and $\cdot$ in (10.37) denotes an inner product. In (10.37), R_a denotes the binding process of the PGLa peptide in solution to the membrane-bound state p_m. The chemical reaction rates R_a, R_m, and R_i can be computed from reaction mechanism (10.33). Initially the solution of PGLa with concentration a_o is inserted into a flow-cell chamber defined by $\partial \Omega_{\text{in}}$. The initial conditions of (10.35) and (10.36) are given by

$$a|_{t=0} = a_o \text{ in } \partial \Omega_{\text{in}}, \qquad a|_{t=0} = 0 \text{ otherwise}$$
$$p_m|_{t=0} = 0 \text{ in } \partial \Omega_{\text{surf}}, \qquad p_i|_{t=0} = 0 \text{ in } \partial \Omega_{\text{surf}}. \qquad (10.38)$$

The conductance of the membrane depends on the concentration of conducting PGLa pores and the equilibrium number of aqueous pores. A PGLa pore can be viewed as a peptide ion channel (see Chapter 2) in which the pore is formed by the insertion and oligomerization of PGLa molecules. Aqueous pores (water-filled pores) in the membrane result from the random thermal fluctuation of lipids. Denoting G_o as the equilibrium conductance of the aqueous pores (that is, the conductance of the membrane with zero transmembrane potential $V_m = 0$), then the total conductance of the membrane is given by

$$\hat{G}_m(t) = G_o + \kappa_p \int_{\partial \Omega_{\text{surf}}} p_n(t,x)\, dS. \qquad (10.39)$$

In (10.39), κ_p is a proportionality constant relating the mean conductance of the pores to the molar concentration of pores, p_n. The mean conductance G_p of each PGLa pore (Figure 10.8) is equal to κ_p/N_A, where N_A is Avogadro's number. From experimental measurements and theory G_p is expected to be in the range of picosiemens to nanosiemens; see [78, 89].

Summary. Assume that the following are specified for the PMFP device: the initial concentration a_o, the diffusion coefficients D from coarse-grained molecular dynamics, and the governing equations (10.35) and (10.36) with the boundary conditions (10.37) and initial conditions (10.38). Then the conductance of the membrane, $\hat{G}_m(t)$, can be computed using (10.39). Given the experimentally measured membrane conductance, denoted by $G_m(t)$, then least-squares estimates of the reaction rate constants $R_a(t)$, $R_m(t)$, and $R_i(t)$ that describe the pore formation reaction mechanism can be computed by minimizing $(G_m(t) - \hat{G}_m(t))^2$ at each time t. Refer to §10.3.4 for details on parameter estimation methods.

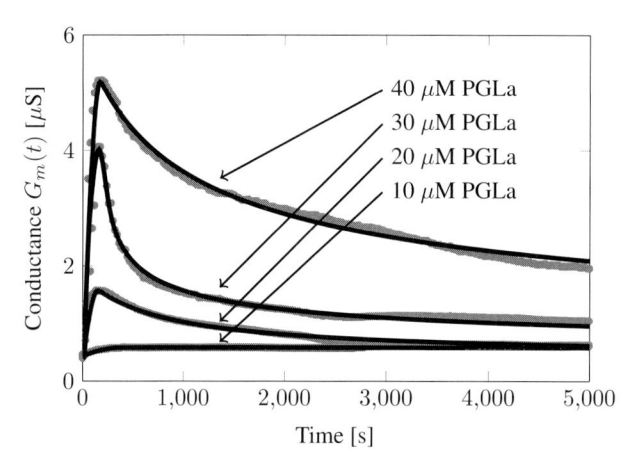

Figure 10.10 Experimentally measured and model-predicted conductance for DphPC tethered membrane with 10, 20, 30, and 40 μM of PGLa. The predictions are made using (10.35) and (10.36) with the reaction mechanism given by (10.33) and simulation parameters provided in Table C.2 on page 414. The experimental results are extracted from [88].

10.5.4 Experimental Results: Reaction Dynamics of PGLa

How accurate are the diffusion-based models for predicting the pore formation dynamics of PGLa? Figure 10.10 displays the experimentally measured and model-predicted conductance (using the model presented in §10.5.3) for varying concentrations of PGLa. The mean-absolute percentage error for the experimentally measured and model-predicted conductance in Figure 10.10 is 3.7 percent; that is, the predicted and measured conductance are in excellent agreement. Initially the conductance increases as a result of PGLa peptides diffusing to the membrane surface, binding, and translocating to the transmembrane configuration. As shown in Figure 10.10, at about 500 s the conductance of the membrane begins to decrease. This suggests that PGLa pores begin to close and prevent the formation of new PGLa pores. Given that PGLa has a positive charge of +4 as a result of the lysine residues, as the membrane becomes saturated with PGLa this causes the overall charge of the membrane to decrease, inhibiting the insertion of PGLa into the membrane. The estimated PGLa pore conductance G_p is in the range of 0.6 to 3.5 pS, in agreement with the expected pore conductance from [78, 89]. As expected the association coefficient k_a (10.33) is four orders of magnitude larger than the protomer formation rate constant k_p. This is expected as the binding of the PGLa to the surface does not require the translocation of the peptide to the transmembrane state. The translocation likely involves the peptide diffusing into a transient aqueous pore, which is a slower process than the peptide directly binding to the membrane surface. The rate of closing, k_c (10.33), is large, suggesting that PGLa pores only form transiently in the uncharged membrane. This provides an explanation for why the transmembrane state of the PGLa has not been observed at physiological temperatures using NMR techniques [7]. The estimated diffusion coefficient of PGLa (D_a in (10.35)) is in the range of 2 to 5 nm^2/ns.

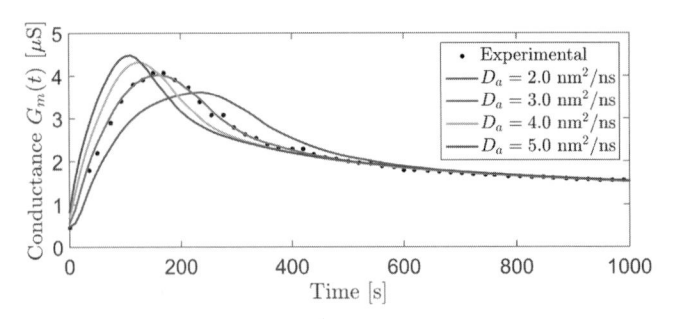

Figure 10.11 Experimentally measured conductance for DphPC tethered membrane with 30 μM of PGLa, and model-predicted conductance with varying D_a values. The predictions are made using the model (10.35), (10.36) with the reaction mechanism given by (10.33), and simulation parameters provided in Table C.2 on page 414. The experimental results are extracted from [88].

How sensitive is the membrane conductance $G_m(t)$ to variations in the analyte diffusion coefficient D_a?[24] To gain insight, Figure 10.11 provides $G_m(t)$ for D_a in the range of 2 to 5 nm^2/ns. From Figure 10.11 it is clear that $D_a = 3$ nm^2/ns provides a resulting $G_m(t)$ that is in agreement with the experimental measurements for the 30 μM PGLa interaction with the DphPC membrane. For large D_a the conductance increases faster; however, the formed PGLa pores also close faster than compared to the $D_a = 3$ nm^2/ns case. The reverse effect is observed for $D_a < 3$ nm^2/ns as seen in Figure 10.11. Note that, although the membrane conductance is strongly dependent on D_a, $G_m(t)$ is negligibly dependent on the surface and transmembrane-bound diffusion coefficients (D_i in (10.36)) of PGLa. This results because $D_i \ll D_a$ such that the reaction-diffusion equation in (10.36) can be approximated by a set of nonlinear ordinary differential equations.

How does the membrane permanent surface charge density[25] affect PGLa pore formation? To gain insight, we constructed several membranes containing different concentrations of charged palmitoyloleoyl-phosphatidylglycerol (POPG) lipids and then measured the current response when PGLa was added. Figure 10.12 displays the response of the 10% tethered membrane with charged (POPG lipids) and uncharged (DphPC) lipids resulting from the addition of 30 μM PGLa. As seen in Figure 10.12 the model-predicted results are in excellent agreement with the experimentally measured conductance. As expected, the negatively charged POPG lipids promote the binding of positively charged PGLa peptides. Also, as the percentage of POPG lipids increases, the membrane conductance increases. Comparing Figure 10.10 with Figure 10.12, it is

---

[24] Recall D_a appears in the mesoscopic model (10.35).

[25] The permanent surface charge density of a membrane results from lipid molecules that contain a nonzero net charge in the lipid head group. Charged lipids are important in biology as they induce electrostatic forces on ions and charged molecules adjacent to the membrane surface. Negatively charged lipids contain a net negative charge in the head group, and positively charged lipids contain a net positive charge in the head group. The POPG lipid specifically contains a single negative charge associated with the phosphate in the head group. For completeness, zwittrionic lipids have zero net charge in the head group from an equal number of positive and negative electrical charges in the head group.

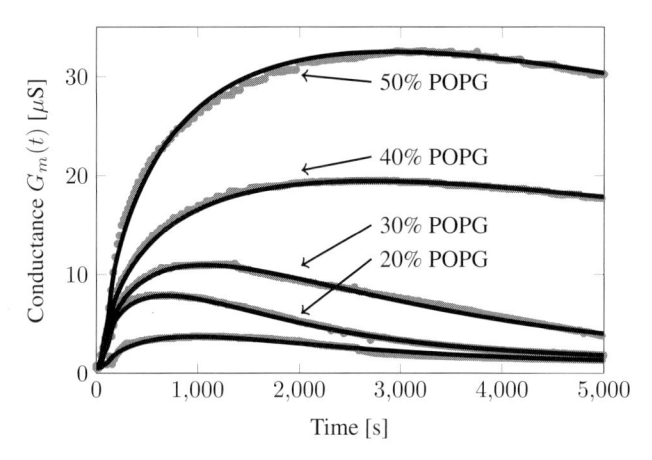

Figure 10.12 Experimentally measured (dotted) and model-predicted (solid line) membrane conductance for tethered membranes composed of 10, 20, 30, 40, and 50 percent POPG with a PGLa concentration of 30 μM. The predictions are made using (10.35), (10.36) with the reaction mechanism given by (10.33), and simulation parameters provided in Table C.3 on page 415. The experimental results are extracted from [88].

clear that PGLa has an affinity for forming pores in biological membranes containing negatively charged lipids typically found in prokaryotic cells. Using the reaction-rate model we find that the rate of protomer formation, k_1 in (10.33), increases as the negative charge of the membrane increases. However, the rate of pore closure (k_c in (10.33)) decreases as the negative charge of the membrane increases. This suggests that as the negative charge of the membrane increases there is an increase in the number of PGLa pores and pore lifetime. This makes PGLa especially effective for disrupting and perforating biological membranes containing a net negative charge.

10.6 Asymptotic Poisson–Nernst–Planck Model and Lumped Circuit Parameters

The fractional-order macroscopic model constructed in Chapter 8 can be linked to the PNP model (10.14) in this chapter using small-signal perturbation analysis. Throughout this section we focus on methods for computing the electrolyte resistance R_e and the double-layer capacitance C_{dl} in the fractional-order model using the results of the PNP model. The goal is to construct analytical expressions for the lumped circuit that do not require numerically solving the PNP model. Figure 10.13 illustrates the continuum PNP model domain and associated lumped circuit elements that are considered in this section. The parameters R_B and R_{BV} are the estimated electrolyte resistance in the case of a blocking electrode or Butler–Volmer electrode, respectively. C_B and C_{BV} are the estimated double-layer capacitance in the case of a blocking electrode and a Butler–Volmer electrode, respectively. As we show, the characteristics of the double-layer capacitance depend on the boundary conditions selected for the PNP model.

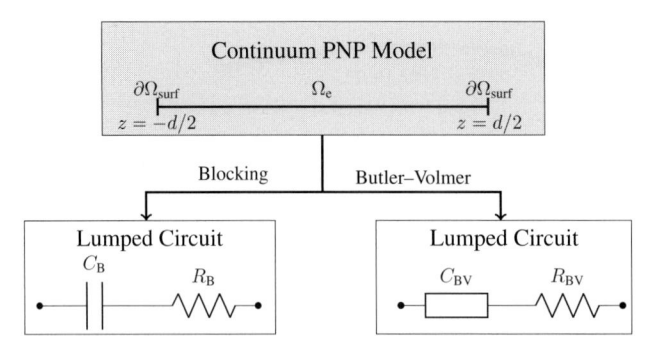

Figure 10.13 Schematic of the one-dimensional spatial domain of the PNP model (10.14) to estimate the double-layer capacitance and electrolyte resistance. $\partial\Omega_\text{surf}$ indicates the location of the electrode and counter electrode, and Ω_e is the electrolyte domain. The distance between the two electrodes is d and the area A of each electrode is equal. If blocking (or no-flux) boundary conditions are used for the PNP model then the resulting lumped circuit capacitance and electrolyte resistance are denoted by C_B and R_B, respectively. If the Butler–Volmer boundary conditions are used for the PNP model then the resulting lumped circuit capacitance and electrolyte resistance are denoted by C_BV and R_BV, respectively.

10.6.1 Double-Layer Capacitance and Electrolyte Resistance for Blocking Electrode

Consider an electrode and counter electrode placed at positions $z = -d/2$ and $z = d/2$, respectively, as illustrated in Figure 10.13. Let us assume that the electrolyte is composed of two symmetric ions c^1 and c^2 (identical mass, diffusion coefficient, and electric mobility), where c^1 has a charge of $+q$, and c^2 has a charge of $-q$. The boundary conditions we consider for the PNP model (10.14) at the domain boundaries $\partial\Omega_\text{surf}$ in Figure 10.13 are

$$J^i = 0 \text{ on } \partial\Omega_\text{surf} \text{ for } i \in \{1, 2\},$$

$$\phi(-d/2, t) = -\frac{V_o}{2}\sin(\omega t),$$

$$\phi(d/2, t) = \frac{V_o}{2}\sin(\omega t), \qquad (10.40)$$

where $V_o > 0$ and $\omega = 2\pi f$ is the angular frequency of the excitation potential. For sufficiently small V_o, the PNP model (10.14) can be represented by the linear approximation[26]

$$\frac{\partial}{\partial t}(c^1 - c^2) = D\frac{\partial^2}{\partial z^2}(c^1 - c^2) + \frac{2c_B q}{k_B T}\frac{\partial^2 \phi}{\partial z^2},$$

$$\frac{\partial^2 \phi}{\partial z^2} = -\frac{q}{\varepsilon}(c^1 - c^2), \qquad (10.41)$$

---

[26] For sufficiently small V_o in (10.40), the concentrations satisfy $c^1(z) = c_B + \eta^1(z)e^{j\omega t}$ and $c^2(z) = c_B + \eta^2(z)e^{j\omega t}$, where $\omega = 2\pi f$, c_B is the bulk concentration, and $\eta^1(z)$ and $\eta^2(z)$ represent the perturbed concentration from the small excitation potential.

10.6 Asymptotic Poisson–Nernst–Planck Model and Lumped Circuit Parameters

where c_B is the bulk concentration in Ω_e, k_B is Boltzmann's constant, and T is the temperature. To simplify the expression (10.41), we introduce the following parameters:

$$c^1(z,t) = c_B + \eta^1(z)e^{j\omega t}, \quad c^2(z,t) = c_B + \eta^2(z)e^{j\omega t},$$
$$\Psi(z,t) = (\eta^1(z) - \eta^2(z))e^{j\omega t}, \tag{10.42}$$

where $j = \sqrt{-1}$ is the unit imaginary number, and $\eta^1(z)$ and $\eta^2(z)$ represent the perturbed concentration from the small excitation potential. Substituting (10.42) into (10.41) results in the linear system of ordinary differential equations

$$\frac{d^2\Psi}{dz^2} = \left(\frac{1}{\lambda_D^2} + j\frac{\omega}{D}\right)\Psi,$$
$$\frac{d^2\phi}{dz^2} = -\frac{q}{\varepsilon}\Psi, \tag{10.43}$$
$$\lambda_D^2 = \frac{\varepsilon k_B T}{2 c_B N_A q^2}.$$

In (10.43), the parameter λ_D is the Debye screening length and gives the characteristic length scale of electrostatic screening.

Using the boundary conditions (10.40) in the complex domain, and the ordinary differential equations (10.43), we can construct a closed-form expression for the impedance of the electrode and electrolyte. The resulting impedance for the blocking electrodes at angular frequency $\omega = 2\pi f$ is

$$Z_B(\omega) = \frac{2\lambda_D}{\varepsilon A \omega \left(j - \frac{\omega \lambda_D^2}{D}\right)} \left[\frac{1}{\sqrt{1 + j\frac{\omega \lambda_D^2}{D}}} \tanh\left(\frac{d}{2\lambda_D}\sqrt{1 + j\frac{\omega \lambda_D^2}{D}}\right) + j\frac{d\omega \lambda_D}{2D}\right], \tag{10.44}$$

where A is the surface area of the electrode. Equation (10.44) gives an accurate description of the electrode and electrolyte impedance for a small sinusoidal excitation potential V_o. The typical frequency used for the excitation potential of engineered tethered membranes is below $f = 10$ kHz. At this frequency, we are in the low-frequency limit of (10.44) where $\omega \lambda_D^2 \ll D$ is satisfied. Given $\omega \lambda_D^2 \ll D$, the resulting impedance from (10.44) is

$$Z_B(\omega) \approx \frac{\lambda_D^2 d}{AD\varepsilon} + \frac{2\lambda_D}{j\omega A \varepsilon} = R_B + \frac{1}{j\omega C_B}, \tag{10.45}$$

where R_B is the electrolyte resistance and C_B is the double-layer capacitance of the electrodes. The electrolyte resistance R_B depends on the temperature T, electrolyte diffusion D, ion charge q, and bulk concentration c_B. As expected, the electrolyte resistance does not depend on the dielectric permittivity ε of the electrolyte solution. For an electrolyte concentration of 150 mol/m³ and two electrodes with area $A = 2.1 \times 10^{-6}$ m, the values are $\lambda_D \approx 0.79$ nm, $R_B \approx 50\ \Omega$, and $C_B \approx 940$ nF. The typical double-layer capacitance of the engineered tethered membrane (if both the electrode and counter electrode have the same surface area A) is C_{dl} in the range of 110 to 200 nF. Therefore, the double-layer capacitance C_B from the PNP model with blocking electrodes tends to overestimate the estimated double-layer capacitance from the fractional-order macroscopic model. Additionally, the double-layer capacitance C_B is an ideal capacitor as it does not contain any fractional-order terms.

10.6.2 Double-Layer Capacitance for Reaction-Limited Electrode

In the fractional-order macroscopic model constructed in §8.2, the double-layer capacitance satisfies a fractional-order power law. However, in the PNP model with blocking electrodes, the double-layer capacitance in C_B (10.45) acts as an ideal capacitor. This results because the PNP model with the no-flux boundary conditions accounts for the charge accumulation at the electrodes but neglects diffusion-limited charge transfer, reaction-limited charge transfer, and ionic adsorption dynamics that may occur at the electrode surface. These double-layer charging effects were accounted for in the fractional-order macroscopic model using the double-layer impedance element[27]

$$Z_{dl}(\omega) = \frac{1}{(j\omega)^p C_{dl}}, \qquad (10.46)$$

where C_{dl} is the double-layer capacitance and p is the fractional-order operator. In this section we use reaction-limited boundary conditions in the PNP model and estimate the resulting double-layer capacitance C_{BV} which accounts for both charge accumulation and reaction-limited charge transfer. Note that we neglect the electrolyte resistance in this section as it does not contribute to the analysis of the electrical double layer at the electrode surface.

Consider an electrode and counter electrode placed at positions $z = -d/2$ and $z = d/2$, respectively, as illustrated in Figure 10.13. If we use a reaction-limited boundary condition at the electrode surface $\partial \Omega_{\text{surf}}$ of the PNP model, then it is possible to construct a double-layer impedance element that satisfies a specific fractional-order power law. At the electrode-electrolyte interface, a commonly used technique to account for cathodic and anodic reactions at the electrode surface is to use the Butler–Volmer equation

$$J = J^1 + J^2 = \frac{I}{A}\left[\exp\left(\frac{\alpha_a q}{k_B T}(V - V_{\text{eq}})\right) - \exp\left(-\frac{\alpha_c q}{k_B T}(V - V_{\text{eq}})\right)\right], \qquad (10.47)$$

where J is the total flux of ions from the electrolyte to the electrode, I is the current through the electrode, A is the surface area of the electrode, α_a is the anodic charge transfer coefficient, α_c is the cathodic charge transfer coefficient, V is the electrode potential, and V_{eq} is the equilibrium electrode potential. The equilibrium potential of the electrode is equal to the potential that results when the anodic and cathodic chemical reactions are at chemical equilibrium. The boundary conditions of the PNP model we

---

[27] Refer to §8.2.1 for the derivation of the double-layer impedance of the fractional-order capacitance C_{dl} in the fractional-order macroscopic model.

10.6 Asymptotic Poisson–Nernst–Planck Model and Lumped Circuit Parameters

consider are

$$J^1 + J^2 = \frac{I}{A}\left[\exp\left(\frac{\alpha_a q}{k_B T}(V - V_{eq})\right) - \exp\left(-\frac{\alpha_c q}{k_B T}(V - V_{eq})\right)\right] \text{ on } \partial\Omega_{\text{surf}},$$

$$\phi(-d/2, t) = -\frac{V_o}{2}\sin(\omega t),$$

$$\phi(d/2, t) = \frac{V_o}{2}\sin(\omega t). \tag{10.48}$$

To construct the resulting electrode and electrolyte impedance when Butler–Volmer kinetics are present at the electrode surface, we neglect the migration term in the PNP model (10.14), assume a low excitation potential V_o, and set $\alpha_a = \alpha_c$. These assumptions with the boundary conditions (10.48) can be used to construct the impedance of the electrode and electrolyte,[28]

$$Z_{\text{BV}}(\omega) = \frac{2k_B T}{Aq^2 N_A c_B \sqrt{j\omega D}}\tanh\left(\frac{d}{2}\sqrt{\frac{j\omega}{D}}\right), \tag{10.49}$$

where N_A is Avogadro's number and A is the size of the electrode. The impedance (10.49) is the finite-length Warburg impedance. If we assume that the distance between the electrodes satisfies $d \gg 1$, then (10.49) can be approximated by

$$Z_{\text{BV}}(\omega) \approx \frac{2k_B T}{Aq^2 N_A c_B \sqrt{j\omega D}} = \frac{1}{C_{\text{BV}}\sqrt{j\omega}}, \tag{10.50}$$

where C_{BV} is the Butler–Volmer double-layer capacitance of the electrodes. Interestingly, the capacitance C_{BV} does not depend on the dielectric permittivity of the electrolyte. The impedance in (10.50) is known as the Warburg impedance commonly used to model the charging dynamics of electrodes. The key result of (10.50) is that imposing the Butler–Volmer boundary conditions and evaluating the impedance of the PNP model results in the double-layer impedance satisfying a fractional-order power law with $p = 0.5$ with a capacitance of C_{BV}.

To construct the general fractional-order impedance of the electrical double layer, Z_{dl} (10.46), with p in the range of 0 to 1 used in the fractional-order macroscopic model requires defining a flux boundary condition that accounts for the diffusion-limited charge transfer, reaction-limited charge transfer, and ionic adsorption dynamics. Though the Butler–Volmer equation (10.47) is commonly used, it only provides an expression for the fractional-order parameter $p = 0.5$. Another possibility is to use fractional integral constitutive equations. That is, we introduce fractional-order partial differential operators into the PNP model (10.14) to account for anomalous diffusion and then construct the associated impedance that results from a small-voltage sinusoidal excitation. For details on this procedure refer to [92, 234, 235].

---

[28] The derivation of (10.49) is identical to the derivation of the impedance, (10.45). For details, refer to [125, 234].

10.7 Complements and Sources

Four important modeling aspects were presented in this chapter: continuum theories to model the dynamics of ions, ICS biosensor arrays, methods for estimating model parameters and performing model identification, and surface chemical binding reactions. Each of these areas has extensive literature; below we outline only a small fraction of this literature that is relevant to engineered membranes.

As described in §10.3, mass-transport dynamics (partial differential equation) coupled with chemical kinetics result in a diffusion partial differential equation with Neumann and Dirchlet boundary conditions [109, 140, 271]. Similar formulations for binding and dissociation between a soluble analyte and an immobilized ligand are studied in [138]. Goldstein et al. discuss the accuracy and theoretical basis of different models for mass-transport effects in the binding of analytes in [140]. Dehghan discusses several different finite-difference methods to solve the advection-diffusion equation and discusses the stability of the numerical methods in [97].

10.7.1 Poisson–Nernst–Planck (PNP) Model

PNP is a widely used method for modeling the dynamics of ions in biological systems, and electrons in electrical circuits [176, 224, 253 255, 265, 280, 312, 353, 371, 483]. In biophysics, the PNP theory is used to model ion transport through ion channels (this was pioneered by Eisenberg and colleagues) and nanopores [80, 167, 176, 224, 265, 280, 371]. For example, in [312], the bioelectronic interface was studied using the steady-state PNP theory in which an electrogenic cell was placed on the top of a field-effect-transistor gate. Interested readers are referred to [128] for an introduction to bioelectronic interfacing using semiconductor devices. Two common methods for the derivation of the PNP theory are to begin from either the electrochemical potential of equilibrium thermodynamics, or using the properties of diffusion and electrostatics [373]. However, the PNP theory is limited in it application as it requires the mean-field approximation [482], and additionally neglects the finite-volume effect of ions and correlation effects. In Chapter 11 we explore how the finite-volume effect of ions and polarization effects of water can be included in a continuum model using similar ideas to the construction of the PNP equations presented in this chapter.

10.7.2 ICS Biosensor Arrays and Multicompartment Models

In §10.4, the mass-transport dynamics of the ICS biosensor were modeled using a system of partial differential equations coupled with surface chemical kinetics. For small analyte concentrations and high binding-site densities, it is possible to construct a multicompartment model which consists of a system of ordinary differential equations. The multicompartment model makes the approximation that the analyte concentration in specific regions of the flow chamber is spatially homogeneous; these regions are called compartments. For example, in [271, 288], mass-transport dynamics are formulated using a two-compartment model where analyte molecules move between the two compartments. For the ICS biosensor, the two-compartment model is constructed

10.7 Complements and Sources

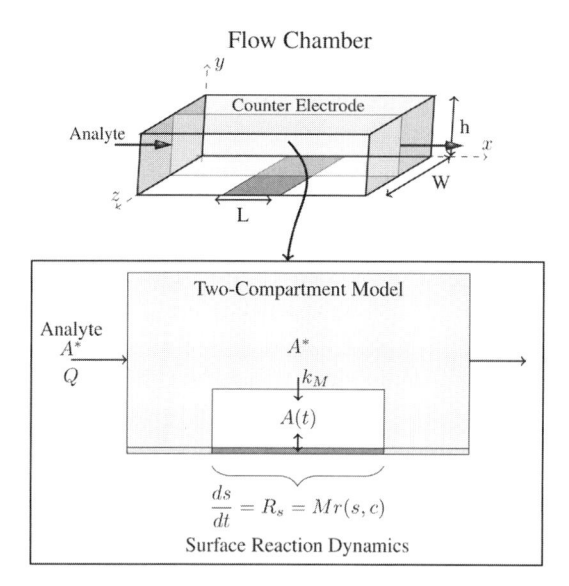

Figure 10.14 A color version of this figure can be found online at www.cambridge.org/engineered-artificial-membranes. Schematic of the flow chamber of the ICS biosensor and the two-compartment model. The biomimetic surface (dark gray region at the bottom of flow chamber with length L) indicates the location of the engineered membrane and gold electrode. The electrolyte solution containing the analyte (target molecules) enters the flow chamber at $x = 0$ and flows along the x axis. The gold counter electrode is indicated by the yellow region. The light gray region indicates a slice of the flow chamber that will be used for constructing the two-compartment model. The first compartment is indicated by the light gray region; the concentration of analyte in this region is equal to the inlet analyte concentration A^*. The second compartment (indicated by the white region) contains the time-dependent but spatially homogeneous analyte concentration $A(t)$. The analyte concentration in the second compartment is time dependent as the analyte in this region depends on the biomimetic surface reaction dynamics of the biosensor, and the advection and diffusion of analyte molecules from the first compartment into the second compartment. The mass-transport coefficient k_M accounts for the advection and diffusion of analyte molecules between the two compartments, and R_s represents the surface reaction dynamics at the biomimetic surface.

as illustrated in Figure 10.14. The first compartment is composed of the spatially homogeneous inlet analyte concentration A^*, and the second compartment is composed of a time-dependent but spatially homogeneous analyte concentration $A(t)$ that is in the region adjacent to the surface of the membrane where the surface-binding reactions occur. The mass-transport coefficient k_M accounts for the advection and diffusion of analyte molecules between the two compartments. The interaction between the two compartments can be described by a system of ordinary differential equations. For detailed discussion of the multicompartment model of the ICS biosensor and ICS biosensor arrays refer to [2, 3, 221].

10.7.3 Parameter Estimation and System Identification

Model parameter estimation and system identification methods are key to constructing a useful dynamic model. Excellent references on these topics include [250, 325].

Additionally, we propose to use the Langmuir–Hinshelwood equation that is classically used to describe the dynamics of adsorption processes at surfaces. However, several other binding mechanisms may also be possible. Some excellent references for chemical reactions at solid surface interfaces include [270, 396]. It is also possible to envisage more sophisticated surface reactions that can take place to form aqueous pores compared to the Hill-type approximation [392] of the aggregation and binding processes of the reaction mechanisms presented in [424] for α-hemolysin pores, and in [432] for cytolysin A pores used in this chapter.

An issue with the construction of complex reaction dynamic networks is the identifiability of such networks using experimental data derived from tethered membranes and the existence and uniqueness of solutions to the reaction-diffusion PDEs. Guaranteeing the existence and uniqueness of solutions for reaction-diffusion PDEs requires knowledge of Sobolev spaces, embedding, and PDE theory [115, 346]. Note that the source term in the reaction-diffusion equation is composed of a system of reaction rates. Existence and uniqueness of the reaction-diffusion problem are only guaranteed for sufficiently smooth and bounded reaction rates (which depend on the concentration of the species). If these conditions are not satisfied then the reaction-rate source terms in the PDEs can cause a singularity to form in finite time; that is, the source term can cause a discontinuity (shock), or force a concentration value to infinity. Assuming that the reaction-rate terms in the reaction-diffusion PDEs satisfy these conditions, what requirements on the function class of reaction rates are necessary to ensure identifiability of the functional form of the reaction rates? Specifically, consider a reaction-diffusion equation

$$\frac{\partial c^i}{\partial t} = \nabla \cdot D_i \nabla c^i + R^i(\theta, c), \qquad (10.51)$$

where $c = [c^1, c^2, \ldots, c^n]'$ is a column vector with element c^i denoting the concentration of species $i \in \{1, \ldots, n\}$, D_i is the diffusion coefficient of species i, and $R^i(\theta, c)$ is the functional form of the reaction rates with parameters θ.

The model-predicted current response $I(t)$ using the reaction-diffusion model can be compared with the experimentally measured current response from the engineered tethered membrane to estimate both D^i and $R^i(\theta, c)$. How many different values of D^i and functions $R^i(\theta, c)$ can be used such that the computed current response is in agreement with the experimentally measured current response? In general, several values of D^i and functions $R^i(\theta, c)$ can be used in the reaction-diffusion model such that the computed and measured current are in agreement [250]. However, if the function $R^i(\theta, c)$ is fixed with only θ allowed to vary, then estimation of the parameters D_i and θ can be performed using finite-element discretization and linear least squares [279], stochastic approximation expectation maximization [143], and Bayesian inference techniques [469]. Note that the function $R^i(\theta, c)$ can be estimated using sophisticated NMR techniques [32]. In silico methods, such as coarse-grained molecular dynamics and all-atom molecular dynamics, can also be used to estimate $R^i(\theta, c)$ as illustrated in Chapters 14 and 15.

In this chapter we have illustrated how tethered membranes can be used to estimate the reaction pathway of pore-forming peptides and proteins. However, other experimental techniques that can be used include lytic experiments, gel electrophoresis, site-directed mutagenesis, and cryoelectron microscopy [23, 110, 259, 309]. These methods can be used in combination with the PFMP dynamic model to alleviate the nonidentifiability issue associated with estimating the specific reaction pathway leading to the formation of peptide ion channels.

10.8 Closing Remarks

This chapter has focused on constructing dynamic models of the ICS biosensor and the pore formation measurement platform. The dynamic models were constructed using the Poisson–Nernst–Planck system of equations coupled with surface reaction-diffusion equations. In the case of the ICS biosensor, the binding dynamics and surface reaction dynamics are known; however, in the case of the PFMP these reaction dynamics are unknown and must be estimated using experimental measurements.

In this chapter low-voltage impedance measurements were utilized to estimate the tethered-membrane conductance, which allowed the electro-osmotic flow of ions in the PNP model to be neglected. Additionally, the use of low-voltage excitations implies that electroporation does not occur. For the design of novel drug delivery methods it is desirable to have a model that allows the design of excitation potentials to control the pore formation reaction dynamics of analytes. The dynamic models presented in this chapter provide a basis for constructing a model which can simultaneously account for the electroosmotic flow, the surface reaction-diffusion dynamics, and the process of electroporation.

11 Electroporation Models in Engineered Artificial Membranes

11.1 Introduction

A lipid membrane that contains negligible defects and no ion channels will not allow ions to pass through the membrane. Indeed a membrane can be viewed as an electric capacitor. However, if an excitation is applied to the membrane this will cause the temporary breakdown of the membrane, causing water-filled pores to form that are large enough to allow the passage of ions and molecules through the membrane – this process is called electroporation.

Electroporation is the process of aqueous pore formation in biological membranes when a voltage potential (e.g., ion gradient) is applied across the membrane. Several microbiology techniques are based on the use of electroporation to increase the permeability of the cell membrane to allow the passage of macromolecules, drugs, DNA, and antimicrobial peptides into the cell. These microbiology techniques are focused on how to design the excitation potential to allow the passage of these molecules through the membrane while ensuring the membrane is not irreversibly damaged. In this chapter we construct mesoscopic[1] diffusion models of electroporation in engineered tethered membranes. The formulation in this chapter relies heavily on the Smoluchowski–Einstein equation, which is a probabilistic model for the number of aqueous pores formed. Additionally, we use statistical mechanics and electrodiffusive dynamic models to estimate lumped circuit parameters such as membrane conductance, membrane capacitance, and double-layer capacitance in the presence of electroporation. Recall that Chapters 4 and 6 dealt with the construction and experimental measurement methods for the electroporation measurement platform. Put simply, this chapter gives mathematical models and associated insight at the mesoscopic level to explain the experimental results of Chapter 6.

11.1.1 Applications of Electroporation

Electroporation has been used in a variety of in vitro and in vivo biotechnical applications for antitumor treatment, protein insertion, cell fusion, and gene and drug delivery.

---

[1] Recall that the mesoscopic level lies between the macroscopic reaction-rate models and microscopic atomistic models. The mesoscopic models deal with the physical phenomena on micrometer length scales and microsecond times scales.

The process of electroporation results in the formation of pores in the membrane that allow molecules and ions to pass through the hydrophobic bilayer membrane. During electroporation the lipid molecules remain intact; they merely undergo a collective configurational change to form nanometer-sized aqueous pores.[2] These aqueous pores allow ions and molecules to pass through the membrane. There are two types of electroporation:

Reversible electroporation. Reversible electroporation occurs if, after the excitation potential is removed, the lipids can rearrange to form an impermeable membrane.
Irreversible electroporation. Irreversible electroporation occurs if, after the excitation potential is removed, the lipids cannot rearrange to form an impermeable membrane.

For the insertion of macromolecules, DNA, and drugs, typically reversible electroporation is used because the goal is not to destroy the cell membrane. However, irreversible electroporation is useful if the goal is to destroy the cell membrane. The first research involving electroporation occurred on human Jurkat cells[3] in 2003 [33]. If the molecules to be transported across the membrane are chemotherapeutic agents (for treating cancer), the process is known as electrochemotherapy. Additionally, if DNA is being transported across the membrane, the process is referred to as gene electrotransfer. Gene electrotransfer is a promising approach that may lead to the cure for Alzheimer's disease, Parkinson's disease, and brain cancer as it allows chemotherapeutic agents to be implanted into specific neurons. The first successful treatment of malignant cutaneous tumors using irreversible electroporation was completed in 2007 on mice (with complete tumor removal in 12 out of the 13 mice) [9]. However, a major challenge when performing irreversible electrochemotherapy is to ensure that only specific cancerous tissues are irreversibly electroporated. Not all cell membranes have the same resistance to electroporation; therefore, care must be taken when designing irreversible electrochemotherapy protocols.

11.1.2 What Is Electroporation?

Electroporation is the process of aqueous pore formation resulting from changes in the potential applied across the membrane (transmembrane potential). Electroporation is a multistep process with several distinct phases. As illustrated schematically in Figure 11.1, electroporation involves the movement of lipids and the formation of aqueous pores. The first step of the electroporation process is the application of an excitation potential across the membrane. From Chapter 10, when an excitation potential is applied, this causes the electromigration of ions from the electrolyte solution to the surface of the engineered tethered membrane. This causes an increase in the electrostatic

---

[2] Aqueous pores are water-filled pores that traverse the engineered artificial membrane.
[3] An immortalized human white blood cell. The Jurkat cell line contains specifically an immortalized line of human T lymphocyte cells obtained from the peripheral blood of an adolescent boy in 1970 with T cell leukemia.

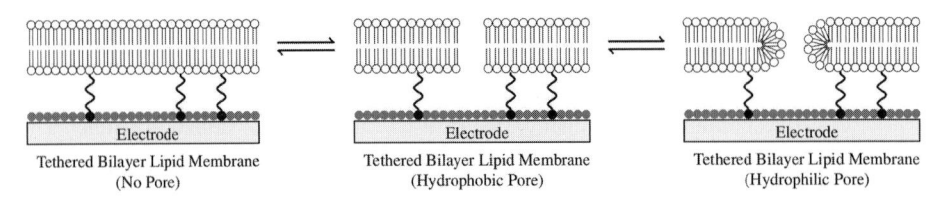

Figure 11.1 Schematic of the process of electroporation in an engineered membrane. Thermal fluctuations of lipids promote the creation of a hydrophobic pore. This pore can subsequently close (return to the no-pore state) or form a hydrophilic pore. This process of pore creation and pore destruction occurs in all biological membranes.

pressure on the membrane surface, which promotes the formation of aqueous pores. Two types of aqueous pores can form as a result of electroporation: hydrophobic and hydrophilic. A hydrophobic pore can spontaneously form in the membrane as a result of thermal fluctuations, as depicted in Figure 11.1. Typically the size of the hydrophobic pore is in the range of 3 to 5 Å. The hydrophobic pore is unstable since the hydrophobic tails of the lipids are in direct contact with the water molecules. Therefore, the hydrophobic pore either closes or rapidly forms a hydrophilic pore. The transition of the hydrophobic pore to the hydrophilic pore involves rearrangement at the lipid head groups to create the surface of the hydrophilic pore. The hydrophilic pore has a higher stability than the hydrophobic pore as the hydrophilic head groups of the lipids line the interior of the pore. This minimizes the interaction of the hydrophobic lipid tails with water. The size and dynamics of the hydrophilic pore are governed by the transmembrane potential. If the transmembrane potential is sufficiently high then the hydrophilic pore will expand until irreversible electroporation occurs, destroying the membrane. However, for sufficiently small transmembrane potentials the hydrophilic pore will expand until an equilibrium pore radius is reached, or contract and reseal for zero transmembrane potentials. If the hydrophilic pore closes leaving the tethered membrane intact then the electroporation process is reversible.

Note that both the hydrophobic and hydrophilic pores are aqueous pores as they contain water-filled pores. The process of aqueous pore formation and destruction occurs naturally in all membranes; however, the application of a transmembrane potential will promote the creation of aqueous pores. In other words, application of a transmembrane potential promotes electroporation.

11.1.3 Mesoscopic Model of Electroporation

For a mesoscopic model of the electroporation process to be useful it must account for the electrolyte dynamics and the formation and dynamics of aqueous pores. The mesoscopic model of the electroporation process presented in this chapter is illustrated schematically in Figure 11.2. We describe three continuum models to account for the electrolyte dynamics and aqueous pore dynamics.

Aqueous pore dynamics. The number N and radius r of aqueous pores is modeled using the Smoluchowski–Einstein equation. The Smoluchowski–Einstein equation is derived from statistical mechanics and the energy required to form an aqueous pore. The energy to create an aqueous pore is commonly constructed by assuming the membrane is a dielectric and elastic continuum. This energy term accounts for the electromechanical properties of lipids in the engineered membrane. The Smoluchowski–Einstein equation is an advection-diffusion partial differential equation.

Pore conductance model. The membrane conductance G_m is dependent on the number and size of aqueous pores present with each aqueous pore having a conductance $G_c(r)$. A continuum model for estimating $G_c(r)$ is constructed in this chapter that accounts for steric effects of ions, Coulomb correlations between ions, and polarization and screening effects of ions and water. The conductance model is composed of a system of partial differential equations that model the concentration dynamics of ions in the engineered membrane.

Capacitance model. For low excitation potentials, the double-layer capacitance C_{dl} of the engineered membrane is approximately constant. However, in the process of electroporation it may be desirable to use large excitation potentials (above 500 mV), in which case the steric effects of ions, Coulomb correlations between ions, and polarization and screening effects of ions and water at the bioelectronic interface are non-negligible. In such cases the double-layer capacitance C_{dl} is dependent on the double-layer potential V_{dl}. In this chapter we construct a system of partial differential equations to model the concentration dynamics of ions at the bioelectronic interface of the engineered membrane.

The mesoscopic model described in this chapter consists of the three continuum models all coupled to the fractional-order macroscopic model presented in Chapter 8. The fractional-order macroscopic model accounts for the combination of diffusion-limited charge transfer, reaction-limited charge transfer, and ionic adsorption dynamics that are present at the bioelectronic interface. The mesoscopic model is used to estimate the current response $I(t)$ of the engineered membrane resulting from the excitation potential $V_s(t)$ as illustrated in Figure 11.2.

The mesoscopic model consists of a system of partial differential equations coupled to a system of fractional-order differential equations. The Smoluchowski–Einstein equation models the pore population and radii $n(r, t)$ as a function of time t and is represented by a nonlinear second-order partial differential equation (PDE); see Appendix A for a short description of PDEs. Assuming surface tension is not dependent on pore density, the Smoluchowski–Einstein equation can be represented by a second-order parabolic partial differential equation. The pore conductance $G_c(r)$ and double-layer charging dynamics C_{dl} are estimated using a fourth-order linear PDE coupled with a second-order nonlinear PDE when Coulomb correlations, polarization effects, and screening effects are non-negligible. However, if these and steric effects are negligible, then a second-order elliptic PDE coupled with a system of second-order mixed-type PDEs can be used. In this chapter all PDEs and associated boundary conditions are defined to model the current response from engineered membranes that undergo electroporation.

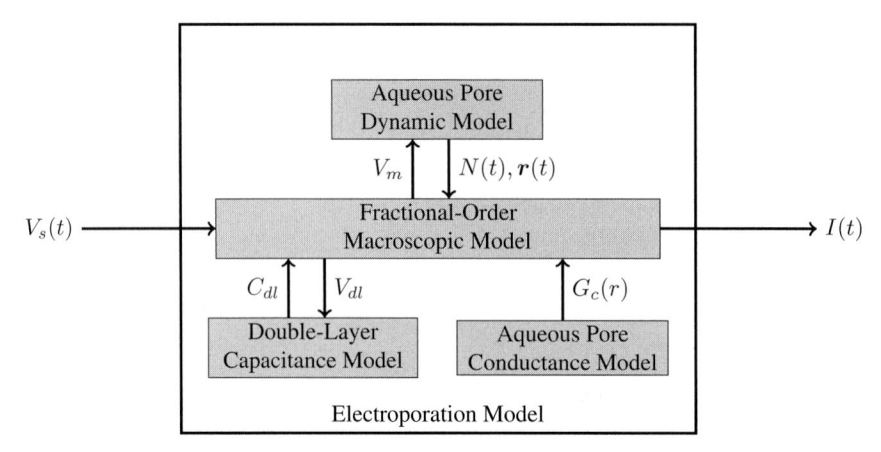

Figure 11.2 Schematic of the electroporation model in the engineered membrane. The model consists of the fractional-order macroscopic model coupled to continuum models of the number and size of aqueous pores (Smoluchowski–Einstein equation), double-layer capacitance (Poisson–Fermi–Nernst–Planck (PFNP) equation), and the conductance of an aqueous pore (generalized Poisson–Nernst–Planck (GPNP) equation). $V_s(t)$ is the excitation potential, $I(t)$ is the resulting current response, V_m is the transmembrane potential, C_d is the double-layer capacitance, V_{dl} is the double-layer potential, $N(t)$ is the number of pores, $r(t) = [r_1, r_2, \ldots, r_{N(t)}]'$ are the pore radii, and $G_c(r)$ is the conductance of an aqueous pore.

11.1.4   Organization of This Chapter

This chapter uses concepts in electrochemistry and statistical mechanics to model engineered membranes. The chapter begins with the construction of the mesoscopic model and a description of how parameters of the engineered membrane can be estimated from the continuum models. §11.2 presents the Smoluchowski–Einstein equation that models the dynamics of aqueous pores that result from changes in the transmembrane potential. The Smoluchowski–Einstein equation encodes all the mechanical properties of the engineered membrane using the energy required to form an aqueous pore. §11.3 presents the mesoscopic model of the engineered membrane, illustrated in Figure 11.2. The fractional-order macroscopic model of the membrane is presented in §11.3.1. If the fractional-order macroscopic model was coupled directly with the Smoluchowski–Einstein equation, then estimating the current response would require the solution to a nonlinear partial differential equation coupled with a fractional-order nonlinear integrodifferential equation. Note that we cannot use a two–time scale singular perturbation approximation here as the Smoluchowski–Einstein equation and fractional-order macroscopic model both operate on the same time scale (for electroporation this is typically milliseconds). However, as we show in §11.3.2, it is possible to use singular perturbation approximations on parameters of the energy required to form an aqueous pore. This allows us to model the dynamics of aqueous pores using asymptotic approximations to the Smoluchowski–Einstein equation which results in a system of ordinary differential equations (ODEs). The system of ODEs that models the dynamics of aqueous pores and current response is dependent on several electromechanical parameters.

In §11.4 we use continuum models to estimate the value of these important parameters, specifically the following:

Pore parameters. §11.4.1 presents the generalized Poisson–Nernst–Planck (GPNP) model to estimate the pore conductance $G_p(r)$ and potential energy term $W_{es}(r, V_m)$ which is contained in the energy required to form an aqueous pore.
Double-layer capacitance. §11.4.2 presents the Poisson–Fermi–Nernst–Planck (PFNP) model to estimate the double-layer charging capacitance C_{dl} as a function of the double-layer potential V_{dl}. Additionally, §11.6 illustrates how to detect and model faradic reactions at the bioelectronic interface.
Parameter estimation. The GPNP and PFNP model the concentration of ions in an electrolyte as a function of time. However, we are interested in computing the parameters $G_p(r), C_p(r), W_{es}(r, V_m)$, and C_{dl}.

Notice that, once the parameters $G_p(r), W_{es}(r, V_m)$, and C_{dl} are estimated, they can be directly plugged into the fractional-order macroscopic model. The reason is that the steady-state value of these parameters is reached on a time scale of microseconds, whereas the dynamics of electroporation and charge accumulation (accounted for by the fractional-order macroscopic model and Smoluchowski–Einstein equation) are on the time scale of milliseconds. Therefore, we have implicitly used a two–time scale approximation when constructing the mesoscopic model. That is, the electrodiffusive dynamics of the ions is on a significantly faster time scale than the dynamics of pore formation and charge accumulation at the bioelectronic interface and membrane surface.

11.2 Smoluchowski–Einstein Equation

The Smoluchowski–Einstein equation [27, 126, 318] provides a useful probabilistic model for the number and size of aqueous pores in an engineered membrane, thereby allowing us to model electroporation. In a membrane, the creation and destruction of large numbers of aqueous pores (see Figure 11.1) occurs as a result of thermal fluctuations. The Smoluchowski–Einstein equation takes into account the mechanical properties of the membrane and how the membrane reacts to changes in the transmembrane potential.

The Smoluchowski–Einstein equation is an advection-diffusion partial differential equation (see Appendix A for a short description of PDEs) that governs the probability distribution of the number of aqueous pores as a function of their radius r and time t. Denoting $n(r, t)$ as the unnormalized pore probability density function,[4] the Smoluchowski-Einstein equation asserts

$$\frac{\partial n}{\partial t} = D\frac{\partial}{\partial r}\left[\frac{n}{k_B T}\frac{\partial W(r, n)}{\partial r} + \frac{\partial n}{\partial r}\right] + S(r, n), \tag{11.1}$$

---

[4] Here $n(r, t)$ is an unnormalized probability density with respect to radius r, meaning that $n(r, t) \geq 0$ and $\int_{\mathbb{R}_+} n(r, t)dr = N(t)$, where $N(t)$ is the normalization constant of $n(r, t)$.

where D is the diffusion coefficient of pores, k_B is the Boltzmann constant, T is the temperature, $W(r, n)$ is the change in energy of the membrane to create a pore of radius r, and $S(r, n)$ models the creation and destruction rate of pores. Equation (11.1) describes the time evolution of the unnormalized probability density function of the number and size of pores as a function of temperature and the energy required to form the pore. Given the solution to (11.1), computing the number of pores with a radius in the interval r to $r + \Delta r$ at time t requires evaluating $n(r, t)\Delta r$.

Two important parameters that can be computed from $n(r, t)$ are $N(t)$, which is the total number of pores in the membrane at time t, and $R(t)$, which is the average size of pores in the membrane at time t. Formally, $N(t)$ and $R(t)$ are given by

$$N(t) = \int_0^\infty n(r, t) dr,$$

$$R(t) = \boldsymbol{E}[r] = \frac{1}{N(t)} \int_0^\infty \eta n(\eta, t) d\eta. \tag{11.2}$$

Here, $N(t)$ is the normalization constant of the unnormalized pore probability density function $n(r, t)$, $\boldsymbol{E}$ denotes the expectation operator, and $R(t)$ is the expected value of the pore radii from the normalized pore probability density function $n(r, t)/N(t)$.

Notice that the right-hand side of (11.1) models both the source of pores (creation and destruction of pores) and the dynamics of created pores. If $D = 0$, then only the source term $S(r, n)$ would contribute to the dynamics of the unnormalized pore probability density function $n(r, t)$. In such cases, pores could only be created or destroyed, but they would not change size. In the case that $S(r, n) = 0$ and $D \neq 0$, then (11.1) takes the form of the homogenous Smoluchowski–Einstein equation.[5] In this case the population of pores does not change over time: $N(t) = N(0)$ where $N(0)$ is the initial population of pores. Only the size of the pores changes over time. Both the source term and size dynamics of the pore are dependent on the temperature T and the transmembrane electric potential V_m. Intuition tells us that if we increase the temperature of the membrane then this should promote the formation of pores. Interestingly from (11.1) we see that an increase in temperature T causes a decrease in the contribution from the energy required to form a pore; however, the source term $S(r, n)$ for the creation of pores must increase as T increases. Typical initial and boundary conditions for (11.1) can be found in [297–299, 318]. A common assumption is that initially for a membrane with zero transmembrane potential $V_m = 0$, all pores have radius r_*, and any pores for which $r < r_*$ close rapidly. Therefore, the solution to (11.9) can be estimated by solving (11.1)–(11.8) with the initial condition $n(r, 0) = G_m(0)/G_p(r_*)$ and absorbing boundary condition $n(r, t) = 0$ if $r < r_*$. The parameter r_* can be interpreted as the radius at which a hydrophobic pore transitions to a hydrophilic pore.

---

[5] If the source term $S(r, n) = 0$, then (11.1) is an inhomogeneous PDE, and if $S(r, n) \neq 0$, then (11.1) is a nonhomogeneous PDE.

11.2.1 Source Term and Energy Term of the Smoluchowski–Einstein Equation

Below we discuss the source term $S(r, n)$ and hydrophilic aqueous pore energy term $W(r, n)$ that arise in the Smoluchowski–Einstein equation (11.1).

The source term $S(r, n)$ in the Smoluchowski–Einstein equation (11.1) accounts for the creation and destruction of aqueous pores in the membrane. The phenomenological equation commonly used for $S(r, n)$ is

$$S(r, n) = \underbrace{\frac{v_c h_m}{k_B T} \frac{\partial U}{\partial r} e^{U/k_B T}}_{\text{Pore Creation}} - \underbrace{v_d n \mathbf{1}\{r < r_*\}}_{\text{Pore Destruction}}, \tag{11.3}$$

which has units of s^{-1} to account for the rate of change of the number of aqueous pores. In (11.3), U is the energy of a nonconducting hydrophobic pore, v_c is the "attempt-rate density" (e.g., the fluctuation rate per unit volume), v_d is the destruction-rate density (e.g., the frequency of lipid fluctuations), r_* denotes the minimum radius for the transition from a hydrophobic pore to a hydrophilic pore, and $\mathbf{1}\{r < r_*\}$ is the indicator function.[6] $S(r, n)$ assumes the formation and destruction of pores is a two-step process, namely, pore creation and pore destruction. Below we discuss each of these important processes.

Pore creation. It is assumed that all created pores are nonconducting hydrophobic pores, as illustrated in Figure 11.1 on page 214. The rate of creation of pores with energy U to $U + dU$ is given by [453]

$$\frac{v_c h_m}{k_B T} e^{U/k_B T} dU. \tag{11.4}$$

Notice that (11.4) resembles the Arrhenius equation[7] used to compute chemical reaction rates. The Arrhenius equation is based on the physical principle that, as temperature rises, molecules move faster and collide more often, which in turn causes an increase in the likelihood of molecular rearrangements. In (11.4), U is the activation energy associated with the creation of a nonconducting hydrophobic pore, and $k_B T$ is the average kinetic energy. The energy U is generally expressed in terms of modified Bessel functions. Typically for engineered membranes, however, the energy U is approximated by a quadratic function

$$U(r) = W_* \left(\frac{r}{r_*}\right)^2 + W_{es}(r, V_m), \tag{11.5}$$

where W_* is the energy of a pore of radius r_* (the minimum radius for the transition from a hydrophobic pore to a hydrophilic pore), and $W_{es}(r, V_m)$ is the electrical energy required to form the hydrophobic pore of radius r_*.

The creation rate of pores with radii in the range of r to $r + dr$ can be computed by substituting the relation $dU = (\partial U/\partial r)dr$ into (11.4). Additionally, we can compute

---

[6] The indicator function $\mathbf{1}\{r < r_*\} = 0$ if $r_* < r$, and 1 otherwise. This is also known as the Heaviside step function.

[7] The Arrhenius equation relates reaction rates to temperature and is given by $k = Ae^{-U/k_B T}$, where k is the reaction rate, A is the frequency factor or preexponential factor, and U is the activation energy.

$\eta(r_1, r_2)$, the average creation rate of hydrophobic pores with radii in the range of r_1 to r_2. The average creation rate $\eta(r_1, r_2)$ is

$$\eta(r_1, r_2) = \frac{1}{r_2 - r_1} \int_{r_1}^{r_2} \left. \frac{v_c h_m}{k_B T} \frac{\partial U(\xi)}{\partial r} \right|_{r=\xi} e^{U(\xi)/k_B T} d\xi. \tag{11.6}$$

Typical values of the parameters in (11.6) are $v_c = 2 \times 10^{38}$ s/m^3, $r_* = 0.5$ nm, $h_m = 4.0$ nm, and $W_* = 45\, k_B T$ for $V_m = 0$. Hydrophobic pores can be created with any radii. However, as illustrated by (11.6), the creation rate of pores with radii $r \ll r_*$ is negligible (e.g., $\eta(r_1, r_2) \approx 0$). Therefore, most hydrophobic pores are created with radii near r_*.

Pore destruction. Equation (11.4) provides the creation rate of aqueous pores; however, we also require the destruction rate of aqueous pores. We assume that any created hydrophobic pore with radius $r \geq r_*$ will spontaneously convert to a hydrophilic pore, as illustrated in Figure 11.1 on page 214. If a pore is created with $r < r_*$, then the hydrophobic pore will only remain open on the time scale of the thermal fluctuations of the lipids. Denoting v_d as the frequency of lipid fluctuations, the fraction of aqueous pores $r < r_*$ that are destroyed in time dt is $v_d dt$. Therefore, the rate of pore destruction for pores with radius in the range of r to $r + dr$ is given by

$$v_d n \mathbf{1}\{r < r_*\} dr, \tag{11.7}$$

where n is the unnormalized probability density function of the Smoluchowski–Einstein equation (11.1) and $\mathbf{1}\{r < r_*\}$ is the indicator function. The step (indicator) function $\mathbf{1}\{r < r_*\}$ accounts for the assumption that nonconducting pores with $r < r_*$ have a lifetime on the order of the lipid fluctuations. Typical values of the parameters in (11.7) are $r_* = 0.5$ nm and $v_d = 10^{11}$ s^{-1}.

The source term $S(r, n)$, defined in (11.3), in the Smoluchowski–Einstein equation (11.1) is constructed by the difference of the creation rate given by (11.4) and the destruction rate given by (11.7).

Hydrophilic pore energy. Having constructed the source term $S(r, n)$ (11.3), we now construct the free energy $W(r, n)$ of hydrophilic aqueous pores. The classical model for free energy $W(r, n)$ of a hydrophilic aqueous pore, in (11.1), consists of four energy terms: the pore edge energy γ, the membrane surface tension $\sigma(n)$, the electrostatic interaction between lipid heads, and the transmembrane potential energy contribution $W_{es}(r, V_m)$. The pore energy $W(r, n)$ in (11.1) can be modeled using

$$W(r, n) = 2\pi \gamma r - \pi \sigma(n) r^2 + \left(\frac{C}{r}\right)^4 + W_{es}(r, V_m) + W_m, \tag{11.8}$$

with the energy contribution from the mechanobiological properties of the tethers included as W_m. The surface tension $\sigma(n)$ is dependent on n because the creation of pores will decrease the area occupied by the membrane. If we denote A_m as the area of the membrane, then the effective area of the membrane is $A'_m = A_m - A_p$, where A_p is the area of the membrane occupied by the pores and is dependent on the pore density distribution function n. As A_p increases, the surface tension of the membrane is

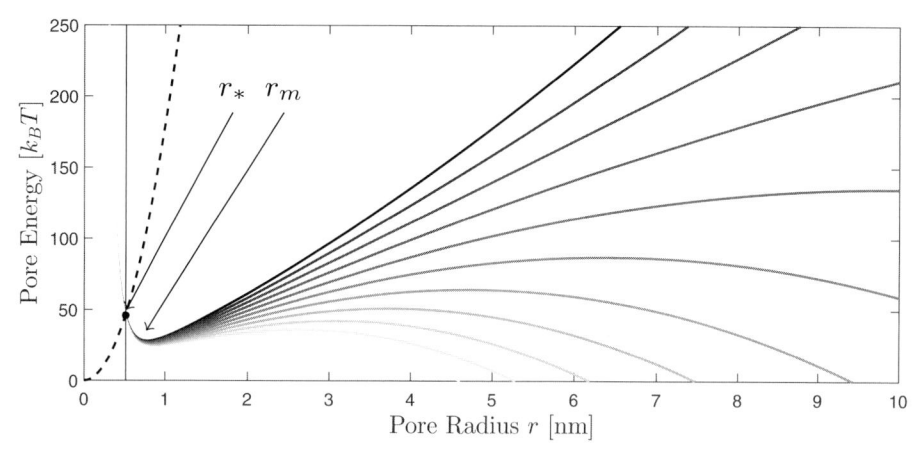

Figure 11.3 The pore energy $W(r, n)$ (11.8) as a function of the pore radius r. Computed hydrophilic (solid lines) and hydrophobic (dashed black line) pore energy for transmembrane potentials V_m in the range of 0 to 1 V. The hydrophilic energy over the range 0–1 V is represented by different shades of gray. The hydrophilic energy at $V_m = 0$ V is given by the darkest black line, and the hydrophilic energy at $V_m = 1$ V is given by the lightest gray line. The parameter r_* is the minimum radius for the transition from a hydrophobic pore to a hydrophilic pore, and r_m is the equilibrium radius of a hydrophilic pore. The curves are computed using (11.5) and (11.8) with the parameters defined in Table C.6 on page 418.

expected to decrease. However, if $A_p \ll 1$ then the surface tension dependence on n can be neglected with $\sigma(n) = \sigma$. The linkages of the tethers to the membrane are analogous to "springs" and act to restrain the enlargement of aqueous pores. This is similar to the experimentally measured results in [67], which suggest that irreversible electroporation cannot create pores that are larger than the cytoskeletal network anchors. Modeling the mechanical properties of the membrane as an elastic continuum and assuming a permanent tethered network anchorage, the effect of the tethers is accounted for via the energy required to deform the Hookean springs – formally, the energy contribution can be modeled using $W_m = 0.5 K_t r^2$ with K_t denoting the spring constant of the tethers.

Insight into the dynamics of pore creation, destruction, and dynamics can be gained from the hydrophobic and hydrophilic pore energy. The aqueous pore energy is illustrated in Figure 11.3. The parameter r_*, which is the minimum pore radius at which a hydrophobic pore will spontaneously convert to a hydrophilic pore, does not vary significantly even for transmembrane potentials of $V_m = 1$ V. Any pore created with a radius $r > r_*$ will spontaneously convert to a hydrophilic pore. This follows from the difference in energy between a hydrophilic and hydrophobic pore, as illustrated in Figure 11.3. That is, for $r > r_*$, the transition from a hydrophobic pore to a hydrophilic pore will have an energy difference of $U(r) - W(r, n) > 0$. The equilibrium pore radius of hydrophilic pores, r_m, also does not vary significantly with increased transmembrane potential V_m. However, as V_m increases beyond approximately 600 mV, pores that grow in radius beyond a threshold value will continue to grow until the membrane ruptures. This is an example of irreversible electroporation. The threshold of pores is located at the local maxima of the hydrophilic pore energy. For example, for a transmembrane potential $V_m = 600$ mV, any pore that has a size larger than approximately 7 nm will continue to grow until irreversible electroporation takes place.

11.2.2 Summary

In this section we have presented the Smoluchowski–Einstein equation (11.1) which provides a probabilistic model for $n(r, t)$ that defines the number and size of aqueous pores in the engineered membrane. The Smoluchowski–Einstein equation depends on the relationship between the pore radius and energy, as well as r_* (the minimum radius for the transition from a hydrophobic pore to a hydrophilic pore). For hydrophobic pores, the pore energy $U(r, n)$ can be approximated by (11.5), and for hydrophilic pores, the pore energy $W(r, n)$ can be approximated by (11.8). Assuming that all hydrophobic pores created with $r > r_*$ spontaneously convert to hydrophilic pores, the pore energy will contain a local maxima at r_* and a local minima at $r_m > r_*$ (equilibrium minimum energy radius r_m of hydrophilic pores at $V_m = 0$). The majority of hydrophilic pores will have a radius around r_m as this is a potential well of the hydrophilic pore energy $W(r, n)$.

11.3 Multiphysics (Mesoscopic) Model of Electroporation

This section constructs a multiphysics model[8] of the electroporation process. First we construct a macroscopic equivalent circuit model of the engineered membrane when electroporation is present. The equivalent circuit model parameters depend on the number of aqueous pores present and the conductance, capacitance, and electrical energy of these pores. The main aim is to use the continuum models, including the Smoluchowski–Einstein equation of §11.2, to determine the circuit model parameters from first principles. The continuum models are used to account for steric and screening effects of ions.

11.3.1 Equivalent Circuit Model of Electroporation

The equivalent circuit model of the tethered membrane is given by the fractional-order macroscopic model presented in §8.2; however, the membrane conductance G_m and capacitance C_m depend on the number and size of aqueous pores in the membrane. When these dependencies are taken into account, the equivalent circuit model for electroporation is given by Figure 11.4.

We now discuss how expressions for the individual circuit elements can be obtained using mesoscopic modeling.

---

[8] By multiphysics model, we mean a multi–time scale model, where each level of abstraction operates on a different time scale. Here we combine the physical models of the movement of electrolyte ions and the process of electroporation. This is a multi–time scale and multi–spatial scale model that involves both continuum theory and reaction-rate theory.

11.3 Multiphysics (Mesoscopic) Model of Electroporation

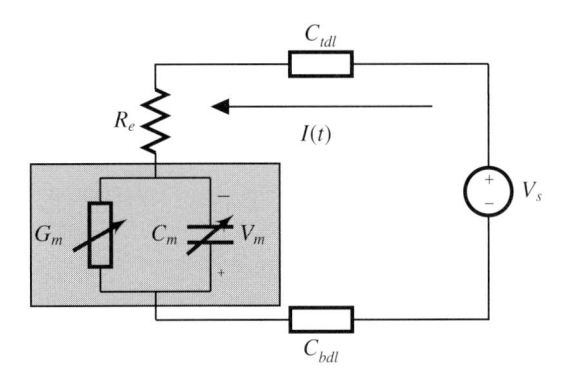

Figure 11.4 Fractional-order macroscopic model of the tethered membrane with electroporation. The membrane conductance G_m and capacitance C_m depend on the number of aqueous pores in the membrane. The other circuit parameters are defined in §8.2 and given by (8.1) on page 141.

Membrane conductance G_m. From the Smoluchowski–Einstein equation (11.1), one can construct expressions for the membrane conductance G_m. The Smoluchowski–Einstein equation yields the pore distribution function $n(r, t)$ over all possible pore radii r as a function of time t and transmembrane potential V_m. To compute the total membrane conductance $G_m(t)$, we require an estimate of the conductance of the aqueous pores. Denoting $G_p(r)$ as the conductance of an aqueous pore of size r, the total membrane conductance G_m can be computed as the expected value of the number of pores of radius r multiplied by the associated conductance $G_p(r)$ of each pore. That is, with E denoting mathematical expectation,

$$G_m(t) = E[G_p(r)] = \frac{1}{N(t)} \int_0^\infty G_p(r) n(r, t) dr. \qquad (11.9)$$

Membrane capacitance C_m. From the Smoluchowski–Einstein equation defined by (11.1)–(11.8), one can construct expressions for the membrane capacitance C_m. The membrane capacitance C_m is modeled as a set of parallel capacitors. The total tethered membrane area is given by A_m, and the capacitance of an aqueous pore of size r is given by $C_p(r)$; then the total membrane capacitance is given by

$$\begin{aligned} C_m(t) &= \frac{\varepsilon_m \varepsilon_0}{h_m} \left(A_m - E[\pi r^2] \right) + E[C_p(r)] \\ &= \frac{\varepsilon_m \varepsilon_0}{h_m} \left(A_m - \frac{1}{N(t)} \int_0^\infty \pi r^2 \, n(r, t) dr \right) + \frac{1}{N(t)} \int_0^\infty C_p(r) \, n(r, t) dr. \end{aligned} \qquad (11.10)$$

In (11.10), ε_m denotes the relative dielectric permittivity of the membrane, and ε_0 is the dielectric permittivity of a vacuum. The pore capacitance $C_p(r)$ is a function of pore radius r as a result of the screening effects of water. Equation (11.10) is valid assuming that the membrane area A_m remains constant. If this is not the case then the relation between the capacitive current I_c and transmembrane voltage V_m is given by

$$I_c = \frac{dq_m}{dt} = C_m \frac{dV_m}{dt} + V_m \frac{dC_m}{dt}, \tag{11.11}$$

where $q_m = C_m V_m$ is the charge on the membrane surface. The capacitance C_m depends on several parameters, including anion and cation sizes, ionic valences, surface charges on the membrane resulting from charged lipids, and the transmembrane voltage V_m. A common model for the voltage-dependent membrane capacitance is $C_m = C_0(1 + \alpha V_m^2)$, where C_0 is the equilibrium capacitance at $V_m = 0$, and α is a constant. For low transmembrane potentials $|V_m| < 300$ mV, typically $dC_m/dt \approx 0$ and the membrane capacitance can be approximated by $C_m(t)$ in (11.10).

Double-layer capacitance C_{dl}, p. In the fractional-order macroscopic model of the engineered membrane, illustrated in Figure 11.4 where $C_{dl} = [C_{tdl}^{-1} + C_{bdl}^{-1}]^{-1}$, the fractional-order capacitance C_{dl} accounts for the electrical double-layer charging dynamics at the gold electrode interface. That is, C_{dl} models the charge accumulation in the Stern and diffuse[9] double layers adjacent to the electrode and counter electrode of the engineered membrane.

As is evident from (11.9) and (11.10), both the membrane conductance $G_m(t)$ and membrane capacitance $C_m(t)$ are dependent on the population and radii of aqueous pores. The overall electrical response of the tethered membrane for a given excitation V_s can be modeled by substituting (11.9) and (11.10) into the fractional-order macroscopic model (Figure 11.4). Evaluating the overall electrical dynamics of the membrane using (11.9), (11.10), and Figure 11.4 is computationally demanding as this requires the simultaneous solution of a nonlinear partial differential equation coupled to a fractional-order nonlinear integrodifferential equation. These equations cannot be solved independently because the pore distribution function $n(r, t)$ and the fractional-order macroscopic model are both dependent on the transmembrane potential V_m and evolve at approximately the same characteristic time scales. However, it is possible to construct a useful approximation by using the tools of singular perturbation theory and the principle of minimum energy to construct a system of fractional-order ordinary differential equations which can describe the electrical dynamics of the engineered membrane.

11.3.2 Singular Perturbation Approximation and Electrical Dynamics

In this section we construct an asymptotic (in time) approximation to the Smoluchowski–Einstein equation that converts the original advection-diffusion partial differential equation in (11.1) into a system of nonlinear ordinary differential equations. From a computational point of view, such ordinary differential equations are easier to handle. The asymptotic Smoluchowski–Einstein equation is then coupled to the

---

[9] The Stern layer represents the charge accumulation of a single layer of ions adjacent to the electrode surface. The diffuse layer represents the charge accumulation at the electrode surface that does not include the Stern layer.

fractional-order model which can be used to estimate the current response of the membrane that results from a given excitation potential.

We start by constructing a dynamic model for the total pore population $N(t)$ at time t. $N(t)$ can be obtained by integrating the pore distribution function $n(r,t)$ (11.1) over all possible pore radii. Using the tools of singular approximations [297], the asymptotic ordinary differential equation for $N(t)$ is given by

$$\frac{dN}{dt} = \alpha e^{(\frac{V_m}{V_{ep}})^2} \left(1 - \frac{N}{N_o} e^{-q(\frac{V_m}{V_{ep}})^2}\right), \quad (11.12)$$

where the non-negative scalar α is the pore creation-rate coefficient, V_{ep} is the characteristic voltage of electroporation, N_o is the equilibrium pore density at $V_m = 0$, and $q = (r_m/r_*)^2$ is the squared ratio of the equilibrium minimum energy radius r_m of hydrophobic pores at $V_m = 0$ with r_* the minimum energy radius of hydrophilic pores.

Three assumptions are used to apply the singular perturbation approximation to construct the ordinary differential equation (11.12) that describes the pore population $N(t)$ from the Smoluchowski–Einstein equation:

Assumption 1. The energy of nonconducting hydrophilic pores in (11.3) can be approximated by the quadratic function

$$U(r) \approx E_* \frac{r^2}{r_*^2}, \quad (11.13)$$

where E_* is the energy barrier for creating a conducting pore, and r_* is the minimum radius of a conducting pore (i.e., the transition radius from hydrophilic pore to a hydrophobic pore).

Assumption 2. The pore distribution function $n(r,t)$ in (11.1) adjusts instantaneously to temporal variations in pore energy W (11.8).

Assumption 3. The electrical energy required to form a pore is $W_{es}(r, V_m) = ar^2 V_m^2$, where the proportionality constant a depends on the properties of the electrolyte and the membrane.

Assumption 1 is valid for typical physiological parameters in engineered tethered membranes. Assumption 2 is a quasistatic approximation and is a reasonable assumption assuming that changes in V_m, the transmembrane potential, occur on a time scale of approximately 5 μs. However, minor violations of this will only impact the equilibrium pore population N_0 and therefore may introduce a small deviation in the estimated $N(t)$. Assumption 3 is valid if steric effects and polarization effects are negligible in the electrolyte solution.

Given the above three assumptions, the singular perturbation approximation can be applied to the Smoluchowski–Einstein equation (11.1) to construct the first-order differential equation

$$\frac{dN}{dt} = K \left(1 - \frac{N}{N_{eq}}\right) \quad (11.14)$$

with parameters K and N_{eq}, which describe the pore population dynamics. The parameters K and N_{eq} are related to the parameters $\{v_d, r_*, r_m, \gamma, \sigma, C, K_t, k_B, T\}$ in the Smoluchowski–Einstein equation (11.1), and the parameters $\{\alpha, V_{ep}, q, N_o\}$ in (11.12) via the following:

$$K = \left(\frac{v_d}{r_*^2}\left(\frac{|W'_{r_*}|}{U'_{r_*} + |W'_{r_*}|}\right)e^{-W_*/k_BT}\right)e^{(V_m/V_{ep})^2} = \alpha e^{(V_m/V_{ep})^2},$$

$$N_{eq} = \left(\frac{v_d(k_BT)^{3/2}}{r_*^2 D}\sqrt{\frac{2\pi}{W''_{r_m}}}\left(\frac{1}{U'_{r_*} + |W'_{r_*}|}\right)e^{-W_m/k_BT}\right)e^{q(V_m/V_{ep})^2} = N_o e^{q(V_m/V_{ep})^2},$$

$$V_{ep} = \frac{1}{r_*}\sqrt{\frac{k_BT}{a}},$$

$$W_* = 2\pi\gamma r_* - \pi\sigma r_* + \left(\frac{C}{r_*}\right)^4 + 0.5 K_t r_*^2,$$

$$W_m = 2\pi\gamma r_m - \pi\sigma r_m + \left(\frac{C}{r_m}\right)^4 + 0.5 K_t r_m^2,$$

$$W'_{r_*} = \left.\frac{\partial W}{\partial r}\right|_{r=r_*}, \quad U'_{r_*} = \left.\frac{\partial U}{\partial r}\right|_{r=r_*}, \quad W''_{r_m} = \left.\frac{\partial^2 U}{\partial^2 r}\right|_{r=r_m}.$$

Notice that the parameters $\{\alpha, V_{ep}, q, N_o\}$ are only dependent on the pore energy W (11.8) and the energy of nonconducting hydrophilic pores, U (11.13), at the two radii r_m and r_*. Recall that r_m is the minimum energy radius of a hydrophobic pore at $V_m = 0$, and r_* is the minimum energy radius of a hydrophilic pore.

Equation (11.12) can be used to model the population dynamics of aqueous pores in the membrane; however, we also need a model to describe the dynamics of the generated pore radii. The Smoluchowski–Einstein equation (11.1) can be represented as an advection-diffusion partial differential equation where it is apparent that the advection and diffusion terms contribute to the pore size dynamics. The advection-diffusion form of the Smoluchowski–Einstein equation is

$$\frac{\partial n}{\partial t} = D\frac{\partial}{\partial r}\left[D\frac{\partial n}{\partial r} + \left(\frac{D}{k_BT}\frac{\partial W(r,n)}{\partial r}\right)n\right] + S(r,n)$$

$$= D\frac{\partial}{\partial r}\left[D\frac{\partial n}{\partial r} + vn\right] + S(r,n), \qquad (11.15)$$

where v is the drift velocity or the radii of pores. In (11.15), the diffusion coefficient D is related to the random increase or decrease in pore radii as a result of random thermal fluctuations. The advection term is a consequence of the energy minimization principle: namely, a hydrophilic pore will always attempt to minimize the pore energy W defined in (11.8). It is possible to determine if the advection transport rate or diffusion transport rate dominates the pore radii dynamics by using the Péclet number[10] P_e. The Péclet

---

[10] From Chapter 9, the Péclet number P_e defines the ratio of diffusive flux to convective flux.

number of (11.15) is

$$P_e = \frac{Lv}{D} = \frac{L}{k_B T} \frac{\partial W}{\partial r},$$

where L is the characteristic length scale associated with the pores. For a typical length scale of $L = 1$ nm, and using the pore energy W defined in (11.8) with the parameters defined in Table C.6 on page 418, $P_e \approx 0.004$. Since $P_e \ll 1$, the advective flux dominates the dynamics of the pore radii. Therefore, for a conducting hydrophilic pore, the pore radius r will evolve according to the ordinary differential equation

$$\frac{dr}{dt} = v = -\frac{D}{k_B T} \frac{\partial W(r, V_m)}{\partial r}, \tag{11.16}$$

where $W(r, V_m)$ is the pore energy given by (11.8) and v is the drift velocity from (11.15). Any created pores are assumed to have an initial radius of r_* (minimum energy radius of the hydrophilic pore) and evolve according to the ordinary differential equation (11.16).

Using (11.12) and (11.16), the membrane conductance $G_m(t)$ and membrane capacitance $C_m(t)$ can be modeled by

$$G_m(t) = \sum_{i=1}^{\lfloor N(t) \rfloor} G_p(r_i),$$

$$C_m(t) = \frac{\varepsilon_m \varepsilon_0}{h_m} \left(A_m - \sum_{i=1}^{\lfloor N(t) \rfloor} \pi r_i^2 \right) + \sum_{i=1}^{\lfloor N(t) \rfloor} C_p(r_i),$$

$$\frac{dr_i}{dt} = -\frac{D}{k_B T} \frac{\partial W(r, V_m)}{\partial r_i} \quad \text{for} \quad r_i \in \{1, 2, \ldots, \lfloor N(t) \rfloor\},$$

$$\frac{dN}{dt} = \alpha e^{\left(\frac{V_m}{V_{ep}}\right)^2} \left(1 - \frac{N}{N_o} e^{-q\left(\frac{V_m}{V_{ep}}\right)^2} \right), \tag{11.17}$$

where $G_p(r)$ and $C_p(r)$ are the associated conductance and capacitance of a conducting pore of radius r. The complete dynamics of the tethered membrane are given by simultaneously solving the system of ordinary differential equations (11.17) and the fractional-order macroscopic model illustrated schematically in Figure 11.4. The fractional-order macroscopic model is defined by (8.1) on page 141.

11.4 Continuum Model of Electroporation: Aqueous Pore Conductance and Double-Layer Capacitance

In the previous section, we constructed an equivalent circuit model of electroporation given by the system of ordinary differential equations (11.17) and the fractional-order macroscopic model (Figure 11.4). These ordinary differential equations are dependent on the following parameters:

- pore conductance $G_p(r)$,
- pore capacitance $C_p(r)$,
- the double-layer capacitance C_{dl}, and
- the electrical potential energy W_{es} for pore formation.

The parameters $G_p(r), C_p(r), C_{dl}$, and W_{es} in turn depend on

(i) electric potential ϕ,
(ii) charge density ρ,
(iii) ionic flux J, and
(iv) dielectric permittivity ε in proximity to an aqueous pore.

In this section we construct two continuum models for computing ϕ, ρ, J, and ε; see the concept chart in Figure 11.2.

Continuum model 1: Generalized Poisson–Nernst–Planck (GPNP). In §11.4.1 the GPNP model is constructed. The GPNP model is similar to the Poisson–Nernst Planck (PNP) model in §10.3.3; however, the GPNP model accounts for the finite-size effects (or steric effects) of ions in the electrolyte. These steric effects contribute to the movement of ions in the electrolyte solution; indeed, the potential ϕ, charge density ρ, and ionic flux J depend on the steric effects of ions in the electrolyte. As such, the parameters $G_p(r), C_p(r), C_{dl}$, and W_{es} depend on the steric effects of ions. Note that the GPNP model assumes the dielectric permittivity ε of the electrolyte and membrane are constant. If this assumption does not hold, then the PFNP model should be used to estimate ϕ, ρ, J, and ε.

Continuum model 2: Poisson–Fermi–Nernst–Planck (PFNP). In §11.4.2 the PFNP model is constructed. The PFNP model accounts for the steric effects of ions and water, the correlation effect of ions (Coulomb correlations), and the polarization and screening effect of water. The PFNP model can be used to estimate ϕ, ρ, J, and ε where the dielectric permittivity of the electrolyte is spatially dependent. The correlation effect of ions and the polarization and screening effect of water are non-negligible when large electric potentials are present. For example, if the double-layer potential V_{dl} is large (above a few hundred millivolts), dielectric permittivity adjacent to the bioelectronic interface is spatially dependent.

The GPNP model can be used to estimate ϕ, ρ, and J assuming that the dielectric permittivity ε of the electrolyte and membrane are constant. For large excitation potentials, the correlation effect of ions and the polarization and screening effect of water molecules cause the dielectric permittivity ε to be spatially dependent. In such cases, the PFNP model can be used to estimate ϕ, ρ, J, and ε. In §11.5 we illustrate how the pore conductance $G_p(r)$, pore capacitance $C_p(r)$, double-layer capacitance C_{dl}, and the electrical potential energy W_{es} can be computed from the GPNP and PFNP models.

11.4.1 Continuum Model 1: Generalized Poisson–Nernst–Planck (GPNP) Equation

Here we construct the GPNP model for the ionic flux and potential gradients in aqueous pores within the engineered membrane. These parameters are important since they can be used to compute the pore conductance $G_p(r)$, pore capacitance $C_p(r)$, double-layer capacitance C_{dl}, and the electric potential energy $W_{es}(r, V_m)$ of an aqueous pore. In engineered membranes the asymmetric electrolyte solution[11] is composed of multiple ionic species, and when an external voltage is applied there exist Stern and diffuse electrical double layers at the surface of the electrodes and membrane. These electrical double layers contain high concentrations of species in which the steric effects of the ions must be accounted for. The GPNP continuum model [449] can be used to account for these electrodiffusion dynamics and steric effects.

Consider an electrolyte composed of n ionic species indexed by $i \in \{1, 2, \ldots, n\}$. For example, $i = 1$ could denote Na$^+$ ions and $i = 2$ could denote Cl$^-$ ions. Then the GPNP model is given by

$$\frac{\partial c^i}{\partial t} = -\nabla \cdot [J^i] = \nabla \cdot \left[D^i \nabla c^i + \frac{D^i q z^i c^i}{k_B T} \nabla \phi - D^i c^i \nabla \ln \left(1 - \sum_{j=1}^{n} N_A a_j^3 c^j \right) \right], \tag{11.18a}$$

$$\nabla \cdot (\varepsilon \nabla \phi) = -\rho = -\sum_{i=1}^{n} F z^i c^i, \quad i \in \{1, 2, \ldots, n\}. \tag{11.18b}$$

For ion species i, c^i is the concentration, D^i is the diffusivity tensor, z^i is the charge valency, q is the elementary charge, N_A is Avogadro's number, and a_i is the effective ion hydration size of chemical species i. Note that the expression in square brackets in (11.18a) is the concentration flux J^i. In Poisson's equation (11.18b), ϕ denotes the potential field, and F is Faraday's constant. The GPNP model (11.18) is composed of a system of nonlinear partial differential equations coupled with an elliptic partial differential equation (see Appendix A for a short overview of partial differential equations). Note that if $a_j = 0$ in (11.18), then (11.18) reduces to the standard PNP model presented in §10.3.3.

---

[11] An asymmetric electrolyte is composed of several ion species that have different properties such as ionic radius and/or diffusion coefficient.

Derivation of the GPNP model. The main idea underpinning the GPNP model (11.18) is the inclusion of a steric constraint on the maximum concentration of the analyte species. We now show that the GPNP model (11.18) can be obtained by starting with the PNP model (10.10) and then considering an additional excess chemical potential added to the mass conservation equations to account for steric effects. The PNP model with the added excess chemical potential is given by

$$\frac{\partial c^i}{\partial t} = \nabla \cdot \left[\frac{D^i}{k_B T} c^i \nabla \mu^i \right]$$

$$= \nabla \cdot \left[\frac{D^i}{k_B T} c^i \nabla (\mu_{id}^i + \mu_{ex}^i) \right]$$

$$= \nabla \cdot \left[\frac{D^i}{k_B T} c^i \nabla \left(q z^i \phi + k_B T \ln \left(\Lambda_i^3 c^i \right) + \mu_{ex}^i \right) \right]$$

$$= \nabla \cdot \left[D^i \nabla c^i + \frac{D^i q z^i c^i}{k_B T} \nabla \phi + \frac{D^i}{k_B T} c^i \nabla \mu_{ex}^i \right]$$

$$\nabla \cdot (\varepsilon \nabla \phi) = -\sum_{i=1}^{n} F z^i c^i \qquad \text{for } i = \{1, 2, \ldots, n\}, \tag{11.19}$$

where Λ_i (on the third line of the equation) is the de Broglie wavelength[12] of species i. In (11.19) the total chemical potential of species i is

$$\mu^i = \mu_{id}^i + \mu_{ex}^i. \tag{11.20}$$

Here, μ_{id}^i is the ideal chemical potential[13] used in the PNP equations, and μ_{ex}^i is the excess chemical potential which accounts for the steric interactions of species i.

There are several methods which can be used to estimate the excess chemical potential μ_{ex}^i to account for steric interactions: density functional theory, the distance of closest approach, and the local packing fraction. The excess chemical potential μ_{ex}^i in (11.19) is

$$\mu_{ex}^i = k_B T \ln(\gamma^i)$$

$$= k_B T \ln \left(\frac{1}{1 - \sum_{i=1}^{n} \frac{c^i}{c_{max}^i}} \right), \tag{11.21}$$

where γ^i is the activity coefficient of species i, and c_{max}^i is the maximum allowable concentration of species i. Here we assume a for species i given by $c_{max}^i = 1/N_A a_i^3$.

---

[12] The de Broglie wavelength is defined as the ratio of Planck's constant h to the momentum p of the particle and is given by $\Lambda_i = h/p$. The de Broglie wavelength accounts for the quantum-mechanical effect that every moving particle (e.g., ion, water) is associated with a wavelength. Just as light can act as both a particle and a wave (wave-particle duality), molecules can act as both a particle and a wave. The definition of the de Broglie wavelength for light and particles is identical. In the electrolyte solution of engineered tethered membranes Λ_i is a constant such that $\nabla \Lambda_i = 0$.

[13] The ideal chemical potential μ_{id}^i is commonly referred to as the electrochemical potential in the thermodynamic literature.

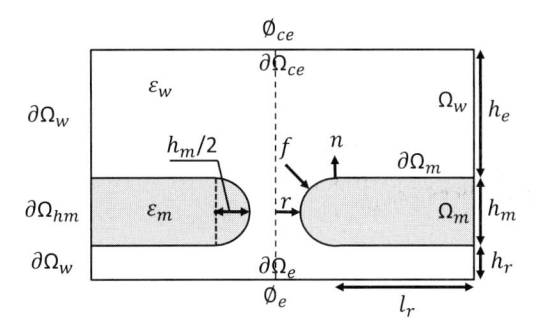

Figure 11.5 Schematic of an aqueous pore in an engineered membrane. The geometry and boundary conditions for (11.18) are specified in (11.22). The variables $\partial\Omega$ specify the various boundary surfaces in (11.22) for estimating the electrostatic potential ϕ, charge density ρ, and ionic flux J^i of species i. Recall that these parameters are used to estimate the pore conductance $G_p(r)$, the pore capacitance $C_p(r)$, and the electric potential energy $W_{es}(r, V_m)$. The aqueous pore is assumed to have a toroidal structure. The pore shape is modeled as axisymmetric with dimensions given by r, l_r, h_r, h_m, and h_e. The potentials at each electrode are defined by ϕ_e and ϕ_{ec}, respectively. The electrolyte has an electrical permittivity ε_w and the membrane an electrical permittivity ε_m.

Note that if v_i is the volume of a particle of species i, then $c^i_{max} = 1/N_A v_i$. Substituting (11.21) into (11.19) yields the GPNP equation (11.18).

Notice that with the steric effects included the GPNP model (11.18) is able to account for anisotropic diffusion, asymmetric electrolytes, multiple ionic species with different valences, and the Stern and diffuse electrical double layers present at the surface of the electrodes and membrane. Note that (11.21) only provides an approximation to the steric effects of ions in a solution. A first-principles method for construction of the excess chemical potential μ^i_{ex} from ab initio molecular dynamics or quantum mechanics simulations is to use the Widom insertion method from statistical mechanics. However, using first-principles quantum mechanics to compute μ^i_{ex} is a computationally prohibitive task, with only a few results currently available in the literature for simple planar geometries and dilute electrolytes in thin slits [134, 158]. For further details the reader is referred to [135].

Boundary and initial conditions, and geometry. The boundary and initial conditions and simulation geometry for the GPNP model (11.18) need to be specified in order to compute the electrical potential ϕ, charge density ρ, and ionic flux J^i for species $i \in \{1, 2, \ldots, n\}$. Recall that these parameters are required to compute the pore conductance $G_p(r)$, pore capacitance $C_p(r)$, double-layer capacitance C_{dl}, and electric potential energy $W_{es}(r, V_m)$ of an aqueous pore. Regarding the simulation geometry of the aqueous pore, typically a toroidal pore is assumed. One justification for the toroidal pore structure is that it can be obtained from first-principles molecular dynamics simulations as will be described in Chapter 15. In this chapter, we use the toroidal pore structure illustrated in Figure 11.5 to compute ϕ, ρ, and J^i for species $i \in \{1, 2, \ldots, n\}$.

The boundary and initial conditions of (11.18) for computing ϕ, ρ, and J^i are specified as

$$n \cdot J^i = 0 \text{ in } \partial\Omega_m \cup \partial\Omega_e \cup \partial\Omega_{ec}, \tag{11.22a}$$

$$\phi_m - \phi_w = 0 \text{ in } \partial\Omega_m,$$

$$\varepsilon_m \nabla \phi_m \cdot n - \varepsilon_w \nabla \phi_w \cdot n = 0 \text{ in } \partial\Omega_m, \tag{11.22b}$$

$$C_s(\phi_e - \phi) + \varepsilon_w n \cdot \nabla \phi = 0 \text{ in } \partial\Omega_e,$$

$$C_s(\phi_{ec} - \phi) + \varepsilon_w n \cdot \nabla \phi = 0 \text{ in } \partial\Omega_{ec}, \tag{11.22c}$$

$$c^i = c_o^i \text{ in } \partial\Omega_w,$$

$$n \cdot \nabla \phi = 0 \text{ in } \partial\Omega_{hm} \cup \partial\Omega_w. \tag{11.22d}$$

Equation (11.22a) states that the membrane surface $\partial\Omega_m$ is assumed to be perfectly polarizable so that the ionic flux J^i in the direction n that is normal to the membrane surface is zero. Also, since there are no surface reactions present at the bioelectronic interface ($\partial\Omega_e$ and $\partial\Omega_{ec}$), a no-flux boundary condition is present at the gold bioelectronic interface. To ensure the well-posedness of the Poisson equation (11.18b), the internal boundary conditions on the membrane-electrolyte interface are specified by (11.22b). At the bioelectronic interface ($\partial\Omega_e$ and $\partial\Omega_{ec}$), a compact Stern layer exists with a capacitance per unit area given by C_s. The Stern layer adjacent to the electrodes is modeled using (11.22c), where ϕ_e and ϕ_{ec} are the specified potentials at the gold electrode $\partial\Omega_e$ and gold counter electrode $\partial\Omega_{ec}$, respectively. Equation (11.22d) provides the ambient boundary conditions away from the pore and c_o^i denotes the initial concentration of ionic species $i \in \{1, 2, \ldots, n\}$.

Summary. This section has discussed the GPNP continuum model for electroporation. Given the boundary conditions (11.22), the electrostatic potential ϕ, charge density ρ, and ionic flux J^i of species $i \in \{1, 2, \ldots, n\}$ can be evaluated by solving the GPNP system (11.18). These are used in §11.5 to compute the lumped circuit parameters $G_p(r)$, $C_p(r)$, and C_{dl}, and the electrical potential energy W_{es} in the electrical circuit model of the engineered membrane with electroporation; see Figure 11.4.

11.4.2 Continuum Model 2: Poisson–Fermi–Nernst–Planck (PFNP) Equation

In this section we describe the second continuum model, namely, the PFNP model of electrodiffusive dynamics in engineered membranes. An important property of the PFNP model compared to the GPNP model is that it accounts for the correlation effect of ions (Coulomb correlations) and the polarization and screening effects of water.[14] For species $i \in 1, 2 \ldots, n$ (which includes ionic species and water), the PFNP equation

---

[14] Refer to §3.6 for a detailed discussion of these important electrolyte dynamics.

is given by

$$\frac{\partial c^i}{\partial t} = -\nabla \cdot [J^i] = \nabla \cdot \left[D^i \nabla c^i + \frac{D^i q z^i c^i}{k_B T} \nabla \phi - D^i c^i \nabla \ln\left(1 - \sum_{j=1}^{n} N_A a_j^3 c^j\right) \right],$$

$$\varepsilon\left(1 - l_c^2 \nabla^2\right) \nabla^2 \phi = -\rho = -\sum_{i=1}^{n} F z^i c^i. \tag{11.23}$$

Here l_c^2 denotes the Coulomb correlation length with all other parameters defined in the GPNP model (11.18). Coulomb correlation describes the correlation between the spatial position of electrons due to their Coulomb repulsion. Note that this is a quantum-mechanical effect that results from the interaction of electrons. For highly concentrated electrolytic solutions, there are no water molecules present and the dielectric permittivity is equal to the permittivity of the ions (Na^+ and/or Cl^-), which have a small dipole moment. The dipole moment of Cl^- is in the range of 0.35 to 0.85 C·m (estimated from ab initio molecular dynamics) whereas water has a dipole moment of 1.85 D (units of Debye or approximately 3.3×10^{-30} C·m) [130]. However, Coulomb correlations are dominant in these regions where the effective permittivity increases as a result of the alignment of the dipole-dipole pairs between ions. These Coulomb correlations lead to a wavelength-dependent permittivity on the length scale associated with these dipole-dipole pairs. The Coulomb correlation length is equal to length scale associated with these dipole-dipole pairs. Note that at low concentrations of ionic species the permittivity is dominated by the polarization of the dipolar water molecules, which is a constant for electrolytes.

It is instructive to compare the GPNP model (11.18) of the previous section with the PFNP model (11.23). The major difference between the GPNP (11.18) and PFNP (11.23) models is that the Poisson equation (11.18b) has been replaced with the Poisson–Fermi equation. The Poisson–Fermi equation is a fourth-order Cahn–Hilliard-type partial differential equation. The dielectric permittivity operator $\varepsilon(1 - l_c^2 \nabla^2)$ approximates both the dielectric permittivity and the linear response of correlated ions [31, 362]. This allows the Poisson–Fermi equation to account for screening effects caused by polarizable water molecules in the engineered membrane.

Derivation of PFNP. A derivation of the Poisson–Fermi equation using nonequilibrium thermodynamics can be found in [362]. The main assumption is that a linear dielectric response exists for the chemical species in the electrolyte solution. The linear dielectric response comprises a constant permittivity ε plus a nonlocal contribution from Coulomb correlations. The parameter ε accounts for both the electronic polarizability of ions and the dielectric relaxation of water. Additionally, the total charge variations are assumed to occur on a length scale larger than l_c. In typical physiological conditions these assumptions are satisfied. It was also verified in [361] that (11.23) provides an exact description of the dielectric for weak and strong coupling, and reasonable approximation at intermediate coupling with l_c set to the Bjerrum length given by $q^2/\varepsilon k_B T$.

Boundary and initial conditions, and geometry. Our aim is to compute the ionic flux J^i of species i in the electrolyte and the associated electrical potential ϕ. These important parameters are required for computing the pore conductance $G_p(r)$, pore capacitance $C_p(r)$, electrical energy required to form a pore, $W_{es}(r, V_m)$, and the double-layer capacitance C_{dl}. Computing J^i and ϕ requires specifying the boundary and initial conditions and the simulation geometry of the PFNP model (11.23). The parameters J^i and ϕ are required to evaluate the pore conductance $G_p(r)$, capacitance $C_p(r)$, and electrical energy $W_{es}(r, V_m)$ required to form a pore. The toroidal pore geometry we use for the PFNP model is identical to that used for the GPNP model in §11.4.1 and is illustrated in Figure 11.5. The material and boundary conditions of (11.23) for computing J^i and electric field ϕ are given by

$$n \cdot J^i = 0 \text{ in } \partial\Omega_m \cup \partial\Omega_e \cup \partial\Omega_{ec}, \tag{11.24a}$$

$$\phi_m - \phi_w = 0 \text{ in } \partial\Omega_m,$$

$$\varepsilon_m \nabla\phi_m \cdot n - \varepsilon_w \nabla\phi_w \cdot n = 0 \text{ in } \partial\Omega_m,$$

$$\varepsilon_w l_c^2 n \cdot \nabla(\nabla^2\phi) = 0 \text{ in } \partial\Omega_m, \tag{11.24b}$$

$$C_s(\phi_e - \phi) + \varepsilon_w n \cdot \nabla\phi = 0 \text{ in } \partial\Omega_e,$$

$$\varepsilon_w l_c^2 n \cdot \nabla(\nabla^2\phi) = 0 \text{ in } \partial\Omega_e,$$

$$C_s(\phi_{ec} - \phi) + \varepsilon_w n \cdot \nabla\phi = 0 \text{ in } \partial\Omega_{ec},$$

$$\varepsilon_w l_c^2 n \cdot \nabla(\nabla^2\phi) = 0 \text{ in } \partial\Omega_{ec}, \tag{11.24c}$$

$$n \cdot \nabla\phi = 0 \text{ in } \partial\Omega_{hm} \cup \partial\Omega_w,$$

$$\varepsilon_w l_c^2 n \cdot \nabla(\nabla^2\phi) = 0 \partial\Omega_{hm} \cup \partial\Omega_w, \tag{11.24d}$$

$$c^i = c_o^i \text{ in } \partial\Omega_w.$$

Let us parse (11.24). Equation (11.24a) states that the membrane surface $\partial\Omega_m$ is perfectly polarizable so that the ionic flux is zero in the direction that is normal to the surface of the membrane. Also, since there are no surface reactions present at the gold bioelectronic interface ($\partial\Omega_e$ and $\partial\Omega_{ec}$), no-flux boundary conditions are also present at the gold bioelectronic interface. To ensure the well-posedness of the Poisson–Fermi equation, for the internal boundary conditions on the membrane-electrolyte interface we require that there is no net charge accumulation on the membrane surface, $\partial\Omega_m$. Assuming the membrane surface acts as an ideal metal surface (e.g., $n \cdot (\varepsilon(l_c^2\nabla^2 - 1)\nabla\phi) = q_s$, where q_s is the charge accumulation on the surface of the electrode), the total charge accumulation resulting from the electrolyte solution (q_w) and membrane solution (q_m) is given by

$$n \cdot \left(\varepsilon_w(l_c^2\nabla^2 - 1)\nabla\phi_w\right) = q_w,$$

$$n \cdot (\varepsilon_m\nabla\phi_m) = -q_m.$$

Note that electrostatic correlations of molecules in the membrane are neglected in this formulation. For $q_w = q_m$ on the surface of the membrane $\partial\Omega_m$ the boundary conditions (11.24b) must hold. The boundary condition $\varepsilon_w l_c^2 n \cdot \nabla(\nabla^2\phi) = 0$ in $\partial\Omega_m$ merely states that the charge density on the surface of the membrane is flat. At the electrode surface a compact Stern layer exists with a capacitance per unit area given by C_s. This Stern layer results from the hydrogen bonding of water directly adjacent to the electrode surface, which has been observed experimentally. The Stern layer adjacent to the electrodes is modeled using (11.24c) with ϕ_e and ϕ_{ec} the prescribed potentials at the respective electrodes. Equation (11.24d) provides the ambient boundary conditions away from the pore with c_o^i the initial concentration.

Summary. This section has described the second continuum model, namely, PFNP for electroporation. Given the boundary conditions (11.24), the electrostatic potential ϕ, charge density ρ, dielectric permittivity ε, and ionic flux J^i for species $i = \{1, \ldots, n\}$ can be evaluated by solving (11.23). Despite the apparent complexity in the structure of the equations, the continuum models are used in §11.5 to compute the lumped circuit parameters $G_p(r)$, $C_p(r)$, and C_{dl}, and potential energy W_{es} in the electrical circuit model of the engineered membrane with electroporation; see Figure 11.4. Thus the continuum models can be viewed as a first-principles physics-based approach for determining the macroscopic parameters of the engineered membrane. Of course, the continuum model itself has certain parameters that need to be specified; these are estimated from experimental results or computed from first principles via molecular dynamics in Chapters 14 and 15.

11.5 Computing Engineered Tethered-Membrane Parameters from Continuum Theory

Suppose an aqueous pore of radius r is present in an engineered membrane. The main outcome of this section is that we can directly relate the continuum models of the previous section to the electrical lumped circuit model of an aqueous pore. Figure 11.6 illustrates this abstraction from a continuum model to a lumped circuit model. Specifically, we describe how the GPNP and PFNP continuum models of the previous section can be used to compute the pore conductance $G_p(r)$, pore capacitance $C_p(r)$, double-layer capacitance C_{dl}, and the electrical energy $W_{es}(r, V_m)$ required to form an aqueous pore. In a nutshell, the GPNP model (11.18) or the PFNP model (11.23) provides expressions for the electrostatic potential ϕ, charge density ρ, dielectric permittivity ε, and ionic flux J^i for species $i \in \{1, \ldots, n\}$ in the pore geometry illustrated in Figure 11.6; these expressions are then used to evaluate the lumped circuit parameters. We also provide a method to select which continuum model (GPNP model or PFNP model) to use to estimate these parameters.

11.5.1 Computing Pore Conductance

Consider a pore of radius r in the engineered membrane as illustrated in Figure 11.5. The conductance $G_p(r)$ of a pore of radius r can be computed from the total current I_p that passes through the pore, and the transmembrane potential V_m. The pore conductance $G_p(r)$ is given by

$$G_p(r) = \frac{I_p(r)}{V_m} \quad \text{where} \quad I_p(r) = F \sum_i \int_0^r 2\pi \xi J^i(\xi) \cdot n_e \, d\xi. \tag{11.25}$$

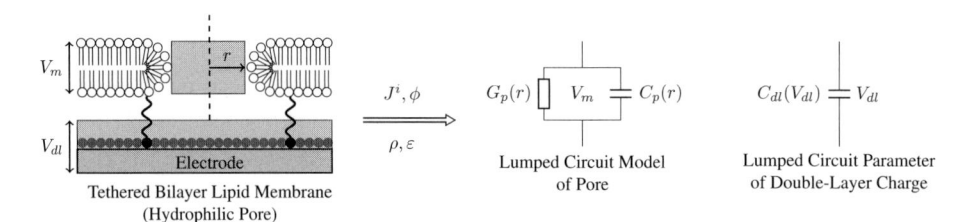

Figure 11.6 Schematic of how the continuum model will be used on the hydrophilic pore structure to estimate the lumped circuit parameters $G_p(r)$, $C_p(r)$, and C_{dl}. The continuum model is used to compute the ionic flux J^i, electrostatic potential ϕ, charge density ρ, and dielectric permittivity ε for different transmembrane potentials V_m, and double-layer potentials V_{dl}. Given the computed parameters from the continuum models, the pore conductance $G_p(r)$, pore capacitance $C_p(r)$, and double-layer capacitance C_{dl} can be computed. The gray rectangles depict the regions that constitute the lumped circuit parameters in the engineered membrane model.

Here n_e is the normal vector to the electrode surface, ξ is the distance from the center of the pore, J^i is the flux of species i, and F is the Faraday constant. The evaluation of J^i and V_m from the GPNP or PFNP models is that they are both time dependent, which results in the pore conductance also being a function of time. However, after a period of several microseconds the steady-state pore conductance $G_p(r)$ is reached. Note that the steady-state pore conductance occurs when the ratio of the ionic current I_p and the transmembrane potential V_m becomes a constant.

11.5.2 Electrical Potential Energy for Pore Formation

The electrical potential energy $W_{es}(r, V_m)$ involved in the formation of an aqueous pore accounts for any electric forces that arise from applying a transmembrane potential V_m. More specifically, if a transmembrane potential V_m is applied across an aqueous pore of radius r, then an electric force is induced on the interface of the solvent and the membrane surface. Note that this force is associated with the electrostatic pressure P_e (force per unit area) caused by the applied transmembrane potential V_m on the membrane surface. The aim below is to compute $W_{es}(r, V_m)$ in terms of the electrostatic pressure P_e. Qualitatively, $W_{es}(r, V_m)$ accounts for the electric energy difference between a membrane with no aqueous pore and a membrane with an aqueous pore of radius r. Denoting $U_{es}(r)$ as the total electric potential energy of a membrane with an aqueous pore of radius r, the electrical potential energy $W_{es}(r, V_m)$ involved in the formation of an aqueous pore is

$$W_{es}(r, V_m) = U_{es}(0) - U_{es}(r) = \frac{1}{2}\int_\Omega \rho_0 \phi_0 \, d\Omega - \frac{1}{2}\int_\Omega \rho_r \phi_r \, d\Omega, \tag{11.26}$$

where ρ_0 and ϕ_0 are the charge density and electrostatic potential when no aqueous pore is present, ρ_r and ϕ_r are the charge density and electrostatic potential when an aqueous pore of radius r is present, and Ω is the entire volume of the membrane and aqueous pore. The expression (11.26) can be used to evaluate $W_{es}(r, V_m)$; however, here we provide an alternative method for evaluating $W_{es}(r, V_m)$ based on the electrostatic pressure P_e.

An alternative method to evaluate $W_{es}(r, V_m)$ is to integrate the electrostatic force required to expand a pore from radius 0 to radius r. To compute this electrostatic force, we assume that the pore is at local mechanical equilibrium, and that pore expansion only occurs in the radial direction. Therefore, for an aqueous pore of radius r, we only need to consider the force acting on the surface $S(r)$ of the pore. The electrostatic pressure P_e at the interface of two dielectrics can be computed from the Maxwell stress tensors. The Maxwell stress tensor relates electrostatic pressure to mechanical momentum. Using the Maxwell stress tensor, the aqueous pore electrical energy is given by

$$W_{es}(r, V_m) = -\int_0^r \left(\int_{S(\xi)} n \cdot (T_w(S(\xi)) - T_m(S(\xi))) n \, dS \right) d\xi,$$

where $T_w = \varepsilon_w \left(\frac{1}{2} |\nabla \phi_w|^2 I - \nabla \phi_w \otimes \nabla \phi_w \right)$,

$$T_m = \varepsilon_m \left(\frac{1}{2} |\nabla \phi_m|^2 I - \nabla \phi_m \otimes \nabla \phi_m \right). \tag{11.27}$$

In (11.27), T_w and T_m are known as the Maxwell stress tensors, which are evaluated on the surface $S(r)$ of the aqueous pore of radius r; I denotes the identity matrix, n denotes the normal vector to the membrane surface (Figure 11.5), $S(r)$ the surface of the pore of radius r, and $\otimes$ is the dyadic product (i.e., $\nabla \phi \otimes \nabla \phi = \nabla \phi \nabla \phi^T$). The expression $P_e = (T_w - T_m)n$ in the first equation denotes the electrostatic pressure on the surface of the pore resulting from the transmembrane potential V_m. In simple terms (11.27) integrates the electrostatic pressure on the surface of an aqueous pore with radius in the range 0 to r nm to yield the electric potential energy of the pore, $W_{es}(r, V_m)$.

11.5.3 Computing Pore Capacitance

The capacitance of an aqueous pore of radius r, $C_p(r)$, illustrated in Figure 11.5, can be computed by relating the electric energy stored in a capacitor to the electrical potential energy $W_{es}(r, V_m)$ stored in the aqueous pore.

Let us assume that we have two charged plates, one with charge $+Q$ and the other with charge $-Q$ with a potential difference between them of V. Then, from the charge-voltage relation of capacitance, $C = Q/V$. The electric energy stored in the capacitor is given by the difference in energy associated with two uncharged plates, and the charged plates. Denoting $U_{es}(q)$ as the electric energy required to move a charge of q from one plate to the other plate, the electric energy stored in the capacitor is

$$W_c(V) = U_{es}(0) - U_{es}(q) = -\int_0^Q \frac{\xi}{C} d\xi = 1\frac{1}{2}CV^2. \tag{11.28}$$

From (11.28), the electric energy stored in the capacitor is only a function of the capacitance C and the squared potential difference across the capacitor, V^2. Given that the electric energy stored in the aqueous pore, $W_{es}(r, V_m)$ (11.27), is equal to the electric energy stored in the capacitor, $W_c(V_m)$ (11.28), the pore capacitance can be evaluated from

$$C_p(r) = -\frac{2W_{es}(r, V_m)}{V_m^2}. \tag{11.29}$$

From (11.29), the aqueous pore capacitance $C_p(r)$ is proportional to the electric energy stored in a capacitor, $W_{es}(r, V_m)$. Notice that if $W_{es}(r, V_m)$ is not proportional to V_m^2, then the pore capacitance $C_p(r)$ will depend on the transmembrane potential V_m.

11.5.4 Double-Layer Capacitance

Here we discuss how the double-layer capacitance C_{dl} can be evaluated. We focus on using the PFNP model (11.23); however, the GPNP model can also be used to evaluate C_{dl} if the correlation effect of ions (Coulomb correlations) are negligible, causing the dielectric permittivity in proximity to the bioelectronic interface to be spatially dependent.

To compute C_{dl}, we can use the double-layer voltage-to-charge relation $Q_{dl} = C_{dl}V_{dl}$. Since capacitance is the ratio of charge to voltage, the double-layer capacitance C_{dl} can be evaluated as

$$C_{dl} = \frac{Q_{dl}}{V_{dl}} = \frac{\int_0^{h_{dl}} \rho \, d\ell}{\int_0^{h_{dl}} \nabla\phi(\ell) \cdot n_z \, d\ell}, \tag{11.30}$$

where ρ is the charge density, ϕ is the electrostatic potential, n_z is the normal vector perpendicular to the electrode surface, and h_{dl} is the height of the double layer. The height of the electrical double layer, h_{dl}, is the distance from the electrode surface at which the potential gradient $\nabla\phi$ is a constant. From Poisson's equation (10.3) on page 180, this is the distance from the electrode surface at which the charge density $\rho = 0$. In (11.30), the parameters ρ, ϕ, and h_{dl} depend on the hydration ion sizes, diffusion coefficients, charges, and double-layer potential V_{dl}. Notice that (11.30) does not explicitly depend on the dielectric permittivity of the electrolyte; however, if correlation effects are present in the electrolyte then the PFNP model should be used to estimate ρ and ϕ in (11.30).

11.5.5 Detection Tests for Ionic Correlation Effects

Detection of Coulomb correlation effects is important in deciding whether to use the GPNP model (11.18) or the PFNP model (11.23) to compute the electrostatic potential ϕ, charge density ρ, and ionic flux J^i of species $i \in \{1, \ldots, n\}$. A good rule of thumb for selecting between the GPNP model and PFNP model is that if high ionic concentrations are present, this will lead to strong ion-ion correlations, in which case Coulomb correlations are non-negligible and the PFNP model should be used. Here we consider two methods to detect if Coulomb correlations are present. The first is based on computing the approximate dielectric permittivity $\bar{\varepsilon}$, and if $|\nabla\bar{\varepsilon}| \approx 0$, then Coulomb correlations are negligible and the GPNP model (11.18) can be used to compute ϕ, ρ, and J^i. The second is based on estimating the characteristic length scale of electrostatic screening, namely, the Debye length. If the Debye length is comparable to the dimensions of the ions (a few angstroms), then steric and Coulomb correlations are non-negligible and the PFNP model should be used.

Dielectric Permittivity and the Maxwell–Garnett Equation

To test if the magnitude of the spatial gradient of the dielectric permittivity $\bar{\varepsilon}$ is approximately zero (e.g., $|\nabla \bar{\varepsilon}| \approx 0$) requires that we have a method to compute $\bar{\varepsilon}$. Here we approximate $\bar{\varepsilon}$ using the Maxwell–Garnett equation.[15]

Consider an electrolyte solution comprising n ionic species indexed by $i \in \{1, 2, \ldots, n\}$ with each ion having an effective size a_i. In the construction of the GPNP model and FPNP model we assumed a for species i given by $c^i_{max} = 1/N_A a_i^3$. Let us define the dielectric permittivity ε_i as the permittivity in the volume occupied by ions associated with species i. Then, the dielectric permittivity is given by the Maxwell–Garnett equation

$$\bar{\varepsilon} = \varepsilon_w \left[\frac{1 + 2f \sum_{i=1}^{n} \frac{c^i}{c^i_{max}} \frac{\varepsilon_i - \varepsilon_w}{\varepsilon_i + 2\varepsilon_w}}{1 - f \sum_{i=1}^{n} \frac{c^i}{c^i_{max}} \frac{\varepsilon_i - \varepsilon_w}{\varepsilon_i + 2\varepsilon_w}} \right]. \qquad (11.31)$$

Here f is the packing coefficient,[16] and ε_w is the dielectric permittivity of water. We cannot directly evaluate (11.31) because the dielectric permittivity ε_i inside an ion of size a_i is unknown. However, the key parameter in (11.31) is the ratio of the concentration c^i to the maximum concentration c^i_{max}. If $c^i \ll c^i_{max}$, then (11.31) states that $\bar{\varepsilon} \approx \varepsilon_w$ and $|\nabla \bar{\varepsilon}| \approx 0$. However, if c^i and c^i_{max} have similar magnitudes, then the dielectric permittivity $\bar{\varepsilon}$ in (11.31) is spatially dependent. Therefore, the Maxwell–Garnett equation (11.31) states that, for high analyte concentrations near the steric limit, the dielectric permittivity is spatially dependent such that $|\nabla \bar{\varepsilon}| > 0$.

The test for Coulomb correlations can be performed by computing the concentration of ions, c^i, from the GPNP model. Then, if there exist concentrations c^i with comparable magnitude to c^i_{max}, this suggests that Coulomb correlations are present in these regions. For engineered membranes this typically occurs at the surface of a gold electrode bioelectronic interface.

Dielectric Permittivity Estimation from Continuum Model

Though the Maxwell–Garnett equation (11.31) gives an approximate expression for the dielectric permittivity $\bar{\varepsilon}$, it requires knowledge of the local permittivity inside the volume occupied by each ion which is typically unknown. Below we illustrate how the dielectric permittivity $\bar{\varepsilon}$ can be estimated from the PFNP model (11.23).

---

[15] The Maxwell–Garnett equation can be derived using several methods starting from Maxwell's equations and the Lorentz force; see [266].
[16] The packing coefficient is the fraction of volume occupied by ions in a unit volume. For example, $f = 1$ for perfect packing, and $f = \pi/6$ for simple cubic packing.

Let us denote the spatially dependent dielectric permittivity by $\bar{\varepsilon}$. The dielectric permittivity $\bar{\varepsilon}$ is dependent on the polarization charge density,[17] denoted by $\Psi(x)$, and is given by

$$\Psi(x) = -\varepsilon_w \nabla^2 \phi_w(x) - \rho(x), \qquad (11.32)$$

where $\rho(x)$ and $\phi_w(x)$ are the spatially dependent charge density and electric potential from the PFNP model (11.23). The parameter $\Psi(x)$ is also known as the *concentration polarization*. The dielectric constant must satisfy Poisson's equation (10.3), namely, $\bar{\varepsilon} \nabla^2 \phi_w = -\rho$. With $\nabla \cdot (\bar{\varepsilon} \nabla \phi_w) \approx \bar{\varepsilon} \nabla^2 \phi_w$, and the polarization charge density $\Psi(x)$ in (11.32), the spatially dependent dielectric constant $\bar{\varepsilon}(x)$ is given approximately by

$$\bar{\varepsilon}(x) \approx \frac{\varepsilon_w}{1 + \Psi(x)/\rho(x)}. \qquad (11.33)$$

Note that the PFNP model (11.23) gives the time-dependent charge density $\rho(x, t)$ and electric potential $\phi_w(x, t)$. Therefore, the spatially dependent dielectric permittivity will also be time dependent. Physically, this time dependence results as the ions in the electrolyte move in response to the applied electric field, which also results in the charge density $\rho(x, t)$, electric potential $\phi_w(x, t)$, and polarization all being time dependent. Since the dielectric is dependent on both the charge density and polarization, it will also be time dependent. In engineered membranes, experimental measurements are typically performed on the time scale of milliseconds, where the charge density is at its steady-state value. Therefore, we only consider the steady-state spatially dependent dielectric constant $\bar{\varepsilon}(x)$ in (11.33).

Poisson–Boltzmann Equation and Ionic Correlation Effects

Instead of directly estimating the dielectric permittivity, we can also estimate the interaction strength between ions to test for the presence of Coulomb correlation effects. Here we use the GPNP model (11.18) and nondimensionalization to estimate the characteristic length scale of ionic screening effects. If the characteristic length of ionic screening is on the order of the size of the ions then Coulomb correlation effects are non-negligible and the PFNP model (11.23) should be used.

Poisson's equation provides a relation between the electric potential ϕ and the concentration of species i via

$$\nabla \cdot (\varepsilon \nabla \phi) = -\sum_{i=1}^{n} F z^i c^i, \qquad (11.34)$$

where ε is the dielectric permittivity, F is Faraday's constant, z^i is the valency if species i, and c^i the concentration of species i. We require an expression for c^i to evaluate the electric potential ϕ. In the GPNP model (11.18) this was provided by the generalized Nernst–Planck equation (11.18a). We are interested in the equilibrium concentration c^i

---

[17] The polarization charge density $\Psi(x)$ is related to the electric polarization $\boldsymbol{P}$ by $\Psi(x) = -\nabla \cdot \boldsymbol{P}$. The electric polarization $\boldsymbol{P}$ is a vector field that accounts for the dipole moments (both static and induced by an external electric potential ϕ) in a dielectric material such as water. The total charge density is given by the sum of the free charge density ρ and the polarization charge density $\Phi(x)$.

of ions adjacent to the bioelectronic gold interface where ionic correlation effects are expected to be present. The closed-form expression for the equilibrium concentration c^i can be computed by solving for the steady-state solution of (11.18a) with the boundary conditions (11.22). The equilibrium concentration c^i is

$$c^i = \frac{c_o^i \exp\left(-\frac{z^i q\phi}{k_B T}\right)}{1 + \sum_{j=1}^{n} \frac{c_o^j}{c_{max}^j}\left(\exp\left(-\frac{z^j q\phi}{k_B T}\right) - 1\right)} \qquad (11.35)$$

where c_o^i is the bulk concentration of species i, c_{max}^i is the maximum concentration of species i, k_B is Boltzmann's constant, and T is the temperature. Substituting (11.35) into (11.34) results in the Poisson–Boltzmann equation for the electrostatic potential at the bioelectronic interface:

$$\nabla \cdot (\varepsilon \nabla \phi) = -\sum_{i=1}^{n} Fz^i \left[\frac{c_o^i \exp\left(-\frac{z^i q\phi}{k_B T}\right)}{1 + \sum_{j=1}^{n} \frac{c_o^j}{c_{max}^j}\left(\exp\left(-\frac{z^j q\phi}{k_B T}\right) - 1\right)} \right]. \qquad (11.36)$$

The Poisson–Boltzmann equation (11.36) is a nonlinear partial differential equation that gives the electrostatic potential ϕ of the GPNP model (11.18) adjacent to the bioelectronic interface at steady state.

In engineered tethered membranes typically the bulk concentration $c_o^i \ll c_{max}^i$. Additionally, if we assume that the potential ϕ is sufficiently small such that $z^i q\phi \ll k_B T$, then (11.36) becomes

$$\nabla^2 \phi = \left(\frac{1}{\lambda_D^2}\right)\phi, \qquad \lambda_D^2 = \frac{\varepsilon k_B T}{\sum_{i=1}^{n} c_o^i N_A (qz^i)^2}. \qquad (11.37)$$

Equation (11.37) is known as the linear Poisson–Boltzmann equation and provides the expression for the electrostatic potential of point-like ions in a uniform dielectric with permittivity ε. The parameter λ_D is known as the Debye length and it is the characteristic length scale of the screening effects of the ions. If λ_D is small, then Coulomb correlations are likely present. The Debye length λ_D is related to the Bjerrum length λ_B, which is the length scale for Coulomb correlations. The Bjerrum length λ_B is the length at which the Coulomb energy balances the thermal energy and is given by

$$\lambda_B = \left[4\pi \lambda_D^2 \sum_{i=1}^{n} c_o^i N_A (z^i)^2\right]^{-1}. \qquad (11.38)$$

Therefore, as λ_D decreases, the Coulomb correlations increase. Recall that the Bjerrum length λ_B is equal to the Coulomb correlation length l_c in (11.23) if the ions are treated as point-like (no steric effects).

The detection test for Coulomb correlations using only the bulk concentration of species relies on evaluating the Debye length λ_D in (11.37). If λ_D is significantly larger than the size of the ions then Coulomb correlations are negligible and $\lambda_B \approx 0$, suggesting that $l_c \approx 0$ in (11.23). In engineered tethered membranes the typical concentration of electrolyte ions used is 150 mol/m^3 of NaCl. The Debye length associated with this electrolyte concentration is $\lambda_D = 0.76$ nm, which is comparable to the size of the electrolyte ions. Therefore, Coulomb correlations may be non-negligible in the engineered tethered membrane.

11.6 Faradic Reactions at the Bioelectronic Interface

In an engineered tethered membrane there exist charge accumulation, diffusion-limited charge-transfer, and reaction-limited charge-transfer phenomena at the bioelectronic interface.[18] In the reaction-rate model of Chapter 8, these processes were modeled using a double-layer capacitance C_{dl} and fractional-order operator p. In this section, we generalize the no-flux boundary conditions of the PFNP model (11.23) at the electrode surface to account for reaction-limited charge-transfer processes, namely, the faradic reaction process. The boundary conditions are constructed based the faradic reaction-rate equation for charge transfer at the surface of the bioelectronic interface.

What is a faradic reaction? Faradic currents are a result of electrochemical reactions at the electrode surface. Specifically, a faradic reaction occurs when an electron is transferred from the ions to the electrode (adsorption), or from the electrode to the ions (desorption). This transfer of electrons between the electrode surface and ions can be accounted for by the boundary condition

$$n \cdot J^i = -R \text{ in } \partial \Omega_e \cup \partial \Omega_{ec},$$

where $\partial \Omega_e$ is the electrode surface, $\partial \Omega_{ec}$ is the counter electrode surface, and R is the electron transfer reaction. Notice that this would replace the no-flux boundary condition (11.24a) used when no faradic reactions are present at the electrodes. If faradic reactions are present, they must be accounted for as the electrostatic potential ϕ, charge density ρ, dielectric permittivity ε, and ionic flux J^i for species $i = \{1, \ldots, n\}$ depend on the specified boundary conditions of the PFNP model (11.23).

11.6.1 Faradic Reactions and Double-Layer Charging at the Bioelectronic Interface

Assume that the excitation potential V_s at the gold bioelectronic interface is sinusoidal. Therefore, we only need to consider the steady-state impedance of the bioelectronic interface. In the reaction-rate regime, the transport phenomena at the bioelectronic interface can be accounted for by the total electrode impedance Z, which is the sum of the double-layer impedance Z_{dl}, the faradic impedance Z_F, and a kinetic impedance Z_k. The kinetic impedance Z_k accounts for the inertia of the ions in the electrolyte

---

[18] §3.6 on page 49 discusses these transport phenomena in detail.

solution. For a sinusoidal excitation potential V_s, it takes finite time for the ions to change speed in response to changes in the excitation potential. This affect is accounted for by the kinetic impedance Z_k. In engineered tethered membranes, the frequency of excitation is typically in the hertz to kilohertz range, in which case $Z_k \approx 0$. The double-layer impedance Z_{dl} accounts for the charge accumulation, diffusion-limited charge transfer, and reaction-limited charge transfer at the surface of the bioelectronic interface. In Chapter 8, the double-layer impedance Z_{dl} was equal to the impedance of the double-layer capacitor C_{dl}; that is,

$$Z_{dl} = \frac{1}{(j2\pi f)^p C_{dl}}, \quad (11.39)$$

where f is the frequency of the sinusoidal excitation (in hertz), $j = \sqrt{-1}$ is the imaginary number, and $p \in (0, 1]$ is the fractional-order parameter. The double-layer impedance Z_{dl} does not account for charge adsorption or desorption at the bioelectronic interface. That is, if the electrode is assumed to be perfectly polarizable then the charging dynamics at the bioelectronic interface can be accounted for by the double-layer impedance Z_{dl}. However, if the bioelectronic interface is nonblocking then these transport phenomena can be accounted for by the faradic impedance Z_F.

Typically for engineered tethered membranes, the faradic impedance Z_F at the gold electrode, which is coated by polyethylene glycol, is significantly larger than the double-layer impedance Z_{dl} such that the faradic reaction processes are negligible. Gold is nonreactive to most biochemical species, and polyethylene glycol prevents protein adsorption to the surface. In engineered tethered membranes, faradic reactions may result if the double-layer potential is sufficiently large, or if a particular electrolyte species directly reacts with the gold or polyethylene glycol coating the surface. If the double-layer potential increases beyond 900 mV (the oxidation potential of gold) then this will cause a faradic reaction to occur. For every electrode there is an intrinsic charge injection limit, restricting the voltage that can be generated at the electrode surface – this limit is the oxidation potential of the electrode. Once this voltage is breached, the purely double-layer capacitive charging can no longer be maintained, and faradic reactions will occur. Above 900 mV, adsorption of hydrogen from water on the surface of the gold electrode occurs. Another source of faradic reactions may occur if particular organic residues form on the surface of either the gold or the polyethylene glycol – this will polarize the electrode.

How can we detect if faradic reactions are present in the engineered tethered membrane? Let us assume that we have only the gold electrode bioelectronic interface present with no membrane present. Then, if we apply a constant excitation potential $V_s(t)$, if only double-layer charging effects (nonfaradic processes) are present, then the current response will decay rapidly to zero on a time scale of a few microseconds. The fractional-order differential equations that model the current response of the bioelectronic interface with no membrane present are

$$\frac{d^p V_{dl}(t)}{dt^p} = \frac{1}{R_e C_{dl}} (V_s(t) - V_{dl}(t)), \qquad I(t) = \frac{1}{R_e} (V_s(t) - V_{dl}(t)).$$

Here C_{dl} is the double-layer capacitance, V_{dl} is the double-layer potential, p is the fractional-order operator, and $I(t)$ is the current response. At steady state for a constant excitation potential $V_s(t)$, the total charge in the double layer, q_{dl}, satisfies $q_{dl} = C_{dl}V_s$. If faradic processes are present at the bioelectronic interface, they typically occur at a significantly slower rate than the double-layer charging dynamics. This will cause a background faradic current $I_F(t)$ to result from the surface reactions. If a faradic current is present, then the associated charge accumulation in the double layer is $q_{dl}(t) = C_{dl}(V_s - R_e I_F(t))$. Therefore, to detect for the presence of faradic reactions at the bioelectronic interface requires detecting if nonzero currents are present after the double-layer charging of C_{dl} has reached the steady state.

11.6.2 Faradic Reaction Boundary Conditions for the PFNP Continuum Model

If faradic reactions are present at the bioelectronic interface, then two processes occur simultaneously at the electrode surface to induce a current. The first is the nonfaradic accumulation of charge in the double layer and can be accounted for using the double-layer impedance Z_{dl}. The second is a faradic reaction process[19] in which cations react with electrons from the metal electrode. In this section we use the general theory of reaction rates (based on nonequilibrium thermodynamics and transition-state theory) to construct boundary conditions of the PFNP model (11.23) to account for faradic reactions. The results are based on including steric effects (refer to §3.6) in the generalized Frumkin–Butler–Volmer equation for particle adsorption on surfaces. The generalized Frumkin–Butler–Volmer equation is given by

$$\frac{dq}{dt} = J_F = k_R c_s^i e^{-\alpha_r z^i \lambda_s \nabla \phi \cdot \mathbf{n}} - k_O c_e^i e^{\alpha_o z^i \lambda_s \nabla \phi \cdot \mathbf{n}}, \tag{11.40}$$

where q is the charge at the electrode surface resulting from faradic reactions, J_F is the faradic flux at the electrode-electrolyte interface, c_s^i is the electrolyte ion concentration at the Stern layer, c_e^i is the adsorbed electrolyte ion concentration at the electrode surface, k_R and k_O are reaction-rate constants, α_r and α_o are the transfer-rate constants, z^i is the number of electrons transferred per electrolyte ion, λ_s is the thickness of the Stern layer, and $\nabla \phi \cdot \mathbf{n}$ is the electric field strength across the Stern layer. Note that $\lambda_s \nabla \phi \cdot \mathbf{n}$ is the potential drop across the Stern layer. Notice that the generalized Frumkin–Butler–Volmer equation (11.40) does not account for the steric size of ions. Below we illustrate how the steric size of ions can be included in the faradic flux J_F at the electrode-electrolyte interface.

In a general faradic reaction, there are z^i electrons transferred from the electrode to the electrolyte cation ion i to produce the surface-adsorbed reduced state. The concentration

---

[19] A faradic reaction process governs the exchange of charge between a metal electrode and an electrolyte solution. It is an example of an electrochemical reaction. The excess charge in the metal is confined near the surface of the electrode, and a balancing charge exists in the electrolyte side of the electrode which extends into the electrolyte solution (e.g., the Stern and diffuse layers). Note that faradic reactions do not account for the charge accumulation in the electrode as a result of the movement of electrolyte ions and reorientation of water dipoles because electrons are not directly transferred between the electrode and electrolyte.

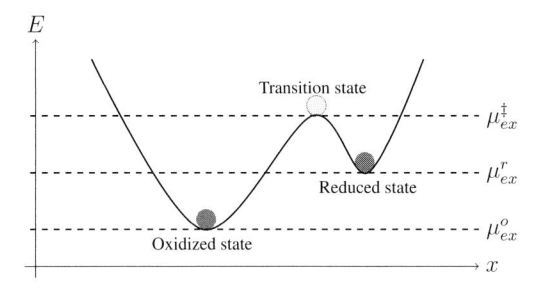

Figure 11.7 Energy landscape of the faradic reaction (11.41), where the vertical axis E denotes energy and the horizontal axis x denotes the reaction coordinate. The figure shows that there are two possible equilibrium states for the cation: oxidized and reduced. The cation spends negligible time in the transition state. As a voltage potential is applied to the electrode the energy levels of the oxidized and reduced states change. This variation promotes the transfer of cations from the oxidized state to the reduced state, or from the reduced state to the oxidized state.

of the cation ion is c^i and the reduced-state concentration of the ion is r^i. Formally, this reaction is given by

$$c^i + z^i q \underset{k_R}{\overset{k_O}{\rightleftharpoons}} r^i. \tag{11.41}$$

Note that q is the electron charge, c^i is the surface concentration of the electrolyte in units of mol/m^3, and r^i is the surface-bound reduced state of the electrolyte with units of mol/m^2. The reaction rates k_R for the reduction process and k_O for the oxidation process are dependent on the local concentration of ions, the voltage potential, and the steric properties of the electrolyte and surface-bound reduced electrolyte.

To construct expressions for k_R and k_O we use the theory of excess chemical potential. In the thermodynamically consistent formulation of reaction kinetics, the reaction complex explores a landscape of excess chemical potential $\mu_{ex}(x)$ between local minima μ_{ex}^o and μ_{ex}^r with an activation barrier $\mu_{ex}^\ddagger$. This energy landscape is illustrated in Figure 11.7. From the energy landscape in Figure 11.7, the net reaction rate R in the change in reduced electrolyte is given by

$$\frac{dr}{dt} = R = v\left[e^{-(\mu_{ex}^\ddagger - \mu_{ex}^o)/k_B T} - e^{-(\mu_{ex}^\ddagger - \mu_{ex}^r)/k_B T}\right]. \tag{11.42}$$

The parameter v in (11.42) is a frequency prefactor[20]. The reaction rate (11.42) is valid assuming that the oxidized and reduced states of the ion have a sufficiently long lifetime such that $\mu_{ex}^\ddagger - \mu_{ex}^o \gg k_B T$ and $\mu_{ex}^\ddagger - \mu_{ex}^r \gg k_B T$. For the reaction (11.41), the excess

---

[20] The frequency prefactor, or preexponential factor, represents the frequency of collisions between the reactant molecules. Here the reactants are the oxidized state and electrons, as described by the reaction in (11.41).

chemical potentials $\mu_{ex}^{\ddagger}$, μ_{ex}^{o}, and μ_{ex}^{r} are given by

$$\mu_{ex}^{o} = k_B T \ln(a^o) + z^o q \phi - z^o q \phi_s + E^o$$
$$= k_B T \ln(c^i) + k_B T \ln(\gamma^o) + z^o q \phi - z^o q \phi_s + E^o,$$
$$\mu_{ex}^{r} = k_B T \ln(a^r) + z^o q \phi + z^o q \phi_s + E^r$$
$$= k_B T \ln(r^i) + k_B T \ln(\gamma^r) + z^o q \phi - z^o q \phi_s + E^o,$$
$$\mu_{ex}^{\ddagger} = k_B T \ln(\gamma^{\ddagger}) + \alpha z^o q \phi + (1-\alpha) z^o q (\phi - \phi_s) + E^{\ddagger}. \tag{11.43}$$

The parameters a^o and a^r denote the absolute chemical activity[21] of c^i and r^i, respectively; c^i is the concentration of ions in the oxidized state in the electrolyte solution, and r^i is the concentration of ions in the reduced state on the surface of the electrode; ϕ is the potential, and ϕ_s is the potential drop across the Stern layer; $\alpha \in [0, 1]$ is the transfer coefficient; and E^o, E^r, and $E^{\ddagger}$ are the reference energies of the oxidant, reactant, and transition state, respectively. The steric contribution of the oxidized species i (with concentration c^i) is given by

$$\gamma^o = \frac{1 - N_A \sum_{i=1}^{n} v_i c_o^i}{1 - N_A \sum_{i=1}^{n} v_i c^i}$$
$$\approx \frac{1}{1 - N_A \sum_{i=1}^{n} c^i v_i}$$
$$\approx \frac{1}{1 - \sum_{i=1}^{n} c^i / c_{max}^i} \tag{11.44}$$

and the steric contribution of the reduced species r is given by

$$\gamma^r = \frac{1}{1 - \sum_{i=1}^{n} r^i / r_{max}^i}. \tag{11.45}$$

The parameter r_{max}^i is the maximum surface-bound concentration of reduced species r, and γ^o is identical to the steric chemical overpotential used to construct (11.23) where v_i is the volume of species i and c_o^i is the bulk concentration. Note that the approximation in (11.44) results as $\gamma^o = 1$ if $c^i = c_o^i$. That is, no steric contributions are present if the electrolyte is at the bulk concentration level. Substituting (11.43) into (11.42), the resulting reaction rate R is given by

$$R = k_r c^i \gamma^o e^{-\alpha z^o q \phi_s / k_B T} - k_o c^r \gamma^r e^{(1-\alpha) z^o q \phi_s / k_B T}$$
$$= k_r c^i \gamma^o e^{-q \phi_s / 2k_B T} - k_o c^r \gamma^r e^{q \phi_s / 2k_B T}, \tag{11.46}$$

where k_r is the rate constant of the reduction reaction, k_o is the rate constant of the oxidation reaction, and the last equality holds for cations with $z^o = 1$, and $\alpha = 0.5$. The faradic reactions that occur on the electrode surface are governed by the following

---

[21] The absolute chemical activity of a species (also known as the thermodynamic activity) is a measure of an "effective concentration" of a species in a mixture. Formally, the absolute chemical activity is defined by $a^r = \exp(\mu^r / N_A k_B T)$, where μ^r is the chemical potential, N_A is Avogadro's number, k_B is Boltzmann's constant, and T is the temperature.

system of nonlinear ordinary differential equations:

$$\frac{dr}{dt} = R, \quad \frac{dc^i}{dt} = -R, \qquad (11.47)$$

where the reaction rate R is given by (11.46) which is dependent on the Stern layer potential ϕ_s, concentration of electrolytes, steric effects, and the surface concentration of reactants.

Using (11.47), we can now construct the boundary conditions of the PFNP formula (11.23). The only modification in the boundary conditions (11.24) to include the faradic reactions is to change the cation interface flux at both electrode surfaces to satisfy

$$n \cdot J^i = -R \text{ in } \partial\Omega_e \cup \partial\Omega_{ec}. \qquad (11.48)$$

Using these boundary conditions and the PFNP formula (11.23) allows us to model the effects of faradic reactions at the electrode surface. Note that if $k_o = k_r = 0$ then no faradic reactions are present at the surface of the electrodes.

11.7 Complements and Sources

The main results presented in this chapter were the Smoluchowski–Einstein equation, the GPNP model, and the PFNP model. These models can be used with experimental measurements to estimate important parameters such as the number and size of aqueous pores in the engineered membrane in response to an excitation potential. The models and engineered tethered membrane can be used to study the process of electroporation in a controlled environment. Recall that electroporation has applications in electrochemotherapy for antitumor treatment, protein insertion, cell fusion, debacterialization, and gene and drug delivery.

For further information on the derivation of the Smoluchowski–Einstein equation from statistical mechanics, and the use of singular perturbation analysis to construct the ODE that describes the population of aqueous pores, see [27, 126, 297, 318]. Notice that the parameters of the Smoluchowski–Einstein equation are based on experimental results and making physically plausible assumptions about the physical processes present. For example, aqueous pore energy models are constructed by assuming that the membrane is a dielectric and elastic continuum [6, 69, 100, 189, 297, 298, 300, 411, 413, 475, 478–480]. The energy of nonconducting hydrophilic pores in (11.3) can be approximated by the quadratic function. Typical physiological parameters are used; refer to [170, 214, 286, 393]. Additionally, for engineered artificial membranes, we make the assumption that the energy contribution of tethers can be modeled using $W_m = 0.5K_t r^2$ with K_t denoting the spring constant of the tethers [162, 163, 411]. Note that the energy model for tethers is identical to that of the cytoskeletal network presented in [194, 348, 351, 419]. It is, however, possible to use lower-level models such as coarse-grained molecular dynamics and all-atom molecular dynamics, to estimate the aqueous pore energy. Note, however, that even without using these models, the asymptotic approximation of the Smoluchowski–Einstein equation has been successfully used to model DNA translocation into cells [95, 96, 170, 214, 286, 393].

The continuum GPNP and PFNP models [162, 163, 165, 247, 248] that were used extensively in this chapter are dependent on the pore shape, the chemical potential used to construct the model, and the boundary conditions of the associated partial differential equations. Although different hydrophobic pore shapes can be considered, typically a toroidal pore is assumed [26, 297, 300]. Note that the toroidal pore structure corresponds to the estimated hydrophobic pore shape obtained from molecular dynamics simulations [99, 463]. The derivation of the continuum models in this chapter relies on defining the free energy of the system and assuming the mean-field approximation holds. A complete derivation of the PFNP model from first principles can be found in [362], and applications are provided in [31, 246, 361]. A similar construction is used for the GPNP model. Notice that it is also possible to use the results from atomistic models to construct associated continuum models. For example, estimates of the excess chemical potential μ_{ex}^i to account for steric interactions are now being computed using density functional theory[22] [30, 135, 333]. Finally, the boundary conditions in the PNP, GPNP, and PFNP models are selected to match the expected physical phenomena present or to ensure the well-posedness of the solution to the PDEs. For example, to ensure the well-posedness of the Poisson equation (11.18b), the internal boundary conditions on the membrane electrolyte interface are satisfied by (11.22b) [483]. For the adsorption dynamic boundary conditions of the PFNP, we included steric effects in the generalized Frumkin–Butler–Volmer equation for faradic reactions (or particle adsorption) on surfaces [5, 117, 228, 258]. This was based on the general theory of reaction rates (based on nonequilibrium thermodynamics and transition-state theory). Additionally, in the boundary conditions we assume that there exists a Stern layer that results from the hydrogen bonding of water directly adjacent to the electrode surface as experimentally observed [440, 462]. The properties of water and ions directly adjacent to a charged electrode are still not well understood from an experimental or theoretical point of view. As such, care must be taken when selecting which type of boundary conditions to employ for the continuum models presented in this chapter.

11.8 Closing Remarks

This chapter has discussed dynamic models for electroporation in engineered membranes. A two-level predictive model, consisting of a reaction-rate (macroscopic) model and a continuum (mesoscopic) model, was developed to relate the pore dynamics to the current-voltage response of the membrane. The reaction-rate model had fractional-order macroscopic dynamics present, as illustrated in Figure 11.4 where both membrane

---

[22] This is an approximate method for solving the ab initio quantum mechanics used to estimate electron shell interactions. Note the results from this model have higher accuracy compared to molecular dynamics; however, the simulation sizes and time horizons that are computationally feasible are smaller than that of molecular dynamics simulations.

conductance G_m and capacitance C_m are dependent on the population of aqueous pores present. The population and dynamics of aqueous pores in the membrane was modeled using a continuum model. This continuum model was obtained using asymptotic approximations to the Smoluchowski–Einstein equation (11.1) of electroporation that is dependent on the pore conductance G_p, pore capacitance C_p, and the energy required to form aqueous pores, W_{es}. The GPNP and PFNP continuum models were used to estimate these important biological parameters in engineered membranes. Recall that the GPNP is able to account for the steric effects of ions by including an excess chemical potential μ_{es}^i for each analyte species i in the original PNP equation. The PFNP can account for both steric effects of analyte species and the polarization effects of water by using the excess chemical potential μ_{es}^i and using the Poisson–Fermi equation to model the electrostatic potential in the tethered membrane.

Future research on the process of electroporation can include accounting for dynamic temperature effects, designing excitation potentials to control the population and dynamics of pores, and including sophisticated density functional theory results into the excess chemical potential of the GPNP and PFNP models. These are challenging research problems because the dynamic temperature effects will impact the dynamics of pore creation, the dynamics of the pore radii, and the pore conductance, capacitance, and electrical energy required to form the pores as these are all temperature dependent. The major challenge of constructing excitation potentials to control electroporation is that the structural integrity of the tethered membrane and disulfide gold supports must be maintained. Therefore, very large excitation waveforms, or large charge accumulation in either the membrane surface or electrode surface, must not occur. Constructing density functional theory simulations that include accurate electrode-to-solvent and solvent-to-lipid interactions that are tractable will require careful assumptions about the electronic properties of the electrode, solvent, and lipid molecules.

12 Electroporation Measurements in Engineered Membranes

12.1 Introduction

The electroporation measurement platform (EMP) discussed in Chapters 4 and 6 is a synthetic biological device built out of artificial membranes to study electroporation in a controlled environment. In this chapter we apply the continuum models for electroporation developed in Chapter 11 to predict and interpret the response of the EMP device. Specifically we evaluate how accurately the continuum models predict the response of the EMP to membranes containing different tether densities, lipid types, and sterols and excited using different waveforms. Thus, this chapter (which validates electroporation models with experimental data in precisely controlled environments) together with the previous chapter (which formulates continuum models) gives a complete treatment of electroporation at the mesoscopic level.

Before proceeding, the reader should recall that electroporation is the process of aqueous pore formation resulting from changes in the transmembrane potential. A schematic of the electroporation process is given in Figure 12.1. To ensure only the process of electroporation is present, the following test is performed for all experiments involving the electroporation measurement platform (see Chapter 6 for details of such experiments). An excitation potential V_s is applied and the resulting current is recorded; then the negative potential $-V_s$ is applied and the resulting current recorded. If the current response resulting from V_s is related to the current response of $-V_s$ by a sign change then we can conclude that only the process of electroporation is present. This conclusion follows from the dynamic model of engineered membranes given by Figure 11.4 and (11.17). Recall that the process of electroporation is transmembrane potential symmetric; that is, the polarity or sign of the applied transmembrane potential does not change the dynamics the electroporation process. All experimental measurements reported below satisfied this test.

Organization of Chapter

Given the mesoscopic models in Chapter 11, in this chapter we evaluate important parameters of the engineered membrane using both in silico and in vitro techniques. More importantly, we give significant insight into the process of electroporation by discussing several aspects.

12.1 Introduction

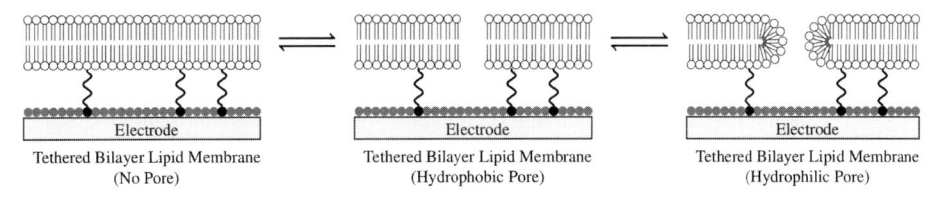

Figure 12.1 Schematic of the process of electroporation in an engineered membrane. Thermal fluctuations of lipids or applying a potential across the membrane promotes the creation of a hydrophobic pore. This pore can subsequently close (return to the no-pore state) or form a hydrophilic pore. This process of pore creation and pore destruction occurs in all biological membranes and also in tethered artificial membranes.

Figure 12.2 illustrates the aspects of electroporation that we discuss, which comprises three topics. The first is the aqueous pore conductance and dynamics (number and size of aqueous pores). The second topic includes methods for modulating the aqueous pore dynamics by adjusting the tether density and composition of the engineered membrane. The third topic involves the bioelectronic interface charging dynamics, which impact the electroporation process. Several sections are dedicated to each of these topics as discussed below. In summary, we work from pore to electrode in Figure 12.2, starting with aqueous pores, then how to modulate the dynamics of aqueous pores, and closing with insight into the charging dynamics of the gold electrode bioelectronic interface.

Aqueous Pore Dynamics and Conductance

The aqueous pore dynamics are described by the Smoluchowski–Einstein equation ((11.1) on page 217), which depends on the aqueous pore conductance $G_p(r)$, capacitance $C_p(r)$, and electric potential energy $W_{es}(r, V_m)$ stored in an aqueous pore. Given

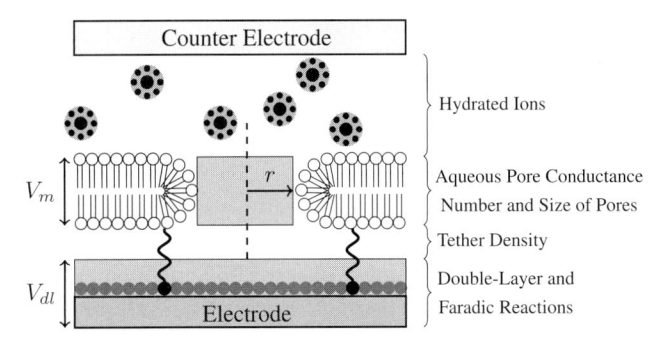

Figure 12.2 Schematic of the electroporation processes studied in Chapter 12 using the mesoscopic continuum models constructed in Chapter 11 and experimental measurements from the electroporation measurement platform (EMP). The hydrated ions are represented by the black dots surrounded by 8 smaller black dots, the electrical double layer is represented by the gray region with a double-layer voltage potential of V_{dl}, and the toroidal pore region of radius r is represented by the gray region with a transmembrane potential of V_m. The excitation potential between the electrode and counter electrode is V_s.

these parameters, it is possible to evaluate the population and size of aqueous pores in an engineered tethered membrane. In §12.2 and §12.3 we use experimental measurements from the EMP and the continuum models constructed in Chapter 11 to gain insight into aqueous pore dynamics and conductance in response to an excitation potential. Details on the results presented in these sections are provided below.

Conductance, spreading conductance, and capacitance of pore. In silico techniques are used to compute important parameters of the engineered membrane without using experimental measurements. Specifically, in §12.2 the generalized Poisson–Nernst–Planck (GPNP) model is used to compute the pore conductance $G_p(r)$ and electrical energy required to form a pore, $W_{es}(r, V_m)$. §12.2 shows that, near the aqueous pore, significant nonlinear potential gradients are present that restrict the current flowing through the pore. This effect is denoted as the "spreading conductance" and is dominant for pore radii significantly larger than the membrane thickness, causing the pore conductance to scale proportionally to the pore radius (i.e., $G_c(r) \propto r$). Additionally in §12.2, we illustrate that the electrical potential energy to form a pore, namely, W_{es}, is proportional to $V_m^{2.2}$ where V_m is the transmembrane potential. Given the relation between W_{es} and the pore capacitance $C_p(r)$ in §11.5.4, this suggests that the pore capacitance also depends on the transmembrane potential V_m.

Population and size of pores. In §12.3 the mesoscopic model is used to estimate the population and size of pores in an engineered membrane based on the experimentally measured current response from the engineered membrane. Recall from Chapter 11 that the mesoscopic model is given by the system of ordinary differential equations ((11.17) on page 227) and the fractional-order macroscopic model illustrated schematically in Figure 11.4 on page 223.

Modulating Aqueous Pore Dynamics

How sensitive are the dynamics of aqueous pores to changes in the tether density or composition of the engineered membrane? In §12.4–§12.7 we use experimental measurements from the EMP and the continuum models constructed in Chapter 11 to investigate how to modulate the aqueous pore dynamics by varying the tether density and membrane lipid composition. Details on the results presented in these sections are provided below.

Sensitivity of current response. In §12.4, we evaluate the sensitivity of the model-predicted current response to variations in the electroporation model parameters. It is illustrated that the electroporation model is sensitive to variations in the double-layer capacitance C_{dl}, fractional-order parameter p, voltage of electroporation, V_{ep}, tether spring constant K_t, equilibrium membrane conductance G_o, and membrane capacitance C_m.

Effect of tether density on pores. §12.5 illustrates how the mesoscopic model can be used to estimate the dynamics of aqueous pores in engineered membranes containing tether densities in the range of 1 to 100 percent.

Heterogeneous membrane mixtures. In §12.6 and §12.7 experimental measurements and the mesoscopic model are used to estimate the current response of engineered membranes composed of DphPC lipids, lipids from *Escherichia coli* and

Saccharomyces cerevisiae membranes, and membranes containing cholesterol concentrations in the range of 0 to 50 percent.

Bioelectronic Interface Charging Dynamics

When an excitation potential is applied to an engineered tethered membrane, it causes the electrolyte ions to form regions where charge accumulates. The location of this charge accumulation is either on the membrane surface or on the surface of the gold electrode (bioelectronic interface). Both these charge accumulation regions contribute to the dynamics of aqueous pores. In §12.8 and §12.9 we use experimental measurements from the EMP and the continuum models to study the charging dynamics of the bioelectronic interface, which includes both nonfaradic and faradic charging dynamics. Details on the results presented in these sections are provided below.

Hydration radius of ions and faradic reaction rates. §12.8 discusses the in vitro techniques that use both the mesoscopic models in Chapter 11 and experimental measurements from engineered membranes to estimate important biological parameters. Specifically, the Poisson–Fermi–Nernst–Planck (PFNP) model and experimental measurements are used to estimate the hydration size (the steric size) of ions, and electrode surface adsorption reaction rates (faradic reaction rates) of ions can be estimated from the current response of engineered membranes. Recall that the faradic reaction rates discussed in §11.6 account for the reaction-limited charge-transfer phenomena at the bioelectronic interface (this includes all processes that involve ion absorption and desorption from the electrode-to-electrolyte interface). In the lumped circuit model of the engineered membrane, these processes are accounted for by the double-layer capacitance C_{dl} and fractional-order parameter p. We illustrate how the hydration radius of ions can be estimated using experimental measurements and the PFNP model.

Double-layer charging dynamics. In §12.9 we use the PFNP model to compute the double-layer capacitance C_{dl}. The results suggest that C_{dl} is dependent on Coulomb correlations, polarization and screening, and the steric effects of ions. Therefore, these properties of the electrolyte must be considered when estimating C_{dl}.

Large excitation potentials. In §12.10 the mesoscopic model and experimental measurements are used to estimate the dynamics of the membrane conductance G_m, membrane capacitance C_m, and double-layer capacitance C_{dl} in the presence of large excitation potentials V_s in the range of -1 to 1 V. The results show that the membrane capacitance C_m is insensitive to these large excitation potentials. However, both the membrane conductance G_m and the membrane capacitance C_m depend on the concentration dynamics of cations and anions present at the electrode surface. These dynamics are accounted for by allowing C_{dl} to depend on the double-layer potential V_{dl}, as discussed in §12.9.

12.2 Aqueous Pore Conductance, Capacitance, and Electrical Energy

In electroporation, aqueous pores form in the membrane when an excitation potential is applied across the membrane. In this section we compute the aqueous pore conductance

$G_p(r)$ (11.25), aqueous pore capacitance $C_p(r)$ (11.29), and the aqueous pore electrical energy W_{es} (11.27) using the Poisson–Nernst–Planck (PNP) and GPNP continuum models constructed in Chapters 10 and 11. The surprising result of this section is that the pore conductance $G_p(r)$ increases linearly with pore radius. This is unusual since typically the conductance increases with area of the cross section of a cylinder, which is proportional to the square of the radius. As described below, this unusual behavior occurs because the pore is a cylinder of small height. So edge effects (called *spreading conductance* below) dominate the expression for the conductance of the pore; these edge effects have a linear dependency on the pore radius. Also we show that the electrical energy required to form a pore, namely, W_{es}, is proportional to $V_m^{2.2}$ where V_m is the voltage across the membrane. Since W_{es} is proportional to $V_m^{2.2}$, this also means that the aqueous pore capacitance $C_p(r)$ depends on the transmembrane potential V_m. This is unusual since the electrical energy in a cylinder[1] is proportional to the square of voltage.

12.2.1 Aqueous Pore Conductance

Continuum models (PNP, GPNP, and PFNP) can be used to compute the electrostatic potential ϕ and ionic flux J^i of species i in proximity to an aqueous pore. Given ϕ and J^i, the aqueous pore conductance can be evaluated as

$$G_p(r) = \frac{I_p(r)}{V_m} \text{ where } I_p(r) = F \sum_i \int_0^r 2\pi \xi J^i(\xi) \cdot n_e \, d\xi,$$

where n_e is the normal vector to the electrode surface, ξ is the distance from the center of the pore, J^i is the flux of species i, F is the Faraday constant, and V_m is the transmembrane potential.

A useful model for the conductance of aqueous pores in engineered membranes must account for the "spreading conductance" [8, 27, 156, 238, 301] that limits the flow of ions to the confined geometry of an aqueous pore. Formally, the spreading conductance results from the convergence of the electric current flux lines as illustrated in Figure 12.3. The larger the rate of convergence of the flux lines, the larger the effect of the spreading conductance on the aqueous pore conductance G_p. In the engineered membrane, the aqueous pore conductance results from the sum of the bulk spreading conductance G_{bs}, the conductance in the aqueous pore, G_a, and the spreading conductance G_{rs} from the tethering reservoir. Each of these parameters is dependent on the structure of the aqueous pore and tethering reservoir as illustrated in Figure 12.3. Insight into the difference between the spreading conductance and conductance in the aqueous pore can be obtained if we assume the electrolyte solution can be modeled by a homogeneous conductance σ. Additionally, if we assume that the membrane is not close to the electrode (i.e., if it is more than 100 μm away from the electrode), then the aqueous

---

[1] For example, the electric energy stored in a cylindrical capacitor is $CV^2/2$, where C is the capacitance and V is voltage across the capacitor.

12.2 Aqueous Pore Conductance, Capacitance, and Electrical Energy 255

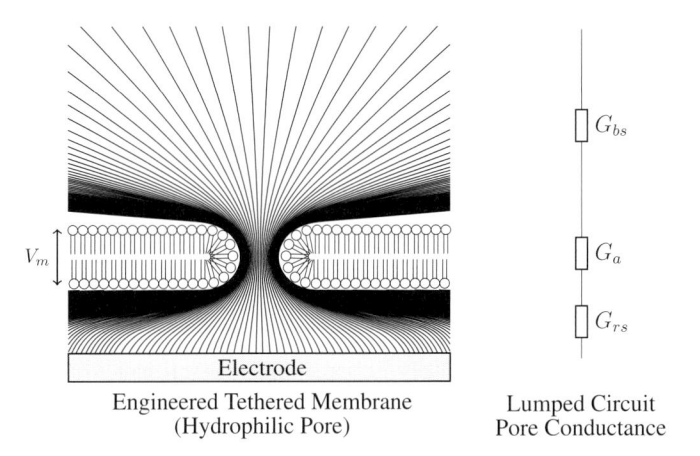

Figure 12.3 Computed total electric current flux $\sum_{i=1}^{n} J^i$ of the n ions, illustrated by the black lines, from the GPNP model in §11.4.1 with parameters defined in Table C.5 on page 417 for an aqueous pore of radius $r = 3$ nm. The parameter V_m is the transmembrane voltage, G_{bs} is the bulk spreading conductance, G_a is the conductance in the aqueous pore, and G_{rs} is the reservoir spreading conductance. The pore conductance is $G_p = G_{bs} + G_a + G_{rs}$. For an aqueous pore of size $r = 3$ nm, the associated pore conductance from the GPNP model is $G_p = 4.32$ nS.

pore conductance is

$$G_p = 2G_{bs} + G_a = 8r\sigma + \frac{\sigma\pi}{h_m}r^2. \tag{12.1}$$

In (12.1), we see that the spreading conductance term is proportional to r, and the conductance in the aqueous pore is proportional to r^2. From (12.1), as r increases, the conductance in the aqueous pore will dominate the pore conductance such that $G_p \approx G_a$. Therefore, for large pore radii the pore conductance will satisfy $G_p \propto r^2$. As we illustrate below, however, this r^2 proportionality relation will not hold if the membrane is in proximity to the electrode surface as illustrated in Figure 12.3. To compute G_p when the membrane is in proximity to the electrode surface we utilize the results of the PNP and GPNP continuum models.

Model-based predictions versus experimental measurements. Figure 12.4 presents the numerically computed pore conductance G_p (11.25) and electrical energy W_{es} (11.27) required to form a pore evaluated using the GPNP and PNP models in §11.4.1. As seen, the model-predicted conductance values for pores with $r < 3$ nm are in agreement with the experimentally measured conductance of a single pore obtained from planar bilayer lipid membranes (BLMs) using patch-clamp and linearly rising current protocols [193, 211, 275]. This is expected because in this region the pore conductance is dominated by both the bulk spreading conductance G_{bs} and the conductance of aqueous pores, G_a. That is, the tethering reservoir spreading conductance G_{rs} does not significantly contribute to G_p. For $r > 3$ nm, the conductance of the aqueous pores in the BLMs (which are not in proximity to an electrode surface) satisfies $G_p \propto r^2$. This is expected because in (12.1), for large pore radii, $G_p \propto r^2$. However, this relation does not hold for aqueous pores in proximity to an electrode surface. We use the PNP model

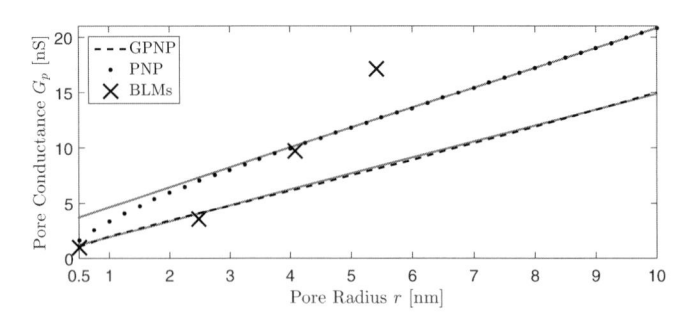

Figure 12.4 GPNP and PNP model-predicted conductance G_p (11.25) with parameters defined in Table C.5 on page 417. The experimentally measured conductance (denoted by X) for freestanding bilayer lipid membranes (BLMs) is obtained from [193, 211, 275]. For the GPNP model, the aqueous pore conductance is approximately equal to $G_p = 1.44r$ nS as indicated by the gray line.

to estimate the pore conductance where steric effects are assumed negligible, and we use the GPNP model to compute the pore conductance that includes steric effects. The computed conductance in Figure 12.4 between the PNP and GPNP models is a result of the steric effects. Recall, from §11.4.1, that for $\sum_{j=1}^{N} N_A a_j^3 c^j \ll 1$ the steric effects are negligible and the estimated conductance using the GPNP and PNP models would be identical. From Figure 12.4, the computed conductance approximately follows a $G_p \propto r$ relationship for pore radii up to $r = 10$ nm. This is expected for small pore radii where the spreading conductance of the bulk dominates. However, why does $G_p \propto r$ result even for large pore radii? It must be the case that the spreading conductance G_{rs} caused in the $h_r = 4$ nm tethering reservoir above the bioelectronic interface dominates the pore conductance. Therefore, the conductance of aqueous pores in tethered membranes depends on the distance between the membrane and the electrode surface.

12.2.2 Aqueous Pore Electrical Energy

In simple terms the aqueous pore acts as a capacitor and stores electrical energy. The electrical energy W_{es} stored in the aqueous pore results from the force exerted on the dielectric interface between the electrolyte and the lipids in expanding the pore from a radius of 0 to r. Insight into this energy W_{es} can be gained if we first consider a cylindrical aqueous pore in a freestanding membrane (no bioelectronic interface) and do not account for the dynamics of multiple ionic species with different properties such as hydration size and/or diffusion coefficient. For this case we show that $W_{es} \propto V_m^2$. However, if we consider a toroidal pore geometry with a membrane in proximity to the bioelectronic interface, as illustrated in Figure 11.5, and account for the dynamics of multiple ionic species, we show that $W_{es} \propto V_m^{2.2}$. The key is that both the properties of the electrolyte and the geometry of the aqueous pore contribute to the electrical energy stored in the pore.

Electrical Energy in a Cylindrical Pore

Consider the two cylindrical segments, one representing a membrane with a cylindrical lipid pore and the other representing a membrane with a cylindrical aqueous pore,

12.2 Aqueous Pore Conductance, Capacitance, and Electrical Energy

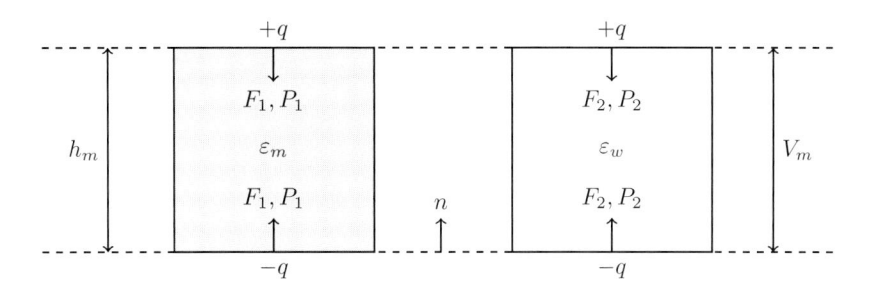

Figure 12.5 Schematic of two cylindrical pores with identical radius r in an engineered tethered membrane. The surface of the membrane is illustrated by the dashed lines and has a thickness of h_m. The lipid-filled pore is illustrated by the gray region and has a dielectric permittivity of ε_m. The water-filled pore is illustrated by the white region and has a dielectric permittivity of ε_w. F_1 and F_2 denote the electrostatic force acting on the two cylinders, and P_1 and P_2 denote the electrostatic pressure acting on the surfaces of the two cylinders. V_m is the transmembrane potential, n is the normal vector to the membrane surface, and q is the total charge on the surface of each cylindrical pore.

illustrated in Figure 12.5. The change in electrical energy between the cylindrical lipid pore and cylindrical aqueous pore is equal to the electrical energy W_{es}. We assume that the total charge q on both surfaces of the cylinder is uniform with no other charges present. Then the electric field E_i in each cylinder is uniform and is given by

$$E_i = \frac{q}{\pi r^2 \varepsilon_i} n, \tag{12.2}$$

where r is the radius of the cylinder, ε_i is the dielectric permittivity inside the cylinder, and n is the unit normal vector perpendicular to the membrane surface. The electrostatic force[2] F_i and electrostatic pressure P_i exerted on both surfaces as a result of the charge q are given by

$$F_i = -\frac{q}{2\pi r^2} E = -\frac{q^2}{2\pi r^2 \varepsilon_i}, \quad P_i = \frac{q^2}{2\varepsilon_i}. \tag{12.3}$$

Using either the electrostatic force or the electrostatic pressure, and the capacitance $q = C_i V_m$, we can compute the total electrical energy contained in the pore as a function of the transmembrane potential V_m. Both methods are illustrated below:

$$W_i = 2\int_{\xi=0}^{r} P_i 2\pi \eta \, d\xi = \frac{\pi r^2 \varepsilon_i V_m^2}{2h_m}, \quad W_i = -\int_{\xi=0}^{h_m} F_i \cdot n \, d\xi = \frac{\pi r^2 \varepsilon_i V_m^2}{2h_m}. \tag{12.4}$$

As seen from (12.4), the electrical energy in the cylinder is equal to the total electrostatic pressure acting on the surface of the cylinder, or the energy required to move the charged surfaces a distance of h_m from each other.

Having evaluated the electric energy W_i (12.4) for a uniform charge q on the surface of the cylinders in Figure 12.5, we can compute the electric energy W_{es} stored in a water-filled cylinder (assuming uniform charge on both surfaces of the cylinder and

---
[2] Also known as the Lorentz force.

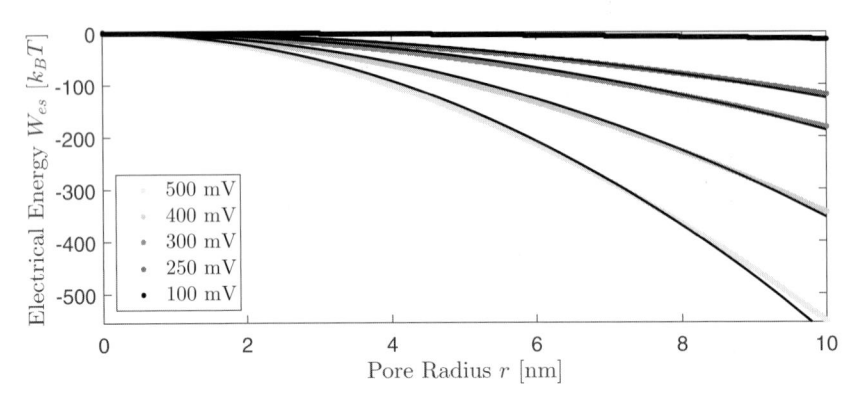

Figure 12.6 Model-predicted aqueous pore electrical energy W_{es} (11.27) computed using the GPNP model in §11.4.1 with the parameters defined in Table C.5 on page 417. The solid black lines indicate the fit with $W_{es} = 0.11 r^2 V_m^{2.2}/k_B T$.

equal transmembrane potential V_m). With these assumptions, the total electrical energy stored in a water-filled cylinder is

$$W_{es} = W_1 - W_2 = \frac{\pi r^2}{2h_m}(\varepsilon_m - \varepsilon_w)V_m^2, \qquad (12.5)$$

where W_1 is the electrical energy stored in the lipid cylinder (dielectric permittivity ε_m) and W_2 is the electrical energy stored in the water cylinder (dielectric permittivity ε_w). Given $\varepsilon_m > \varepsilon_w$, the water cylinder stores more charge per unit area and has a higher electrostatic pressure compared with the lipid cylinder. As such, the electrostatic pressure difference between the lipid cylinder and water cylinder results in the expansion of the water cylinder in the radial direction r such that, as V_m increases, r is expected to increase. The same methodology can be used to estimate the electrical energy W_{es} contained in an aqueous pore; however, the expressions for the electrostatic pressure on the surface of the membrane depend on the electrodiffusive dynamics and steric effects of ions in the electrolyte solution.

Electrical Energy in a Toroidal Pore

Here we compute the electrical energy in the aqueous pore illustrated in Figure 11.5 using the GPNP continuum model defined in §11.4.1. The GPNP model accounts for the toroidal pore geometry with a membrane in proximity to the bioelectronic interface, as well as the dynamics of the ion in the electrolyte solution.

Figure 12.6 presents the predicted W_{es} using the GPNP defined in §11.4.1. The results compare favorably with previously computed $W_{es}(r, V_m)$ using simplified governing equations that do not include electrodiffusive effects [1, 300]. From Figure 12.6, the proportionality between W_{es} and V_m follows a fractional power law. Typically the assumption of a squared law is used to compute W_{es} [1, 300]. This assumption is valid for narrow cylindrical pores where $\partial \phi/\partial r \approx 0$ and $\partial \phi/\partial z \approx V_m/h_m$ on the surface of the pore. The computed ϕ using the GPNP model does not satisfy these conditions on

12.2 Aqueous Pore Conductance, Capacitance, and Electrical Energy

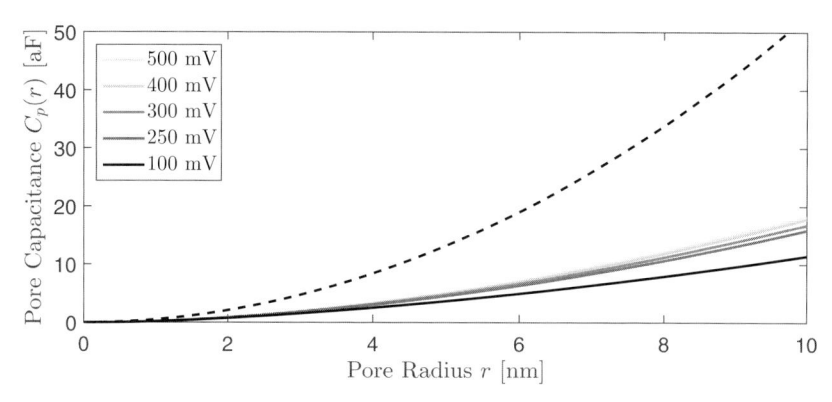

Figure 12.7 Model-predicted aqueous pore capacitance $C_p(r)$ (11.5.3) at different transmembrane potentials V_m computed using the GPNP model in §11.4.1 with the parameters defined in Table C.5 on page 417. The transmembrane potential V_m is indicated by the gray line, with $V_m = 500$ mV the darkest, and $V_m = 100$ mV the lightest. The dashed black line indicates the pore capacitance of a cylindrical pore computed using (12.5) and (11.28) with $\varepsilon_m = 2$, $\varepsilon_w = 78$, and $h_m = 4$ nm.

the pore surface, illustrating the importance of including the electrodiffusive dynamics of the electrolyte.

12.2.3 Aqueous Pore Capacitance

Here we compute the capacitance $C_p(r)$ (see §11.5.3) for the aqueous pore illustrated in Figure 11.5 using the GPNP continuum model defined in §11.4.1. The GPNP model accounts for the toroidal pore geometry with a membrane in proximity to the bioelectronic interface, as well as the dynamics of the ion in the electrolyte solution.

Figure 12.7 illustrates the model-predicted aqueous pore capacitance $C_p(r)$ computed using the GPNP defined in §11.4.1. As seen, $C_p(r)$ depends on both the radius of the pore and the transmembrane potential V_m. The model-predicted aqueous pore capacitance for a cylindrical pore that does not account for electrolyte dynamics or the bioelectronic interface is illustrated by the dashed black line. A possible explanation for the cylindrical pore capacitance being larger than the model-predicted toroidal pore capacitance is that the electrical energy stored in a cylindrical pore of radius r is larger than the electrical energy stored in a toroidal pore of radius r. From (11.28), this will cause the cylindrical pore to have a larger capacitance compared to the toroidal pore. The magnitude of $C_p(r)$ is on the order of attofarads. Interestingly, the magnitude of the capacitance of an aqueous pore is comparable to that of a metal-oxide-semiconductor transistor – the most basic element in the design of large-scale integrated circuits. The typical capacitance of an engineered tethered membrane is in the range of 14 to 18 nF. Therefore, the ratio of the capacitance of the membrane to a single aqueous pore is on the order of 1×10^9. Therefore, unless a significant number of aqueous pores are present in the engineered membrane, the total membrane capacitance can be assumed constant.

12.3 Pore Radii and Membrane Conductance Dynamics

In this section we apply the electroporation model illustrated schematically in Figure 11.2 to estimate the time-dependent pore radii $r_i(t)$ of each of the $i \in \{1, 2, \ldots, N\}$ aqueous pores, the membrane conductance $G_m(t)$, the transmembrane potential $V_m(t)$, and the double-layer charging potential $V_{dl}(t)$ resulting from an applied excitation potential $V_s(t)$. All these parameters depend on the pore conductance $G_p(r)$, pore capacitance $C_p(r)$, and electrical energy required to form a pore, W_{es}, computed in §12.2. The main result of this section is that, for excitation potentials with a characteristic time scale of milliseconds, all the aqueous pores in the engineered artificial membrane have the same radius. Therefore, the change in membrane conductance $G_m(t)$ from electroporation is dominated by the creation and destruction of aqueous pores and not the size dynamics of the aqueous pores.

The experimental measurement and predicted voltages, pore radii, membrane resistance, and current are presented in Figure 12.8 for the 10 percent tether density zwittrionic C20 diphytanyl-ether-glycero-phosphatidylcholine (DphPC) membrane. As can be seen in Figure 12.8(a), the experimentally measured and model-predicted currents are in excellent agreement. Figure 12.8(b) shows that, the application of a voltage excitation causes an immediate increase in the double-layer voltage V_{dl} due to an increase in the charge at the electrode surface. The transmembrane potential V_m simultaneously increases as a result of the excitation potential. Increasing V_m results in the formation of pores. As seen in Figure 12.8(c), a significant change occurs in the resistance of the membrane after applying an excitation potential. In Figure 12.8(d), the maximum radius r_{max} and mean radius $\bar{r}$ are provided to illustrate how much the pore radii can vary. As V_m increases, pores are generated and expand according to (11.17). From (11.17), the radii of all pores diffuse to the minimum-energy pore radius given by $\partial W/\partial r_i = 0$ with an advection velocity proportional to $D/k_B T$. As seen in Figure 12.8(d), generated pores rapidly expand to the minimum-energy pore radius because the spread between r_{max} and $\bar{r}$ is negligible. This allows the number of pores, N, from equation (11.17) to be computed using the relation $N = 1/(R_m G_p(r_{max}))$ with G_p given in Figure 12.4. The major result of Figure 12.8 is that the radii of pores in the membrane are approximately homogeneous, and the change in membrane conductance is dominated by the population $N(t)$ of conducting aqueous pores.

12.4 Sensitivity of Current Response to Model Parameters

How sensitive is the fractional-order macroscopic model-predicted current response $I(t)$ to variations in the lumped circuit parameters (e.g., membrane conductance G_m and membrane capacitance C_m)? In this section, we vary each parameter in the fractional-order macroscopic model, illustrated in Figure 11.4 and defined (11.17), and compute the associated model-predicted current response. Variations in the predicted current response $I(t)$ for different values of the model parameters provide insight into the

12.4 Sensitivity of Current Response to Model Parameters

(a) Measured and model-predicted current response $I(t)$.

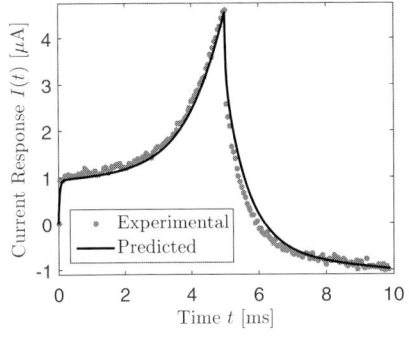

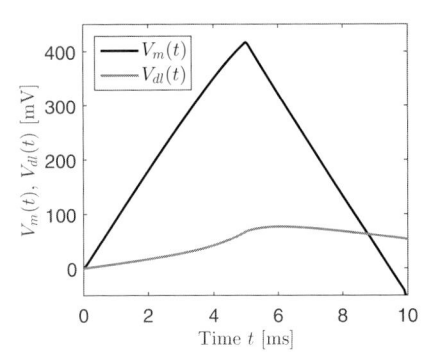

(b) Model-predicted transmembrane potential V_m and double-layer potential V_{dl}.

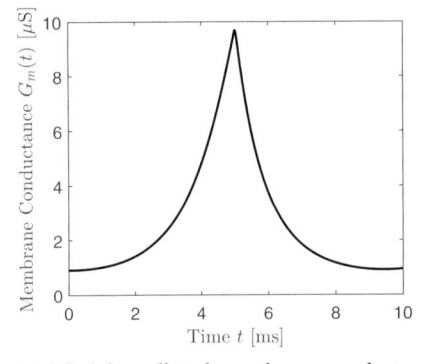

(c) Model-predicted membrane conductance $G_m(t)$.

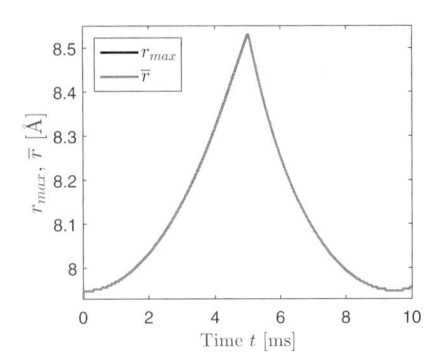

(d) Model-predicted maximum pore radius r_{max}, and mean pore radius $\bar{r}$.

Figure 12.8 The measured and model-predicted current $I(t)$, membrane and double-layer voltage potentials, membrane conductance $G_m(t)$, and pore radii for a 10 percent tether density DphPC bilayer membrane. The excitation potential $V_s(t)$ consists of a linearly increasing potential at a rate of 100 V/s for 5 ms, and -100 V/s for 5 ms. All model predictions are computed using the fractional-order electroporation model depicted in Figure 11.4 together with (11.17) with the parameters defined in Table C.6 on page 418. The aqueous pore conductance in (11.17) is evaluated using the GPNP model.

sensitivity of the fractional-order macroscopic model to variations in the model parameters. The sensitivity of the fractional-order macroscopic model is important for using the model and experimental measurements from engineered membranes to estimated important biological parameters such as membrane conductance G_m and membrane capacitance C_m.

In Chapter 8 impedance measurements were used to estimate the sensitivity of the fractional-order macroscopic model to variations in C_m, G_m, C_{dl}, p, and R_e. Here we focus on the variations in $I(t)$ that result from time-dependent excitation potentials $V_s(t)$ that cause electroporation. To gain insight into the sensitivity of the predicted current response to variations in model parameters, Table C.6 on page 418 provides the uncertainty associated with each parameter. The uncertainty is computed by finding the range in which the parameter can vary and still have the predicted current response, from Figure 11.4 and (11.17), be in good agreement with the measured current response. Specifically, the electrolyte resistance R_e has a negligible effect on the current response since the membrane conductance and capacitive charging dominate the current flow. The initial jump in current at the start of the triangular excitation potential is dominated by C_m. The slope of the current preceding the initial jump at time 0.5 ms in Figure 12.8(b) is proportional to the equilibrium membrane conductance G_m at zero transmembrane potential. At time 2 ms, there is a deviation from the linear current response due to the electroporation process. The double-layer capacitance C_{dl} dominates the current response as the triangular excitation potential decreases. As expected, C_m, R_m, and C_{dl} can be determined accurately. The electroporation parameters α and N_o have a large uncertainty because dN/dt, given by (11.17), is exponentially dependent on V_m, V_{ep}, and q and linearly dependent on N_o and α. Given that G_m is dependent on N, it is expected that the uncertainty of α and N_o is larger than that of V_{ep} and q. The tether spring constant K_t has a large uncertainty because effects caused by K_t are only pronounced in the current response if large pores (i.e., $r \gg r_m$) are present. As seen in Figure 12.8, since the pore radii are only slightly larger than r_m, the current is dominated by the nucleation and destruction of pores.

12.5 Effect of Tether Density of Membrane Electroporation Dynamics

Here we use experimental measurements of engineered membranes composed of DphPC lipids[3] and the electroporation model defined in (11.17) to investigate how variations in the tether density impact the equilibrium membrane conductance, membrane thickness, and characteristic voltage of electroporation. The experimental measurements are obtained from the EMP device.

As the tether density of the engineered membrane is increased, the equilibrium membrane conductance is expected to decrease. This is because the tethers anchor the lipids to the gold electrode, and so an increase in tether density will reduce the density of aqueous pores formed by random thermal fluctuations. Figure 12.9 presents the experimentally measured and model-predicted current response of the EMP for the 1 percent tether density and 10 percent tether density DphPC bilayer membrane, and the 100 percent tether density DphPC monolayer membrane. As seen in Figure 12.9, the resistance begins to change after approximately 1 ms when the transmembrane potential is sufficiently high to cause the nucleation of aqueous pores. As expected, the maximum of the current response of the three membranes results at 5 ms when the excitation potential is at its maximum value of 500 mV. At 5 ms, the current response decreases for increasing tether density. This suggests that the equilibrium membrane conductance decreases for increasing tether density. Using the electroporation model (11.17), the estimated equilibrium conductance of the 1, 10, and 100 percent tether density membranes are 1.00,

---

[3] DphPC bilayers are used extensively to construct engineered tethered membranes given their high stability in harsh environments; see §2.3.1 for details.

12.5 Effect of Tether Density of Membrane Electroporation Dynamics

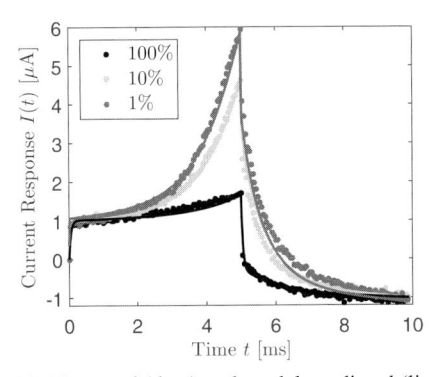

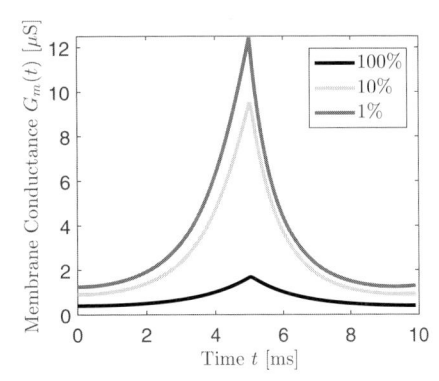

(a) Measured (dots) and model-predicted (lines) current response $I(t)$.

(b) Model-predicted membrane conductance $G_m(t)$.

Figure 12.9 Experimentally measured and model-predicted current $I(t)$, and membrane conductance $G_m(t)$, for tethered DphPC membranes resulting from a triangular excitation potential $V_s(t)$ at a rate of 100 V/s for 5 ms. All predictions are computed using the model specified by Figure 11.4 and (11.17) with the parameters defined in Table C.6 on page 418. The aqueous pore conductance in (11.17) is evaluated using the GPNP model.

0.66, and 0.33 μS, respectively. For the 100 percent tether density membrane, pore expansion is hindered by the tethers; therefore, changes in resistance are primarily a result of pore nucleation and destruction governed by (11.17). For the 100 percent tether density membrane, it may be the case that all pores in the membrane are hydrophilic because the tethers may prevent the transition from hydrophilic to hydrophobic structure. If only hydrophilic pores are present, the membrane resistance is dominated by the nucleation of pores and not the dynamics of the pores.

Using the electroporation model (11.17) and the experimental measurements in Figure 12.9, it is possible to estimate the thickness of the engineered membrane as a function of the tethering density. Using the electroporation model, the estimated membrane capacitance of the 1, 10, and 100 percent tether density membranes are 16, 13, and 12 nF, respectively. Using these values of membrane capacitance, the thickness of the membrane can be estimated using $h_m = \varepsilon_m A_m / C_m$ with $A_m = 2.1$ mm^2, the area of the membrane surface, and ε_m and C_m. For the 1, 10, and 100 percent tether density membranes we obtain membrane thicknesses of 4.63, 4.28, and 3.48 nm, respectively. These values are in excellent agreement with neutron reflectometry measurements of similar DphPC-based tethered membranes [152]. As expected, the thickness of the tethered DphPC membrane decreases for increasing tether density. This results from the tethers suppressing the thermal fluctuations of lipids in the engineered membrane. The dramatic decrease in membrane thickness for the 100 percent tether density membrane compared with the 1 and 10 percent tether density membranes results from the combined effect of an increased tether density and the dibenzyl group that binds the phytanyl tails in the 100 percent tether density DphPC monolayer.

An important parameter for the application of electroporation therapies is the characteristic voltage of electroporation, V_{ep}. The characteristic voltage of electroporation

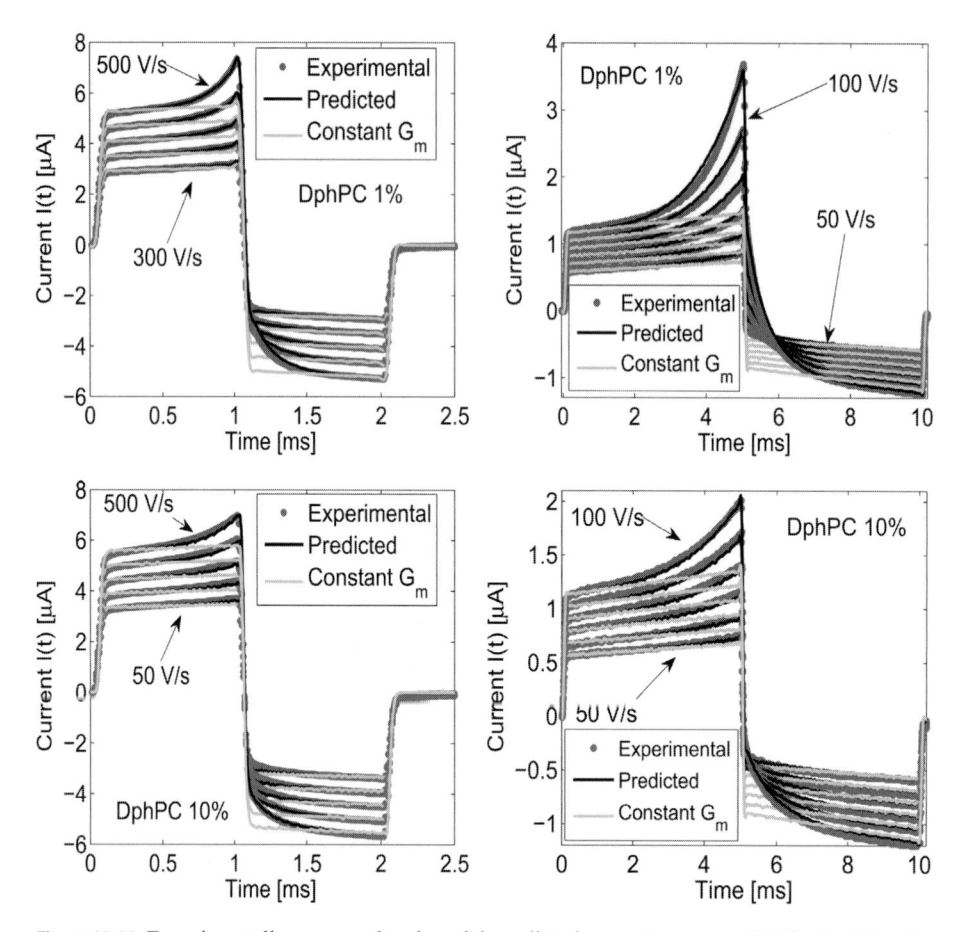

Figure 12.10 Experimentally measured and model-predicted current response $I(t)$ for the 10 and 1 percent tether density DphPC membrane. The triangular excitation potential V_s is defined by 300–500 V/s for 2 ms, and 50–100 V/s for 5 ms. All predictions are computed using Figure 11.4 and (11.17) with the parameters defined in Table C.6 on page 418. The aqueous pore conductance in (11.17) is evaluated using the GPNP model.

provides a measure of the magnitude of the transmembrane potential necessary to cause the formation of aqueous pores. From the electroporation model (11.17), and V_{ep} increases, a larger excitation voltage is required to cause the nucleation of aqueous pores in the engineered membrane. Using the results in Figure 12.9 and the electroporation model, the characteristic voltage of electroporation, V_{ep}, can be estimated as a function of the tether density. For the 1, 10, and 100 percent tether density membranes we obtain characteristic voltages of electroporation of 375, 510, and 650 mV, respectively. As expected, V_{ep} increases with increasing tether density. Therefore, when designing electroporation therapies it is essential to account for the tethering density of the membrane (which is equivalent to the density of cytoskeletal supports in cell membranes).

In Figure 12.10, the experimentally measured and predicted current $I(t)$ (using the fractional-order model illustrated in Figure 11.4) is provided for several different linearly increasing and decreasing excitation potentials. As seen from Figure 12.10, excellent agreement between the experimentally measured and predicted current is obtained. For small-magnitude excitation potentials one would expect the membrane resistance to remain constant because the effects of electroporation, governed by Figure 11.4 and (11.17), are negligible. Indeed this is the case because electroporation is only present in the 1-ms drive for potentials above 300 V/s, and at 5 ms for potentials above 50 V/s. The reason the 5-ms rise (see Figure 12.10) has larger relative electroporation effects compared with the 1-ms rise is because the nucleation and dynamics of pore radii evolve for a longer period of time. As expected, the magnitude of the current response for the 10 percent tether density membrane is less than the current response for the 10 percent tether density membrane as a result of the tethers hindering the nucleation and expansion of pores. The estimated electrical double-layer capacitance used to compute the current for the 10 percent membrane is $C_{dl} = 65$ nF, and that for the 1 percent tether density is $C_{dl} = 39$ nF. In reference to Table C.6 on page 418, the expected value of C_{dl} is in the range of 118 to 137 nF. Despite this minor discrepancy, the estimated current, using the model in (11.17) and illustrated in Figure 11.4, is in excellent agreement with the experimentally measured current.

12.6 Heterogeneous Membrane Mixtures

In this section we use engineered membranes as a hardware platform, along with the mathematical model (11.17), illustrated schematically in Figure 11.4, to estimate important biological parameters for the process of electroporation in unique membrane architectures. That is, we construct specialized engineered tethered membranes composed of lipids from archaea, yeast (*S. cerevisiae*), and *E. coli* bacteria. Note that these heterogeneous engineered membranes represent an in vitro model of cell membranes from various classes of single-celled microorganisms, namely, archaea, eukaryota, and bacteria.

Figure 12.11 presents the experimentally measured and model-predicted current response for DphPC, *E. coli*, and *S. cerevisiae* lipid membranes with tether densities of 1 and 10 percent. As seen, excellent agreement is obtained between the predicted and measured current response. As expected the resistance to electroporation increases as the tether density increases.

As is evident from Figure 12.11, the resistance to electroporation from the highest curve to the lowest curve is given by DphPC, *E. coli*, and *S. cerevisiae*. The difference in the resistance to electroporation of DphPC compared to that of *E. coli* and *S. cerevisiae* is a result of the larger attractive nonbonded interactions between DphPC lipids compared with the nonbonded interactions of *E. coli* and *S. cerevisiae* lipids. Additionally, DphPC lipids are known to have a lower diffusion coefficient compared to common phospholipids found in prokaryotic and eukaryotic membranes [378]. It is suggested in [377] that the stability of the DphPC membrane is closely related to the slow conformational motion of the phytanyl chains in the DphPC lipids. Note that an increase in tether density for all three membrane architectures will result in a decrease in the

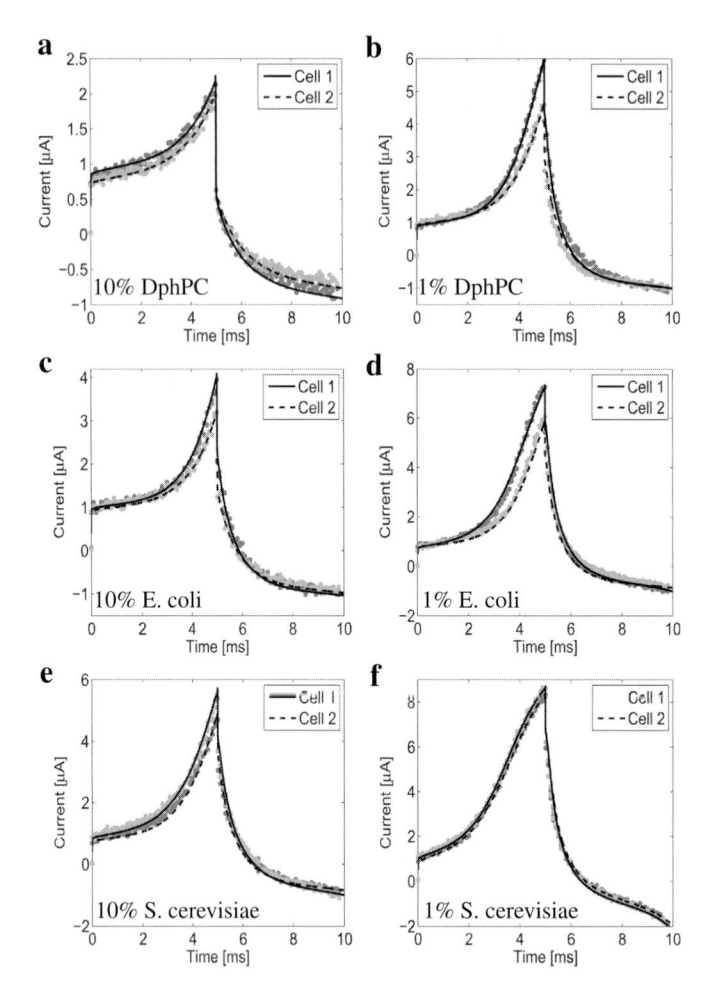

Figure 12.11 The measured and predicted current response of the 1 and 10 percent tether density DphPC, *S. cerevisiae*, and *E. coli* membranes. The excitation potential V_s is defined by a linear ramp of 100 V/s for 5 ms followed by a -100 V/s for 5 ms. Cell 1 and cell 2 denote the flow-cell number of the tethered membrane in which the measurement was made. All predictions are computed using the dynamic models, Figure 11.4 and (11.17), with the parameters defined in Table C.6 on page 418.

conformational motion of the lipids in each associated membrane. Therefore, we conclude that the difference in the resistance to electroporation between the DphPC and *E. coli* and *S. cerevisiae* membranes is a result of the larger attractive nonbonded interactions of the DphPC lipids compared with *E. coli* and *S. cerevisiae* lipids.

Does increased membrane thickness imply that fewer aqueous pores form due to electroporation? From Chapter 8, the thickness of the *E. coli* membrane is $h_m = 3.29$ nm, and the thickness of the *S. cerevisiae* membrane is $h_m = 4.30$ nm. Given the *S. cerevisiae* membrane is approximately 1 nm thicker than the *E. coli* membrane, we expect *S. cerevisiae* to have a higher resistance to electroporation compared with the

E. coli membrane. However, a surprising observation is that the *E. coli* membrane is more resistant to electroporation than that of *S. cerevisiae*. Therefore, only comparing the membrane thickness between membranes cannot be used to determine which membrane has an increased resistance to electroporation. A possible mechanism for the difference in electroporation between *E. coli* and *S. cerevisiae* is that aqueous pores in the *E. coli* membrane are primarily formed by the flip-flop of specific lipid molecules [324]. From molecular dynamics simulations [324] it is reported that the resistance to electroporation is a result of the reduced mobility of lipopolysaccharides, which comprise approximately 50 percent of the *E. coli* membrane, such that phospholipids primarily stabilize the aqueous pores. In comparison, the *S. cerevisiae* membrane primarily contains the phospholipids dipalmitoylphosphatidylcholine, dioleoylphosphatidylcholine, palmitoyloleoylphosphatidylethanolamine, palmitoyloleoylphosphatidylamine, and palmitoyloleoylphosphatidylserine with cholesterol [185]. Since lipopolysaccharides are not present in the *S. cerevisae* membrane, this reduces the energy for lipid flip-flop and therefore decreases the resistance to electroporation compared to the *E. coli* membranes.

12.7 Membranes with Sterol Inclusions

How can we control the mechanical properties of the engineered membrane? In this section we illustrate how including cholesterol in an archaebacterial membrane can be used to control the following properties of the membrane: conductance, thickness, diffusion of lipids, line tension, and surface tension. Since archaebacterial membranes are used for constructing membrane-based biosensors (refer to Chapter 3), knowledge of how cholesterol affects the membrane properties is important for the design of biosensors.

Using the fractional-order macroscopic model, formulated in (11.17) and illustrated in Figure 11.4, together with experimental measurements allows us to estimate (train) the parameters of the macroscopic model via least-squares fitting. We can then use the resulting trained model to predict the membrane conductance G_m that is dependent on the diffusion coefficient D, surface tension σ, line tension γ, and membrane capacitance C_m (which is dependent on the membrane thickness h_m).

Figure 12.12 provides the experimentally measured and model-predicted current response of the tethered archaebacterial membrane with cholesterol ranging from 0 to 50 percent (50 percent indicates that 50 percent of the membrane is composed of archaebacterial lipids, and the other 50 percent is composed of cholesterol). As seen the computed current is in excellent agreement with the experimentally measured current response of the tethered archaebacterial membrane. This allows the experimental measurements and dynamic model, Figure 11.4 and (11.17), to be used to estimate important biological parameters of interest. The fractional-order operator p is in the range of 0.95 to 0.98, suggesting that a diffusion-limited process is present at the bioelectronic gold interface of the tethered archaebacterial membrane. The associated capacitance C_{dl} is in the range of 120 to 180 nF. The membrane capacitance C_m, conductance G_m, and characteristic voltage of electroporation, V_{ep}, are provided in Table 12.1. How does the

Table 12.1 Macroscopic model parameters for archaebacterial membrane (see Figure 11.4 for the equivalent circuit model)

% Cholesterol	C_m	G_m	V_{ep}
0	34.4 nF	0.16 μS	270 mV
10	32.4 nF	0.48 μS	290 mV
20	31.4 nF	0.59 μS	300 mV
30	31.0 nF	1.86 μS	330 mV
40	35.0 nF	0.09 μS	345 mV
50	41.0 nF	0.15 μS	350 mV

concentration of cholesterol affect the characteristic voltage of electroporation, denoted by V_{ep}, and the membrane conductance G_m in the engineered membrane? Table 12.1 provides the experimentally determined membrane capacitance C_m, membrane conductance G_m, and characteristic voltage of electroporation, V_{ep}. From Table 12.1, we see that as the concentration of cholesterol increases there is an increase in V_{ep}. Recall from §11.3 that an increase in V_{ep} will cause a decrease in the rate of pore formation. Therefore, an increase in V_{ep} is associated with an increase in the stability of the membrane. As the concentration of cholesterol increases from 0 to 30 percent, the associated conductance of the membrane increases from 0.16 to 1.86 μS. However, for 40 and 50 percent cholesterol membranes, the associated conductance is 0.09 and 0.15 μS. The change in V_{ep} and G_m as a function of cholesterol concentration is a result of the interaction of the cholesterol with the lipids in the membrane. Recall from §8.4 that in an engineered membrane containing cholesterol, two possible domains exist. The first is the lipid domain, which contains low concentrations of cholesterol, and the second

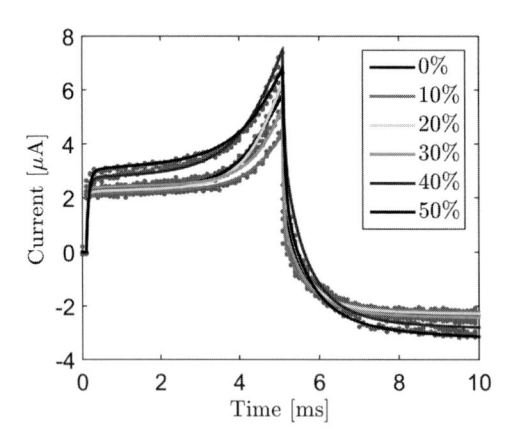

Figure 12.12 A color version of this figure can be found online at www.cambridge.org/engineered-artificial-membranes. Experimentally measured and model-predicted current response for archaebacterial tethered membranes containing cholesterol. The experimentally measured current response is represented by the gray dots, and the predicted current response from Figure 11.4 and (11.17) by the lines. The parameters for the numerical predictions are given in Table 12.1. The excitation V_s is given by a linearly increasing voltage with a rate of 100 V/s for 5 ms, then a linearly decreasing voltage with a rate of -100 V/s for 5 ms.

12.8 Estimating Hydration Ion Size and Faradic Reaction Rates

is the lipid raft domain, in which high concentrations of cholesterol are present. As the concentration of cholesterol increases, this causes an increase in the area occupied by the lipid raft domain compared with that of the lipid domain. A possible explanation for the conductance increase observed when the cholesterol concentration increases from 0 to 30 percent is that there is an insufficient amount of cholesterol to promote the formation of lipid rafts. Since lipid molecules pack together with other lipids in a tighter configuration than with cholesterol, this will cause an increase in the membrane conductance. As the cholesterol concentration is increased beyond 40 percent, a sufficiently large amount of cholesterol is present to allow the formation of tightly packed lipid rafts. The main source of leakage then results from the interface of the lipid raft domain and the lipid domain – as such, the conductance of the membrane decreases for cholesterol concentrations above 40 percent.

12.8 Estimating Hydration Ion Size and Faradic Reaction Rates

The hydration size of ions is important for estimating the electric potential ϕ, charge density ρ, and ionic flux J^i of species i used to evaluate the aqueous pore conductance $G_p(r)$, capacitance $C_p(r)$, and electrical potential energy stored in an aqueous pore, W_{es}. Faradic reactions account for reaction-limited charge-transfer phenomena that may be present at the bioelectronic interface. If faradic reactions are present at the bioelectronic interface, they will also contribute to the evaluation of ϕ, ρ, and J^i. Note that all these parameters, G_p, C_p, and W_{es}, contribute to the dynamics of electroporation in the engineered tethered membrane.

In this section, we describe how experimental measurements from the electrical response of the electrode surface can be used to determine both the hydration radius[4] of ions and the faradic reaction rates that may be present at the electrode surface. In engineered membranes, the electrostatic potential ϕ, charge density ρ, dielectric permittivity $\bar{\varepsilon}$, and ionic flux J^i for species $i \in \{1, \ldots, n\}$ all depend on the hydration radius of ions and water, and faradic reaction rates. As such, the lumped circuit parameters G_p, C_p, C_{dl}, and W_{es} also depend on the hydration radius of ions and faradic reaction rates (if present).

Remark: The atomic radii of ions are known; however, the precise hydration radii of ions are not known. The typical values for the hydration radius of sodium and chloride ions lie in the range of 0.2 to 2 Å. Experimental methods used to estimate the hydration radius of ions include infrared spectroscopy, X-ray and neutron diffraction, and nuclear magnetic resonance spectroscopy. Additionally, quantum mechanics, molecular dynamics, and statistical mechanics can be used to estimate the hydration radius of ions. For

---

[4] The hydration radius of an ion is the size of both the ion and the water molecules that bind with the ion via hydrogen bonding and van der Waals forces. For example, the slightly positive hydrogen atoms in water are attracted to Cl^- ions, while the slightly negative oxygen atom in water is attracted to Na^+ ions. This causes water molecules to surround both the Cl^- and Na^+ ions. The process of water associating with ions is known as *solvation*.

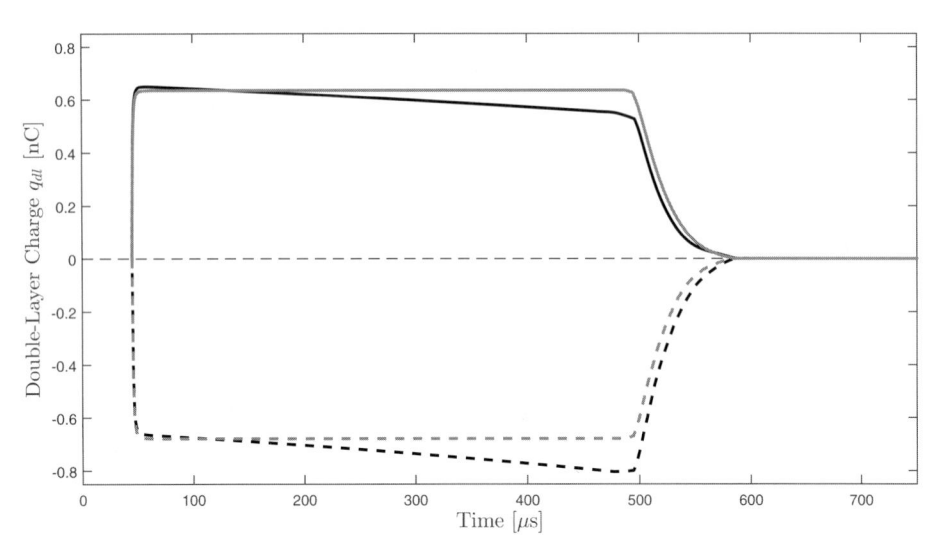

Figure 12.13 PFNP model-predicted double-layer charge dynamics resulting from a voltage step from 0 to 300 mV at $t = 45$ μs, then a linear decrease to 0 mV in 10 μs at $t = 500$ μs. The solid line indicates the charge for a positive voltage step, and the dashed line indicates the response for a negative voltage step. The gray lines indicate the response with no faradic reactions present, and the black lines indicate the response with faradic reactions present.

Cl$^-$ and Na$^+$ ions, there are typically two or three hydration shells[5] present. However, the hydration radii of these ions range from 0.2 to 2 Å because the hydration radius is dependent on several factors including temperature and ionic concentration.

Experimental setup. To estimate the hydration radius of ions and the faradic reaction-rate constants, we use experimental measurements from a gold surface bioelectronic interface with no membrane, in combination with the PFNP model (11.23) with the faradic surface adsorption boundary conditions (derived in §11.6). Recall that these faradic surface boundary conditions are composed of a generalized Frumkin–Butler–Volmer equation (11.40) with steric effects included. The excitation potential applied to the gold electrode bioelectronic interface is a voltage step of magnitude V_s. This is followed by a linear decrease in applied potential from V_s to 0 V in a period of 20 μs. If no faradic reactions are present ($k_r = k_o = 0$) then the associated current response will only result from the steric effects of the analyte species and from the screening effects of polarizable water.

Model-predicted hydration radius of ions and faradic reactions. Prior to using the PFNP model and experimental measurements to estimate the hydrated ionic radius and faradic reaction rates, we consider the charge accumulation at the electrode surface that results with no faradic reactions present and with faradic reactions present. For an excitation potential of $V_s = 300$ mV the model-predicted charge q_{dl} at the electrode surface is illustrated in Figure 12.13. The difference in the double-layer charge q_{dl} which

---

[5] A hydration shell is a monolayer of water that surrounds the ion.

results from the positive and negative excitation potentials, respectively, is a result of the difference in the hydrated ionic radius of Na$^+$ ions and Cl$^-$ ions. The computed ionic radius of Na$^+$ is 0.95 Å, and that for Cl$^-$ is 1.81 Å. In the case that faradic reactions are present, then the double-layer charge q_{dl} on the electrode is expected to change not only rapidly as a result of the electrolyte double-layer charging dynamics, but also on a slow time scale as a result of the faradic reactions that occur on the surface.[6] Figure 12.13 also illustrates how the charge q_{dl} varies when $k_r = 5 \times 10^{-8}$ m/s and $k_o = 5 \times 10^{-8}$ 1/s. As seen, there is an approximately linear response in the q_{dl} which results from the faradic reactions. As expected, the negative excitation potential causes a larger change in q_{dl} as a result of an increase in the adsorption of Na$^+$ ions at the surface of the electrode.

Experimental measurements of the hydration radius of ions and faradic reactions. How does the model-predicted charge response in Figure 12.13 (using the PFNP model (11.23) with faradic surface adsorption boundary conditions) compare with real-world experimental measurements? Figure 12.14 displays the experimentally measured and model-predicted charge response of a gold electrode bioelectronic interface. From the charge response q_{dl}, we see that there is a difference in the ionic radius of Na$^+$ ions and Cl$^-$ ions, and that faradic reactions are present at the electrode surface. As seen in Figure 12.14(c) the experimentally measured charge response and model-predicted charge response from the PFNP model (11.23) with the faradic boundary conditions (11.48) are in excellent agreement. In this special case where faradic reactions are present at the bioelectronic interface, we must account for both nonfaradic and faradic reactions when estimating the charge accumulation dynamics at the electrode surface. Typically for engineered tethered membranes, faradic reactions are not present because the gold electrode bioelectronic interface is coated with polyethylene glycol, which resists reaction with proteins and peptides. For the results in Figure 12.14(c), it may be that sections of gold are directly exposed to the analyte solution, allowing for binding of erroneous organic species.

12.9 Electrical Double-Layer Charging Dynamics

In this section we continue to study the properties of individual aqueous pores that form during electroporation. Specifically, we compute the double-layer charging capacitance $C_{dl}(V_{dl})$ and spatially dependent dielectric constant ϵ using the PFNP continuum model and of (11.23), (11.24). Recall that the PFNP model accounts for steric effects and Coulomb correlations; it is these steric effects and Coulomb correlations that determine the double-layer charging capacitance and a spatially dependent dielectric. (The GPNP model of the previous section cannot model Coulomb correlations; hence we use the PFNP model.) Recall from Figure 11.4 that the double-layer capacitance C_{dl} is an important component in computing the current response of the engineered membrane.

---

[6] Details on the fast charging dynamics of nonfaradic processes and slow charging dynamics of faradic processes are discussed in §11.6.

(a) Excitation potential $V_s(t)$.

(b) Measured current response $I(t)$.

(c) Double-layer charge q_{dl} with $I = dq_{dl}/dt$.

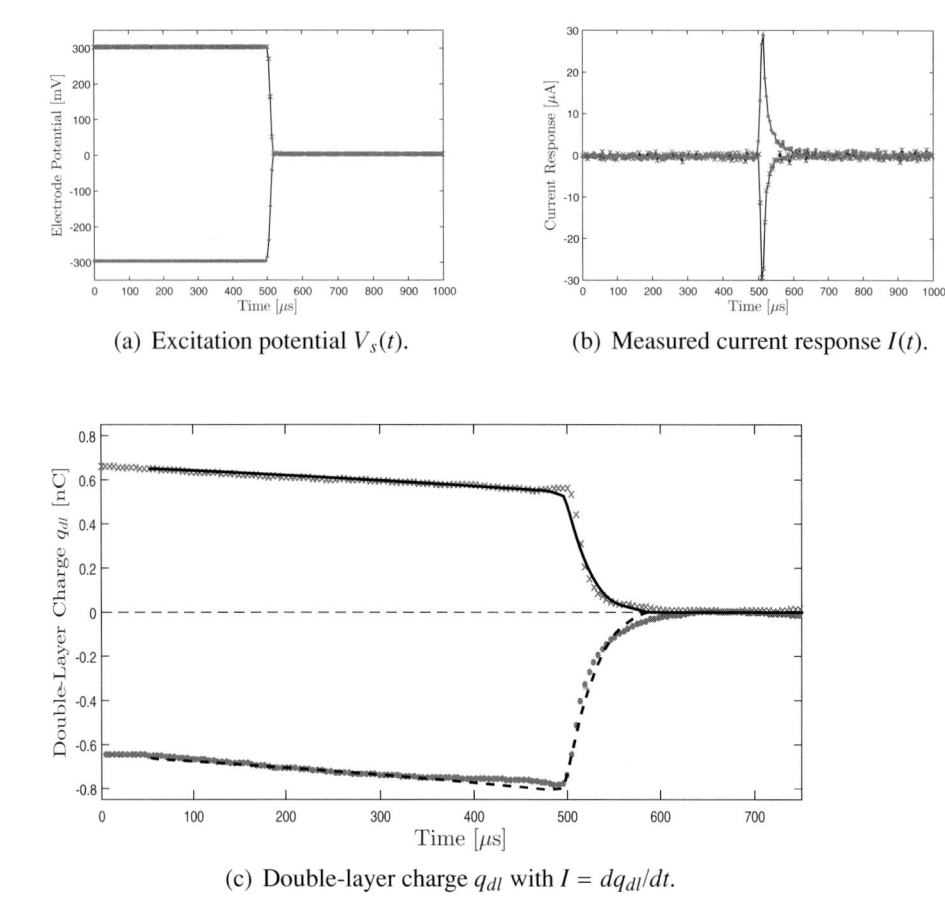

Figure 12.14 Experimentally measured and PFNP model-predicted charge response of the gold electrode bioelectronic interface to a voltage excitation V_s. For the positive voltage excitation (gray × in (a)), we display the associated current response (gray × in (b)) and charge accumulation (gray × in (c)). For the negative voltage excitation (gray dots in (a)), we display the associated current response (gray dots in (b)) and charge accumulation (gray dots in (c)). A negative charge, $q_{dl} < 0$, indicates there is a surplus of Cl^- ions at the electrode surface, and $q_{dl} > 0$ indicates there is a surplus of Na^+ ions at the electrode surface. The model-predicted charge response is given by the solid black line (positive excitation) and dashed black line (negative excitation). All model predictions are computed using the PFNP model (11.23) with the faradic boundary conditions (11.48).

12.9.1 Spatially Dependent Dielectric Constant at the Bioelectronic Interface

Why is the PFNP model (11.23) used to compute the double-layer capacitance C_{dl}? The PFNP model is used because Coulomb correlations[7] are present in proximity to the bioelectronic interface for large double-layer potentials V_{dl}. In this section we apply the

---

[7] Refer to the discussion in §11.4.2 on Coulomb correlations. Coulomb correlations are a quantum-mechanical effect that results from the interactions of the spatial position of electrons in ions and water.

detection test for Coulomb correlations constructed in §11.5.5 to justify the use of the PFNP model for estimating C_{dl}.

Testing for Coulomb correlations. To apply the detection test for Coulomb correlations constructed in §11.5.5, we use the PFNP model (11.23) with the boundary conditions (11.24) and adjust the electrode potential $\phi(0)$ to values in the range of -500 to 500 mV while the counter electrode potential is set to zero. For each electrode potential $\phi(0)$, the electric potential ϕ, charge density ρ, dielectric permittivity $\bar{\varepsilon}$, and height of the electric double layer, h_{dl}, adjacent to the bioelectronic interface is computed.

In engineered membranes, the electrolyte solution is always composed of an asymmetric electrolyte[8] (for example, an NaCl electrolyte solution). Here we consider an electrolyte solution composed of Na$^+$ and Cl$^-$ ions at a concentration of 150 mol/m^3, with the hydration radii of sodium and chloride ions given by $a_{Na} = 0.5$ nm and $a_{Cl} = 0.55$ nm, respectively. Additionally, we include 55,000 mol/m^3 of water molecules each with a radius of $a_w = 0.14$ nm. Figure 12.15 illustrates the computed ϕ, ρ, h_{dl}, and $\bar{\varepsilon}$ for different electrode potentials $\phi(0)$ in the range of -500 to 500 mV. The results in Figure 12.15(a) show that electrical double layers are present in proximity to the electrode surface, where the gradient of the potential is nonconstant. Comparing Figures 12.15(a) and 12.15(b), we see that the double-layer regions are associated with an unequal concentration of sodium and chloride ions in proximity to the electrode surface. As expected, for increasing electrode potentials $\phi(0)$, both the electric potential ϕ and charge ρ increase. Additionally, for electrode potentials below 150 mV, the charge ρ is negative symmetric.[9] However, above 150 mV the charge density ρ is asymmetric between the positive and negative electrode potentials. This asymmetry in the charge accumulation results from the combined effects of Coulomb correlations, polarization, screening, and steric effects of the ions. Figure 12.15(c) illustrates the computed double-layer thickness as a function of the electrode potential. For electrode potentials below 150 mV the double-layer thickness is symmetric and only depends on the magnitude of the electrode potential. However, for excitation potentials above 150 mV the double-layer thickness is asymmetric as a result of the steric effects of the ions. Figure 12.15(d) illustrates the computed spatially dependent dielectric constant $\bar{\varepsilon}$ for different electrode potentials in the range of -500 to 500 mV. Interestingly, the computed dielectric contains regions that have both higher and lower dielectric constants than the bulk electrolyte, which has a relative permittivity of 80. From our discussion in §11.5.5 and the results in Figure 12.15(d), since the condition $|\nabla \bar{\varepsilon}| \neq 0$, the PFNP model will be used to estimate the parameters ϕ and ρ required to compute the double-layer capacitance C_{dl}.

Spatially dependent dielectric. Why is the relative dielectric $\bar{\varepsilon}$ in Figure 12.15(d) spatially dependent? The spatially dependent dielectric $\bar{\varepsilon}$ results from the polarization effects of water. In bulk solution, the oxygen atoms of water face toward the sodium ions, and the hydrogen atoms of oxygen face toward the chloride ions. This reorientation of the polar water molecules causes its own electric field, which cancels out most of the

---

[8] An asymmetric electrolyte contains ions of different sizes, diffusion coefficients, and charges.
[9] If the electrode potential $\phi(0)$ results in a charge density of ρ, then an electrode potential $-\phi(0)$ results in a charge density of $-\rho$.

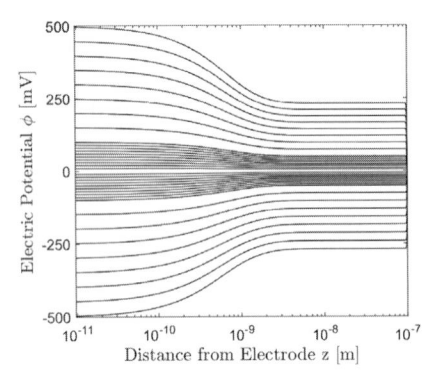

(a) Potential ϕ in the electrolyte solution.

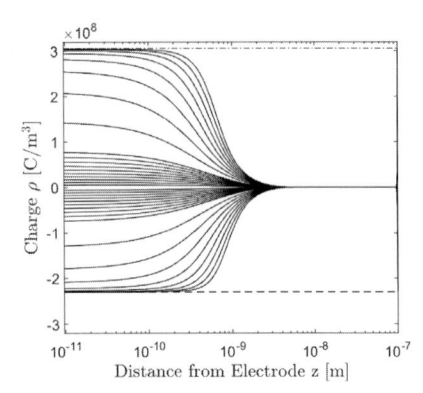

(b) Charge density ρ in the electrolyte solution. The dashed lines indicate the saturated charge ρ_{max} if only sodium ions (upper dashed line) or chloride ions (lower dashed line) are present.

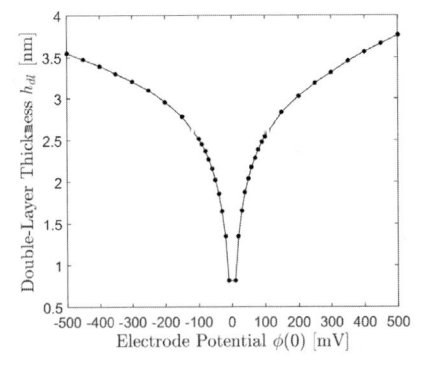

(c) Electrical double-layer thickness h_{dl}.

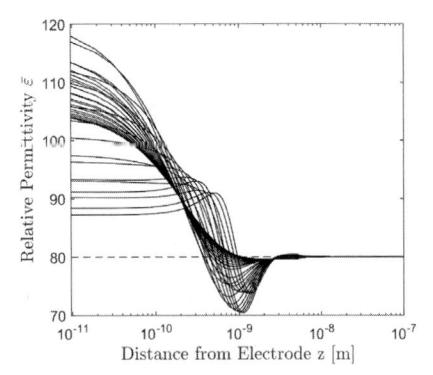

(d) Relative dielectric constant $\bar{\varepsilon}$ in the electrolyte solution.

Figure 12.15 Computed potential ϕ, charge density ρ, electrical double-layer thickness h_{dl}, and relative dielectric $\bar{\varepsilon}$ using the PFNP model (11.23) with the boundary conditions (11.24) and parameters defined in §12.9. The surfaces of the electrodes are at positions $z = 0$ nm and $z = 100$ nm, and the relative dielectric in the bulk electrolyte is $\varepsilon_w = 80$. The hydration radius of 2 sodium ions, chloride ions, and water are given by $a_{Na} = 0.5$ nm, $a_{Cl} = 0.55$ nm, and $a_w = 0.14$ nm, respectively. The multiple curves in (a), (b), and (d) denote the value of the electric potential, charge, and relative dielectric constant at different electrode potentials $\phi(0)$ in the range of -500 to 500 mV. The main conclusion is that for high-magnitude electrode potentials the charge density in proximity to the electrode saturates at the steric limit ρ_{max}. Therefore, for high-magnitude potentials the double-layer capacitance is expected to decrease.

electric field that would exist if the ions were located in a vacuum. As such, the sodium and chloride ions are effectively "shielded" by the water molecules. Therefore, for increasing concentrations of sodium and chloride ions, the relative dielectric constant is expected to decrease. Additionally, a decrease in dielectric can result from the water dipole moments becoming highly oriented at large electrode potentials. If the potential increases, the dipoles become saturated (e.g., further polarization is not possible for

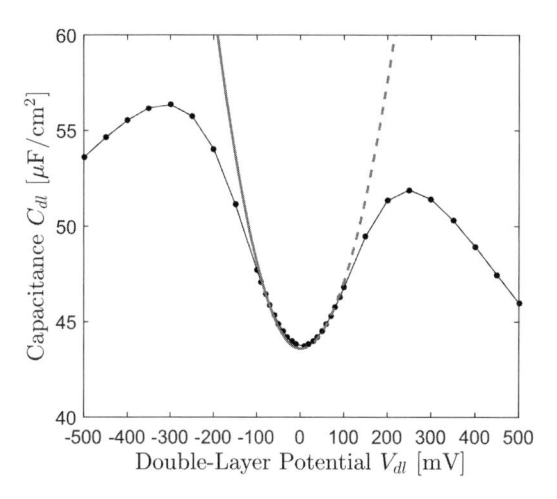

Figure 12.16 Computed double-layer capacitance C_{dl} as a function of the double-layer potential V_{dl} using the PFNP model (11.23) with the boundary conditions given in §12.9. The dashed gray line is a fitted quadratic model $C_{dl} = C_{dl+}(1 + \alpha_+ V_{dl}^2)$ for positive double-layer potentials, and the solid gray line is a fitted quadratic model $C_{dl} = C_{dl-}(1 + \alpha_- V_{dl}^2)$ for negative double-layer potentials. The simulation results are reported at a simulation time of $t = 10\ \mu s$. The main conclusion is that for low double-layer potentials V_{dl} in the range of -200 to 200 mV, the double-layer capacitance C_{dl} is quadratically dependent on V_{dl}. However, for large V_{dl} the double-layer capacitance does not depend quadratically on V_{dl}.

increasing potential), which results in a decrease in the dielectric permittivity [18, 471]. At approximately 1 nm from the electrode surface, Figure 12.15(d) illustrates the decrease in the electrode potential that results from the high concentration of sodium and chloride in this region, and the saturation of the water dipoles. However, why is there an increase in the dielectric in the 0–1 nm region of the electrode surface? The reason is that in this region there is a negligible concentration of water present to shield the electric field produced from the ions. Because the concentration of water decreases while the concentration of ions increases, there will be an increase in the dielectric constant.

12.9.2 Voltage-Dependent Double-Layer Capacitance

Does the double-layer capacitance C_{dl} depend on the double-layer potential V_{dl}? Figure 12.16 illustrates the computed double-layer capacitance C_{dl} as a function of V_{dl}. For double-layer potentials V_{dl} below 150 mV, C_{dl} depends on the magnitude of V_{dl} but not the polarity. This follows from the results in Figure 12.15, which indicate that both the charge density ρ and double-layer thickness h_{dl} for electrode potentials below 150 mV only depend on the magnitude of the electrode potential. A common model of C_{dl} is to assume that C_{dl} depends quadratically on V_{dl}. In Figure 12.16, the gray lines indicate the quadratic fit of the capacitance. As seen, the quadratic approximation is reasonable for V_{dl} below 150 mV.

For large double-layer potentials V_{dl}, the capacitance C_{dl} depends on both the magnitude and the polarity of V_{dl}. For large V_{dl}, the double layer becomes saturated with ions, which causes a decrease in the double-layer capacitance. Specifically, at a V_{dl} above 150 mV, a single layer of ions is not sufficient to balance the polarization charge[10] at the surface of the electrode, and therefore multiple layers of ions develop to balance the electrode charge. This can be seen in Figure 12.15(b) where the charge density is constant adjacent to the electrode surface. Therefore, as V_{dl} increases, the total charge in the electrical double layer, Q_{dl}, decreases, causing the decrease in the capacitance C_{dl}. The range of potentials V_{dl} for which the capacitance decreases depends on the hydration size of the ions. The decrease in C_{dl} occurs below $V_{dl} = -300$ mV and above $V_{dl} = 230$ mV. The asymmetric decrease in capacitance is expected because the hydration radius of sodium ions is $a_{Na} = 0.5$ nm, which is smaller than the hydration radius of chloride ions, $a_{Cl} = 0.55$ nm; therefore, fewer chloride ions can reach the electrode surface when a positive double-layer potential is applied.

12.10 Large Excitation Potentials and Double-Layer Charging Dynamics

In this section we illustrate that for large excitation potentials V_s above a few hundred millivolts, the double-layer charging dynamics depend on the polarity of the double-layer potential. Since the transmembrane potential V_m depends on the double-layer potential V_{dl}, this will cause V_m to depend on the polarity of V_s. Therefore, for large excitation potentials it is expected that the nature of electroporation (the population and size of pores) will be different when the applied potential V_s has different polarities but identical magnitude. This is in contrast to the results in the previous section, where, for low excitation potentials below 500 mV, electroporation was identical for both positive and negative excitation potentials.

In the equivalent circuit model (11.17) of the membrane, illustrated schematically in Figure 11.4, the membrane capacitance C_m depends only on the population and size of conducting pores in the engineered membrane. However, for large excitation potentials (above 500 mV), the membrane capacitance also depends on the size of ions, their valency, and the transmembrane potential V_m. In such cases the induced current from charges accumulating on the surface of the membrane is given by (11.11). Additionally, from the results in §12.9, it is also possible for the double-layer capacitance C_{dl} to depend on the size of ions, their valency, and the transmembrane potential V_{dl}. In this section we illustrate how these important engineered membrane dynamics can be observed from real-world experimental data.

To observe the difference in charging effects requires a large excitation potential V_s to be applied across the engineered membrane. Figure 12.17 displays the current response of an archaebacterial membrane with 1 percent tether density to several excitation potentials. The positive excitation potentials in Figure 12.17(a) are composed of an identical linear initial voltage ramp of 500 V/s, followed by a linearly increasing voltage ramp in the range of 35 to 75 V/s, then followed by a 1-ms linearly decreasing voltage ramp

---

[10] The polarization charge at the electrode surface results from electrons in the electrode attracting counter ions in the electrolyte to create a zero net charge across the electrode-electrolyte interface.

12.10 Large Excitation Potentials and Double-Layer Charging Dynamics

(a) Excitation potential V_s.

(b) Measured current response I.

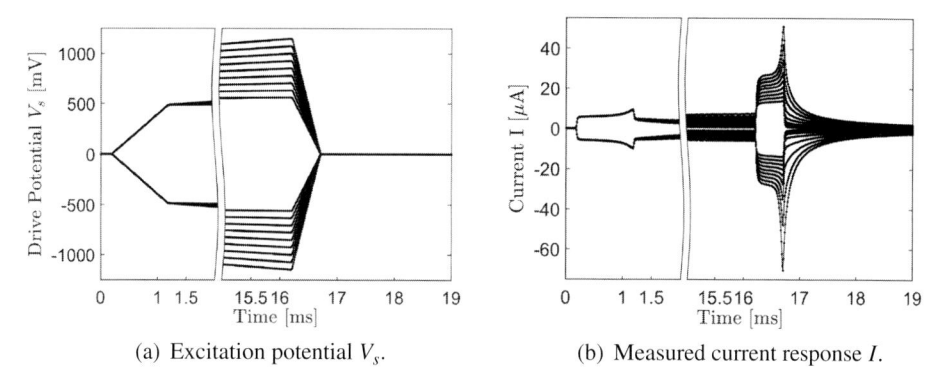

Figure 12.17 Experimentally measured current response of a 1 percent tethered DphPC membrane for various time-dependent excitation potentials V_s with maximum magnitudes in the range of -1000 to 1000 mV as illustrated in (a).

to zero potential. The negative excitation potentials V_s are the negative of the positive excitation potentials. As seen from Figure 12.17(b), large positive excitation potentials have a larger impact on the current response compared with the large negative excitation potentials. Since all physiological parameters are held constant between the positive and negative excitation potentials, this current-response difference is a result of the dynamics of the membrane and electrolyte solution. These effects can be accounted for by allowing the membrane conductance C_m to vary, or for the double-layer capacitance C_{dl} to vary (11.17), depending on the cations and anions present.

To determine if the membrane capacitance C_m or double-layer capacitance C_{dl} in the fractional-order model (see Figure 11.4 on page 223) depends on the properties (size, valency, and diffusion coefficient) of the cations and anions present requires estimating the transmembrane potential V_m and associated double-layer potential V_{dl}. We use the following parametric models for C_m and C_{dl}:

$$C_m = C_{m0}\left(1 + \alpha_m V_m^2\right),$$
$$C_{dl} = C_{dl0}\left(1 + \alpha_{dl}(V_{dl})V_{dl}^2\right)$$
$$\alpha_{dl}(V_{dl}) = \begin{cases} 1.6 & \text{if } V_{dl} > 0 \\ 0.8 & \text{if } V_{dl} \leq 0. \end{cases} \quad (12.6)$$

In (12.6), the scalar α_m is a measure of the voltage dependence of the membrane capacitance C_m on the transmembrane potential V_m. C_{m0} is the membrane capacitance at zero transmembrane potential, $V_m = 0$. The parameter C_{dl0} is the double-layer capacitance at zero double-layer potential $V_{dl} = 0$, and the parameter $\alpha_{dl}(V_{dl})$ is voltage dependent because it is dependent on the dominant species (cations or anions) in the electrical double layer. Using the fractional-order macroscopic model (Figure 11.4) with the voltage-dependent C_m and C_{dl} from (12.6) we obtain the current response in Figure 12.18(b), which results from the excitation potential V_s in Figure 12.18(a). As seen, excellent agreement is achieved between the measured and predicted current response using the

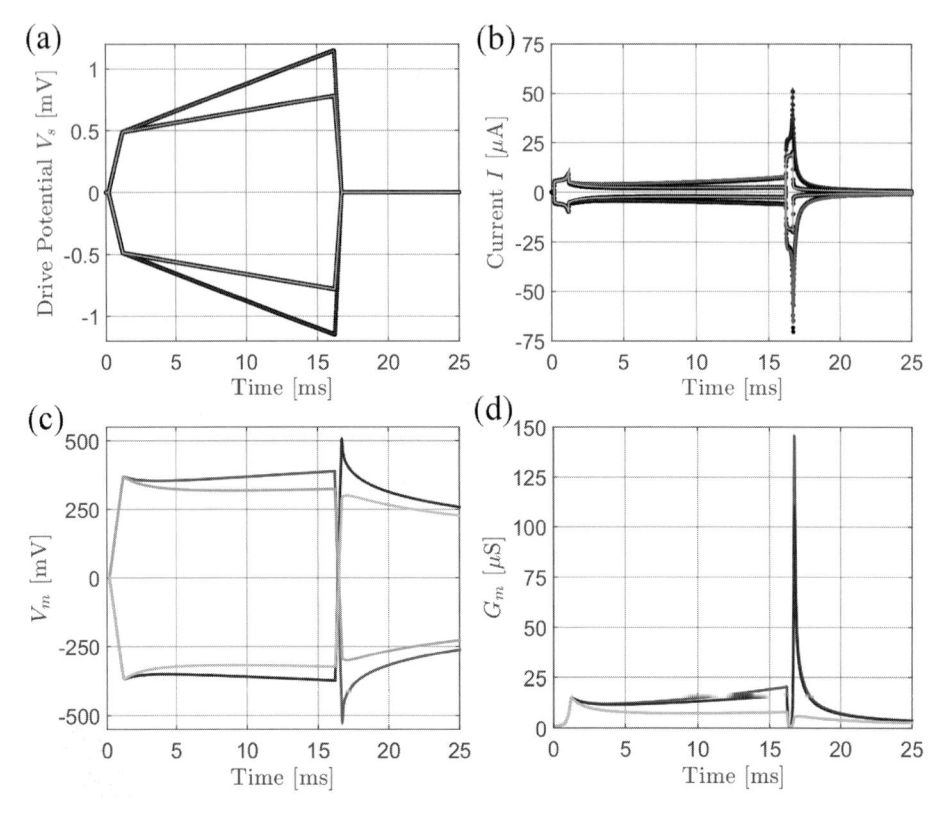

Figure 12.18 Experimentally measured and macroscopic model-predicted current response $I(t)$ for the 1 percent tether density DphPC membrane. The excitation potential V_s is given in (a), the associated current response is given in (b), the model-predicted transmembrane potential V_m is given in (c), and the model-predicted membrane conductance G_m is given in (d). The model-predicted responses $I(t)$, $V_m(t)$, and $G_m(t)$ are computed using the fractional-order electroporation model depicted in Figure 11.4 together with (11.17) with the parameters defined in Table C.6 on page 418 and $C_{dl}(V_{dl})$ defined in (12.6).

modified fractional-order macroscopic model. These results were repeated to ensure that at these high excitation potentials the tethered membrane or disulfide supports did not dissociate from the gold surface.

To determine if the difference between the model-predicted and actual current response is due to the charging dynamics of the membrane, C_m, we observe that the estimated transmembrane voltage V_m for these excitation potentials, illustrated in Figure 12.18(c), remains at approximately 300 mV. However, even though this transmembrane potential could cause a change in area of a freestanding membrane, the tethers in the engineered membrane restrict this change in area. Therefore, for this analysis we set $\alpha_m = 0$, and the difference in the current response must be contained in the double-layer charging dynamics of C_{dl}. Further insight into the charging dynamics of C_{dl} for the excitation potential presented in Figure 12.18 is provided in Figure 12.19. The charging dynamics of the electrical double layer appear to be well approximated by

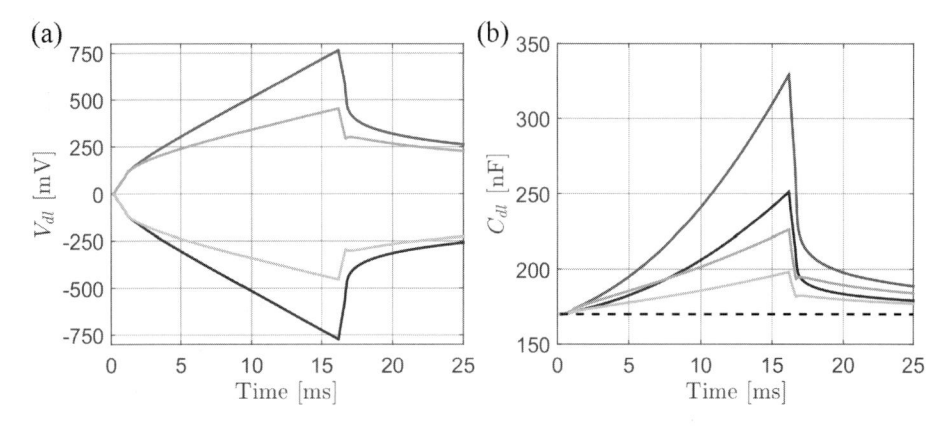

Figure 12.19 Predicted double-layer voltage V_{dl} and double-layer capacitance C_{dl} from the macroscopic model (Figure 11.4) with the parameters defined in Table C.6 on page 418 and the excitation potentials V_s provided in Figure 12.18(a). The main conclusion is that the double-layer capacitance C_{dl} is quadratically dependent on the double-layer potential V_{dl} and the polarity of the double-layer potential V_{dl}.

the parametric expression in (12.6) for C_{dl}. As expected from our analysis in §12.9 C_{dl} has a larger value when anions are the dominant species in the double layer compared with cations. This effect only becomes experimentally measurable at high double-layer charging potentials V_{dl} as seen from the results in Figure 12.19.

12.11 Complements and Sources

In this chapter we have applied the continuum models constructed in Chapter 11, together with experimental measurements from the EMP, to compute important parameters of engineered membranes. Several important insights were determined for how tether density, lipid composition, sterol inclusions, excitation potentials, the hydration size of ions, Coulomb correlations, and faradic reactions impact the current response of engineered tethered membranes. The main result is that accurate models for electrodiffusive dynamics and aqueous pore energy are essential in order to determine important biological parameters from experimental measurements from the EMP.

Voltage-Dependent Double-Layer Capacitance

In §12.9 and §12.10 the PFNP model [247, 248] and experimental measurements were used to illustrate that for large excitation potentials the double-layer capacitance C_{dl} is voltage dependent. Initially the C_{dl} increases with an increase in the magnitude of the voltage, and then C_{dl} begins to decrease after the magnitude of the voltage exceeds the steric charge accumulation capacity. This type of double-layer capacitance phenomenon has been observed in several experimental studies of charge accumulation in metallic nanopores [209] and on the surface of electrodes [30, 31, 150]. In this chapter we illustrated that this process is present at the bioelectronic interface and how the double-layer charge accumulation can be modeled using the PFNP continuum model. Note that these steric charging effects are negligible for small excitation potentials below 500 mV.

Spreading Conductance and Aqueous Pore Electrical Energy

In §12.2, we illustrated using the GPNP model that significant nonlinear potential gradients are present which restrict the current flowing through the aqueous pore. This effect is denoted as the "spreading conductance" and is dominant for pore radii significantly larger than the membrane thickness, causing the pore conductance to scale proportionally to the pore radius (i.e., $G_c(r) \propto r$). This is a surprising result at first glance; however, if we consider the literature on spreading conductance [8, 27, 238, 301], we conclude that the spreading conductance caused in the $h_r = 4$ nm tethering reservoir above the bioelectronic interface dominates the pore conductance. Note that if the engineered tethered membrane was not in proximity to the bioelectronic interface, then the relation $G_c(r) \propto r$ would not hold because the electrodiffusive dynamics in the tethering reservoir would not significantly restrict the movement of ions. As such, if no bioelectronic interface is present then $G_c(r) \propto r^2$.

A common assumption in the literature is to assume that the aqueous pore is composed of a narrow cylindrical pore, where $\partial \phi/\partial r \approx 0$ and $\partial \phi/\partial z \approx V_m/h_m$ on the surface of the pore. In such cases the electrical energy required to form the aqueous pore satisfies the proportionality relation $W_{es} \propto r^2$ [1, 300]. However, using the GPNP model we find that the computed potential in the pore, ϕ, does not satisfy these conditions. Additionally, we find that the electrical energy required to form the pore satisfies the relation $W_{es} \propto r^{2.2}$. This illustrates the importance of including an accurate model of the electrodiffusive dynamics when estimating important biological parameters for electroporation.

Heterogeneous Membrane Mixtures and Sterol Inclusions

In §12.6, the fractional-order macroscopic model and experimental measurements were used to study how the concentration of cholesterol in an engineered membrane changes the voltage of electroporation and conductance of the membrane. The results in Table 12.1 are in agreement with the experimentally measured results for 1-palmitoyl-2-oleoyl-sn-glycero-3-phosphocholine [64], 1,2-dipalmitoyl-sn-glycero-3-phosphocholine [159], DphPC [430], phosphatidylcholine, phosphatidylserine and phosphatidyl-glycerol [192], and egg yolk phosphatidylcholine [210] membranes that contain different concentrations of cholesterol. At an atomistic level, a possible explanation for the different membrane conductance values G_m of the DphPC membrane containing different concentrations of cholesterol is that cholesterol interferes with the packing of the hydrophobic chains in the lipids. For DphPC membranes the phytanyl tails (hydrophobic chains of the lipids) form a tightly packed network with neighboring hydrocarbon chains being interdigitated [174, 375, 377]. It is possible, if the membrane contains a low concentration of cholesterol, that these cholesterol molecules will interfere with the tightly packed hydrocarbon network, which will increase the membrane conductance. To numerically study this process requires the use of atomistic models such as coarse-grained molecular dynamics and all-atom molecular dynamics.

12.12 Closing Remarks

In this chapter we have applied the continuum models constructed in Chapter 11, together with experimental measurements from the EMP, to compute important parameters of engineered membranes. The chapter had seven important results:

(i) The conductance $G_p(r)$ of aqueous pores in engineered membranes is proportional to the radius r of the pore. This unusual behavior occurs because the toroidal aqueous pore is of small height and is adjacent to the bioelectronic interface.

(ii) The double-layer capacitance C_{dl} depends on the effects of Coulomb correlations, the polarization and screening of water, and the steric effects of water and ions. The result is that, for double-layer potentials V_{dl} above a few hundred millivolts, C_{dl} depends on V_{dl}.

(iii) The electrical energy stored in a toroidal aqueous pore satisfies $W_{es} \propto V_m^{2.2}$. This is in contrast to the electrical energy stored in a cylindrical aqueous pore, which satisfies $W_{es} \propto V_m^2$.

(iv) The capacitance of aqueous pores, $C_p(r)$, is of magnitude 10^{-18} F. As such, unless a large number of aqueous pores are present, the total capacitance of the engineered membrane, C_m, can be assumed constant when the process of electroporation is present.

(v) The hydration size of ions was estimated using the PFNP model experimental measurements. It was found that the hydration size of the Cl^- ion is 0.95 Å and the hydration size of the Na^+ ion is 1.81 Å.

(vi) Most aqueous pores in engineered tethered membranes have an identical size; that is, the distribution of pore radii is concentrated at a single value.

(vii) An increase in tether density will cause an increase in the resistance of the membrane to the formation of aqueous pores. This can be viewed as an increase in the stability of the membrane.

13 Electrophysiological Response of Ion Channels and Cells

13.1 Introduction

In this chapter we continue our study of continuum models for engineered membranes. We construct mesoscopic models for the electrophysiological response platform (ERP). The ERP is a synthetic biological device built out of artificial membranes for measuring ion-channel dynamics and the electrophysiological response[1] of cells grown on the surface of the engineered membrane. Recall that the ERP was synthesized in Chapter 3; our aim now is to model how the ERP works mathematically. The ERP can be used for drug screening and diagnosing diseases in which ion-channel functionality is disrupted (i.e., channelopathies). To estimate the electrophysiological response of cells and embedded ion channels requires that the electroporation dynamics of the tethered membrane be accounted for in the mesoscopic model of the ERP.

Motivation: Ion-Channel and Cellular Measurements

For measuring the electrophysiological response of individual ions or groups of ions, patch clamping is the gold standard. Patch clamping involves electrical measurements of the current-voltage response at different electrolyte concentrations. Patch clamping produces information-rich data which are widely used to study and validate ion-channel gating models. However, patch clamping is a labor-intensive process requiring a highly skilled experimenter to micromanipulate a glass pipette under a microscope to record data from one cell or membrane segment at a time.

Regarding measuring the electrophysiological response of cells, there are two classical methods. The first is to use substrate-integrated microelectrode arrays, and the second is to use sharp or patch microelectrodes that puncture the cell membrane [400]. A limitation of these invasive cell-measurement techniques is that, because they use sharp and patch microelectrodes, only a limited number of cells can be measured. Moreover, these techniques can cause the interior of the cell to leak into the electrolyte. Hence, these techniques can only measure the response of cells for short periods of time before they destroy the cell. Substrate-integrated microelectrode arrays provide a

---

[1] An electrophysiological response is defined as the change in membrane permeability as a result of an applied electric field, that is, the dynamics of the membrane conductance in response to an applied electric field.

noninvasive method for measuring the electrophysiological response of cells; however, a major challenge when using these sensors is to ensure sufficient cell adhesion and coverage [46, 400]. An emerging technology to ensure cell adhesion is to use a metal electrode coated with a polycationic film onto which an adhesion protein, such as Glycocalyx or Fibronectin, is used to bind with the cell membrane [400].

Instead of the above methods, this chapter discusses the synthetic biological device called the ERP. The ERP provides a controllable noninvasive tethered-membrane environment for measuring the response of ion channels and cells in a controllable environment. In Chapter 4 we discussed how to construct the ERP and also described methods for embedding ion channels, and methods for growing cells on the surface of the ERP. In this chapter we analyze the electrical response of the ERP using mesoscopic continuum models. We then evaluate the accuracy of these models by comparing their predictions with experimental measurements to perform noninvasive measurements.

To give some perspective on the ERP, note that the area of systems biology studies models for the response of single and multiple cells. The ERP can be viewed as an in vitro noninvasive platform for validating systems biology models. Also, the lumped circuit models we use in this chapter to model the ERP and response of multiple cells can be viewed as a systems biology model.

Organization of Chapter

In this chapter we construct mesoscopic models of the ERP and then use experimental measurements from the ERP to validate the accuracy of the models and also to estimate the electrophysiological response of ion channels and cells grown on the surface of the membrane. The chapter is organized as follows:

(i) In §13.2 we generalize the mesoscopic model (presented in §11.3) to account for the dynamics of embedded ion channels. This generalized model can be used to estimate the conductance dynamics of embedded ion channels while accounting for the formation and dynamics of aqueous pores.
(ii) In §13.3 the generalized mesoscopic model from §13.2 is used to measure the conductance dynamics of embedded ion channels using the experimentally measured current response from the ERP. Specifically we study the conductance dynamics of the voltage-gated sodium (NaChBac) ion channel.
(iii) In §13.4 we generalize the mesoscopic model (presented in §11.3) to account for the dynamics of cells grown on the surface of the tethered membrane. This generalized model can be used to estimate the conductance dynamics of the cell membrane while accounting for the conductance dynamics of the tethered membrane.
(iv) In §13.5 the generalized model from §13.4 is applied to measure the conductance dynamics of skeletal myoblasts (muscle cells) grown on the surface of the tethered membrane using the experimentally measured current response from the ERP. Note that skeletal myoblasts are attractive donor cells for cardiomyoplasty used to regenerate damaged myocardium tissue produced by acute myocardial infarction [260, 319].

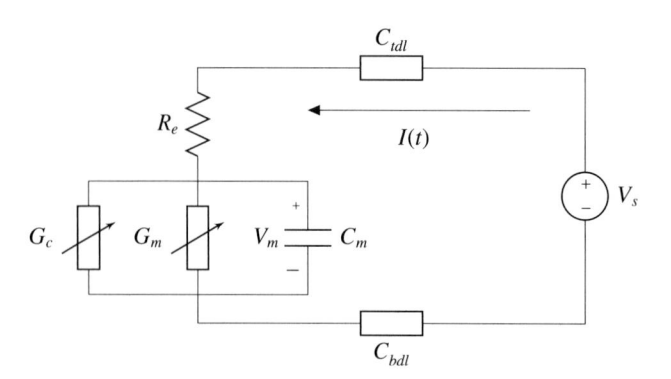

Figure 13.1 Fractional-order macroscopic model of the electrophysiological response platform (ERP) for measuring embedded ion-channel conductance G_c. V_m is the membrane potential, V_s the applied potential, and $I(t)$ the current response. C_{bdl} and C_{tdl} are the gold electrode and gold counter electrode bioelectronic interface capacitances. The circuit parameters are described in §13.2.

The results in this chapter illustrate how the ERP and generalized mesoscopic model can be used to study the dynamics of embedded ion channels and cells grown on the surface of the membrane in a controllable environment (the properties of the tethered membrane, electrolyte, and the applied electric field can all be precisely controlled in the ERP).

13.2   Dynamic Model of Embedded Ion Channels

Consider ion channels embedded in an engineered membrane. The current response of the ERP, denoted by $I(t)$, is a result of two simultaneous processes. The first is the process of electroporation, which causes the formation of aqueous pores within the tethered membrane, and the second is the ion-channel gating dynamics. In Chapter 11 we constructed a mesoscopic model of the electroporation process which comprised the system of ordinary differential equations ((11.17) on page 227) and the fractional-order macroscopic model illustrated schematically in Figure 11.4 on page 223. Here, we include the ion-channel gating dynamics in the mesoscopic model, which allows us to model the current response of the ERP. The ion-channel gating dynamics are accounted for by the conductance G_c that represents the conductance of the ion channels. The total conductance of the membrane results from the aqueous pore conductance G_m and the ion-channel conductance G_c in parallel. The complete fractional-order macroscopic model is provided in Figure 13.1.

To estimate the ion-channel conductance G_c, we minimize the mean-square error between the predicted current response $\hat{I}(t)$ from the generalized mesoscopic model in Figure 13.1 and the experimentally measured current response $I(t)$. The experimental measurements yield a time series $\{I(T_1), I(T_2), \ldots, I(T_K)\}$ of the current response at instants $T_1, T_2, \ldots, T_K$. The least-squares estimator for the parameters in the generalized mesoscopic model is given by

$$\theta^* \in \arg\min_{\theta \in \mathbb{R}_+^n} \left\{ \sum_{i=1}^{K} (I(T_i) - \hat{I}_m(T_i))^2 \right\}. \tag{13.1}$$

In (13.1), the parameter θ^* denotes the solution to the constrained optimization problem, constrained since the elements are non-negative. Note that typically only the parameters of the ion-channel conductance G_c are unknown because we already have low-variance (accurate) estimates of the parameters G_m, C_m, C_{dl}, and R_e of the tethered membrane.

13.3 Electrophysiological Response of a Voltage-Gated Ion Channel

In this section we use the ERP to study the electrophysiological response of the prokaryotic sodium channel NaChBac from *Bacillus halodurans*. The NaChBac channel is a voltage-gated ion channel[2] that selectively allows the passage of specific ions at a rate determined by electric potential gradient across the membrane (i.e., the transmembrane potential). The NaChBac channel was first reported in 2001 [341] and is likely an evolutionary ancestor of the larger four domain sodium channels in eukaryotes [65, 68].

The gating dynamics of NaChBac have been studied using patch clamping [227]; however, given the slow kinetics of NaChBac, it was not possible to detect the gating dynamics at room temperature (i.e., 21°C). In [227] the patch-clamping measurements were therefore performed at an elevated temperature of 28°C. The results in [227] suggest that NaChBac has several closed states with voltage-dependent transitions. It is, however, possible to measure the electrophysiological response of NaChBac at room temperature using the ERP and generalized mesoscopic model illustrated in Figure 13.1. The primary reason that the ERP can be used is that it measures the response of millions of NaChBac channels in the engineered tethered membrane with no other interfering transport phenomena present (e.g., from other ion channels or from non-negligible leakage currents that may be present in performing patch-clamping measurements).

To study the electrophysiological response of the voltage-gated NaChBac ion channel, the excitation potential illustrated in Figure 13.2(a), is applied to the ERP and the associated current response recorded. The excitation potential in Figure 13.2(a) is composed of increasing and decreasing voltage ramps with different slopes. The experimentally measured and model-predicted current response is displayed in Figure 13.2(b). As seen the experimentally measured and model-predicted current response are in excellent agreement, suggesting the estimated conductance G_c is a valid representation of

---

[2] Voltage-gated channels initiate action potentials in nerve, muscle, and other excitable cells and are vital for transcellular communication.

the ensemble conductance of the NaChBac channels. From Figures 13.2(c) and 13.2(d), for positive transmembrane potentials the estimated conductance of NaChBac channels, G_c, increases slightly; however, there is a dramatic increase in conduction for negative transmembrane potentials. Like most voltage-gated channels, NaChBac inactivates for positive transmembrane potentials [227].

Why is there an increase in conductance for a positive transmembrane potential? One possible reason is that there is a population difference in the orientation of the embedded NaChBac ion channels in the tethered membrane. Given that the conductance of a NaChBac channel is approximately 12 pS [341] and assuming the NaChBac is in a closed state for $V_m > 0$, it is possible to estimate the total population of NaChBac channels in each direction from the two peaks occurring at 1.7 and 2.2 μs in Figure 13.2(d). At 1.7 μs the conductance is 23 μS corresponding to 1.9 million conducting channels, and at 2.2 μs the conductance is 170 μS corresponding to 14.1 million conducting channels. Given the tethered membrane has a surface area of 2.1 mm^2, with 14.1 million conducting channels, this corresponds to each channel occupying an area of approximately 150,000 nm^2. From Figure 13.2(d) notice that the contribution of electroporation is negligible compared to that of the conductance dynamics resulting from the NaChBac channel. Note that if the population of NaChBac channels decreases then the effects of electroporation must be accounted for.

13.4 Dynamic Model of Electrophysiological Response of Cells

The electrophysiological measurement of cells allows for the diagnosis of channelopathic diseases such as cystic fibrosis and Bartter syndrome.[3] For example, the *cystic fibrosis transmembrane conducting regulator* protein blocks the flow of chloride and thiocyanate ions, causing a decrease in the cell membrane conductance. The ERP utilizes an engineered membrane for the noninvasive electrophysiological measurement of cells, which can be used to detect such a change in conductance. Typically an adhesion protein, such as Glycocalyx, or Fibronectin is used ensure there is sufficient adhesion of the cell to the sensing surface [400]. Using a suitably designed tethered membrane, cells can achieve sufficient coverage and adhesion to allow for their electrophysiological measurement. It was determined that a 100 percent tether density archaebacterial-based monolayer provides a suitable membrane to promote cell growth for electrophysiological measurement. Remarkably this allows the response to be measured using a noninvasive technique in the proximity of a synthetic membrane that mimics the electrophysiological response of biological membranes.

In this section we construct a macroscopic model of the ERP for measuring the electrophysiological response of cells grown on the surface of the ERP. The model accounts for both the electroporation and the dynamics of ion channels embedded in the cellular membranes, which is of importance for detecting channelopathic diseases.

---

[3] Refer to [20, 156] for a detailed discussion of ion-channel diseases.

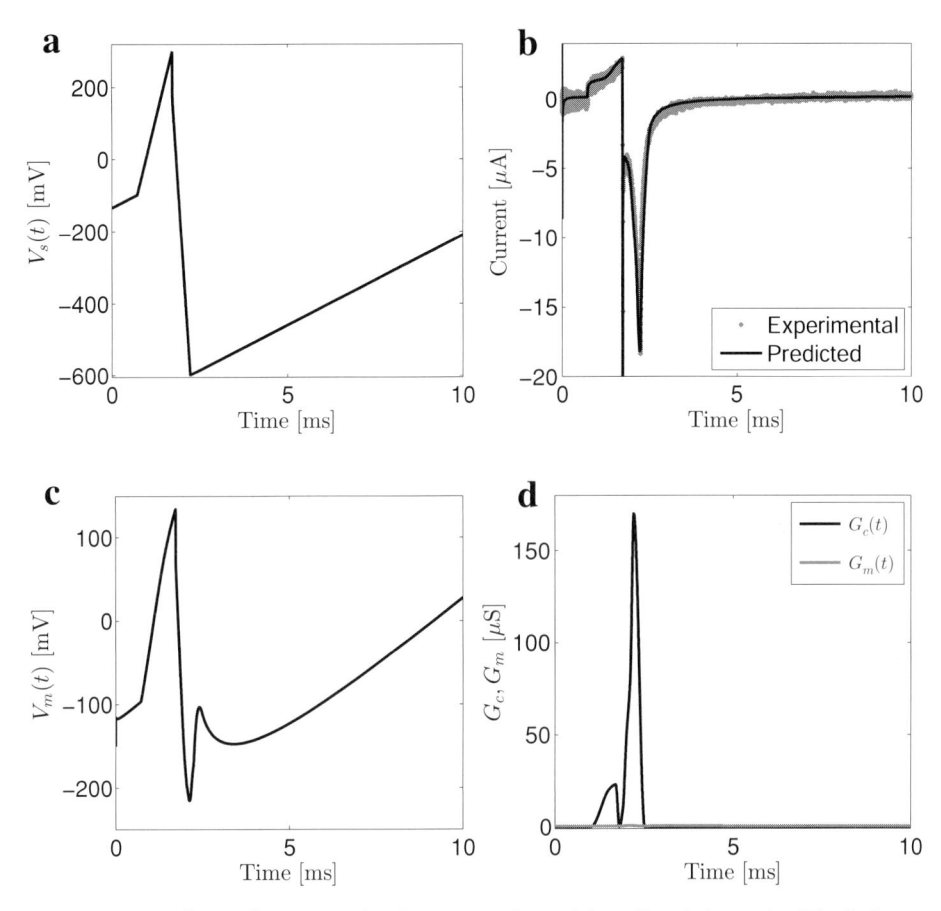

Figure 13.2 Experimentally measured and macroscopic model-predicted electrophysiological response of NaCHBac ion channels in the ERP. The numerical predictions are computed using the model in Figure 13.1 with the parameters defined in Table C.6 on page 418. (a) is the excitation Excitation potential $V_s(t)$. (b) Experimentally measured and model-predicted current response. (c) Predicted transmembrane potential $V_m(t)$. (d) Predicted total aqueous pore $G_m(t)$ and NaChBac conductance $G_c(t)$.

13.4.1 Macroscopic Model of the Electrophysiological Response Platform

Estimating the electrophysiological response of cells using experimental measurements from the ERP requires a dynamic model of the engineered membrane and the cells. In this section, a fractional-order model is provided that can be used to estimate the electrophysiological response of cells using experimental measurements from the ERP.

A schematic of the ERP is illustrated in Figure 13.3. The cellular suspension is modeled by assuming each cell has identical membrane capacitance C_c [22] and cytoplasm resistance R_c [21, 422], the conductive properties of each patch of the cell are uniform, and the side of the cell has negligible area. Given that the cells have uniform characteristics, the entire cell suspension can be represented by an equivalent circuit as illustrated in Figure 13.4. Note that R_L models the leakage resistance caused by current flowing

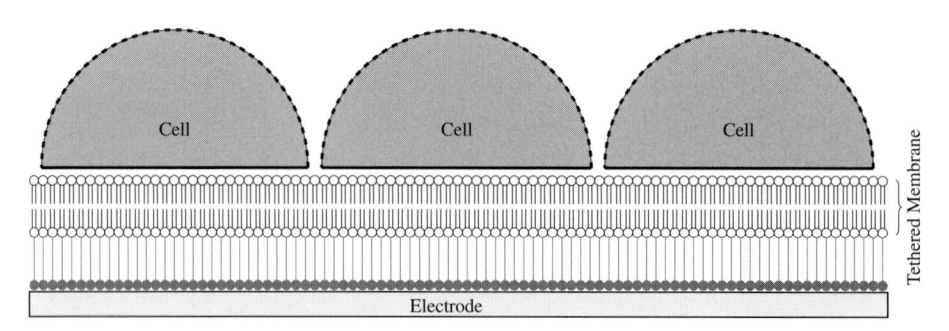

Figure 13.3 Schematic of the ERP. Cells are grown on top of an engineered membrane. The aim of the ERP is to measure the electrophysiological response of the cells. The bottom of these cells is indicated by the solid black line, and the top of the cells is indicated by the dashed black lines. The engineered membrane is composed of a DphPC membrane with a 100 percent tether density. Note that there are two types of membranes in the setup: the membranes of the individual cells and the engineered membrane on which the cells are grown.

directly from the tethered-membrane surface to the electrolyte solution [46]. The circuit parameters C_{tdl}, R_e, G_m, C_m, and C_{bdl} are defined in Figure 11.4 and (11.17). To account for the different polarizations of the top and bottom of the cell membrane we have defined the parameters C_{tm}, G_{tl}, and G_{tm} for the capacitance, leakage conductance, and variable conductance of the top surface of the cell adjacent to the bulk electrolyte, and C_{bm}, G_{bl}, and G_{bm} for the bottom surface of the cell adjacent to the tethered-membrane surface. R_{cm} is the total cytoplasmic resistance of the cellular suspension. In Figure 13.4, V_r denotes the resting potential of the cell membrane maintained by leakage ion channels.

The membrane capacitances C_{tm} and C_{dm} in the circuit model of Figure 13.4 are directly proportional to the fractional surface area of the top and bottom cell membranes (refer to Figure 13.3). Denoting A_t as the surface area of the top of the cell and A_b as the surface area of the bottom of the cell, then the total cell membrane capacitance satisfies $C_{tot} = A_t C_c + A_b C_c$. Additionally, C_{tot} is related to the membrane capacitances C_{tm} and C_{bm} as follows:

$$C_{tm} = A_f C_{tot}, \qquad C_{bm} = (1 - A_f) C_{tot},$$
$$A_f = A_t / (A_t + A_b). \qquad (13.2)$$

Equation (13.2) allows us to account for the geometry of the cells in a meaningful way. Note that for a completely symmetric cell the areas of the top and bottom of the cell are equal such that $A_f = 0.5$. Typically, cells resting on a flat surface will have $A_b > A_t$, which will result in $A_f > 0.5$ [421].

To estimate the membrane conductances G_{bm} and G_{tm}, the experimentally measured current $I(t)$ is compared with the model-predicted current from the lumped circuit in Figure 13.4. When agreement between the model-predicted and measured currents is reached, then G_{bm} and G_{tm} denote the electrophysiological response of the cell membrane.

13.4 Dynamic Model of Electrophysiological Response of Cells

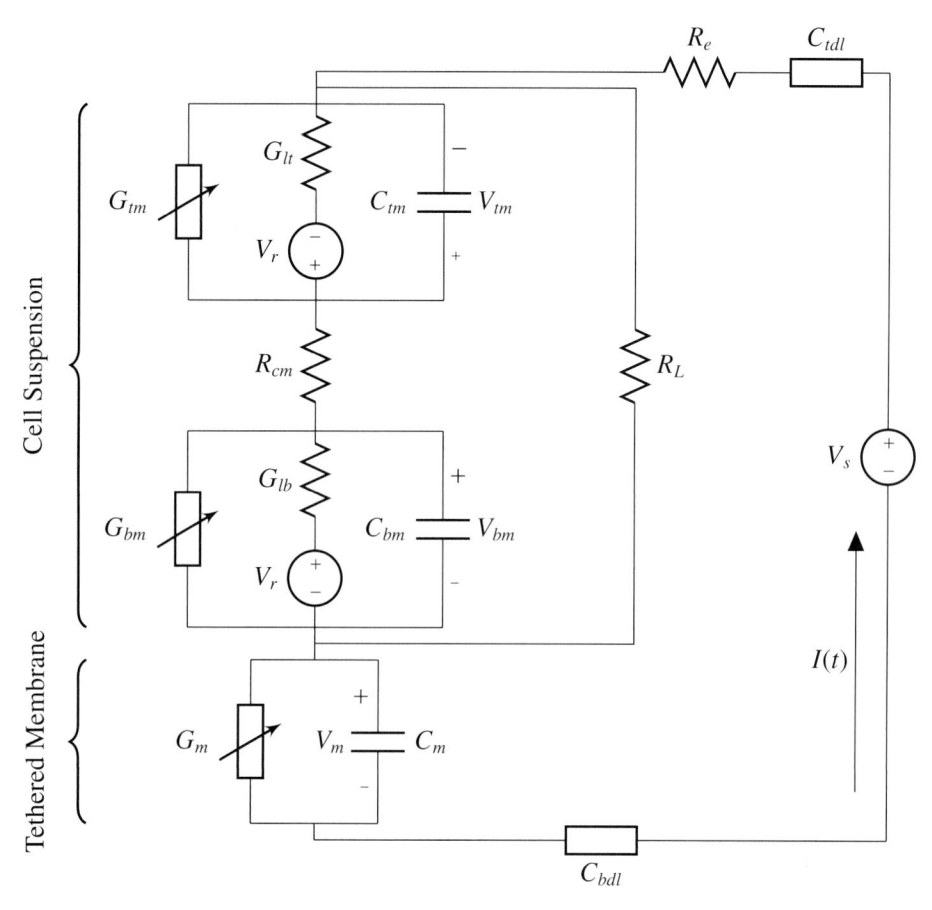

Figure 13.4 Lumped circuit model of the ERP, which comprises the cells grown on top of the engineered membrane. V_s is the driving voltage, V_{tm} is the potential across the cell membrane adjacent to the bulk electrolyte, V_{bm} is the potential across the cell membrane adjacent to the tethered membrane surface, and V_m is the tethered-membrane potential. The circuit parameters are described in §8.2.

13.4.2 Cellular Membrane Conductance and Charging Dynamics

The conductance and charging dynamics of cellular membranes depends on the dynamics of ion channels contained in the membrane. The dynamics of these ion channels depends on the polarity of the transmembrane potential and the charge accumulation at the surface of the membrane. Each of these important processes is discussed briefly here.

Resting potential and polarization. Recall from Figure 13.3 that there are three membranes (the engineered membrane, the bottom cell membrane, and the top cell membrane) present with the cellular membranes containing voltage-gated ion channels. The transmembrane potential of the engineered membrane is V_m, the top cellular membrane potential is V_{tm}, and the bottom cellular membrane potential is V_{bm}. Note that there are four regimes of transmembrane potentials of importance for cellular membranes: resting potential, depolarization, repolarization, and hyperpolarization. The resting potential is associated with the dynamics of ion channels in the cellular membrane that regulate the transmembrane potential. Typically the resting potential, denoted by V_r, of a cellular membrane is maintained at $V_r = -70$ mV. Depolarization occurs when the transmembrane potential increases above the membrane resting potential. During depolarization, voltage-gated sodium ion channels are activated, causing positively charged ions to diffuse back into the cell. After depolarization, a repolarization event occurs where the voltage-gated sodium ion channels close and potassium ion channels open as a result of the increased transmembrane potential. These channels remain open until the transmembrane potential returns to the resting potential. Note that hyperpolarization occurs when the transmembrane potential decreases below the resting potential, which will result in the closing of all voltage-gated ion channels and activation of sodium and potassium ion pumps. Note that these are the main dynamics of the ion channels in the cellular membranes; however, the cellular membrane is composed of thousands of different ion channels and pumps. The electrophysiological response of the cells accounts for the ensemble conductance response of all the ion channels in the cellular membrane.

Charging dynamics. Given the cellular membrane dynamics (resting potential, depolarization, repolarization, and hyperpolarization), we now consider the charging dynamics of the membranes resulting from the linearly increasing potential $V_s(t)$ producing the current response provided in Figure 13.5(a). For the linearly increasing potential $V_s(t)$, the top membrane adjacent to the bulk electrolyte experiences depolarization as the transmembrane voltage V_{tm} increases above the membrane rest potential $V_r = -70$ mV. This depolarization event promotes the activation of voltage-gated sodium ion channels on the top membrane surface. As the top membrane depolarizes, the bottom membrane adjacent to the tethered membrane undergoes hyperpolarization as the transmembrane voltage V_{bm} decreases below -70 mV, which causes all voltage-gated ion channels to close in the bottom membrane. Therefore, for a linearly increasing potential, we expect G_{tm} to vary by contributions from leakage ion channels (i.e., nonselective ion channels which are always open) and electroporation effects while G_{bm} includes effects caused by leakage ion channels, electroporation, and voltage-gated ion channels. For the current response in Figure 13.5(a), the transmembrane potential V_{tm} is in the range of -70 to 50 mV, and V_{bm} is in the range of -70 to -160 mV. For these transmembrane potentials, there were no voltage-gated ion-channel dynamics detected. This suggests that only the process of electroporation is present, and the dynamics of ion-channel pumps and leakage ion channels do not contribute significantly to the electrophysiological response of the membrane for a linearly increasing excitation potential. The associated conductance for the three membranes resulting from the excitation potential producing the results in Figure 13.5(a) is provided in Figure 13.5(b). As expected, since the magnitude of V_{bm} is larger than V_{tm}, the effects of electroporation are more pronounced on G_{bm} compared to that of G_{tm}.

13.5 Electrophysiological Response of Skeletal Myoblasts

In §13.4 we constructed a macroscopic model of the ERP to estimate the electrophysiological response of cells from experimental measurements from the ERP. Here, we perform electrophysiological measurements of skeletal myoblasts grown on the surface of the ERP and use the macroscopic model to estimate the electrophysiological response of these skeletal myoblasts. As mentioned in §13.1, the tethered membrane provides a stable and noninvasive platform for measuring the response of cells grown on it.

Experimental setup. To measure the electrophysiological response of the skeletal myoblasts, a sawtooth voltage excitation waveform was used with a slope of 250 V/s for 2 ms, and another with a slope of -250 V/s for a 2 ms. Figure 13.5 provides the numerically computed and experimentally measured current response of the skeletal myoblasts to the two sawtooth voltage excitation waveforms. Referring to Figures 13.5(a) and 13.5(c), it can be observed that the current response is asymmetric. Therefore, we can conclude that the leakage resistance R_L is sufficiently large, and $A_f > 0.5$ (13.2). Recall that if R_L is sufficiently large then the measured current response $I(t)$ would be dominated by the current flowing through the leakage resistance R_L, and the electrophysiological response of cells would be unobservable. Using the measured current response in Figures 13.5(a) and 13.5(c) with the dynamic model (Figure 13.4), we estimate that $A_f = 0.75$. The estimated A_f is consistent with values computed for neurons on electrode surfaces [421].

Experimental measurements. We now consider the charging dynamics of the ERP and cellular membranes. A linearly decreasing potential $V_s(t)$ (which we subsequently call excitation potential) is applied across the membrane, which results in the current response displayed in Figure 13.5(c). For this excitation potential, the bottom membrane depolarizes, which activates voltage-gated sodium ion channels. Comparing the current response in Figure 13.5(a) and Figure 13.5(c), we see that they are not related by a sign change even though the excitation potential $V_s(t)$ is related by a sign change. This is expected as the current response in Figure 13.5(a) is dominated by electroporation, while the current response in Figure 13.5(c) is a result of both electroporation and the activation of voltage-gated sodium ion channels.

Summary. The current response of the ERP resulting from the positive excitation potential is dominated by electroporation, while the current response from the negative excitation potential is a result of electroporation and the dynamics of voltage-gated ion channels. Therefore, the experimental results from the ERP in combination with the macroscopic fractional-order model (Figure 13.4) can be used to estimate the electrophysiological response of cells and ensembles of voltage-gated ion channels. Given a series of experimental measurements from the ERP using different excitation potentials, an experimenter could perform parameter estimation of an ensemble of ion-channel gating models (i.e., coefficient and exponents in ordinary differential equations of Hodgkin–Huxley-type models or transition probabilities between states in discrete-state Markov models) [57]. This allows the ERP and dynamic model to be used for the ex vivo validation of ion-channel gating models. Additionally, the ERP and dynamic

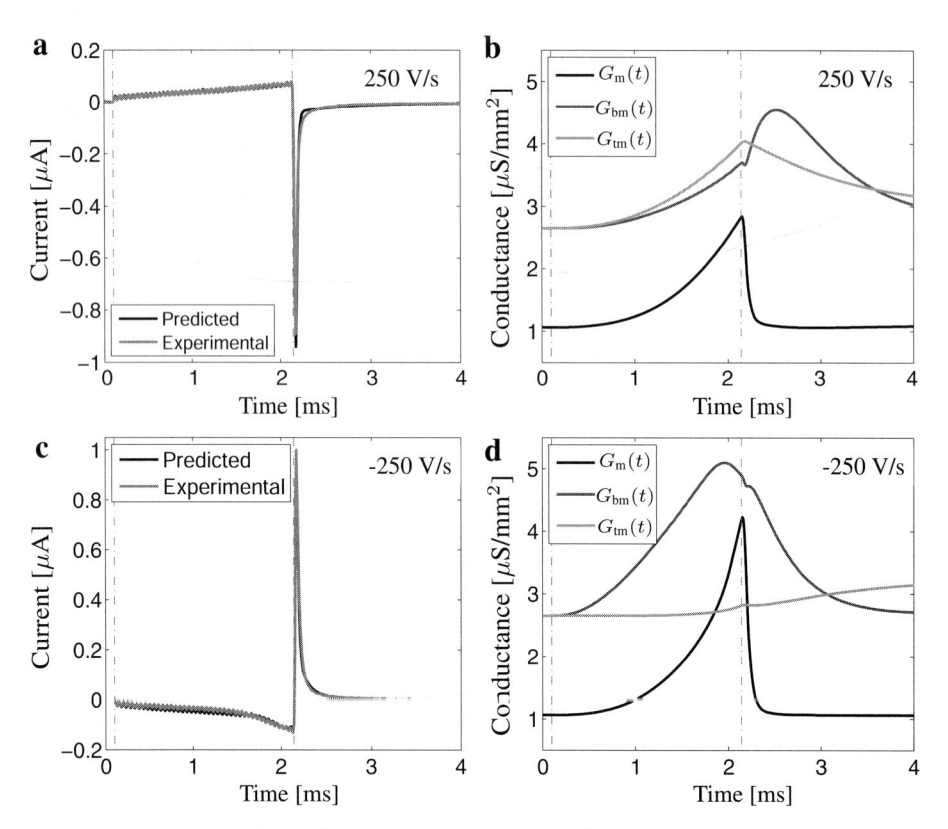

Figure 13.5 (a) The experimentally measured and model-predicted current response $I(t)$ for a sawtooth pulse excitation potential $V_s(t)$ with a slope of 250 V/s for 2 ms. (b) displays the computed membrane conductance G_m, the cell-membrane conductance adjacent to the tethered membrane, G_{bm}, and the conductance of the cell membrane facing the bulk electrolyte solution, G_{tm}, computed using the current response in (a). (c) presents the experimentally measured and model-predicted current response $I(t)$ for a sawtooth pulse excitation potential $V_s(t)$ with a slope of -250 V/s for 2 ms. (d) shows the computed membrane conductance G_m, the cell-membrane conductance adjacent to the tethered membrane, G_{bm}, and the conductance of the cell membrane facing the bulk electrolyte solution, G_{tm}, computed using the current response in (c). All predictions are computed using Figure 11.4 and (11.17) with the parameters defined in Table C.6 on page 418.

model can be used to detect channelopathic diseases as these would cause an anomalous electrophysiological response compared with healthy cells.

13.6 Complements and Sources

The remarkable feature of the ERP is that it provides a noninvasive method for simultaneous, long-term cellular measurement and stimulation from many cells under in vitro and in vivo conditions in a controllable environment. Additionally, the ERP can be used to measure the electrophysiological response of an ensemble of ion channels embedded in the membrane.

The gold standard method for measuring the electrophysiological response of ion channels is patch clamping [156, 222, 281, 305, 356], which provides information-rich data of use for validating ion-channel gating models. However, a limitation with patch clamping is that it is labor intensive and requires sophisticated experimental setup. Common methods for measuring the electrophysiological response of cells is to use either substrate-integrated microelectrode arrays [46, 400] or sharp or patch microelectrodes that puncture the cell membrane [400]. The ERP provides a third method that can be used to measure the electrophysiological response of cells. The major challenge with the use of the ERP is to ensure that sufficient cell adhesion is achieved between the cell membrane and the engineered tethered membrane. The design of adhesion techniques for the ERP is still an active area of research.

The paper [120] describes a patch clamp on a chip, namely, how a planar microstructured quartz chip can be used for whole-cell patch-clamp measurements without micromanipulation or visual control. The work [382] describes potential applications of this technology including massively parallel screening of drugs.

Future applications of the ERP will include using multielectrode arrays to perform multisite cellular measurement and stimulation. This allows the study of the communication network between cells to be elucidated using a multielectrode ERP. As our knowledge of cell biological principles increases, we find there are two major applications of the ERP. The first is the design of multielectrode ERPs with donor cells to construct stable neuron-electrode interfaces that mimic our central nervous system. Since the interface is constructed using the same cells as the central nervous system of the host, it provides the highest resolution of communication possible between exterior electrical equipment and the nervous system. The second is the design of cell-based integrated transistor arrays. Essentially, the multielectrode ERP can be designed such that cells interact cooperatively to perform a specific task. This is analogous to the well-established multiplexing capabilities of ultra-large-scale integrated transistor arrays which employ field-effect transistor arrays to perform logical operations; see also §3.7 for a discussion of future applications.

13.7 Closing Remarks

In this chapter we constructed mesoscopic models of the ERP, which is a novel synthetic biological device built out of engineered membranes. A key aspect of the ERP is that it forms a stable platform on which cells can be grown and measured in a noninvasive manner. If the electrical response of cells is different than what is expected, it could be due to certain genetic diseases; thus, the ERP provides the basis for noninvasive screening of cells. The mesoscopic models and experimental measurements from the ERP were used to estimate important parameters of ion channels embedded in engineered membranes and of cells grown on the surface of the ERP.

A major challenge with future cell-based biosensors is the design of electrode assemblies that comprise thousands of individual microelectrode interfaces. Key to the design of such devices is how to route the electrical signal from each microelectrode to an external sensor efficiently (e.g., using a small number of connections). Methods such as monolithic integration with on-chip circuitry and active switches may be used to time-multiplex the electric signals from the microelectrodes. However, several questions still remain of how to optimally design both the engineered membrane and electrolyte solution to promote cell growth on the membrane surface. The design of such bioelectronic interfaces for synthetic biological devices is an active area of research involving several aspects of physics, chemistry, biology, and engineering.

14 Coarse-Grained Molecular Dynamics

14.1 Introduction

We now move on to a lower level of model abstraction compared to the continuum models of the previous chapters. This chapter describes how coarse-grained molecular dynamics (CGMD)[1] can be used to study important properties of engineered artificial membranes (see Figure 14.1 for perspective on the levels of modeling abstraction). CGMD provides a dynamic simulation model of the engineered tethered membrane that is close to atomic resolution. As explained below, CGMD simulations can be used to estimate important parameters such as the diffusion coefficient of lipids and thereby facilitate building and designing membrane devices such as biosensors (discussed in Part II).

Why use CGMD simulations when one can use continuum models (at a lower spatial resolution) or all-atom molecular dynamics (MD) simulations (at a higher spatial resolution)? In simple terms, CGMD strikes a compromise between full-atom MD (which is often computationally intractable) and continuum theory (which can be inaccurate at a molecular spatial scale). All-atom MD (described in Chapter 15) deals with the trajectory of individual atoms, whereas in continuum theory the probability distribution of the atoms is propagated over time. Since an atom's trajectory is dependent on the trajectory of all other atoms due to intermolecular forces,[2] the computational cost[3] of evaluating all the atom trajectories in an MD simulation is restricted to systems on the nanometer length scale and nanosecond time scale. The main idea of CGMD is to construct "pseudoatoms" (also known as beads) that represent groups of atoms. CGMD then involves evaluating the trajectory of these interacting pseudoatoms instead of specific atom trajectories. An example is the atomistic representation of the DphPC lipid in an MD simulation compared with the coarse-grained representation of the DphPC

---

[1] CGMD is a huge area of research. Our presentation here is highly stylized and focuses exclusively on modeling engineered membranes. For a general comprehensive treatment of CGMD refer to [445] and references therein.
[2] Intermolecular forces are the forces which mediate interaction between molecules or atoms, including forces of attraction or repulsion which act between molecules or atoms and other types of neighboring molecules and atoms.
[3] For n atoms, the evaluation of the forces between all atoms requires $O(n^2)$ force-field evaluations.

Levels of Abstraction				
Ab Initio Molecular Dynamics	Classical Molecular Dynamics	Coarse-Grained Molecular Dynamics	Continuum Theory	Macroscopic Theory
nm fs	nm ns	nm μs	μm μs	m s

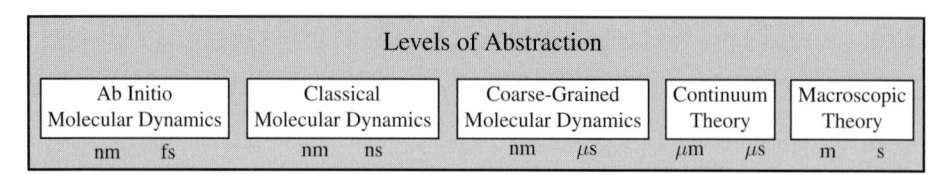

Figure 14.1 Schematic of the levels of abstraction for models of engineered artificial membranes. Abinitio is the lowest level of abstraction while macroscopic is the highest level.

lipid. In the MD simulation, the DphPC lipid is represented by 154 atoms, while a coarse-grained representation of the lipid ranges from 1 to 11 beads (or pseudoatoms) depending on the coarse-grained force field used. There are 23,716 atom-to-atom interactions between two DphPC lipids in the MD simulation, while in the CGMD simulation there are between 1 and 169 bead interactions – this significantly reduces the computational cost of CGMD simulations compared with MD simulations. Note that since CGMD only models the trajectory of the beads and not specific atoms, the CGMD model should not be used when atom trajectories are required such as in the estimation of the charge density across an engineered membrane. In Chapter 15 we show how MD simulations can be used to model engineered membranes at the per-atom atomic level.

From the model abstraction hierarchy (Figure 14.1), even though CGMD can be used to study larger systems at longer time scales than all-atom MD, CGMD can only be used to study small portions of the engineered membrane (nanometers) for a short duration of time (microseconds). Therefore, it is common to combine the results from CGMD simulations with models with higher levels of abstraction that include the continuum and macroscopic models to construct multiphysics models of engineered membranes. A multiphysics model of the engineered membrane, composed of CGMD, continuum, and reaction-rate theories, can be interpreted as an atomistic-to-observable model. A schematic of this atomistic-to-observable model is illustrated in Figure 14.2. The CGMD results provide key biological parameters such as the diffusion tensor and particle density of water, membrane thickness, lipid diffusion, defect density, free energy of lipid flip-flop, and membrane dielectric permittivity. Using continuum and reaction-rate theory, the CGMD results can be directly compared with the experimentally measurable current response from engineered membranes.

CGMD models also yield important insight into the design of membrane biosensors. For example, in [245] the diffusion coefficient of lipids in the distal layer (the lipids adjacent to the electrolyte bath) is larger than for the lipids in the proximal layer (adjacent to the bioelectronic interface). This result is expected as the tethers present in the proximal layer impede the movement of the lipids in the proximal layer. Additionally, in [451] the formation dynamics of an engineered membrane are studied. The results suggest that the lipid concentration used in step 3 to form an engineered membrane (refer to §4.2.1) is critical to minimize the formation of defects in the membrane (if the concentration is too low) or the formation of micelles on the membrane surface (if the concentration is too high).

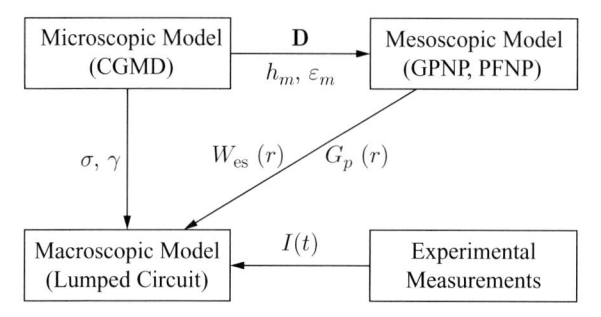

Figure 14.2 Schematic of the engineered tethered membrane from the atomistic-to-observable macroscopic model. The microscopic model is composed of CGMD, which is used to estimate the thickness of the membrane, h_m, the dielectric permittivity of the membrane, ε_m, the lipid diffusion coefficient **D**, the surface tension σ, and the line tension γ. The mesoscopic model is either the generalized Poisson–Nernst–Planck (GPNP) or Poisson–Fermi–Nernst–Planck (PFNP) and is used to compute the pore, conductance $G_p(r)$ and electrical energy required to form a pore $W_{es}(r)$. The macroscopic model is composed of the lumped circuit model of the engineered tethered membrane. The macroscopic model relates the results from microscopic and mesoscopic models to the experimentally measured current response $I(t)$ from the engineered tethered membrane.

Organization of Chapter

This chapter contains four major topics for engineered tethered membranes and also a brief mathematical description of stochastic diffusion processes. The first topic deals with the basics of constructing CGMD simulations of engineered tethered membranes. The second topic is focused on the properties of water at the bioelectronic interface, namely, the water density and the diffusion tensor of water. The third topic is how to evaluate parameters of membrane biomechanics from CGMD simulations, which include defect density r_{ss}, surface tension σ, line tension γ, free energy of lipid flip-flop, free energy of lipid desorption, deuterium order parameter S_{CD}, and lipid transport phenomena (anomalous and Fickian diffusion). The fourth topic is how to evaluate geometric properties such as tethering reservoir thickness, membrane thickness h_m, and area per lipid from CGMD simulations. The chapter closes with an application example of how CGMD simulations can be used to model the binding and pore formation of antimicrobial peptides in an engineered tethered membrane.

Since this chapter contains a variety of results and interpretations stemming from CGMD simulations, for the reader's convenience, we present a short summary of the main results. This serves as a road map for the chapter; we recommend that the reader consult this outline frequently to maintain perspective while reading this chapter. The main results are as follows:

(i) Construct an atomistic-to-observable model of tethered membranes by combining CGMD, continuum models, and reaction-rate models, that is, construct of a multiphysics model of the engineered membrane.
(ii) Estimate the density profile of water at the bioelectronic interface. The water density is a function of the distance from the gold electrode surface.

(iii) Estimate the diffusion tensor of water at the bioelectronic interface. The diffusion tensor plays an important role in the conductance of aqueous pores in the engineered membrane (refer to Chapter 11 for details).

(iv) Compute the potential of mean force (PMF) of DphPC lipids in an engineered membrane as a function of the tether density. The PMF is a key thermodynamic property that can be used to compute the energy of lipid flip-flop and to estimate the aqueous pore density in the engineered membrane. Recall from Chapter 11 that the aqueous pore density is a key parameter in the Smoluchowski–Einstein equation for modeling the dynamics of electroporation in engineered membranes.

(v) Compute the line tension and surface tension of the engineered tethered membrane. These parameters are used in the energy required to form an aqueous pore and can also be used to measure the stability of the engineered membrane.

(vi) Compute the transition times of the lipid diffusion characteristics to be in the ballistic, subdiffusion, or Fickian diffusion regimes. Note that all the continuum models in this book assume that molecules move according to the Fickian diffusion model (e.g., Brownian motion). However, in the CGMD model the trajectory of beads can satisfy any of the three diffusion regimes. Therefore, the results from CGMD can be used to estimate the approximate transition time from one regime to the other.

(vii) Estimate the structural properties of engineered membranes such as tethering reservoir thickness and membrane thickness, which are used to estimate the conductance of an aqueous pore in Chapter 11.

(viii) Estimate how the inclusion of sterols (cholesterol) into the engineered membrane impacts the structural properties and lipid dynamics in the engineered membrane. As was illustrated in Chapter 11, cholesterol is an important design parameter which can be used to control the conductance properties of the engineered membrane.

(ix) Estimate the aliphatic-chain deuterium order parameter of lipids in the engineered membrane. The aliphatic-chain deuterium order parameter provides a measure of the angle of a C-D bond vector with respect to the bilayer normal. Qualitatively it provides a measure of dynamics of the lipid tails with respect the bilayer normal. For a lipid with lipid tails that only rotate around the normal this order parameter would be unity, and zero if the lipid tails on average moved through all angles relative to the bilayer normal.

(x) Estimate the diffusion coefficient of the antimicrobial peptide PGLa and study the surface binding, translocation, and oligomerization of PGLa, which leads to the formation of PGLa pores. Recall that in Chapter 10 a continuum model was utilized to study these reaction pathways that lead to PGLa pore formation.

To summarize, CGMD simulations provide a viable first-principles approach to compute parameters that arise in both the continuum models and reaction-rate models of engineered membranes. This is in contrast to Chapters 8–13, where the parameters of the continuum and reaction-rate models were directly estimated from experimental data. Finally, we remind the reader that the CGMD codes are available at the website of the book. To get a better grasp of the material, we strongly recommend that the reader implement at least a basic-level CGMD simulation of the tethered bilayer.

14.2 Basics of Coarse-Grained Molecular Dynamics

CGMD is a computer simulation technique that allows one to model the time evolution of a system of interacting particle groups (known as pseudoatoms or beads) using Newtonian (classical) mechanics. Given n beads, the equation of motion of the ith bead is given by

$$m_i \ddot{x}_i = F_i, \qquad F_i = -\frac{\partial}{\partial x_i} U(x), \qquad \text{for } i \in \{1, \ldots, n\}. \qquad (14.1)$$

In (14.1), m_i is the mass of the ith bead, $\ddot{x}_i$ is the acceleration of the bead,[4] F_i is the total force acting on the bead, $U(x)$ is the potential energy, and n is the total number of beads in the system. Notice that the potential energy is dependent on the position and type of all n beads in the system, and consequently f_i is also a function of the position and type of all the beads in the system.

Newton's equation of motion (14.1) comprises a system of nonlinear ordinary differential equations. It is for this reason that careful numerical implementation techniques are required to evaluate the trajectory of the beads, especially to preserve the total energy (kinetic and potential energy) of the system. Note that although (14.1) is deterministic and describes the motion of the beads, the trajectories of the beads are random as a result of the initial conditions of the beads generated from a Maxwell–Boltzmann distribution and Monte Carlo simulations. The Maxwell–Boltzmann distribution is used for the initial velocity of the beads, and Monte Carlo simulations are used to select the initial position of the beads in the CGMD simulation.

Note that real-world experimental parameters such as temperature, pressure, and volume are not specified by the potential energy $U(x)$ in (14.1). To specify these parameters and ensure that they are conserved requires that CGMD simulations are performed in specific statistical ensembles.[5] Another important consideration when performing CGMD simulations is the definition of the initial conditions of the beads and the boundary conditions of the CGMD simulation used to solve (14.1). To summarize, conducting CGMD simulations requires specifying (1) the potential energy function $U(x)$ in (14.1), (2) the conserved macroscopic quantities of the system (i.e., the statistical ensemble), and (3) the initial and boundary conditions of the CGMD simulation to be specified, and a numerical method for evaluating (14.1).

MARTINI is a popular CGMD force field used to define the potential energy $U(x)$ for engineered membranes. The MARTINI force field will be used for all CGMD simulations presented in this book. In this section we briefly discuss how CGMD force

---

[4] The notation $\ddot{x}_i$ denotes the second-order time derivative of the position x_i of the ith bead.
[5] A statistical ensemble is the collection of all possible system states which have different microscopic states (for example, bead position and velocity) but identical macroscopic or thermodynamic properties (for example, temperature or pressure).

fields are constructed for engineered membranes. For details on the statistical ensemble and initial and boundary conditions in the CGMD simulation[6] the reader is referred to §15.2.

14.2.1 From an Atomistic to a Mesoscopic Coarse-Grained Description of Engineered Membranes

The main purpose of CGMD simulations is to compute the time-evolving trajectory of groups of interacting beads. The trajectory of the beads should approximate the trajectory that would have resulted if the all-atom MD simulation results were used to estimate the bead positions. Additionally, the CGMD models should also account for the electrostatic interaction between groups of atoms. In the following we present some mesoscopic modeling approaches to construct CGMD force fields and beads that provide an approximate representation of the underlying atomistic nature of the molecules (lipids, water, proteins, and peptides) in engineered membranes.

The results of MD simulations provide the trajectories and macroscopic parameters (temperature, pressure, and energy) that should be reproduced from the trajectory of the bead positions from the CGMD simulation. Note that a CGMD model will generate an equilibrium probability distribution function of the momenta that is consistent with a given MD model if no atom is involved in the definition of more than one bead and if the mass of each bead is defined using the center of mass of the associated atoms in the bead. There are several methods that can be used to construct a CGMD force field given the bead definitions and MD simulation results. These include Boltzmann inversion, iterative Boltzmann inversion, inverse Monte Carlo, force matching, and relative entropy-matching methods.

Inversion methods. Boltzmann inversion, iterative Boltzmann inversion, and inverse Monte Carlo methods are designed such that the empirical probability distributions of various parameters from the CGMD simulation match those of the MD simulation. Examples of such distributions include the probability distribution of the bond length, angle between bead pairs, and dihedral angles between beads, or for nonbonded interactions, the radial distribution function[7] between nonbonded beads. Note that the Boltzmann inversion, iterative Boltzmann inversion, and inverse Monte Carlo methods can only be used to construct potential energy functions that account for pairwise interactions between beads.

Force matching. Force matching seeks to match the potential between atoms in the MD simulation with the potential used between the beads in the CGMD simulation. Typically the estimated interaction potential between the beads is a nonlinear function of the distance between the beads. In such cases, it is common to approximate

---

[6] CGMD and MD simulations have equivalent requirements for the statistical ensemble and initial and boundary conditions used. The difference between CGMD and MD simulations is in the force field used and the particle (atom or bead) definitions.

[7] The radial distribution function describes how the density of a particle varies as a function of radial distance from a reference bead. Formally, the radial distribution function defines the probability of finding a bead at a radial distance r from another reference bead.

the pairwise interaction between beads using a set of linear splines. Note that the force-matching method is not focused on reproducing specific empirical probability distributions from the results of MD simulations.

Relative entropy. The relative entropy-matching method is designed to construct the CGMD force field so that the difference between the configurational phase space[8] of the CGMD simulation and MD simulation is minimized. The relative entropy-matching method provides a solution to the inverse molecular-thermodynamic problem of the optimization of a CGMD model (both bead definitions and bead interaction potentials) to reproduce the properties of a target MD model. The relative entropy-matching method is equivalent to the iterative Boltzmann inversion method when pairwise bead interaction potentials are used. Additionally, the relative entropy-matching method is equivalent to the force-matching method when the n-body bead interaction potential is utilized, with $6n$ the number of degrees of freedom[9] in the CGMD model.

In practical application of CGMD, the results from Boltzmann inversion, force-matching, and relative entropy-matching methods can all be used to construct CGMD models of engineered membranes from the results of MD simulations.

14.3 Atomistic-to-Observable Model of Tethered Membranes

With the above short review of CGMD, we are now ready to construct an atomistic-to-observable simulation model of the engineered tethered membrane. Recall that the construction and experimental measurements of the tethered membrane were detailed in Chapters 4–7. The atomistic-to-observable model is designed to link the results from atomistic simulations to the results obtained from experimental measurements of the macroscopic device. The model is composed of CGMD, a continuum model, and a fractional-order macroscopic model as illustrated in Figure 14.3. The CGMD simulation model is constructed using the MARTINI force field which combines the speed-up benefits of a simplified model with the resolution obtained by atomistically detailed models such as molecular dynamics. Using CGMD simulations allows the diffusion tensor $\mathbf{D}$, geometric properties of the membrane (e.g., thickness of the membrane, h_m), surface tension σ, line tension γ, and the number of pores in the membrane to be estimated. Recall that the fractional-order macroscopic model of the tethered membrane is given by (8.1) and (11.17). In (11.17), the pore energy $W(r, V_m)$ (11.8) is dependent on the surface tension σ and the line tension γ, which can be obtained from the CGMD results. However, the fractional-order macroscopic model is also dependent on the aqueous pore conductance G_p and the electrical energy required to form a pore, W_{es}. Given the diffusion tensor $\mathbf{D}$ and the geometric properties of the membrane, a continuum model can be used to estimate G_p and W_{es}. Two suitable continuum models are the generalized

---

[8] For a system of n atoms, the phase space is the $6n$-dimensional space of all the possible positions x_i and momenta p_i of the atoms, $i = 1, 2, \ldots, n$. At time t, the state of the system, given by the position and velocity of each atom, is given by a unique point in the phase space. Hence, MD and CGMD generate samples from this $6n$-dimensional phase space.

[9] The degrees of freedom of a CGMD simulation is the number of beads in the CGMD simulation.

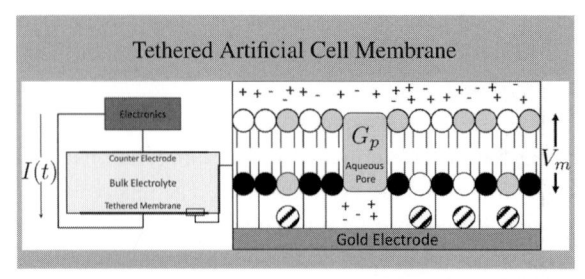

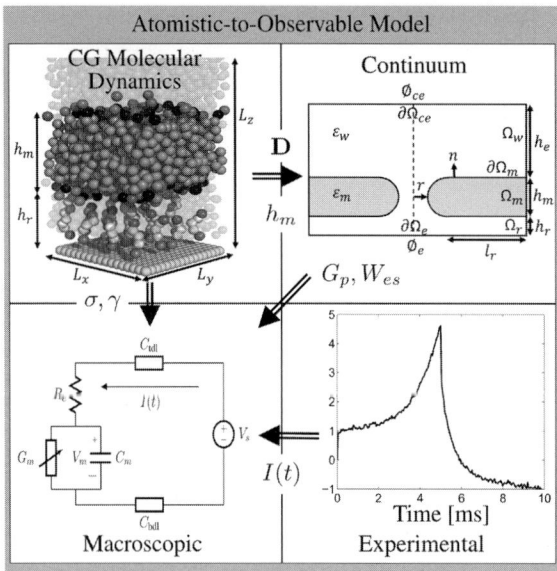

Figure 14.3 A color version of this figure can be found online at www.cambridge.org/engineered-artificial-membranes. Schematic of the tethered membrane and atomistic-to-observable model. The "Electronics" block in the upper most figure represents the electrical equipment used to measure the current response $I(t)$ of the membrane that results from an excitation potential $V_s(t)$. G_p is the conductance of the aqueous pore. In the atomistic-to-observable model the coarse-grained molecular dynamics (CGMD) is used to compute the diffusion **D**, thickness of the membrane h_m, surface tension σ, and line tension γ. The continuum model is used to compute the pore conductance G_p and electrical energy required to form a pore W_{es}. The macroscopic model is used to relate G_p, W_{es}, σ, and γ to the experimentally measured current $I(t)$. In the CGMD panel, the yellow beads model the bioelectronic interface, the translucent blue beads the water, and the green, pink, orange, and dark blue beads the tethered membrane. The brown beads are the disulphide beads, the white are the ethylene glycol components of the tethers, and the red beads are the terminus of the spacers.

Poisson–Nernst–Planck (GPNP) model presented in §11.4.1, and the Poisson–Fermi–Nernst–Planck (PFNP) model presented in §11.4.2. If only steric effects are important then the GPNP model can be used; however, in the case of high-voltage excitation potentials the PFNP should be used as it accounts for both the steric effects of ions and the polarization (or screening) effects of water. As seen in Figure 14.4, the fractional-order macroscopic model is used to link the GPNP continuum model with the experimentally measured current response of the tethered membrane.

14.3 Atomistic-to-Observable Model of Tethered Membranes

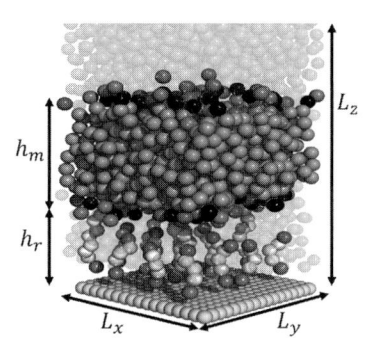

Figure 14.4 A color version of this figure can be found online at www.cambridge.org/engineered-artificial-membranes. Schematic of the CGMD model of the tethered bilayer lipid membrane. The green beads represent the phytanyl tail of the GDPE and DphPC lipids, the NC_3 bead (choline group) is displayed in blue, the PO_4 beads (phosphate group) in orange, the OH beads (hydroxide group) in red, the COC beads (etherglycol) as pink, the polyethylene glycol beads as white, the benzyl disulfide beads as brown, gold beads as yellow, and the water beads W as a translucent blue. L_z is the height of the simulation cell, L_y and L_x are the length and width of the simulation cell, h_r is the tethering reservoir height, and h_m is the membrane thickness.

We now give a detailed CGMD model construction of the lipid bilayer. It is strongly recommended that the reader implement the CGMD code on the book website to get a better understanding. The molecules that comprise the tethered membrane include the zwittrionic C20 diphytanyl-ether-glycero-phosphatidylcholine (DphPC) lipid, C20 diphytanyl-diglyceride ether (GDPE) lipid, benzyl disulfide connected to an eight-oxygen ethylene glycol group terminated by a C20 hydrophobic phytanyl chain (tether), benzyl disulfide connected to a four-oxygen ethylene glycol group terminated by an OH (spacer), and the gold surface. The engineered membrane contains mobile lipids, and physically tethered lipids that comprise DphPC and GDPE lipids.

The CGMD model of the tethered bilayer lipid membrane is constructed using the MARTINI force field.[10] That is, all the molecules that comprise the engineered membrane are mapped into a set of beads where the bead interactions are defined by the MARTINI force field. The MARTINI force field contains four bead categories, namely, Q for charged molecules, P for polar molecules, N for nonpolar molecules, and C for apolar molecules.[11] In total there are 18 beads that comprise the MARTINI force field (four are of type Q, five of type P, four of type N, and five of type C), and a total of 10 unique Lennard-Jones potentials that describe the nonbonded interaction potential between two beads. All bonded interactions (bonds, angles, dihedrals, and impropers) in the MARTIN force field are designed to match the results from molecular dynamics simulations.

---

[10] The MARTINI force field typically maps four atoms to one coarse-grained bead. The bonded interactions (bonds, angles, proper dihedrals, and improper dihedrals) between beads are constructed from all-atom molecular dynamics simulations. The nonbonded interactions are constructed based on experimental data (e.g., oil-water partitioning coefficients).

[11] A polar molecule consists of any group of atoms that contain a permanent electric dipole moment, while an apolar molecule consists of any group of atoms that do not contain a permanent electric dipole moment.

The CGMD simulation model of the engineered membrane accounts for the effect of the tethers which anchor the membrane to the gold surface. For example, a 25 percent tether density membrane is presented in Figure 14.4 where a 25 percent area tether density reflects one spacer molecular for every three tethering molecules. Recall that the engineered membrane can be composed of a tether density in the range of 1 to 100 percent.

Lipids. The phosphatidylcholine head group of the DphPC lipid is represented by two beads: the positive choline (NC_3) by the Q_o bead, and the negative phosphate (PO_4) by the Q_a bead. The ether glycol (COC) is represented by an SN_a bead, and each of the phytanyl tails by four C_1 beads. The Q_o bead models the charged choline head group of the lipid which has no hydrogen bonding, the Q_a bead models the phosphate head group of the lipid which is a hydrogen-bond acceptor, the SN_a bead models the nonpolar ether glycol molecule which is a hydrogen-bond acceptor, and the C_1 bead models a section of the apolar phytanyl tail. The phytanyl and ether glycerol moieties of GDPE are represented by the same mapping as the DphPC; however, the hydroxyl head group (OH) of GDPE is represented by a P_4 bead. The P_4 is the second highest polar coarse-grained bead in the MARTINI force field (the highest is P_5, which is used to model highly polar molecules such as acetamide). In total the DphPC lipid is composed of 12 coarse-grained beads, and the GDPE lipid by 11 coarse-grained beads.

Sterol inclusions. The inclusion of sterol components such as cholesterol in CGMD requires construction of a coarse-grained model of the associated molecule. To construct such a CGMD model of a molecule, one first runs an all-atom MD simulation of the molecule, then selects a coarse-grained topology (e.g., beads) and fits the nonbonded and bonded interactions to the results from the detailed molecular dynamics simulations. The construction of the CGMD simulation model of cholesterol is provided in [91].

Tethers and spacers. The eight ethylene glycol molecules of the tethers are represented by eight polyethylene glycol (PEG) beads. The interaction of the PEG beads is provided in [232, 451]. The benzyl disulfide group is represented by a C_5 bead which has the highest apolar affinity of the MARTINI beads. The interaction between the C_5 bead and P_4 polar bead of water have Lennard-Jones parameters $\sigma = 0.47$ nm and $\varepsilon = 3.1$ kJ/mol. Therefore, the equilibrium distance (the minimum of the Lennard-Jones potential) between the benzyl disulfide group and water is 0.53 nm. This indicates that the benzyl disulfide group is strongly resistant to water, which is represented by the P_4 polar bead. The polyethylene glycol segment of the tethers is represented by four C_1 beads. The spacers are mapped using an identical method as the tethers; however, the hydroxyl group (which is polar) is represented by the P_4 bead.

Gold surface. The gold surface is composed of a square lattice with custom P_f beads spaced 0.3 nm apart. The interaction of the P_f beads is based on the bead-to-bead Lennard-Jones interaction of the P_4 MARTINI bead.[12] The interaction between

---

[12] The parameters of the Lennard-Jones potential that describes the interaction between P_4 and P_4 (self-interaction) are given by $\varepsilon = 5.0$ kJ/mol and $\sigma = 0.47$ nm, where ε is the depth of the Lennard-Jones potential well, and σ is the distance at which the Lennard-Jones potential is zero. The Lennard-Jones parameters for other bead interactions of the P_4 bead are provided in [268].

P_f and P_4 is 1/3 the value of ε (the depth of the Lennard-Jones potential well) between P_4 and P_4, and the interaction between P_f and other bead types is about 12 percent of the MARTINI value between P_4 and respective bead types. The following interactions are excluded: between P_f beads, between the C_5 beads of the tethers and spacers, and between P_4 and Q_o beads of the lipids. The interactions and spacing of the P_f beads are selected to reduce the effects of excess adsorption of beads to the surface; that is, the surface mimics a hydrophobic gold surface. The P_f beads are held in place using a harmonic potential. Note that from the ab initio molecular dynamics results in [77] there is a wetting layer of water at the gold surface; however, beyond 3.2 Å one of the hydrogen molecules in water points toward the gold surface, which is typical for many hydrophobic surfaces. Given that in the CGMD simulation model each water bead represents four water molecules, the water bead interaction with the gold bead surface is selected to be hydrophobic. Note that this type of electrode surface does not account for any of the charging dynamics that occur at the surface of an electrode; therefore, it is assumed that a zero excitation potential is applied.

14.4 Aside: The Fokker–Planck Equation

We make a brief diversion to discuss three important concepts in continuous-time stochastic processes that model the evolution of ions and molecules: (i) stochastic diffusion processes, (ii) the Kolmogorov and Fokker–Planck equations, and (iii) mean first passage times.

These three concepts are widely used in applied probability theory. We use these concepts in §14.5.3 to estimate the spatially dependent diffusion coefficient of water adjacent to the bioelectronic interface. Also the stochastic diffusion process is used in the Langevin equation in §14.8.1 and Brownian dynamics in §15.6.1 to model the dynamics of molecules and ions as they permeate through ion channels in the membrane.

Let t denote continuous time in the interval 0 to T. A stochastic diffusion process $\{x(t)\}$ that evolves in m-dimensional Euclidean space is represented by the stochastic integral equation[13]

$$x(t) = x(0) + \int_0^t f(x(s), s) \, ds + \int_0^t \sigma(x(s), s) \, dw(s), \qquad x(0) \sim p_0(\cdot). \quad (14.3)$$

Here $w(t)$ is a vector of independent Brownian motions; that is, each element $w_i(t)$, $i = 1, 2, \ldots, m$, is an uncorrelated zero-mean Gaussian process with $\mathbf{E}[w_i(t)w_i(s)] = |t - s|$. The second integral in (14.3) is interpreted as an Itô integral and p_0 is the initial probability density function. The functions $f(x, t)$ and $\sigma(x, t)$ are called the drift and

---

[13] The diffusion process (14.3) can also be represented by the stochastic differential equation

$$\frac{dx(t)}{dt} = f(x(t), t) + \sigma(x(t), t) v(t), \qquad x(0) \sim p_0(\cdot), \quad (14.2)$$

where $v(t) = dw(t)/dt$ is continuous-time white noise. This is a heuristic definition since Brownian motion is not differentiable. White noise $\{v(t)\}$ is a zero-mean random process with correlation $\mathbf{E}[v(t)v(\tau)] = \delta(t - \tau)$, where $\delta(\cdot)$ denotes the Dirac delta function.

volatility of the diffusion process; see [195] for regularity conditions on f and σ that ensure x is a diffusion process.

For example, Langevin's equation discussed in §14.8.1 is of the form of (14.3), where $x(t)$ is the vector of positions of multiple interacting ions in an electrolyte.

Remark: The notation in this section is independent of other sections; $x(t)$ denotes a stochastic process here and not necessarily the position of a particle.

14.4.1 Kolmogorov and Fokker–Planck Equations

How does the probability density function of the diffusion process (14.3) evolve over time? The answer is given by the Fokker–Planck and Kolmogorov equations.

Forward Kolmogorov equation (Fokker–Planck equation). Denote the conditional probability density function of the diffusion process as

$$p_{t|0}(x|x_0) = p(x(t) = x|x(0) = x_0).$$

That is, $p_{t|0}(x|x_0)$ is the conditional density of the m-dimensional random variable $x(t) = x$ at time t given the initial state $x(0) = x_0$. Then the conditional density $p_{t|0}(x|x_0)$ of the diffusion process (14.3) evolves versus time t according to the Fokker–Planck equation (forward Kolmogorov equation):

$$\frac{dp_{t|0}(x|x_0)}{dt} = L^* p_{t|0}(x|x_0), \quad x \in \mathbb{R}^m, t \geq 0, \quad p_{0|0}(x|x_0) = \delta(x - x_0). \tag{14.4}$$

In (14.4) the partial differential operator L^* is defined as follows: For any twice-differentiable function ϕ with $\nabla \cdot$ denoting divergence operator, $\mathrm{tr}[\cdot]$ denoting trace (sum of diagonal elements), and superscript $'$ denoting transpose,

$$L^*(\phi) = \frac{1}{2} \mathrm{tr}\left[\nabla^2(Q\phi)\right] - \nabla \cdot [f\phi], \quad \text{where } Q = \sigma\sigma'. \tag{14.5}$$

The differential gradient operator ∇ is with respect to the variable x in (14.4). Note that the initial condition in (14.4) is specified by the Dirac delta function $\delta(x - x_0)$.

The Fokker–Planck equation is often expressed in terms of the *unconditional* density as follows. Using (14.4) and the initial distribution $p_0(x)$ of the diffusion process, the unconditional density $p_t(x(t) = x)$ can be obtained as

$$p_t(x(t) = x) = \int_{\mathbb{R}^m} p_{t|0}(x|x_0) p_0(x_0) \, dx_0.$$

So the unconditional density $p_t(x)$ for the diffusion process (14.3) evolves as

$$\frac{dp_t(x)}{dt} = L^* p_t(x), \quad x \in \mathbb{R}^m, t \geq 0, \quad \text{with initial condition } p_0(x), \tag{14.6}$$

which is another version of the Fokker–Planck equation.

Backward Kolmogorov equation. Instead of the forward Kolmogorov equation (14.4), an alternative characterization of the evolution of the probability density of a diffusion process is the backward Kolmogorov equation. Let $p_{T|t}(x_T|x) = p(x(T) = x_T|x(t) = x)$ denote the conditional probability density of $x(T) = x_T$ at terminal time T

given the state $x(t) = x$, where t lies in the interval 0 to T. The backward Kolmogorov equation for the diffusion process (14.3) reads

$$\frac{dp_{T|t}(x_T|x)}{dt} = Lp_{T|t}(x_T|x), \quad x \in \mathbb{R}^m, t \leq T, \quad p_{T|T}(x_T|x) = \delta(x_T - x). \quad (14.7)$$

Notice that the backward Kolmogorov equation (14.7) is specified by a terminal time condition $p_{T|T}$. Also, the second-order differential operator[14] L in (14.7) is

$$L(\phi) = \frac{1}{2} \mathrm{tr}\left[Q\nabla^2\phi\right] + f'\nabla\phi, \quad (14.8)$$

where the differential operator ∇ is with respect to the variable x in (14.7).

Summary and example. To summarize, the Fokker–Planck equation (14.4) or (14.6) and the backward Kolmogorov equation (14.7) are advection-diffusion partial differential equations. Either equation can be used to describe the evolution of the probability density function of a stochastic diffusion process.

As a simple example, consider the scalar stochastic linear diffusion process

$$x(t) = x(0) + a \int_0^t x(s)ds + w(t).$$

Then the Fokker–Planck equation (14.6) reads

$$\frac{dp_t}{dt} = \frac{1}{2}\frac{d^2 p_t}{dx^2} - ax\frac{dp_t}{dx} - ap, \quad \text{initialized by } p_0.$$

The backward Kolmogorov equation for time $t \in [0, T]$ reads

$$\frac{dp_{T|t}}{dt} = \frac{1}{2}\frac{d^2 p_{T|t}}{dx^2} + ax\frac{dp_{T|t}}{dx}, \quad \text{initialized by } p_{T|T} = \delta(x - x(T)).$$

It is well known for this linear case that if the initial condition p_0 is Gaussian, then $p_{t|0}$ is Gaussian for all time t. When $a = 0$, the Fokker–Planck equation (and backward Kolmogorov equation) specialize to the heat equation.

Stationary distribution. Consider the stochastic diffusion process $x(t)$ in (14.3). Under suitable conditions, the solution p_t of the Fokker–Planck equation (14.6) converges to the stationary density p_∞ as $t \to \infty$. Then setting the time derivative on the left-hand side of (14.6) to zero, the stationary density $p_\infty(x)$ is the solution of the partial differential equation[15]

$$L^* p_\infty = 0, \quad (14.9)$$

where L^* is the partial differential operator defined in (14.5). The stationary density $p_\infty(x)$ is also called the equilibrium or invariant density.

---

[14] The operator L^* in the Fokker–Planck equation has a deeper interpretation: it is the adjoint of the operator L. This means $\int_{\mathbb{R}^m} \alpha(x) L(\beta(x))dx = \int_{\mathbb{R}^m} L^*(\alpha(x)) \beta(x)dx$ for any two functions α, β. This can be verified using integration by parts.

[15] Equivalently, the stationary density p_∞ is the normalized eigenvector of the operator L^* corresponding to the eigenvalue at 0.

(a) Potential energy $U(x)$.

(b) Stationary probability density.

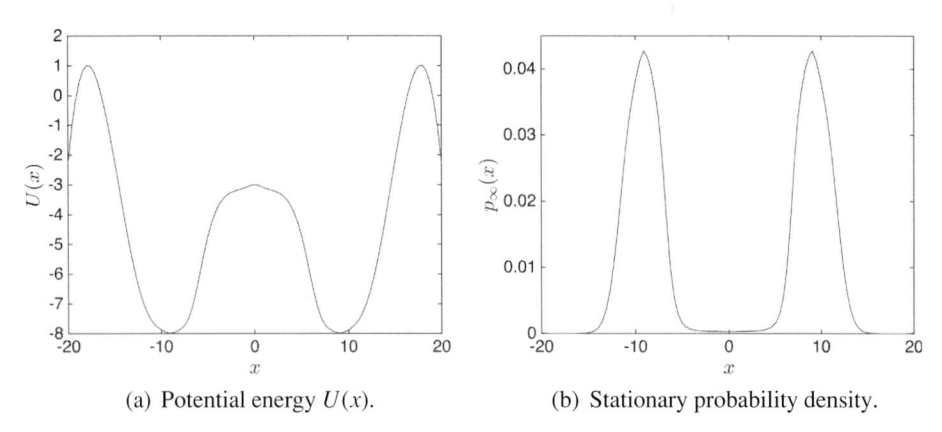

Figure 14.5 Potential energy $U(x)$ and stationary probability density $p_\infty(x)$. There are potential wells at positions $a = -10$ and $c = 10$ and a potential barrier at $b = 0$. As expected, the stationary probability density is much larger at the potential wells. The potential energy is in units of $k_B T$.

14.4.2 First-Passage Time and the Arrhenius Equation

The final concept we discuss is that of the mean first-passage time. Suppose the position of an ion evolves according to the diffusion process (14.3). What is the expected time that the ion takes to reach a particular region for the first time? This expected time is called the mean first-passage time. Let τ denote the first time the diffusion process $\{x(t)\}$ reaches the boundary ∂D of a region D when it starts at $x(0) = x \in D$. Then the mean first-passage time $\mathbf{E}[\tau]$ is the solution $\phi(x)$ of the following partial differential equation:[16]

$$L\phi(x) = -1, \quad \text{with boundary condition } \phi(x) = 0 \text{ if } x \in \partial D, \quad (14.10)$$

where the partial differential operator L is defined in (14.8).

Example: Arrhenius reaction rates and ion permeation. We end this section on diffusion processes with a simple yet insightful example of how ions jump from one binding site to another in an ion channel. The study of how ions travel through the channel is termed permeation and §15.6 gives a short overview of the area.

Consider an ion that moves in one dimension x over the potential energy landscape $U(x)$ illustrated in Figure 14.5(a) in units of $k_B T$, where T denotes temperature. The landscape contains two deep potential wells at positions $a = -10$ and $c = 10$ with a potential barrier at $b = 0$ between them. With D denoting the diffusion constant, the ion moves with dynamics

$$x(t) = x(0) + \int_0^t \nabla U(x(s)) \, ds + \sqrt{2D} \, w(t). \quad (14.11)$$

We now consider two questions regarding the diffusion process $x(t)$ in (14.11).

---

[16] The result follows straightforwardly using the well-known Dynkin formula from stochastic calculus. Dynkin's formula says that the expected value of any function of the diffusion satisfies $\mathbf{E}[\phi(x(t))] = \phi(x(0)) + \mathbf{E}[\int_0^t L\phi(x(s))ds]$. Then choosing $\phi(x(\tau)) = 0$ and $L\phi(x(s)) = -1$ for $0 \le s \le \tau$, Dynkin's formula yields (14.10) for the mean first-passage time.

14.4 Aside: The Fokker–Planck Equation

1. *What is the stationary (equilibrium) distribution of the ion?* One would expect that the ion has a large equilibrium probability of being at the two potential wells. The stationary distribution p_∞ satisfies (14.9), which reads

$$L^* p_\infty = D \nabla^2 p_\infty + \nabla p_\infty \cdot \nabla U + p_\infty \nabla^2 U = 0. \tag{14.12}$$

The reader should verify by substitution that the solution of (14.12) is

$$p_\infty(x) \propto \exp\left(-\frac{U(x)}{D}\right) \tag{14.13}$$

The stationary probability density $p_\infty(x)$ of (14.13) is displayed in Figure 14.5(b), where $D = k_B T$. (Since $U(x)$ is in units of $k_B T$, in Figure 14.5(b) $D = 1$.)

2. *Suppose the ion is at potential well a. What is the mean expected time to reach potential well c?* One would expect that an ion spends most of its time in potential well a or c and occasionally jumps between them. Let us work out the mean first-passage time. Using (14.10) for the mean passage time yields

$$L\phi = D\nabla^2 \phi + \nabla U \cdot \nabla \phi = -1 \tag{14.14}$$

with boundary conditions $\phi(c) = 0$, $\nabla \phi(-\infty) = 0$.

The reader can verify that the solution to (14.14) for the mean passage time starting from any position x to reach potential well c is

$$\phi(x) = \mathbf{E}[\tau_{x \to c}] = \frac{1}{D} \int_x^c \left[\exp\left(\frac{U(y)}{D}\right) \int_{-\infty}^y \exp\left(-\frac{U(s)}{D}\right) ds \right] dy$$

$$\approx \frac{1}{D} \int_x^c \exp\left(\frac{U(y)}{D}\right) dy \int_{-\infty}^c \exp\left(-\frac{U(s)}{D}\right) ds. \tag{14.15}$$

The above approximation holds when $U(b)$ is large and D is small.

We now glean further insight into (14.15) and obtain the Arrhenius equation. For small values of D, Laplace's integral asymptotics can be applied to approximate the two integrals in (14.15). Laplace's method says that, given a function $f(x)$ with $x \in [\alpha, \beta]$ which attains a global maximum at $x^* \in [\alpha, \beta]$, then

$$\int_\alpha^\beta \exp\left(Mf(x)\right) dx \approx \sqrt{\frac{2\pi}{M |\nabla^2 f(x^*)|}} \exp(Mf(x^*)) \text{ as } M \to \infty. \tag{14.16}$$

Similarly, if $M \to -\infty$, then the approximation (14.16) holds with x^* being the global minimum. For sufficiently small D, applying these asymptotics yields

$$\int_x^c \exp\left(\frac{U(y)}{D}\right) dy \approx \sqrt{\frac{2\pi D}{|\nabla^2 U(b)|}} \exp\left(-\frac{U(b)}{D}\right),$$

$$\int_{-\infty}^c \exp\left(-\frac{U(s)}{D}\right) ds \approx \sqrt{\frac{2\pi D}{|\nabla^2 U(a)|}} \exp\left(-\frac{U(a)}{D}\right).$$

Substituting these approximations into (14.15) yields the Arrhenius equation:

$$\phi(a) = \mathbf{E}[\tau_{a \to c}] \approx \frac{2\pi}{\sqrt{|\nabla^2 U(a)| \, |\nabla^2 U(b)}} \exp\left(\frac{U(b) - U(a)}{D}\right). \tag{14.17}$$

Note that $U(b) - U(a)$ is the energy barrier that the ion has to jump over. So the Arrhenius equation (14.17) says (i) the mean first-passage time increases exponentially with the energy barrier and (ii) since the diffusion constant $D = k_B T$, then the mean passage time decreases exponentially with increasing temperature T.

To summarize, we have shown how the Arrhenius equation (14.17) can be obtained from the mean first-passage time of a diffusion process.

Summary. This section was a brief mathematical interlude in applied stochastic processes. We introduced the stochastic diffusion model (14.3), the Fokker–Planck equation (14.6) which characterizes the probability density of the diffusion, the stationary distribution (14.9) of a diffusion process, the mean first-passage time and the equation it satisfies, (14.10), and finally a derivation of the Arrhenius equation (14.17).

14.5 Coarse-Grained Molecular Dynamics Model for the Bioelectronic Interface and Water

Recall that the bioelectronic interface (described in §2.6) is a key ingredient in measuring the electrical response of an engineered membrane; it interfaces the ionic flow of electrical current in the membrane with the electron flow of current in electrical instrumentation. In this section we use CGMD simulation results and statistical mechanics to compute the density profile of water and the diffusion tensor of water in proximity to the bioelectronic interface. Specifically, using CGMD simulations and the Percus–Yevick equation, we show that an uncharged bioelectronic interface (zero double-layer potential V_{dl}) negligibly interacts with the surface of the engineered membrane. Also, using CGMD simulations and the Fokker–Planck equation (discussed in §14.4), we show that the diffusion tensor and particle density of water in the bioelectronic interface are spatially dependent. Recall that the diffusion tensor is important since it provides the diffusion coefficients used in the advection-diffusion, Poisson–Nernst–Planck, GPNP, and PFNP continuum models discussed in Chapters 10 and 11.

The key outcomes of this section are an analytical expression for the radial distribution function of hard-sphere water and the water density at a hard-wall interface. Additionally, the Fokker–Planck equation and mean first-passage time of a Markov process are used to construct an expression for the diffusion tensor of water in proximity to the bioelectronic interface. We evaluate the water density and diffusion tensor of water in proximity to the bioelectronic interface from CGMD simulations. The results suggests that an uncharged bioelectronic interface contributes negligibly to the dynamics of lipids in the engineered tethered membrane.

14.5 Coarse-Grained Molecular Dynamics Model

Motivation: Modeling Water in Confined Spaces

Modeling water in small confined spaces (on the order of nanometers) is an important area in biophysics. It has important consequences in the structure-function relationship of ion channels and bioelectronic interfaces. Water has very different properties in small confined regions compared to its bulk behavior. For example, in several ion channels (such as gramicidin), water molecules traverse the channel in a single file.[17] As another example, in Brownian dynamics simulations, the effect of water molecules is replaced by Brownian motion as a consequence of the fluctuation-dissipation theorem. Our discussion in this section focuses on modeling water in the bioelectronic interface, which is a key component of engineered membranes.

14.5.1 Percus–Yevick Equation and Water Density at the Bioelectronic Interface

In this section we use the Percus–Yevick equation to construct a closed-form expression for the density profile ρ of water at the bioeletronic interface using equilibrium thermodynamics and statistical mechanics. By density profile we mean the spatial variation of the density of water.[18] Knowledge of the density profile of water is important as it can be used to compute the free energy of water, F_w, at the bioelectronic interface, and additionally it can be used to compute the chemical potential μ of water at the bioelectronic interface. In §14.5.3 we use F_w to evaluate the spatially dependent diffusion coefficient of water in proximity to the bioelectronic interface. To compute the water density profile, first we apply the "integral equation approach" from statistical mechanics[19] to construct an analytical expression for the water density ρ with no bioelectronic interface present. Then, we illustrate how the method can be used to construct the analytical expression for the density profile of water adjacent to the bioelectronic interface.

The Radial Density Function of Hard-Sphere Water

CGMD simulations provide Monte Carlo samples of the bead trajectories from a stochastic process that is described by Newton's equations of motion, (14.1), with the initial velocity of the beads being random. The probability of the position and momentum of the beads at any time is defined by a nonequilibrium phase-space probability density function given by $f(\boldsymbol{x}, \boldsymbol{p}, t)$, where $\boldsymbol{x}$ is the position all n beads, $\boldsymbol{p}$ is the momentum of all n beads, and t is the time. The CGMD simulations generate samples from the distribution $f(\boldsymbol{x}, \boldsymbol{p}, t)$. For the density profile of water, we are only interested in sampling from the steady-state distribution $f(\boldsymbol{x}, \boldsymbol{p})$. Using the ergodic hypothesis[20], it is possible to directly estimate $\rho_w(x)$ from samples from the CGMD simulation. Here, however,

---

[17] For the reader familiar with permeation in ion channels, our setup for the bioelectronic interface is different. In an ion channel, water is confined to a narrow cylindrical pore; in the bioelectronic interface, water is confined to a reservoir of very small height.

[18] From elementary physics one knows that the density of bulk water is 10^3 kg/m^3. However, in the bioelectronic interface we are dealing with a narrow region of approximately 4 nanometers of water molecules; in such cases water behaves very differently to its bulk density.

[19] For details on the integral equation approach, refer to [94, 368].

[20] Refer to §15.2 for details on the ergodic hypothesis.

we construct an analytical expression for the radial density function of water, $\rho_w(x)$, assuming the beads interact via a hard-sphere potential.

If the CGMD simulation is performed in the NVT statistical ensemble,[21] then after a sufficiently long time the bead positions from the CGMD simulation are consistent with samples from a Boltzmann distribution. The Boltzmann probability density function is[22]

$$f(\boldsymbol{x}) \propto e^{-\frac{1}{k_B T} U(\boldsymbol{x})}, \qquad (14.18)$$

where k_B is the Boltzmann constant, T is the temperature, and $U(\boldsymbol{x})$ is the potential energy function defined in (14.1) for a system containing n water beads. The water density ρ is related to the pair-correlation function between the beads. Defining $x_{1,2} = \|x_1 - x_2\|$ as the Euclidean distance in three-dimensional space between beads 1 and 2, the pair-correlation function between these beads is

$$g(x_{1,2}) = \frac{1}{Z_n} \int_0^\infty f(\boldsymbol{x}) \, dx_3 dx_4 \cdots dx_n, \qquad (14.19)$$

where Z_n is a normalization constant. Physically, $g(x_{1,2})$ is the probability of finding beads 1 and 2 a distance of at most $x_{1,2}$ from each other. As $x_{1,2} \to \infty$ the probability $g(x_{1,2}) \to 1$. The time-average density (or radial distribution function) of the beads is given by $\rho(x_{1,2}) = \rho_B g(x_{1,2})$, where ρ_B is the bulk density of water. However, the evaluation of (14.19) is a computationally prohibitive task since it requires evaluating an $(n-2)$-multidimensional integral.

How can we approximate the correlation function $g(x_{1,2})$ without directly evaluating (14.19) or using CGMD simulations? The integral equation approach from statistical mechanics can be used to construct analytical expressions for $g(x_{1,2})$ assuming that simplifying assumptions can be made about the interaction potentials (e.g., hard-sphere potentials). The main idea of the integral equation approach in statistical mechanics is to relate pair-correlation functions using an integral expression and then provide a closure that can be used to solve for one of the pair correlations. The two pair-correlation functions we consider are the total correlation function $h(x_{1,2})$ and the direct correlation function $c(x_{1,2})$. The function $c(x_{1,2})$ accounts for all correlations that result from the direct interaction of bead 1 with bead 2, and $h(x_{1,2}) = g(x_{1,2}) - 1$ accounts for the total correlation of bead 1 with bead 2. As $x_{1,2} \to \infty$ then $h(x_{1,2}) \to 0$ because the correlation between the beads is zero at an infinite distance. To construct the integral equation, we consider another bead, bead 3, that interacts with beads 1 and 2. The total correlation function of bead 1 and bead 3, denoted by $h(x_{2,3})$, accounts for all the indirect correlations that result from bead 1 indirectly interacting with bead 2 via bead 3. Given the direct and indirect correlations between the three beads, we can now construct the

---

[21] The NVT statistical ensemble states that the CGMD simulation was performed with a constant number of N beads, contained in a constant volume V, at a constant temperature T.

[22] We say $f(x) \propto$ (proportional to) since the density specified is unnormalized.

integral equation using conservation of mass and that bead 3 can be located anywhere. Using the correlation functions we have the following:

(i) The water density at x_2 that results from direct correlations from bead 1 is $\rho_B c(x_{1,2})$.
(ii) The water density at x_3 that results from direct correlations from bead 1 is $\rho_B c(x_{1,3})$.
(iii) The water density at x_2 that results from indirect correlations from bead 3 is $\rho_B h(x_{2,3})$.
(iv) The water density at x_2 that results from indirect and direct correlations from bead 1 is $\rho_B h(x_{1,2})$.

Therefore, equating the density at position x_2 and integrating over all possible positions of bead 3 results in the integral equation

$$h(x_{1,2}) = c(x_{1,2}) + \rho_B \int_{-\infty}^{\infty} c(x_{1,3}) h(x_{2,3}) \, dx_3. \tag{14.20}$$

Equation (14.20) is known as the Ornstein–Zernike equation [47, 153, 447], which relates the direct correlation function $h(x_{1,2})$ and indirect correlation function $c(x_{1,2})$ between beads. Since the functions $h(x_{1,2})$ and $c(x_{1,2})$ are unknown, we require a closure relation which provides another relation between $h(x_{1,2})$ and $c(x_{1,2})$. The Percus–Yevick closure relation is

$$c(x_{1,2}) = g(x_{1,2}) \left(1 - e^{\frac{1}{k_B T} U(x_{1,2})}\right), \tag{14.21}$$

where $U(x_{1,2})$ is the pairwise potential between beads 1 and 2 excluding all other interactions between beads. Substituting (14.21) into (14.20) results in the Percus–Yevick equation [455, 456] given by

$$g(x_{1,2}) e^{\frac{1}{k_B T} U(x_{1,2})} = 1 + \rho_B \int_{-\infty}^{\infty} \left(1 - e^{\frac{1}{k_B T} U(x_{1,3})}\right) g(x_{1,3})(g(x_{2,3}) - 1) \, dx_3. \tag{14.22}$$

The Percus–Yevick equation (14.22) provides an integral equation[23] that can be used to solve for the radial distribution function $g(\cdot)$ for a given pair potential $U(\cdot)$. Here we assume that the pairwise potential $U(\cdot)$ between beads results from a hard-sphere potential given by

$$U(x_{i,j}) = \begin{cases} \infty & \text{if } x_{i,j} < 2R, \\ 0 & \text{otherwise,} \end{cases} \tag{14.23}$$

where R is the radius of the spheres. The analytical solution for the radial distribution function $g(x_{i,j})$ of the Percus–Yevick equation (14.22) for the hard-sphere potential

---

[23] The Percus–Yevick equation is a Fredholm integral equation of the first kind and can be solved numerically using methods such as variational iteration, homotopy perturbation, and Adomian decomposition.

(14.23) can be constructed using Fourier methods, as detailed in [447]. Given $g(x_{i,j})$, the radial density function of water around any hard-sphere water bead is given by $\rho_w(x_{i,j}) = \rho_B g(x_{i,j})$.

Note that in the CGMD model we use Lennard-Jones potentials for nonbonded pairwise interactions between beads, and additionally the CGMD model comprises several different interacting beads. Therefore, we only expect a rough agreement between the radial distribution function of water computed from the CGMD model and the radial distribution function of water computed from the Percus–Yevick equation (14.22) for the hard-sphere potential (14.23).

The Radial Density Function of Hard-Sphere Water at a Hard-Wall Bioelectronic Interface

In the previous section we constructed the Percus–Yevick equation (14.22) from the Ornstein–Zernike equation (14.20) to estimate the radial density function of water, $\rho_w(r)$, at a radius r from the center of a water bead. Here we wish to estimate the density function of water, $\rho_w(z)$, a distance z from the bioelectronic interface. Estimating $\rho_w(z)$ is different from estimating $\rho_w(r)$ because $\rho_w(r)$ only depends on the pairwise interaction between water; however, $\rho_w(z)$ depends on the pairwise interaction of water and the bioelectronic interface. To estimate $\rho_w(z)$, the "integral equation approach" from statistical mechanics is applied. First, we construct the Ornstein–Zernike equation (14.20) for a system that contains two types beads (one for water, and the other for the bioelectronic surface), then we use the Percus–Yevick closure (14.21) to construct an integral expression for $\rho_w(z)$.

Let ρ_B denote the bulk density of water, and let ρ_s denote the bulk density of the surface beads. If $\rho_s \to 0$ (infinitely dilute), then the Ornstein–Zernike equation for this mixture of beads is given by

$$h(x_{1,2}) = c(x_{1,2}) + \rho_B \int_{-\infty}^{\infty} c(x_{1,3}) h(x_{2,3}) \, dx_3 \quad \text{if } x_2 \text{ is a water bead,}$$

$$h_s(x_{1,2}) = c_s(x_{1,2}) + \rho_B \int_{-\infty}^{\infty} c_s(x_{2,3}) h(x_{1,3}) \, dx_3 \quad \text{if } x_2 \text{ is a surface bead.} \quad (14.24)$$

Here, $h_s(x_{1,2})$ and $c_s(x_{1,2})$ are the total and indirect correlation functions of a water bead at location x_1 and a surface bead at location x_2. The construction of (14.24) is identical to that of the Ornstein–Zernike equation (14.20) when only water is present. The indirect correlation functions are $c(\cdot)$ and $c_b(\cdot)$, and the total correlation functions are $h(\cdot)$ and $h_b(\cdot)$ in (14.24). How can we compute the radial distribution function of water, $g_s(z)$, a distance z from the surface of the bioelectronic interface using (14.24)? The main idea is to construct the relation between $h_s(z)$ and $c_s(z)$, where z is the distance between the water and surface bead, based on the surface bead having an infinitely large radius. Then, given this relation between $h_s(z)$ and $c_s(z)$, we can consider a finite-sized surface bead. The details of this procedure are provided in [153]. The final result is

$$h(x_{1,2}) = c(x_{1,2}) + \rho_B \int_{-\infty}^{\infty} c(x_{1,3}) h(x_{2,3}) \, dx_3,$$

$$h_s(R_s + z) = c_s(R_s + z) + 2\pi \rho_B \int_0^{\infty} \xi h(\xi) \int_{z-\xi}^{z+\xi} c_b(\eta) \, d\eta \, d\xi, \qquad (14.25)$$

where R_s is the radius of the surface bead. Recall that $h_b(z) = g_b(z) - 1$, and that the water density at the bioelectronic interface is $\rho_w(z) = \rho_B g_s(z)$.

To solve for $g_s(z)$ using (14.25) requires two closure relations to construct the integral equations of $h(\cdot)$ and $h_b(\cdot)$. Here we use the Percus–Yevick closure relations for the pair interactions between water and the water and surface beads. The Percus–Yevick closure relations are

$$c(r) = g(r)\left(1 - e^{\frac{1}{k_B T} U(r)}\right), \qquad c_s(r) = g_s(r)\left(1 - e^{\frac{1}{k_B T} U_s(r)}\right), \qquad (14.26)$$

where $U(r)$ is the pair potential between water beads, and $U_s(r)$ is the pair potential between the water bead and the surface at a distance r. Substituting (14.26) into (14.24) yields the integral expression for the radial distribution function of water, $g(r)$, and the radial distribution function of water at the bioelectronic interface, $g_s(z)$. The pair potential $U(r)$ in (14.26) is given by the hard-sphere potential (14.23). The surface bead and water pair potential is defined as the hard-wall potential given by

$$U_s(z) = \begin{cases} \infty & \text{if } z < (R_s + R), \\ 0 & \text{otherwise.} \end{cases} \qquad (14.27)$$

Given (14.25), (14.26), and (14.27), an analytical expression for $g_s(z)$ can be constructed; see [47, 153, 447] for details.

14.5.2 Density Profile of Water at the Bioelectronic Interface

The dynamics of water at the bioelectronic interface have very different characteristics from those in the bulk solution. The thermodynamic, structural, and dynamic properties of water in proximity to the interface are affected by two primary factors: a smaller number of neighboring water molecules compared to water in the bulk solution, and a change in the potential energy of the fluid as a result of interactions with the electrode surface. The density profile of water at the interface is modeled as a periodic function of space with a period equal to the mean thickness of each water layer in proximity to the interface. This suggests that the spatial variation of the density profile occurs at a similar length scale to the molecular diameter of the water molecules. A key question is this: does the density profile from the CGMD simulation model match that from a hard-sphere fluid at a hard-wall interface?

Long-range interactions (beyond a few nanometers) of the bioelectronic interface with the lipid beads can only occur via water density variations in the CGMD simulation model. The reason is that the nonbonded Lennard-Jones potential is cut off to zero potential at a distance of 1.4 nm to reduce the number of bead-bead interactions requiring evaluation in the numerical method used to evaluate Newton's equation of motion (14.1). This is a reasonable approximation because, at distance greater than a few angstroms, the bead-bead interactions are dominated by the attractive van der Waals force decaying as a power law r^{-6}, where r is the distance between the beads. To estimate the water density profile, we use the Percus–Yevick equation for a hard-sphere and hard-wall potential (derived in §14.5.1) and the CGMD simulation results, and illustrate that the bioelectronic interface interacts negligibly with the lipid membrane.

Figure 14.6 displays the water density profile obtained from the CGMD simulation and from the Percus–Yevick equation as well as the nonbonded potentials. Notice that the two density profiles (radial distribution functions) in Figure 14.6(a) are in agreement. This agreement between the CGMD simulation and Percus–Yevick equation results because the equilibrium radius of the Lennard-Jones potential is approximately 0.52 nm, which is approximately equal to the radius $R = 0.49$ nm of the hard-sphere (14.23) and hard-wall pair potentials (14.27). Figure 14.6(b) illustrates the Lennard-Jones potential from the CGMD simulation and the hard-sphere potential from the Percus–Yevick equation. Interestingly, even though the Lennard-Jones potentials and hard-sphere potentials between the CGMD simulation and Percus–Yevick equation are different, the resulting radial distribution functions of water from using either potential are in agreement. This suggests that the radial distribution function of water is insensitive to minor variations in the nonbonded potentials that describe the water-water cut off to zero potential and water-surface interactions. As seen in Figure 14.6, the distance between the local minima and maxima is approximately R, as expected from the discussion in [147, 484]. The initial drop in the density at approximately 0.44 nm from the surface of the bioelectronic interface results from the Pauli repulsion forces (Pauli repulsion force occurs when the distance between beads decreases below σ; refer to Figure 14.6(b)) between the water beads and the bioelectronic interface beads that model the gold electrode. At 4 nm from the bioelectronic interface there is negligible variation in the density profile. This suggests that the gold bioelectronic interface interacts negligibly with the lipid membrane. For example, the diffusion coefficients of the lipids in the proximal and distal layers of the 0 percent tether density membrane are approximately equal – as expected if the bioelectronic interface does not interact with the membrane.

Remark 1: If the bioelectronic interface is charged (or has a nonzero potential across it), then it will interact with the membrane and ions in solution through Coulomb forces. Recall from §2.1 that Coulomb interaction potentials satisfy a power-law relation r^{-1}, where r is the distance between atoms (and Coulomb forces are inversely proportional to r^2). Therefore, if any nonzero charge exists on the bioelectronic interface, this will contribute to the dynamics of the membrane, water, and ions in the solution. If the potential at the bioelectronic interface is sufficiently large (above 50 mV), for example, then this will cause the formation of aqueous pores in the membrane (e.g., electroporation).

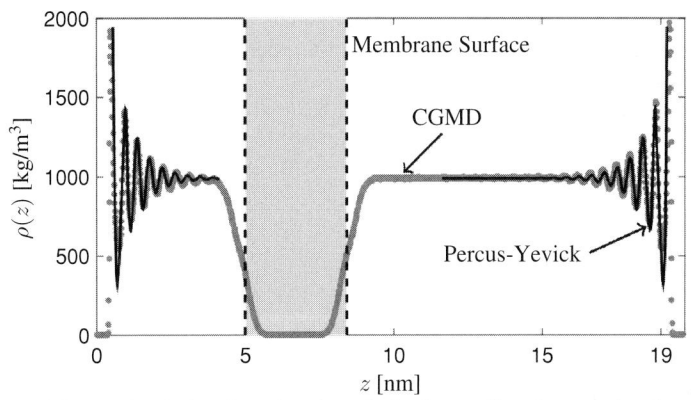

(a) CGMD-simulation based and analytically predicted water density at the bioelectronic interface. The water density from the CGMD simulation results is denoted by the gray dots, and the analytical results from the Percus–Yevick equation for a hard-sphere and hard-wall potential (derived in §14.5.1) with $R = 0.49$ nm and $\rho_B = 1000$ kg/m³ are represented by the black line. The bioelectronic interface is located at $z = 0$ nm and $z = 19.8$ nm, and membrane surfaces are represented by the vertical dotted lines.

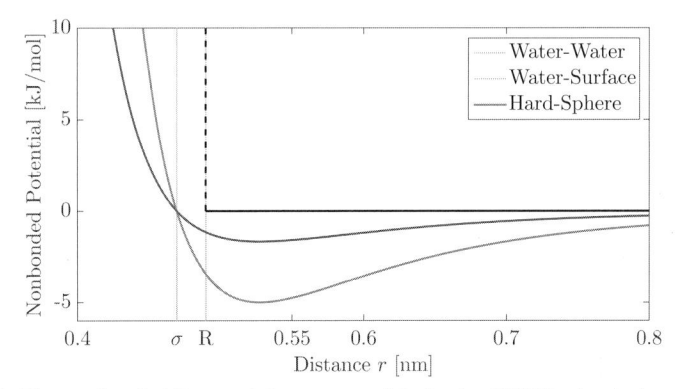

(b) The nonbonded Lennard-Jones potentials in the CGMD simulation that describe the water-water ($\varepsilon = 5.0$ kJ/mol, $\sigma = 0.47$ nm) interaction and the water-surface ($\varepsilon = 1.7$ kJ/mol, $\sigma = 0.47$ nm) interaction, and the hard-sphere potential with $R = 0.49$ nm. The parameter r is the distance between the beads and $U(r)$ is the nonbonded potential. The nonbonded force between the beads is equal to $F = -dU(r)/dr$ and the Pauli repulsion region occurs where $r \leq \sigma$.

Figure 14.6 Numerically predicted radial distribution function of water and the nonbonded potentials of the CGMD simulation and Percus–Yevick equation for water and the bioelectronic interface.

14.5.3 Fokker–Planck Equation: Spatially Dependent Water Diffusion Coefficient

Here we illustrate how the Fokker–Planck equation and CGMD simulations can be used to estimate the spatially dependent diffusion tensor of water adjacent to the bioelectronic interface of the engineered membrane. The complete description of the diffusion of water is given by the second-order diffusion tensor.[24] Specifically, we use the Fokker–Planck equation to construct a relation between the bead trajectories from CGMD simulations to the diffusion tensor of the bead. Here our aim is to compute the diffusion tensor of the water beads from CGMD simulations.

Let Ψ denote the conditional probability density function of observing a bead at position x at time t that was initially at position x_o at time t_o. Assuming the bead position evolves according to Brownian motion, then the time evolution of Ψ is given by the Fokker–Planck equation [249]:

$$\frac{\partial \Psi}{\partial t} = \nabla \cdot \mathbf{D} \cdot [\nabla + \beta(\nabla F)]\Psi$$

$$= \sum_{i=1}^{3}\sum_{j=1}^{3} \frac{\partial}{\partial x_i} D_{ij} \frac{\partial}{\partial x_j} [\Psi + \beta F_w]. \quad (14.28)$$

In (14.28), $\mathbf{D}$ denotes the second-order diffusion tensor with matrix elements D_{ij} that represent the diffusion coefficients in all combinations of directions x, y, and z; $\nabla \cdot \mathbf{D}$ is the divergence of the second-order diffusion tensor; F is the free energy; and β is the inverse of the thermal free energy (i.e., $\beta = 1/k_B T$). The derivation of the Fokker–Planck equation results from the overdamped Langevin equation (a first-order stochastic differential equation) where the bead trajectories contain no average acceleration (e.g., the bead motion is damped). That is, the inertial effects of the beads can be neglected. In CGMD simulations the bead inertia effects[25] are generally neglected so that the equation of motion of the beads can be described by Newton's equation of motion (14.1). Therefore, we use the Fokker–Planck equation to model the conditional probability density function of the bead positions x at time t.

We now argue that the probability density function $\Psi(x)$, which describes the position of particles in the tethering reservoir of the engineered tethered membrane, is independent in each of the coordinate axes (i.e., $\Psi = \Psi_x \Psi_y \Psi_z$), and therefore the diffusion tensor $\mathbf{D}$ is diagonal. Consider the CGMD simulation model of the tethered membrane in Figure 14.4 on page 303. The boundary conditions of the Fokker–Planck equation (14.28) in z are composed of no-flux boundary conditions at the gold bioelectronic interface, denoted by z_b, and the surface of the tethered membrane, denoted by z_t, with the others defined as infinite boundary conditions. From the coordinate axis of the CGMD simulation model (refer to Figure 14.4), the diffusion tensor is diagonal with $D_{xx} = D_{yy} \neq D_{zz}$. The translational invariance parallel to the confining surface ensures that F_w in the parallel direction is constant (i.e., F_w is invariant in the x and y coordinates). The free energy of water along the z axis is related to the equilibrium density of water by $F_w(z) = -k_B T \log(\rho(z)/\rho_o)$ with ρ_o the bulk density. As a result of the no-flux boundary conditions in z, the diffusion coefficients in $\mathbf{D}$, defined following (14.28), will

---

[24] The diagonal elements (D_{xx}, D_{yy}, D_{zz}) of the diffusion tensor $\mathbf{D}$ represent diffusion coefficients measured along each of the x, y, and z coordinate axes. The six off-diagonal terms $(D_{xy}, D_{yz}, \ldots)$ represent the correlation of random motions between each pair of x, y, and z coordinate directions.

[25] If the inertia term for bead motion is non-negligible, then the equation of motion of the beads would be $m_i \ddot{x}_i = F_i + \xi_i \dot{x}_i$ for $i \in \{1, \ldots, n\}$, where ξ is the friction coefficient of the ith bead.

be a function of z. If we discretize the z dimension sufficiently into layers, then in each layer the diffusion coefficients will not vary substantially. This allows Ψ (14.28) to be decoupled such that $\Psi = \Psi_x \Psi_y \Psi_z$ with $\Psi_x(x,t|x_o,t_o)$, $\Psi_y(y,t|y_o,t_o)$ and $\Psi_z(z,t|z_o,t_o)$ denoting the time trajectories of the particle in each respective dimension.

Having constructed the dynamic model (14.28) that relates the bead trajectories to the diffusion tensor, and using the property that $\Psi = \Psi_x \Psi_y \Psi_z$ and $\mathbf{D}$ is diagonal for engineered tethered membranes, the aim is to construct an expression for the diffusion tensor $\mathbf{D}$ as a function of the CGMD simulation results. Assuming that the trajectories of the particle coincide in the x and y coordinates (that is, $\Psi_x(x,t|x_o,t_o) = \Psi_y(y,t|y_o,t_o)$), then the time trajectory of the particle in the x and y directions is given by Fick's second law of diffusion. For the boundary conditions $\Psi_x(x,t|x_o,t_o) \to 0$ as $|x| \to \infty$ and $\Psi_y(y,t|y_o,t_o) \to 0$ as $|y| \to \infty$, the analytical solution of Ψ_x and Ψ_y is given in terms of the Green's function. This allows $D_{xx} = D_{yy}$ to be estimated by evaluating the second central moment (i.e., for D_{xx} the relation $\langle (x - x_o)^2 \rangle = 2 D_{xx}(t - t_o)$ holds where $\langle \cdot \rangle$ denotes the time average). The time evolution of Ψ_z is given by

$$\frac{\partial \Psi_z}{\partial t} = \frac{\partial}{\partial z} \left[D_{zz} e^{-\beta F(z)} \frac{\partial}{\partial z} \left(\Psi_z e^{\beta F(z)} \right) \right]. \tag{14.29}$$

The steady-state solution of (14.29) is equal to the time average of Ψ_z, i.e., $\langle \Psi_z \rangle$. Additionally, the steady-state solution of (14.29) is related to the free energy $F(z)$ via $\beta F(z) = -\ln(\langle \Psi_z \rangle)$. The parameter $\langle \Psi_z \rangle$ can be computed from the CGMD simulation results using the bead density $\rho(z)$ and evaluating

$$\langle \Psi_z \rangle = \frac{\rho(z)}{\int_{z_b}^{z_t} \rho(\xi) d\xi}, \tag{14.30}$$

where z_b and z_t are the locations of the no-flux boundary conditions of (14.29). For the time-dependent case, no analytical solution exists for (14.29) given the no-flux boundary conditions at z_b and z_t. It is, however, possible to relate the diffusion component D_{zz} to the round-trip time, denoted by τ_{rt}, that can be computed from CGMD simulations. D_{zz} can be computed using[26]

$$D_{zz}(z) = \frac{e^{\beta F(z)}}{\partial \tau_{rt}(z)/\partial z} \int_{z_b}^{z_t} e^{-\beta F(z')} dz'. \tag{14.31}$$

In (14.31), τ_{rt} denotes the round-trip time required to start at coordinate z_b, reach position z_t, and then return to z_b. The parameters τ_{rt} and $F(z)$ in (14.30) can be obtained directly from CGMD simulations, allowing (14.31) to be used to estimate $D_{zz}(z)$.

---

[26] For details on the derivation of the mean first-passage time of a Markov process (e.g., the Fokker–Planck equation), see [146, 249, 276, 344].

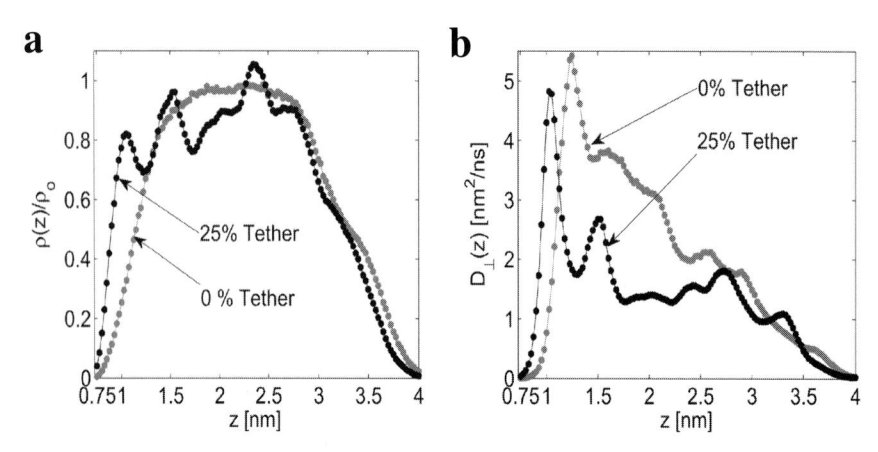

Figure 14.7 Normalized water density and perpendicular diffusion $D_\perp(z)$. The normalized density $\rho(z)/\rho_o$ in (a) and perpendicular diffusion coefficient $D_\perp(z)$ in (b) are computed from the CGMD bead trajectories. The main result is that the perpendicular diffusion coefficient of water approaches the bulk diffusion coefficient of water approximately 2.5 nm from the gold interface located at $z = 0$ nm.

14.5.4 Diffusion Tensor of Water in Tethering Reservoir

CGMD simulations generate time-evolving trajectories of coarse-grained beads. The Fokker–Plank equation can be used to relate the bead positions to the diffusion tensor by using (14.31). In this section we use CGMD simulation results and the Fokker–Plank equation to estimate the spatially dependent diffusion tensor of water for engineered membranes containing a 0 and 25 percent tether density in proximity to an uncharged bioelectronic interface.

The density of water for the 0 and 25 percent tether density membranes is provided in Figure 14.7(a). As a result of the Pauli repulsion force between the water beads and gold beads, the density of water decreases at the hydrophobic gold interface. Comparing the water density of the 0 and 25 percent tether density membranes, it is apparent that the density of water is dependent on the tether density. Given that the water density is position dependent, this strongly suggests that the self-diffusion tensor of the water will also be position dependent. Using (14.31), the position-dependent perpendicular diffusion coefficient $D_\perp(z)$ is provided in Figure 14.7(b). From (14.31), the diffusion coefficient $D_\perp(z)$ is dependent on the round-trip time $\tau_{rt}(z)$ and the density profile $\rho(z)$. Intuitively, the slope of the round-trip time $\tau_{rt}(z)$ is inversely proportional to $D_\perp(z)$. In Figure 14.7(b) the initial peak at 1.05 nm and the second peak at 1.52 nm of the 25 percent tether density membrane diffusion profile match the location of the density peaks at 1.05 and 1.52 nm in Figure 14.7(a). The initial peak at 1.25 nm for the 0 percent tether density membrane diffusion profile results because the round-trip time of a water bead in this region is approximately constant. Additionally, for the 0 percent tether density membrane, the density profile remains approximately constant for $z = 2.0$ nm to $z = 2.5$; however, the round-trip time, illustrated in Figure 14.8, does not remain

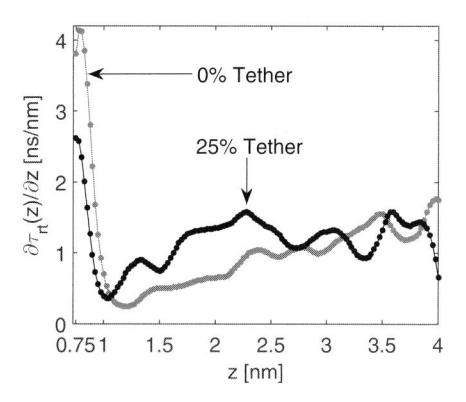

Figure 14.8 The estimated derivative of the round-trip time $\tau_{rt}(z)$ with respect to z numerically computed from the CGMD bead trajectories.

constant. The combined effect of the constant density but varying round-trip time causes $D_\perp(z)$ to be nonconstant in this region.

Finally, to gain additional insight, if no membrane is present, how does the water density and diffusivity evolve as a function of distance from the electrode surface? In Figure 14.9 the water density and spatially dependent diffusion coefficient for the gold interface with no membrane present is provided. From the results in Figure 14.9, the diffusion coefficient $D_\perp(z)$ approximately reaches the bulk diffusion coefficient for $z \geq$ 2.5 nm from the gold interface with no membrane present.

14.5.5 Summary

In this section we discussed the Percus–Yevick equation for the radial distribution function of hard-sphere water and the density of hard-sphere water adjacent to a hard-wall

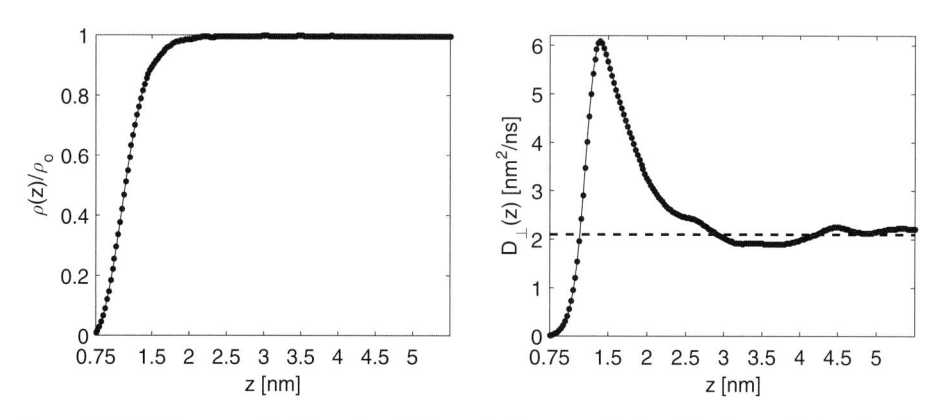

Figure 14.9 (a) The normalized density $\rho(z)/\rho_o$ and (b) perpendicular diffusion coefficient $D_\perp(z)$ are computed from the CGMD bead trajectories at the gold interface with no membrane present.

interface that mimics the bioelectronic interface. Additionally, the Fokker–Planck equation and mean first-passage time of a Markov process were used to construct an expression for the diffusion tensor of water in proximity to the bioelectronic interface. The first major result in this section was that the water density from CGMD simulations is in excellent agreement with the water density evaluated from the Percus–Yevick equation. Second, the diffusion tensor of water adjacent to the bioelectronic interface is spatially dependent. Third, at approximately 2.5 nm from the bioelectronic interface, the diffusion tensor of water and density of water have the same properties as bulk water. This suggests that nonelectrostatic interactions from the bioelectronic interface contribute negligibly to the dynamics of lipids in the engineered tethered membrane.

14.6 Tethered Membrane Dynamics and Energetics

The electroporation measurement platform (EMP) is a novel device constructed out of engineered membranes, as described extensively in Chapter 6. Continuum models of the EMP were also developed in Chapter 11. These continuum models are dependent on the diffusion tensor $\mathbf{D}$, membrane thickness h_m, surface tension σ, and line tension γ of the membrane. All these parameters are crucial for modeling the electroporation processes present in the EMP. This section discusses how CGMD simulations (Figure 14.4) can be used to estimate these important parameters. The implication is that CGMD can be viewed as a first-principles approach of choosing the parameters of the continuum model. Moreover, the continuum model yielded parameters for the macroscopic reaction-rate model of electroporation constructed in Chapter 11. Therefore, this section in concert with Chapter 11 yields a complete atomistic-to-device model for the EMP.

14.6.1 Lipid Energetics and Pore Density

The potential of mean force (PMF) of lipids in a lipid bilayer is a key thermodynamic property that can be used to estimate the pore density and the free energy of lipid flip-flop. Lipid flip-flop occurs when a lipid molecule flips from one side of the lipid bilayer to the other. Here umbrella sampling[27] is used to compute the PMF for moving a single

---

[27] Umbrella sampling (also known as biased MD or importance sampling in statistics) is a numerical technique for estimating the PMF between two states in an MD simulation. Here a state is defined as the position and momenta of all the atoms in the MD simulation. For the NVT ensemble the stationary joint probability density function of the states is given by (15.13 on page 360), and for the NPT ensemble is given by (15.14 on page 360). Performing umbrella sampling requires defining the reaction coordinate between the two states, and defining a biasing potential that can be adjusted to transition the MD simulation from the initial state to the final state through the reaction coordinate. MD simulations are then performed in a finite number of *windows* (intermediate states defined by the reaction coordinate and biasing potential) to generate samples of the position and momenta of all the atoms. Given the sampled position and momenta of the atoms in each window, umbrella integration or the weighted histogram analysis method can be used to estimate the PMF between the initial and final states of the MD simulation. See [49, 197] for details on how to perform umbrella sampling for both CGMD and MD simulations.

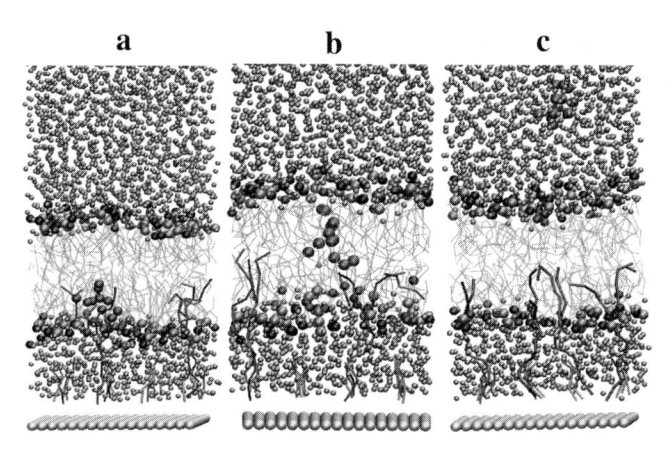

Figure 14.10 A color version of this figure can be found online at www.cambridge.org/engineered-artificial-membranes. Snapshots of restrained DphPC lipid in the 25 percent tether density DphPC membrane for umbrella sampling: (a) the lipid in the equilibrium position, (b) the lipid at the center of the bilayer, and (c) the lipid in bulk water. Water is represented by light blue beads, pulled lipid as magenta spheres, lipid tails as green lines, DphPC and GDPE head groups NC_3, PO_4, and OH as blue, orange and red balls, tethers as violet sticks, and spacers as tan sticks. The gold surface which constitutes the bioelectronic interface, is represented by the yellow beads.

DphPC lipid along the direction normal to the membrane surface for both the 0 and 25 percent tether density membranes. The snapshots for selected umbrella simulation windows for the 25 percent tether density membrane are shown in Figure 14.10.

The computed PMF for both tethering densities is displayed in Figure 14.11(a). The PMF in Figure 14.11(a) can be used to estimate how the free energy of the CGMD simulation changes as a function of the lipid head group PO_4 position relative to the center of the membrane (at $z = 0$ nm). When head group of the lipid is at the surface of the membrane, this is expected to be a local minimum in the PMF because the lipid's hydrophobic tail is in contact with the other lipid's hydrophobic tails, and the hydrophilic lipid head group is in contact with water. As seen in Figure 14.11(a), there are two local minima of the PMFs that occur at -1.82 and 1.82 nm, which corresponds to the equilibrium position of the lipid when the head group is at the surface of the membrane. If the lipid head group is shifted from these two local minima, then the resulting free-energy change will be positive. As such, energy is required to move the lipid head group from these two equilibrium positions. Two important parameters can be estimated from the PMF by computing the difference in energy if the lipid head group is pulled into the solution or if the lipid head group is pulled to the center of the membrane as discussed below.

Free energy of lipid desorption. If the lipid head group is pulled into the water solution from the equilibrium position, this will result in the hydrophobic lipid tail coming into contact with water molecules. As seen in Figure 14.11(a), this results in an increase in the free energy. The energy difference between the equilibrium position of the head group, and the fully solvated position is known as the *free energy of lipid desorption*. The free energy of lipid desorption provides a measure of the membrane stability. If it requires a higher energy to move a lipid from the equilibrium position into the solution, this corresponds to a higher-stability membrane. The free energies of desorption for the 0 and 25 percent tether density membranes are 85.93±0.5 and 91±0.5 kJ/mol, respectively. As expected, the tethers cause the associated energy of lipid desorption to increase as compared with the untethered membrane.

Free energy of lipid flip-flop. As the head group of the lipid is moved to the center of the membrane from the equilibrium position, the steep slope observed in the PMF illustrated in Figure 14.11(a) results from both the lipid head groups interacting with the bilayer interior, and water molecules that are pulled into the bilayer interior by the hydrophilic lipid head group. Note that this effect of pulling water into the bilayer has been observed for other charged or polar molecules that are pulled into the hydrophobic interior of the membrane [36–38, 105]. Figures 14.11(b) and 14.11(c) illustrate this effect of the lipid head group pulling the water into the center of the bilayer, which is typically associated with the formation of an aqueous pore. Using the PMF (Figure 14.11(a)), the free energy required for complete flip-flop of a single lipid is equal to the energy required to move a lipid from its equilibrium position to the center of the bilayer, then to the other leaflet's equilibrium position. From the maxima in the PMFs between the equilibrium position of the head group and the head group at the bilayer center, the free-energy barrier for lipid flip-flop is 89 kJ/mol for the 0 percent tether density membrane and 103.17 kJ/mol for the 25 percent tether density membrane. This is expected from our results in Chapter 12, where the rate of formation of aqueous pores decreases for an increase in tether density.

Remark 2: The computed energy for lipid flip-flop, provided in Figure 14.11(a), is approximately equal to the energy required to move a charged or polar molecule to the center of the membrane [36–38, 105]. However, only the formation of transient aqueous pores is observed from the CGMD simulations. This can be attributed to a lack of electrostatic interaction in the CGMD water model. CGMD water has a zero dipole moment so simulations are run with a dielectric constant of 15 for implicit screening of electrostatic interactions. This means that the interaction of polar or charged molecules is underestimated in hydrophobic environments with the MARTINI water model (e.g., when a lipid (PO_4 group) head group is placed in the interior of a lipid bilayer).

The computed free energies of lipid desorption and lipid flip-flop illustrate that, as the tether density increases, this reduces the probability of formation of aqueous pores in the membrane. Assuming the energy required to form a pore, denoted by ΔG_p, is equal to the energy required for lipid flip-flop, the equilibrium pore density can be computed from $\rho_0 = e^{-\beta \Delta G_p}/A_L$, where A_L is the area per lipid [38]. For DphPC, $A_L = 0.69$ nm^2; therefore, the associated equilibrium pore densities for the 0 and 25 percent tether density membranes are 4375 and 21 pores/m^2, respectively. This is approximately six orders of magnitude less than the experimentally measured pore density, which is in the range of 1.5×10^8 to 3×10^8 pores/m^2. A possible cause for this discrepancy is that the formed membrane contains defects resulting from the manufacturing process, which are not included in the CGMD simulation model. Another possibility is that the membrane contains a large surface tension causing the formation of aqueous pores. Though

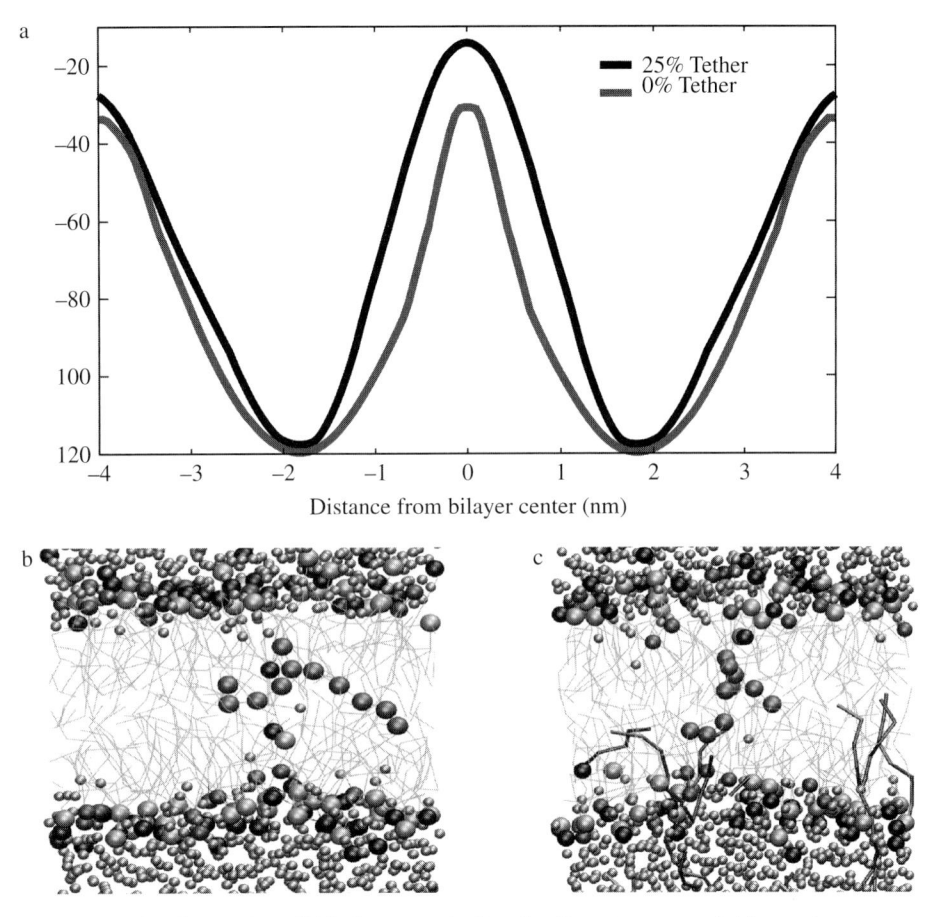

Figure 14.11 A color version of this figure can be found online at www.cambridge.org/engineered-artificial-membranes. (a) PMFs for 0 and 25 percent tether density DphPC lipid membranes. (b), (c) CGMD simulation snapshots of the restrained DphPC lipid at the center of the membrane used to construct the PMF for the 0 and 25 percent tether density DphPC, respectively. Water is represented by light blue beads, pulled lipid as magenta spheres, lipid tails as green lines, DphPC and GDPE head groups NC_3, PO_4, and OH as blue, orange, and red balls, tethers as violet sticks, and spacers as tan sticks.

the CGMD simulation model contains a nonzero surface tension, it is not sufficiently large to cause the formation of aqueous pores.

14.6.2 Line Tension and Surface Tension

We now discuss methods to compute the line tension γ and surface tension σ from CGMD simulations. Recall from §11.2 that γ and σ are important parameters that arise in the energy required to from an aqueous pore in the tethered membrane. The line tension γ is the energy cost per unit length at the boundary between the hydrocarbon lipid tails and water. The surface tension σ is the energy required to increase the surface area of the membrane by a unit area. The ratio γ/σ is defined as the defect density of the engineered membrane. The defect density provides a measure of how the steady-state radii of pores in the engineered membrane change for different line-tension and surface-tension values. Using the stochastic perturbation approximation of the Smoluchowski–Einstein equation and the energy required to form a hydrophilic pore we can compute the steady-state value of the pore radii in the engineered membrane as follows:

$$\frac{dr}{dt} = -\frac{D}{k_B T} \frac{\partial W}{\partial r}\bigg|_{V_m=0}$$

$$= -\frac{D}{k_B T} \frac{\partial}{\partial r}\left[2\pi\gamma r - \pi\sigma r^2 + \frac{C^4}{r^4} + \frac{1}{2}K_t r^2\right]$$

$$= -\frac{D}{k_B T}\left[2\pi(\gamma - \sigma r) + \frac{4C^4}{r^5} + K_t r\right]. \quad (14.32)$$

To solve for the steady-state radius r_{ss}, we solve for the fixed point, namely, the r that yields $dr/dt = 0$. Assuming that the radius $r \ll 1$ and that the spring constant $K_t \ll 2\pi\sigma$, then from (14.32),

$$r_{ss} = \frac{\gamma}{\sigma}. \quad (14.33)$$

The steady-state pore radius r_{ss} is equal to the defect density in the engineered membrane. Notice that, as the ratio γ/σ increases, the average steady-state aqueous pore radius in the membrane will increase.

Line tension. To compute the line tension γ of the membrane we use the procedure described in [183]. The line tension can be computed from the ribbon-like structure (Figure 14.12) using the formula[28]

$$\gamma = \frac{1}{2}\left\langle L_x L_y \left[\frac{P_{xx} + P_{yy}}{2} - P_{zz}\right]\right\rangle, \quad (14.34)$$

where P_{xx}, P_{yy}, and P_{zz} denote the diagonal elements of the pressure tensor, L_x and L_y denote the CGMD simulation cell size in the x and y directions, respectively, and $\langle \cdots \rangle$ denotes the time average.

To construct the lipid structure of Figure 14.12, an intact bilayer containing 320 lipids in a 70 percent DphPC and 30 percent GDPE composition is used. The hydrophilic interior of the bilayer is initially adjacent to the x and z dimensions of the simulation cell. The simulation cell is then expanded in the x direction from 14 to 16 nm and in the y direction from 10 to 13 nm to ensure the membrane forms an edge. Initially a 50-ns equilibration run was performed to allow the edge to form; this was followed by a 250-ns production from which γ (14.34) can be estimated. Simulations are performed

---

[28] The work ΔW required to change the volume of the system in Figure 14.12 by extending the length L_z by ΔL_z is given by $\Delta W = -P_{zz}L_x L_y \Delta L_z$. The work ΔW can also be represented by the sum of the work of changing the system volume minus the work of changing the total length of the ribbon edge such that $\Delta W = -0.5(P_{xx} + P_{yy})L_x L_y \Delta L_z + 2\gamma \Delta L_z$. Equating these two equations for work ΔW and solving for γ yields the instantaneous line tension $\gamma = 0.5L_x L_y [0.5(P_{xx} + P_{yy}) - P_{zz}]$. Taking the ensemble average of the instantaneous line tension gives the relation (14.34). Notice that the surface tension σ does not appear in the expression for line tension γ because the statistical ensemble used for the ribbon, illustrated in Figure 14.12, ensures that $\sigma = 0$.

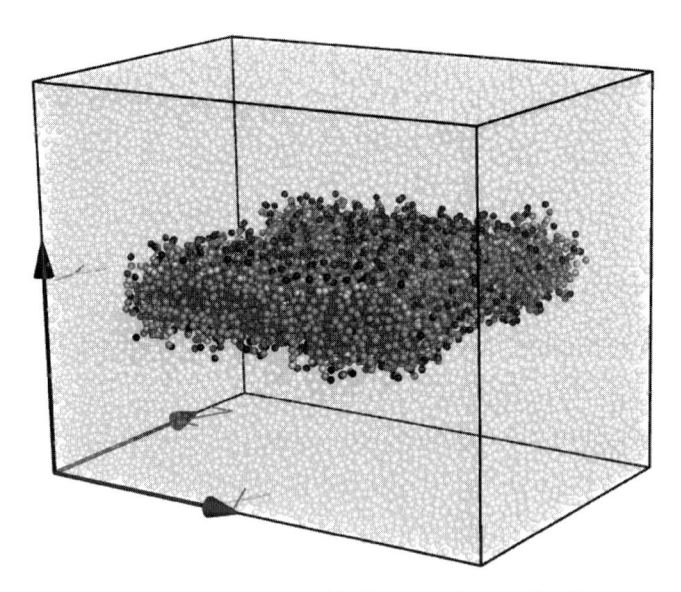

14.6 Tethered Membrane Dynamics and Energetics

Figure 14.12 A color version of this figure can be found online at www.cambridge.org/engineered-artificial-membranes. Ribbon structure of the 0 percent tethered DphPC membrane. Lipid tails are represented by the green beads, the NC_3 bead is displayed in blue, the PO_4 bead in orange, the OH bead in red, the COC bead as pink, and the water beads as a translucent blue. The coloring scheme of the axis is red for x, blue for y, and green for z. Note that this axis is only used for computing the line tension of the membrane as discussed in §14.6.2.

in an $NP_{xy}L_zT$ statistical ensemble[29] at a temperature of 320 K. The temperature is kept constant using the velocity rescaling algorithm with a time constant of 0.5 ps. The pressure is coupled semi-isotropically using the weak-coupling scheme [59] with a time constant of 3 ps, compressibility of 0.3 nm^2/nN, and a reference pressure of 100 kN/m^2.

The computed line tension of the engineered membrane was $\gamma = 12$ pN for both the 0 and 25 percent tether density membranes.

Surface tension. The surface tension of the membrane is computed using [183]

$$\sigma = \frac{1}{2}\left\langle L_z\left[P_{zz} - \frac{P_{xx}+P_{yy}}{2}\right]\right\rangle \qquad (14.35)$$

with the parameters defined following (14.34). The evaluation of (14.35) is performed in the NAP_zT ensemble[30] using a total production run of 250 ns.

The computed surface tensions σ of the 0 and 25 percent tether density membranes are 0 and 15 mN/m, respectively. Notice that an engineered membrane that contains no tethers will not have any induced stress. Therefore, for a 0 percent tether density membrane $\sigma = 0$ mN/m is expected.

---

[29] The statistical ensemble $NP_{xy}L_zT$ represents a molecular dynamics simulation that was performed with the following variables held constant: number of particles, N, pressure P in the x and y directions, simulation cell length L_z in the z direction, and temperature T.

[30] The statistical ensemble NAP_zT represents a molecular dynamics simulation that was performed with the following variables held constant: number of particles, N, pressure P in the z direction, the simulation cell length L_x and L_y in the x and y directions where $A = L_xL_y$ is the cell area, and the temperature T.

14.6.3 Deuterium Order Parameter

The deuterium order parameter,[31] denoted as S_{CD}, is important for characterizing the motional anisotropy of the lipid tails in bilayers [242, 376, 377] for aliphatic chains. It is evaluated as

$$S_{CD} = (3E[\cos^2(\theta)] - 1)/2, \quad (14.36)$$

where θ is the angle between the carbon-to-hydrogen vector and the membrane normal, with $E[\cdot]$ the expectation over all possible angles. $S_{CD} = 1$ when there is perfect alignment of the bond with the bilayer normal, and $S_{CD} = 0$ indicates the bond has a completely random orientation. CGMD simulations are used to compute S_{CD} using

$$S_{CD} = -\frac{S_c}{2}, \quad \text{where } S_c = (3E[\cos^2(\theta_c)] - 1)/2, \quad (14.37)$$

with θ_c the angle between the vector of the two nearest-neighbor beads and the membrane normal.

How do the deuterium order parameters S_{CD} from MD simulations compare with the lipid order parameters computed from CGMD simulations of 0 percent tether density membranes? From Figure 14.13 we see that results from the CGMD model are in reasonable agreement with the deuterium order parameters from molecular dynamics for ester-DphPC and 1,2-dipalmitoyl-sn-glycero-3-phosphocholine (DPPC) membranes. Note that results from MD have illustrated that the order parameters for ether-DphPC lipids have a higher order parameter than the ester-DphPC lipids [375]. Therefore, given that the CGMD model is constructed using ether-DphPC lipids, it is expected that the CGMD simulation would have higher order parameters compared with the ester-DphPC lipids. The results in Figure 14.13 also suggest that ether-DphPC lipids have a higher order parameter then the DPPC lipids. This results as the aliphatic chains in ether-DphPC lipids forming a tightly packed network with neighboring hydrocarbon chains being interdigitated, which results in an increase in their order parameter compared to that of DPPC [375].

14.6.4 Lipid Lateral Diffusion

Lipid diffusion is a key property of cell membranes because it provides the major mechanism by which drugs and other molecules can be absorbed and permeate through the membrane. Most drugs are small amphipathic (both hydrophobic and hydrophilic) molecules, similar to lipids which are also amphipathic (the lipid head group is hydrophilic and the lipid tails are hydrophobic). As such, the diffusion coefficient of lipids is correlated with the diffusion coefficient of absorbed drugs in the membrane.

---

[31] The deuterium order parameter provides a measure of the flexibility of lipids in the bilayer. Formally, it provides the structural orientation between carbon-carbon bonds in the hydrophobic tails of the lipids. The deuterium order parameter can be measured directly using NMR spectroscopic techniques. This allows the model-predicted deuterium order parameter from the CGMD simulation to be compared directly with the experimentally measured order parameter, which is useful for validating the accuracy of the CGMD model.

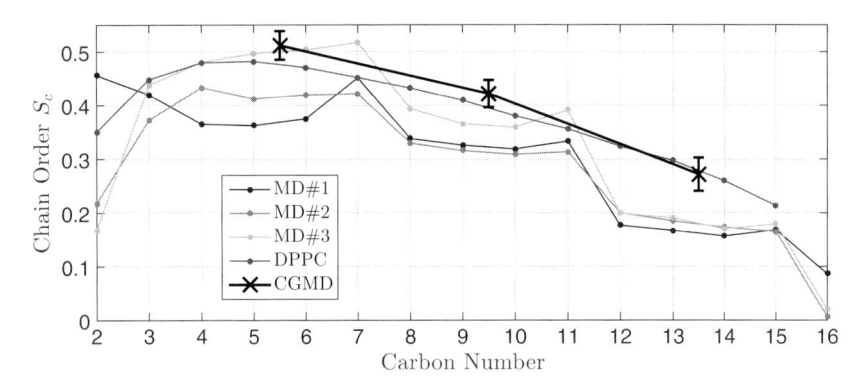

14.6 Tethered Membrane Dynamics and Energetics

Figure 14.13 Computed lipid order parameter from CGMD simulations and all-atom MD simulations of DphPC, ether-DphPC, and DPPC lipids [242, 376, 377]. Here the chain order S_c is illustrated; however, the deuterium order parameter S_{CD} can be computed directly from these results using $S_{CD} = -S_c/2$. The chain order parameter provides a measure of the time-averaged orientation of the covalent bonds of carbon in the lipid tails. As seen, the chain order parameter estimated from the CGMD simulation is in agreement with the results from all-atom MD. The error bars on the CGMD markers indicate the associated error for each data point. For the ester-DphPC molecules (MD#1,MD#2) and ether-DphPC molecules (MD#3), the carbon numbers (3, 7, 11, and 15) are attached to the methyl groups with 16 denoting the terminal methyl, and for DPPC carbon 15 represents the terminal carbon of the aliphatic chain.

In this section we compute the diffusion coefficient of lipids in the engineered tethered membrane using the results from CGMD simulations.

The diffusion of molecules in a homogeneous medium can be described using standard Fickian diffusion where the mean-square displacement is proportional to time. Specifically, the time evolution of the mean-square displacement (variance) of lipids is given by

$$\text{Var}(x(t) - x_o) = \mathbf{E}[(x(t) - x_o)^2] = \int_{-\infty}^{\infty} (\xi - x_o)^2 f(\xi)\, d\xi \propto t^\beta, \quad (14.38)$$

where x_o is the initial position, $x(t)$ is the current position at time t, $\text{Var}(\cdot)$ denotes the variance, the positive real number β is the power-law exponent, and $f(x)$ denotes the stationary probability distribution of the lipid position. Note that (14.38) is the mean-square displacement of a particle that satisfies fractional Brownian motion.[32] For Fickian diffusion, where the mean-square displacement is proportional to time t and $\beta = 1$, the proportionality constant in (14.38) is related to the lipid diffusion coefficient D by the Einstein relation $D = \text{Var}(x(t) - x_o)/4t$. The variance $\text{Var}(x(t) - x_o)$ can be estimated using CGMD simulation results using the ergodic hypothesis, which relates the ensemble average in (14.38) to the time average for the stochastic process of the lipid positions.

---

[32] Brownian motion is a continuous-time Gaussian stochastic process with independent increments having mean zero and variance at time t equal to t.

Table 14.1 Lateral diffusion coefficient $D_\parallel$ (nm^2/μs).

Tether density		0%	25%
Proximal Layer	DphPC	290±23	87±10
	GDPE	256±36	116±16
Distal Layer	DphPC	289±24	115±12
	GDPE	256±38	128±13
Bulk Water		2105±288	2010±233
Tethering Reservoir Water		2730±258	1236±110

The computed lateral diffusion of the DphPC lipids, GDPE lipids, and water for the 0 and 25 percent tether density engineered membranes are provided in Table 14.1. The computed lateral diffusion of the DphPC and GDPE lipids in the proximal layer (i.e., adjacent to the tethering reservoir) and distal layer (i.e., adjacent to the bulk water) are nearly identical. The diffusion of DphPC is related to that of GDPE by a multiplicative factor of 1.13. The effects of the tethers cause the lateral diffusion coefficient of DphPC to decrease by a factor of approximately 3.3 in the proximal layer and 2.5 in the distal layer. Similarly, for GDPE the decrease is 2.2 in the proximal layer and 2.2 in the distal layer. This result is in agreement with the experimental results reported in [289, 446] for different lipids and tethering densities. The lateral diffusion of water in the bulk region is related to the tethering reservoir lateral diffusion by a factor of approximately 1.6 for the 25 percent tethered membrane. It is interesting that the lateral diffusion of water in the tethering reservoir is 2.7 nm^2/ns, which is close to but faster than that in the bulk of 2.1 nm^2/ns for the 0 percent tethered membrane. The interplay between hydrogen-bond breaking and cooperative rearrangement of regions of approximately 1 nm in size causes the lateral diffusion to significantly increase in nanoconfined water regions. Since explicit hydrogen bonds are not included in the CGMD model, we attribute the increased diffusion coefficient to the cooperative rearrangement of water molecules. A unique result of this study is that no anomalous diffusion (where Var$(x(t) - x_o)$ is not proportional to t) was detected for water and lipid head groups near the outer surface of the bilayer lipid membrane. This is in contrast to other MD simulation results of different membranes, which report an anomalous water and lipid diffusivity in the vicinity of the membrane surface.

14.6.5 Geometric Properties of Tethered Membranes

To compute the membrane thickness h_m (see Figure 14.4), the particle density of the lipid head groups is used. Figure 14.14 provides the normalized particle density for the CGMD beads W, PO$_4$, NC$_3$, OH, COC, and the first C$_1$ of the DphPC, and GDPE lipids for the 0 and 25 percent tether density membranes.[33] Recall from §14.3 that the

---

[33] The beads W, PO$_4$, NC$_3$, OH, COC, and C$_1$ are defined in Figure 14.4 on page 303. W represents four water molecules, PO$_4$ represents the phosphate group, NC$_3$ the choline group, OH the hydroxyl group, COC the etherglycol group, and C$_1$ the first bead of the phytanyl tail of the DphPC and GDPE lipids.

14.6 Tethered Membrane Dynamics and Energetics

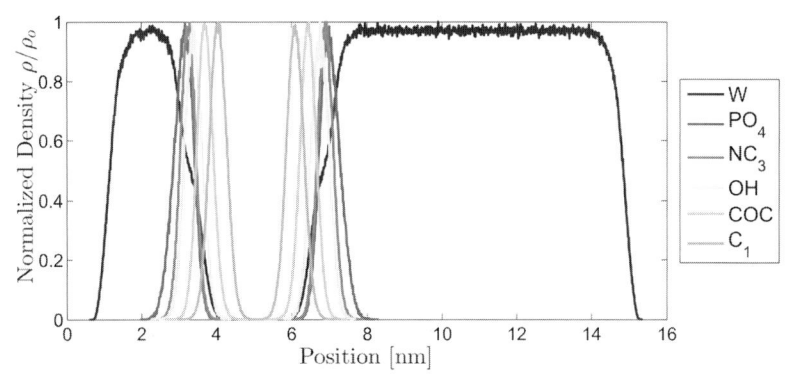

(a) Normalized particle density computed from CGMD for the 0 percent tether density DphPC membrane.

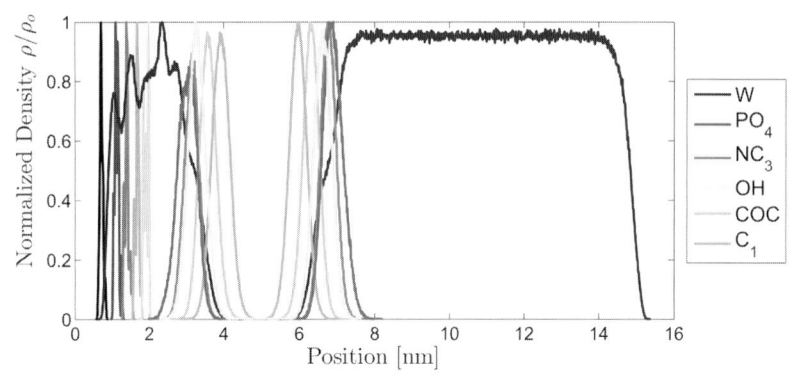

(b) Normalized particle density computed from CGMD for the 25 percent tether density DphPC membrane.

Figure 14.14 Normalized particle density computed from the CGMD bead trajectories. The surfaces of the electrodes are at positions $z = 0$ nm and $z = 16$ nm. The CGMD beads W, PO$_4$, NC$_3$, OH, COC, and C$_1$ are defined in §14.3. The water density in the tethering reservoir depends on the tether density and the thickness of the tethering reservoir.

COC bead is associated with the ether linker in the DphPC and GDPE lipids. In Figures 14.14(a) and 14.14(b) the distinct peaks in the OH and PO$_4$ head group beads of GDPE and DphPC indicate the membrane is intact with negligible defects. As expected, the peak in the particle density in Figure 14.14(b) for water at 2.4 nm occurs between the OH bead of the spacer and the head group of the membrane. The thicknesses of the 25 and 0 percent tether density membranes are $h_m = 3.53$ and $h_m = 3.48$ nm, respectively. The thickness of the phytanyl tails (i.e., hydrocarbon tails) was also computed for the 25 and 0 percent tether density membranes – 2.15 and 2.11 nm, respectively. The reservoir thickness h_r (Figure 14.4) of the 25 and 0 percent tether density membranes is $h_r = 3.30$ nm. These numerically computed values are consistent with the experimentally measured thickness for DphPC-based tethered membranes [152].

14.6.6 Summary

For the reader's convenience, here is a short summary of what was covered in this section. We computed the lateral diffusion of water, lipid diffusion, the energy difference between lipid configurations (lipid desorption and flip-flop) in the membrane, line tension, surface tension, the deuterium order parameter, and the membrane thickness h_m as a function of tether density using CGMD simulations. All these parameters are crucial for modeling the dynamics of engineered membranes. The lipid desorption energy and energy of lipid flip-flop provide a measure of the structural stability of the membrane. As expected, the increase in tether density from 0 to 25 percent resulted in an increase in the lipid desorption energy from 85.93±0.5 to 91±0.5 kJ/mol. Additionally, the energy of lipid flip-flop increased from 89 kJ/mol for the 0 percent tether density membrane to 103.17 kJ/mol for the 25 percent tether density membrane. An increase in tether density was also predicted to cause a reduction in the thickness of the engineered membrane. Specifically, the thicknesses of the 25 and 0 percent tether density membranes are $h_m = 3.53$ and $h_m = 3.48$ nm, respectively. The computed line tension $\gamma = 12$ pN was equal for both the 0 and 25 percent tether density membranes. Note, however, that for high tether densities (e.g., 100 percent tether density), the line tension (force required to form an edge in the membrane) will be tether density dependent. The computed surface tension σ of the 0 and 25 percent tether density membranes was 0 and 15 mN/m, respectively. Therefore, the tethers introduce a small surface tension to the membrane. Not only does the tether density affect the geometry and energetics of lipids, but it also affects the diffusion of lipids and water in the tethering reservoir. For an increase in tether density from 0 to 25 percent, this resulted in a reduction in the lateral diffusion coefficient of water by a factor of 2.2. Additionally, the reduction in the diffusion coefficient of lipids was in the range of 2.0 to 3.3.

14.7 Control of Tethered-Membrane Properties by Sterol Inclusions

Archaebacterial membranes[34] do not contain cholesterol. We have shown in Chapters 8 and 11 how to precisely engineer artificial membranes to mimic such archaebacterial membranes. In this section we show via CGMD simulations that incorporating cholesterol into an archaebacterial membrane controls the properties of the membrane (conductance, thickness, diffusion of lipids, line tension, and surface tension). Since archaebacterial membranes are used for constructing membrane-based biosensors (refer to Chapter 3), knowledge of how cholesterol affects the membrane properties is important for the design of biosensors.

The CGMD simulations of the archaebacterial membranes containing different concentrations of cholesterol are constructed using the method in §14.3. Using the CGMD

---

[34] Recall that archaebacteria are single-celled organisms that form the link between bacteria and eukaryotes. Archaebacteria can tolerate harsh environments with high acidity and salinity. See §2.3.1 for details.

Table 14.2 Lipid and cholesterol diffusion ($nm^2/\mu s$).

Cholesterol Density	DphPC	GDPE	Cholesterol
0 percent	18.0 ± 0.8	14.3 ± 0.7	–
10 percent	12.2 ± 3.5	13.1 ± 3.7	21.8 ± 5.0
20 percent	11.7 ± 1.5	12.8 ± 2.2	18.7 ± 0.4
30 percent	9.5 ± 0.5	10.4 ± 4.7	12.1 ± 0.5
40 percent	6.3 ± 1.5	5.8 ± 1.2	10.0 ± 0.5
50 percent	3.8 ± 2.2	5.3 ± 3.9	7.1 ± 0.1

simulations, we show that cholesterol plays an important role in archaebacterial membrane diffusion dynamics and biomechanics with properties that are unique to lipids with phytanyl tails.[35]

14.7.1 Lateral Diffusion Dynamics of Lipids and Cholesterol

In this section, the results of CGMD simulations are interpreted to gain insight into how the concentration of cholesterol affects the diffusion dynamics of lipids in the archaebacterial membrane. The diffusion of lipids and cholesterol in the membrane is important for the binding and transport of drugs and molecules across the membrane surface.

To gain insight into how the concentration of cholesterol affects the lipid diffusion dynamics, we compute the diffusion coefficient D for DphPC, GDPE, and cholesterol for archaebacterial membranes containing 0 to 50 percent cholesterol densities. The results are provided in Table 14.2. The diffusion coefficient of DphPC for 0 percent cholesterol is in excellent agreement with the experimentally measured diffusion coefficient of 18.1 ± 5.6 $nm^2/\mu s$ [24]. Additionally, the numerically computed diffusion coefficient of cholesterol is in excellent agreement with the experimentally measured diffusion coefficient of cholesterol, which is in the range of 10 to 100 $nm^2/\mu s$ [365, 408]. The cholesterol has a higher diffusion coefficient than DphPC and GPDE, and the diffusion of cholesterol monotonically decreases as the concentration of cholesterol increases. If we consider only the mass and size of cholesterol, it is expected that the lower mass and size of cholesterol compared to DphPC and GDPE will cause cholesterol to have a larger diffusion coefficient. Another contributing factor is that the head groups of DphPC and GDPE both have a larger dipole moment than that of cholesterol, which will also reduce the diffusion coefficient of the lipids compared to that of cholesterol [159]. For DphPC, GDPE, and cholesterol, the diffusion coefficient monotonically decreases as the concentration of cholesterol increases. Cholesterol has a similar effect on the diffusion of 1-stearoyl-2-oleoyl-sn-glycero-3-phosphocholine

---

[35] Eukaryote membranes contain cholesterol to enhance the stability of the membrane, unlike archaebacterial membranes that do not contain cholesterol. As a result of the unique structure of archaebacterial lipids (i.e., the hydrocarbon chains containing methyl groups), the concentration of cholesterol has a noticeably different effect on the membrane properties of archaebacterial membranes compared to that of eukaryotic membranes. Using engineered membranes it is possible to quantify how cholesterol affects the dynamics of archaebacterial membranes.

Table 14.3 Biomechanic parameters of membrane.

	0%	10%	20%	30%	40%	50%
Membrane Thickness (nm)						
PO$_4$	3.80	4.09	4.13	4.17	3.92	3.61
OH	3.36	3.65	3.69	3.76	3.81	3.98
ROH	–	2.92	3.01	3.06	3.09	3.17
Surface Tension (mN/m)						
σ	57.8	73.5	86.1	97.1	68.9	43.9
Line Tension (pN)						
γ	50.8	51.8	54.8	60.5	69.4	70.1

The membrane thickness is computed from the distance between the molecule in the distal layer to the molecule in the proximal layer of the membrane.

(SOPC), 1-stearoyl-2-linoleoyl-sn-glycero-3-phosphocholine (SLPC), 1-stearoyl-2-arachidonoyl-sn-glycero-3-phosphocholine (SAPC), 1-stearoyl-2-docosahexaenoyl-sn-glyerco-3-phosphocholine (SDPC), and DPPC lipids [121, 159].

14.7.2 Biomechanics of Lipids and Cholesterol

How does the concentration of cholesterol affect the archaebacterial membrane biomechanics?[36] Using the results from CGMD simulations, we study how cholesterol content affects the membrane thickness h_m, line tension γ, and surface tension σ. Recall that the fractional-order macroscopic model introduced in Chapter 11 depends on h_m, γ, and σ. Therefore, experimental measurements from the tethered archaebacterial membrane can be used to validate the results from the CGMD simulation. The computed h_m, σ, and γ for the 0 to 50 percent cholesterol membranes is provided in Table 14.3 on the following page.

Table 14.3 displays the predicted membrane thickness (using the CGMD simulation model for the engineered membrane) for concentrations of cholesterol from 0 to 50 percent. The membrane thickness for the 0 percent cholesterol is in agreement with the experimentally measured thickness for DphPC-based tethered membranes which do not contain cholesterol [152]. The position of cholesterol's hydroxyl group (ROH) is always less than the associated head groups (PO$_4$ and OH) of the DphPC and GDPE lipids. This is in agreement with the results from molecular dynamics simulations of DPPC membranes containing 11 and 50 percent cholesterol content [394]. As expected, as the cholesterol content increases from 0 to 30 percent there is a decrease in membrane thickness as a result of the cholesterol forming a complex with the hydrocarbon tails. A similar effect has been observed for cholesterol in DPPC, 1,2-dimyristoyl-sn-glycero-3-phosphocholine (DMPC) SOPC, and 1-palmitoyl-2-oleoyl-sn-glycero-3-phosphocholine (POPC) membranes [119, 173, 429]. Interestingly, for archaebacterial

---

[36] Biomechanics is the study of the mechanical and structural properties of engineered systems. In the context of engineered membranes, we are interested in the thickness h_m, surface tension σ, and line tension γ.

14.7 Control of Tethered-Membrane Properties by Sterol Inclusions

membranes containing 40 and 50 percent cholesterol there is a decrease in membrane thickness. This suggests that high concentrations of cholesterol cause the disentanglement of the tightly connected phytanyl chains of DphPC and GDPE. To validate the CGMD-based membrane-thickness estimate we use experimental measurements from tethered archaebacterial membranes. Recall that the capacitance of the tethered membrane, C_m, is dependent on the dielectric permittivity ε_m, thickness of the membrane, h_m, and surface area of the membrane, A_m. For the tethered bilayer lipid membrane the membrane surface area is fixed at 2.1 mm^2. Therefore, for a constant ε_m, as h_m decreases, the associated capacitance of the membrane must increase. From Table 12.1 on page 268, as the concentration of cholesterol increases from 0 to 30 percent the associated capacitance of the membrane decreases, suggesting that the membrane thickness is increasing. From 40 to 50 percent cholesterol content the tethered membrane capacitance increases, suggesting a decrease in membrane thickness. These results validate the numerically computed archaebacterial membrane thickness provided in Table 14.3 for concentrations of cholesterol from 0 to 50 percent.

As seen from Table 14.3, the cholesterol content affects both the surface tension and the line tension of the membrane. The computed values for the surface tension in Table 14.3 are in agreement with the experimental results in [481] and simulation results in [184, 186, 463] for similar DphPC-based membranes. From MD studies of DPPC membranes, it is observed that as the percentage of cholesterol increases there is a decrease in the surface area of the membrane [159]. Since the surface tension is the energy required to increase the surface area of the membrane by a unit area, if the concentration of cholesterol increases, the surface tension of the membrane is expected to increase. From Table 14.3, for 0 to 30 percent cholesterol the surface tension σ increases with increasing cholesterol content. However, the surface tension decreases for 40 and 50 percent cholesterol content. This results because in the 40 and 50 percent cholesterol membrane there is little free space for the cholesterol to insert, which causes the increased cholesterol content to disentangle the tightly connected hydrocarbon lipid chains of the DphPC and GDPE. From Table 14.3 we see that as the line tension γ of the membrane monotonically increases, the cholesterol content increases. In 1,2-dioleoyl-sn-glycero-3-phosphocholine (DOPC) membranes, cholesterol acts as a two-dimensional surfactant that reduces the line tension [42]. However, the results from the CGMD model suggest that in archaebacterial membranes the cholesterol does not promote separation between lipids that are transitionally disordered and conformationally ordered, and lipids that are both transitionally and conformationally disordered. To validate the computed surface tension σ and line tension γ in Table 14.3, we use experimental results from tethered archaebacterial membranes. Recall that the population of membrane defects increases as the ratio γ/σ increases. The population of membrane defects is given by the equilibrium membrane conductance G_m, which can be measured experimentally from the fractional-order macroscopic model. From the experimentally measured membrane conductance G_m in Table 12.1 on page 268, for 0 to 30 percent cholesterol the ratio γ/σ is expected to decrease, and for 40 to 50 percent, γ/σ is expected to increase. This result is in agreement with the numerically computed σ and γ in Table 14.3.

14.8 Molecular Diffusion and Langevin's Equation

The diffusion coefficient of lipids and macromolecules in the engineered tethered membrane is an important parameter for estimating the diffusion-limited reaction dynamics of molecules on the membrane surface, as in Chapter 10. Additionally, since the diffusion coefficient can be measured using FRAP,[37] it can also be used to validate atomistic models of the engineered membrane. The diffusion of molecules can be described using standard Fickian diffusion models where the mean-square displacement is proportional to time. However, this proportionality relation between mean-square displacement and time is only valid at long time scales where the stochastic process that describes the molecules' positions has attained the steady state (equilibrium). In membranes, this occurs on a time scale of 10 ns. At short time scales, where steady state has not been attained, the law of large numbers does not apply and the time average does not provide a useful estimate of the ensemble average.

Let $f_{X(t)}(x)$ denote the probability density function of the molecules' positions in three-dimensional space x at time t. In this section we study the relation between the mean-square displacement $\text{Var}(x(t) - x_o)$ and time t, where we assume that $f_{X(t)}(x)$ has not reached the steady-state distribution. Formally,

$$\text{Var}(x(t) - x_o) = E[(x(t) - x_o)^2] = \int_{-\infty}^{\infty} (\xi - x_o)^2 f_{X(t)}(\xi)\, d\xi, \qquad (14.39)$$

where $x(t)$ is the position of the molecule, x_o is the initial position, $E[\cdot]$ is the expectation operator, and $f_{X(t)}(x)$ is the probability density function of the molecules' position. First, we derive the proportionality relations and diffusion coefficient for a large molecule using Langevin's equation. Then, we illustrate how to estimate the proportionality relation between $\text{Var}(x(t) - x_o)$ and t from the results of CGMD simulations. Knowledge of these relations and the time necessary for the CGMD simulation to reach steady state is important because diffusion coefficients can only be computed using CGMD trajectories in the Fickian (steady-state) diffusion regime.

14.8.1 Langevin's Equation and Diffusion of Molecules

In this section we use Langevin's equation to derive the relation between the mean-square displacement $\text{Var}(x(t) - x_o)$ and time t for a large molecule in an electrolyte. At the femtosecond time scale we find that $\text{Var}(x(t) - x_o) \propto t^2$, and at the 10-ns time scale that $\text{Var}(x(t) - x_o) \propto t$. These correspond to the ballistic and Fickian diffusion regimes, respectively.

Langevin's equation is a stochastic differential equation that describes the time evolution of the velocity of a large molecule (heavy particle) when it is bombarded by significantly smaller particles (in both size and mass).

---

[37] Fluorescence recovery after photobleaching; refer to §7.3 for details.

14.8 Molecular Diffusion and Langevin's Equation

Langevin's equation for a single particle of mass m with velocity $v(t)$ in three-dimensional space reads

$$mv(t) = mv(0) - m\xi \int_0^t v(s)ds + \sigma w(t), \quad \text{where } \sigma = \sqrt{2m\xi k_B T}. \quad (14.40)$$

In Langevin's equation, ξ is the friction coefficient, $k_B = 1.38 \times 10^{-23}$ J/K is the Boltzmann constant, T is the absolute temperature, and $w(t)$ is a three-dimensional vector of independent Brownian motion; i.e., each element $w_i(t)$ is an uncorrelated zero-mean Gaussian process with $E[w_i(t)w_i(0)] = t$. The three-dimensional random process $\sigma w(t)$ has variance

$$F_o = \sigma^2 E\left[\sum_{i=1}^3 (w_i(t))^2\right] = 3\sigma^2 = 6\xi m k_B T.$$

An important property of Langevin's equation is that the friction coefficient ξ appears both in the velocity integral and in the noise term. Using the Einstein–Smoluchowski relation (fluctuation-dissipation theorem), the friction coefficient can be expressed as

$$\xi = \frac{k_B T}{mD},$$

where D is the diffusion constant.[38] Thus the underlying intuition is that the intensity of the bombardment increases as the energy of the bombarding particles increases (i.e., proportional to D).

Langevin's equation (14.40) is often expressed in terms of continuous-time white noise, where white noise is the time derivative of Brownian motion as (this is heuristic since mathematically speaking Brownian motion is not differentiable)

$$m\ddot{x}(t) = -m\xi \dot{x}(t) + F_R(t), \quad (14.41)$$

where the three-dimensional vector $\ddot{x}(t)$ denotes the acceleration at time t, $\dot{x}$ is the velocity, ξ is the friction coefficient, and $F_R(t)$ is a three-dimensional random force process where each component is modeled as independent continuous-time white noise. So $F_R(t)$ is a zero-mean vector, is uncorrelated with the velocity ($E[F_R(t) \cdot \dot{x}(t)] = 0$), and has variance (where $\cdot$ is the inner product)

$$E[F_R(\tau + t) \cdot F_R(t)] = F_o \delta(\tau), \quad (14.42)$$

where $\delta(\tau)$ is the Dirac delta function and $F_o = 6\xi m k_B T$.

Note that the mean-square velocity of the particles at equilibrium is given by $E[|\dot{x}(t)|^2] = 3k_B T/m$, where $|\cdot|$ denotes the Euclidean norm of a vector. Equation (14.42) is a fluctuation-dissipation theorem that provides the relationship between the

---

[38] In a more sophisticated setting, the diffusion constant $D(x(t))$, is spatially dependent on $x(t)$, implying that ξ would appear inside the integral in (14.40), since position is the integral of velocity. Also, the integral $\int \sigma(x(s))dw(s)$ would then be interpreted as an Itô integral. In a multiple-particle setting, Langevin's equation for individual particles would interact with that of other particles in terms of Coulomb forces [217]. The Mori–Zwanzig formalism (not discussed in this book) gives a detailed justification of Langevin's equation in terms of resolved and unresolved variables.

fluctuation force $F_R(t)$ that changes the velocity $\dot{x}(t)$, and the dissipation in $\dot{x}(t)$ that results from the frictional force $-m\xi\dot{x}(t)$.

Irrespective of whether the Brownian motion model (14.40) or the white noise model (14.41) is used, the noise model is justified as follows: the frictional force on the molecule is assumed to be proportional to the velocity of the particle and the randomly fluctuating force (Brownian motion or white noise), which accounts for the frequent collisions of the molecule with smaller particles such as water molecules. In this sense, explicit water molecules are replaced by implicit water (Brownian motion).

The goal is to use (14.41) to construct a relation between $\text{Var}(x(t) - x_o)$ and time t. Let us set $x(0) = x_o = 0$ as the initial condition for the molecules' position. Multiplying (14.41) by $x(t)$, and using the definition of acceleration and velocity, results in

$$mx(t) \cdot \ddot{x}(t) = -m\xi x(t) \cdot \dot{x}(t) + x(t) \cdot F_R(t),$$

$$\frac{m}{2}\frac{d^2}{dt^2}|x(t)|^2 - m|\dot{x}|^2 = -\frac{\xi m}{2}\frac{d}{dt}|x(t)|^2 + x(t) \cdot F_R(t),$$

$$\frac{d^2}{dt^2}|x(t)|^2 = -\xi \frac{d}{dt}|x(t)|^2 + 2|\dot{x}|^2 + \frac{2x(t)}{m} \cdot F_R(t), \qquad (14.43)$$

where $|\cdot|$ denotes the Euclidean norm of the vector. Given that the larger particles are in thermal equilibrium, the kinetic energy of the large molecule is proportional to the temperature of the significantly smaller particles such that $\boldsymbol{E}[|\dot{x}(t)|^2] = 3k_B T/m$. This relation between the average kinetic energy and the temperature is a result of the equipartition theorem. We now take the ensemble average on both sides of (14.43), and use the relation between kinetic energy and temperature, to obtain the ordinary differential equation

$$\frac{d^2}{dt^2}\boldsymbol{E}[|x(t)|^2] = -\xi \frac{d}{dt}\boldsymbol{E}[|x(t)|^2] + \frac{6k_B T}{m}, \quad \boldsymbol{E}[|x(0)|^2] = 0, \quad \frac{d}{dt}\boldsymbol{E}[|x(0)|^2] = 0. \qquad (14.44)$$

The initial conditions are specified as (14.44) because we set the initial position of the molecule to $x(0) = 0$, and the random fluctuation force $F_R(t)$ is uncorrelated with the molecules' position $x(t)$ such that $\boldsymbol{E}[F_R(t) \cdot x(t)] = 0$. The solution to (14.44) is

$$\boldsymbol{E}[|x(t)|^2] = \frac{6k_B T}{\xi m}\left(t - \frac{1}{\xi} + \frac{1}{\xi}e^{-\xi t}\right). \qquad (14.45)$$

For a large molecule in a bath of significantly smaller particles, (14.45) provides the relation between the mean-square displacement $\text{Var}(x(t))$ and time t. There are two important regimes to consider:

(i) The first is the short-time ballistic regime in which $\xi t \ll 1$ such that (14.45) simplifies to

$$\boldsymbol{E}[|x(t)|^2] \approx \left(\frac{3k_B T}{m}\right)t^2. \qquad (14.46)$$

In this case the mean-square displacement of the large molecule is proportional to t^2 and is in the ballistic regime. In the ballistic regime the molecule collides with very few particles and essentially moves freely as if it was in a vacuum.

(ii) The second is the long-time Fickian regime in which $\xi t \gg 1$ such that (14.45) simplifies to

$$E[|x(t)|^2] \approx \left(\frac{6k_BT}{\xi m}\right) t = Dt, \qquad (14.47)$$

where D is the diffusion coefficient given by Einstein's relation $D = 6k_BT/\xi m$. If the friction coefficient ξ was known, then D could be computed directly from (14.47). In the case of a simple fluid, the friction coefficient ξ is computed using Stokes' law. The key is that the mean-square displacement of the molecule in the Fickian regime is proportional to t.

Although the above derivation provides an accurate estimate for the mean-square displacement of a large molecule in a bath of significantly smaller particles, it cannot be used when the molecule and particles are of similar sizes. If the molecule and particles are of similar sizes, then the Markovian approximation in (14.41) where the frictional force on the molecule is proportional to the velocity ($-m\xi \dot{x}(t)$) does not hold. This results because the molecule cannot instantaneously adjust its velocity and acceleration to the dynamics of the surrounding particles. It is possible to model the dynamics when the molecule and particles are of the same size by introducing a frictional force that is nonlocal in time; that is, the frictional force is a convolution in time which results in a non-Markovian generalization of the Langevin equation (14.41). In such cases, *mode-coupling theory* (MCT) can be used to model the dynamics of the molecules. For example, MCT has recently been applied to model the diffusion dynamics of lipids in membranes [124]. In such cases, the characteristic regimes of the lipid diffusion are composed of ballistic, subdiffusion, or Fickian diffusion regimes. Note that the ballistic and subdiffusion transport phenomena are also commonly referred to as anomalous diffusion.

14.8.2 Nonstationary Lipid Diffusion with Sterol Inclusions

The key question addressed in this section is this: how can one detect if the mean-square error displacement (14.39) of the lipids in an archaebacterial membrane is in the short time ballistic regime ($\text{Var}(x(t) - x_o) \propto t^2$), the subdiffusion regime ($\text{Var}(x(t) - x_o) \propto t^\beta$ with $\beta < 1$), or the Fickian diffusion regime ($\text{Var}(x(t) - x_o) \propto t$) as a function of time t? If these regimes are present, what is the transition time between them? Knowledge of these transition times is important since diffusion coefficients can only be computed using CGMD trajectories in the Fickian diffusion regime.

To start our discussion, we consider the following question: How can we compute the power-law exponent β in $\text{Var}(x(t) - x_o) \propto t^\beta$ given the results of CGMD simulations? The value of β determines if we are in the ballistic, subdiffusion, or Fickian diffusion regime. The value of β can be computed directly from the results of the CGMD simulations. Given $\text{Var}(x(t) - x_o) = \alpha t^\beta$, first we take the logarithm of both sides, then use

Monte Carlo integration to evaluate the expectation operator. The result of these operations is

$$\beta(t) = \frac{\ln(\boldsymbol{E}[|x(t) - x_o|^2])}{\ln(t)} + \ln(\alpha)$$

$$\approx \frac{\ln\left(\frac{1}{n}\sum_{i=1}^{n}|x_i(t) - x_i(0)|^2\right)}{\ln(t)} + \ln(\alpha), \quad (14.48)$$

where $|\cdot|$ denotes the Euclidean norm of a vector, n is the number of molecules in the CGMD simulation, t is the time, and $x_i(t)$ are the coordinates of the ith molecule from the CGMD simulation. Notice that the power-law exponent $\beta(t)$ is time dependent in general. However, we are interested in the ballistic, subdiffusion, or Fickian diffusion regimes where $\beta(t)$ is approximately constant. The conditions for each regime are as follows:

$$\text{ballistic} \quad \text{if} \quad \frac{d\beta}{dt} = 0 \text{ and } \beta = 2,$$

$$\text{subdiffusion} \quad \text{if} \quad \frac{d\beta}{dt} = 0 \text{ and } \beta < 1, \text{ and}$$

$$\text{Fickian} \quad \text{if} \quad \frac{d\beta}{dt} = 0 \text{ and } \beta = 1.$$

From the results in §14.8.1, the ballistic regime occurs at the short time scale of femtoseconds, and the Fickian diffusion regime occurs at the long time scale of several nanoseconds. The subdiffusion regime exists between the ballistic and Fickian diffusion regimes. Note that, from the definition of diffusion[39] from Fick's law, only the Fickian diffusion regime can be used to estimate the diffusion coefficient of lipids. The transport phenomena outside the Fickian diffusion regime (i.e., anomalous diffusion) do not satisfy Fick's law of diffusion, because the ionic transport processes are nonlocal. To model these fractional transport processes, one can use the fractional Langevin equations and the fractional Fokker–Planck equation. Unless fractional processes are present at the time scales of interest (microseconds for engineered tethered membranes), it is not necessary to use these fractional transport models.

Figure 14.15 presents the numerically computed mean-square displacement of DphPC, GDPE, and cholesterol from the CGMD simulation results for membranes composed of no cholesterol (0 percent cholesterol concentration) and membranes where 50 percent of the lipids are cholesterol molecules (50 percent cholesterol concentration). From Figure 14.15, we see that for $t \leq 3$ ns the lipid molecules (i.e., DphPC and GDPE) diffuse in the subdiffusion regime, and for $t \geq 20$ ns the diffusion of the lipids is in the Fickian diffusion regime. This is in agreement with the results predicted using mode-coupling theory for flexible macromolecules [72, 124]. Given that the time step of the CGMD simulation is 20 fs, the ballistic region $\beta > 1$ is not observed in Figure 14.15 for any of the lipids or cholesterol. This is expected because the ballistic region is typically

---

[39] The diffusion coefficient D^i of species i is a proportionality constant between the ionic flux J^i and the gradient of the concentration c^i of species i. Formally $J^i = -D^i \nabla c^i$.

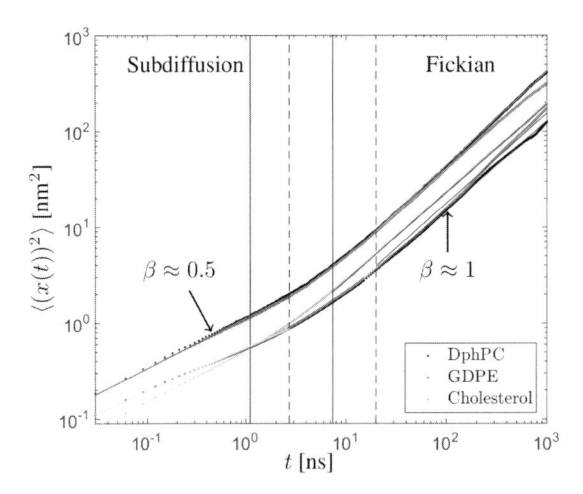

Figure 14.15 A color version of this figure can be found online at www.cambridge.org/engineered-artificial-membranes. Computed mean-square displacement for DphPC, GDPE, and cholesterol in the 0 and 50 percent cholesterol membranes with β defined in (14.48). The red and green lines indicate the subdiffusion regime and Fickian diffusion regime of the lipids in the 0 percent cholesterol membrane. The yellow and blue lines indicate the subdiffusion regime and Fickian diffusion regime of the lipids and cholesterol in the 50 percent cholesterol membrane. Notice that the diffusion dynamics for the lipids are in the subdiffusion regime ($\beta \approx 0.5$) for $t \leq 3$ ns, and for $t \geq 20$ ns the diffusion dynamics are in the Fickian regime ($\beta \approx 1$), as illustrated by the dashed vertical lines. The cholesterol is in the subdiffusion regime for $t \leq 1$ ns and in the Fickian diffusion regime for $t \geq 7$ ns as illustrated by the vertical solid lines. This is in agreement with the mode-coupling theory for flexible macromolecules [72, 124].

observed for $t \leq 10$ fs [124]. Interestingly, the transition times between the subdiffusion and Fickian diffusion regimes of the lipids are not dependent on the cholesterol content. This suggests that the concentration of cholesterol presented contributed negligibly to the caging effect of lipids; that is, only the Fickian diffusion dynamics are strongly dependent on the concentration of cholesterol present. This is a unique feature of the cholesterol on the DphPC and GDPE lipids, as for 1,2-Distearoyl-sn-glycero-3-phosphocholine (DSPC), SOPC, and DOPC it is shown that the cholesterol affects both the subdiffusion and Fickian diffusion dynamics [180]. Given that cholesterol has a substantially different atomistic structure compared to DphPC and GDPE, it is expected that the transition between the subdiffusion and Fickian diffusion regimes will be different than for the lipids. As expected, cholesterol is in the subdiffusion regime for $t \leq 1$ ns and in the Fickian diffusion regime for $t \geq 7$ ns.

14.9 Case Study: Atomistic-to-Observable Model PGLa Pore Formation in Tethered Membranes

We conclude our discussion of CGMD by presenting a detailed case study of how CGMD together with a continuum model can be used to understand the effect of the antimicrobial drug PGLa on the membrane. In Chapters 4 and 6 we described the construction and experimental measurements of a synthetic biological device called the pore formation measurement platform (PFMP), which was built out of artificial membranes. In this section we use the PFMP to study how an antimicrobial peptide forms pores in the membrane. Specifically, we construct a CGMD simulation model of the PFMP (Figure 10.8) to estimate the orientation, conformation, and oligomerization processes involved in creating a conducting PGLa pore in an archaebacterial membrane.

Recall that PGLa is a membrane-active antimicrobial peptide designed for killing "superbugs." PGLa kills the superbug by forming peptide ion channels in the superbug's membrane, which disrupts the operation of other ion channels in the membrane, resulting in cell death. Therefore, understanding the dynamics of pore formation of PGLa at the molecular level is of importance in molecular (in silico) drug screening.

In simple terms, this section gives an atomistic-to-observable model for how PGLa compromises the membrane of bacterial cells. The atomistic-to-observable model that we use is displayed in Figure 14.16. The model is composed of three levels of abstraction: a fractional-order macroscopic model given by (8.1) and (11.17), the generalized reaction-diffusion continuum model presented in §10.5.1, and a CGMD simulation model. The macroscopic model accounts for the diffusion-limited process at the electrode surface and predicts the membrane conductance $G_m(t)$ given the measured current response $I(t)$ from the PFMP. CGMD simulations are used to compute the diffusion coefficient D of surface-bound and transmembrane-bound peptides and are also used to gain insight into the dynamics of binding, translocation, and oligomerization required for pore formation. The computed diffusion coefficients are then used in a generalized reaction-diffusion model. The generalized reaction-diffusion equation is coupled to the surface reaction equations via a "Langmuir–Hinshelwood-like" equation classically used to describe surface binding of molecules. As conducting pores form in the tethered membrane, the conductance of the membrane will increase proportionally to the number of pores. This allows the concentration of surface-bound pores to be used to estimate the membrane conductance $\hat{G}_m(t)$. Validation of a proposed pore formation reaction mechanism is achieved when the experimentally measured conductance $G_m(t)$ and numerically computed conductance $\hat{G}_m(t)$ are in agreement.

14.9.1 Coarse-Grained Molecular Dynamics Simulation of Tethered Membrane Containing PGLa

Recall that in §10.5 we used the PFMP to model the dynamics of PGLa pore formation. However, the PFMP model is nonidentifiable in the sense that it cannot be used to differentiate between the following two possible pore formation reaction mechanisms:

Reaction Mechanism 1.

$$a \underset{k_d}{\overset{k_a^1}{\rightleftharpoons}} m_1 \overset{k_p}{\to} p_1, \qquad np_1 \overset{k_1}{\to} p_n, \qquad mp_n \overset{k_c}{\to} c,$$

14.9 Case Study: Atomistic-to-Observable Model PGLa Pore Formation

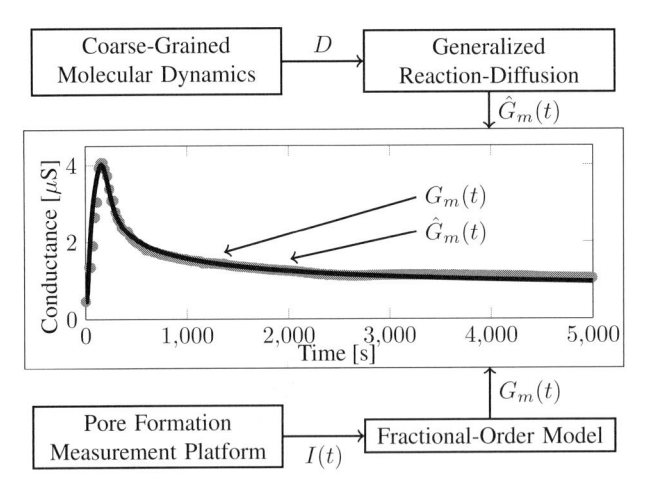

Figure 14.16 Schematic of the mesoscopic-to-macroscopic model. D is the diffusion coefficient of bound PGLa peptides, $\hat{G}_m(t)$ is the predicted conductance, $I(t)$ is the measured current from the PFMP (Figure 10.8), and $G_m(t)$ is the measured conductance.

Reaction Mechanism 2.

$$a \underset{k_d}{\overset{k_a^1}{\rightleftharpoons}} m_1 \overset{k_p}{\to} p_1, \qquad p_1 + p_1 \overset{k_1}{\to} p_2, \qquad mp_2 \overset{k_c}{\to} c,$$

where the reaction-rate parameters are defined following (10.33). Then the PFMP model and experimental measurements from the PFMP cannot be used to differentiate if reaction mechanism 1 or 2 resulted in the formation of the PGLa pores, for example, if all parameters between reaction mechanisms 1 and 2 are identical except for the constant k_1 and $n = 3$. If the rate constant k_1 in reaction mechanism 2 is set to 3/2 the power of the rate constant k_1 in reaction mechanism 1, then the measured current response would be identical for both reaction mechanisms.

CGMD simulations can be used to resolve the above nonidentifiability issue. CGMD (and also MD) can be used to estimate the rate constants of chemical reactions and the change in energy resulting from each step of the chemical reactions in reaction mechanisms 1 and 2. This information can then be included in the PFMP model to estimate if reaction mechanism 1 or 2 is in agreement with the experimental measurements. In this section we use the CGMD simulation model to gain insight into the reaction mechanism and dynamics leading to PGLa pore formation, and the diffusion coefficients D in (10.36). Recall from §10.5 that PGLa is a membrane-active antimicrobial peptide that provides a potential source for new antibiotics against increasingly common multiresistant pathogens (i.e., superbugs) such as methicillin-resistant *Staphylococcus aureus*. The PGLa will preferentially bind and form aqueous pores in the membranes of these superbugs, causing cell death.

Simulating the dynamics of PGLa binding, insertion, and oligomerization requires a spatial dimension of tens of nanometers and a simulation time horizon of several microseconds. With today's computers, all-atom MD simulations are practically limited to system sizes of several nanometers and tens of nanoseconds. However, CGMD simulations can be used to study the dynamics of PGLa pore formation because CGMD allows for a two- to three-order-of-magnitude increase in both system size and simulation time horizon compared to all-atom MD simulations. However, to ensure the CGMD simulations provide reliable trajectory estimates of the beads, we compare some of the results from CGMD simulations to the all-atom MD simulations. Note that the CGMD model is not utilized to study the formation process of aqueous pores (i.e., water-filled pores formed by ionic gradients); for details on the formation process of aqueous pores the reader is referred to Chapter 15, where we study the formation of aqueous pores using all-atom molecular dynamics simulations.

The PGLa peptide contains 21 residues with amino-acid sequence

$$\text{GMASKAGAIAGKIAKVALKAL-NH}_2.$$

As a result of recent advances in ^{2}H-NMR, ^{15}N-NMR, and ^{19}F-NMR spectroscopy it is known that the PGLa peptide has an α-helix configuration[40] when membrane bound. However, the transmembrane state of PGLa (i.e., that the long axis of the α-helix is parallel to the membrane normal) has not been detected experimentally at physiological temperatures. It is expected that PGLa monomers reside in the transmembrane state with the α-helix secondary structure. This is a common trait of similar α-helical monomers including alamethicin, maginin 2, and melittin. Additionally, it is known that amine terminus and carboxyl terminus of antimicrobial peptides such as PGLa are thermodynamically stable (i.e., they have a local energy minimum) when in contact with the surface of the membrane. Therefore, for all CGMD simulations, the secondary structure of the PGLa is constrained to have an α-helix configuration.

To construct the CGMD representation of PGLa, we initially construct the all-atom structure of PGLa in the α-helix configuration. The all-atom structure of PGLa is constructed using the software Molefacture contained in Visual Molecular Dynamics [172]. The secondary structure of the membrane-bound PGLa is defined by a pure α-helix.[41] The all-atom PGLa is coarse grained for use with the MARTINI force field using the protocol described in [284] with each CGMD bead representing approximately four heavy atoms. A schematic of the all-atom and coarse-grained structure of PGLa is provided in Figure 14.17. The membrane is modeled using 512 DphPC CGMD molecules. The parameters of the CGMD simulation model and initial conditions for surface binding, translocation of surface binding to transmembrane binding, and oligomerization are provided in §C.5.

---

[40] The α-helix configuration is the secondary structure of PGLa; see Chapter 2 for details.
[41] The pure α-helix configuration has angle parameters $\phi = -57°$ and $\psi = -47°$ which are associated with the Ramachandran (ϕ, ψ) plot. The Ramachandran plot illustrates the energetically favorable dihedral angles (ϕ, ψ) of the backbone. The angles $\phi = -57°$ and $\psi = -47°$ are energetically favorable for the following reasons: the dipoles of hydrogen-bonding backbone atoms is in near perfect alignment, side chains are separated sufficiently to minimize steric effects, and the side chains are positioned such that the van der Walls interaction of the side chains is favorable increasing the stability of the monomer.

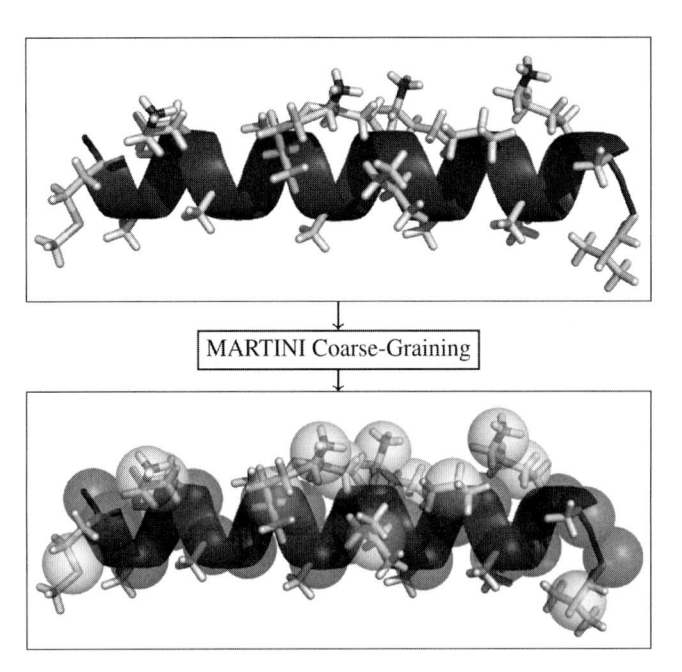

Figure 14.17 A color version of this figure can be found online at www.cambridge.org/engineered-artificial-membranes. Schematic of the all-atom structure of PGLa (GMASKAGAIAGKIAKVALKAL-NH$_2$) and the corresponding MARTINI coarse-grained structure constructed using the protocol in [284]. The PGLa backbone beads are displayed in red, and side chain beads in yellow.

14.9.2 Diffusion of PGLa and Membrane Properties from Coarse-Grained Molecular Dynamics

The surface-bound and transmembrane-bound diffusion coefficients of PGLa play a central role in the dynamics of PGLa pore formation in biological membranes (see (10.35) and (10.36)). To estimate these parameters we use CGMD simulations of PGLa. The diffusion coefficients for surface-bound and transmembrane-bound monomers, dimers, and trimers are provided in Table 14.4. As expected from the Einstein relation,[42] the diffusion coefficient of the PGLa protomers decreases as the number of monomers in each protomer increases. Interestingly the diffusion coefficients for the transmembrane protomers satisfy the "free-drain limit" [316] in which the diffusion coefficient satisfies $D_n = D_i/n$ within the error bounds. This effect has been observed for membrane-bound proteins using single-molecule fluorescence spectroscopy techniques [204]. We do not consider the diffusion of higher-order protomers because the CGMD model results are in agreement with the experimentally measured results for PGLa protomers containing up to three monomers.

---

[42] The diffusion coefficient of a particle is $D = \mu k_B T$, where μ is the mobility, k_B is Boltzmann's constant, and T is the temperature. For the diffusion of spherical particles, Stokes' law gives the mobility as $\mu = 1/(6\pi \xi r)$, where ξ is the viscosity, and r is the radius of the sphere. Larger spheres will have a lower diffusion coefficient.

Table 14.4 Diffusion coefficients of PGLa protomers ($\mu m^2/s$).

	Monomer	Dimer	Trimer
Transmembrane	91.2 ± 19.5	41.8 ± 16.0	26.25 ± 9.0
Surface Bound	127.5 ± 12.4	50.7 ± 14.4	21.0 ± 8.0

14.9.3 Surface Binding and Oligomerization of PGLa from Coarse-Grained Molecular Dynamics

In this section, CGMD simulations of PGLa are used to gain substantial insight into the mechanism of surface binding, the translocation of surface-bound peptides to the transmembrane state, and oligomerization. Surface binding and entry into the transmembrane state are key steps in the pore formation reaction mechanism described in §10.5.1.

Figure 14.18 presents the surface binding, translocation, and oligomerization processes of PGLa in a DphPC membrane from the CGMD model. An explanation of each reaction mechanism is provided below.

Surface binding of PGLa. To initialize the CGMD simulation for surface binding of PGLa, a PGLa monomer with an α-helix secondary structure is placed above the

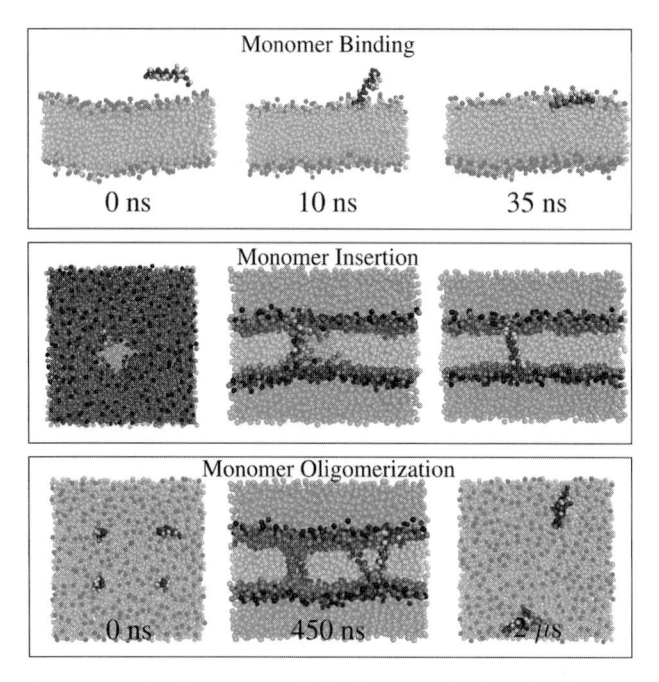

Figure 14.18 A color version of this figure can be found online at www.cambridge.org/engineered-artificial-membranes. Snapshots of CGMD bead positions for the surface binding, translocation, and oligomerization of PGLa in a DphPC membrane. The NC_3 bead is displayed in blue, the PO_4 bead in orange, the lipid tail carbons using green beads, the PGLa backbone beads in red, PGLa side chains using yellow beads, and water using light blue beads. The CGMD setup and parameters are defined in §C.5.

DphPC membrane. Note that in the analyte, the secondary structure of PGLa is a coil; therefore, we assume that the diffusion of the PGLa and the configurational change from the coil to the α-helix configuration has already occurred. As seen in Figure 14.18, the amine terminus of the PGLa monomer first binds to the surface of the DphPC membrane. At 20 ns the PGLa monomer pivots on the amine terminus and begins to embed itself into the membrane surface. The final surface-bound structure of the PGLa monomer is reached at 35 ns with the charged lysine residues pointing into the bulk electrolyte and the hydrophobic region in contact with the hydrophobic phytanyl tails of the DphPC membrane. The monomer remains in the membrane until the simulation horizon is reached at 1 μs. The computed tilt angle, defined as the angle between the helix long-axis vector and the membrane normal, of the PGLa monomer is $90\pm5°$ which is in excellent agreement with the ^{2}H-NMR and MD results of approximately $95°$.

Translocation to transmembrane state. How do the peptides transition from the surface-bound state to the transmembrane state?[43] Both the surface-bound state and the transmembrane state of PGLa are stable.[44] Therefore, random thermal fluctuations are not expected to provide sufficient energy to allow the PGLa to transition from the surface-bound state to the transmembrane state. Indeed, for a 1-μs CGMD simulation, we did not observe the surface-bound monomer (Figure 14.18) transition to the transmembrane state. If, however, a transient aqueous pore is present in the membrane (refer to Chapter 11), then the surface-bound PGLa can diffuse into the transient aqueous pore. An aqueous pore containing a surface-bound PGLa can be viewed as a metastable state between the stable surface-bound and transmembrane-bound states of PGLa. Figure 14.18 illustrates the transition from the surface-bound state to the metastable state where the PGLa has diffused into the aqueous pore. As the transient pore closes, the PGLa transitions from the metastable state to the stable transmembrane state. As expected, the PGLa remains in the transmembrane state for the remainder of the CGMD simulation.

Oligomerization. From the reaction mechanism (10.33), a necessary step for PGLa pore formation is an oligomerization process. Is it possible for PGLa peptides in the transmembrane state to oligomerize in the DphPC membrane? CGMD simulations can explain how PGLa oligomerizes in a DphPC membrane. We consider four PGLa monomers in a transmembrane state as illustrated in Figure 14.18. Initially the PGLa monomers diffuse into the membrane. As time progresses, the monomers form transmembrane dimers. The formation of the transmembrane dimers from the monomers is dependent on the orientation and diffusion dynamics of the peptides; as such the first dimer is formed at 150 ns, and the second at 450 ns. The formation of the two dimers occurs as a result of the amine terminus or carboxyl terminus interacting when two peptides come into close contact. The two transmembrane dimers remain stable for the

---

[43] In the surface-bound state the orientation of the α-helix of the PGLa peptide is perpendicular to the membrane normal. In the transmembrane state, the α-helix of the PGLa peptide is aligned to the membrane normal. These two orientations of the PGLa peptide are illustrated in Figure 10.8.

[44] A stable state corresponds to a configuration (position and momenta of atoms) that is at a local energy minimum.

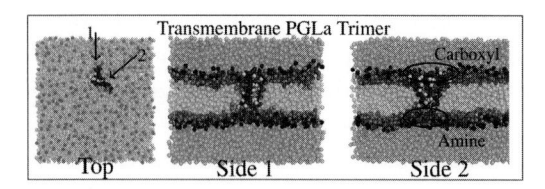

Figure 14.19 Snapshots of CGMD bead positions for the transmembrane PGLa trimer. To illustrate the PGLa trimer structure a top view and two side views are displayed. The CGMD setup and parameters are defined in §C.5.

duration of the CGMD simulation. Can these two dimers bind to form a quadramer complex? Over a 4-μs simulation the two dimers were not observed to form a quadramer complex. A possible explanation for the instability of the quadramer complex is the Coulomb repulsion force that results from the position of the +4-charged lysine residues in PGLa. In the PGLa dimer, the hydrophilic face of each PGLa monomer prefers to face the hydrocarbon interior of the membrane – this is the energy-favorable configuration of the PGLa monomers. The Coulomb energy required to form the quadramer complex is greater than the energy associated with the PGLa dimers. Therefore, the quadramer complex is not an energy-favorable configuration of the PGLa monomers.

What are the possible oligomerization steps necessary to form a transmembrane PGLa trimer? CGMD simulations can give insight into this. Once a PGLa dimer has formed (Figure 14.18), then it is possible for a transmembrane-bound PGLa monomer to bind with the dimer to form a transmembrane PGLa trimer. The resulting trimer is illustrated in Figure 14.19, where the results suggest that the amine terminus of the PGLa monomer stabilizes the trimer while the carboxyl terminus contributes negligibly to the binding of the three PGLa monomers. However, the carboxyl terminus may promote the passage of ions through the membrane by causing local membrane instabilities, increasing the conductance of the membrane. The formed PGLa trimer remains stable for the duration of the simulation.

The results in Figures 14.18 and 14.19 provide snapshots of a possible reaction mechanism (10.33) for PGLa. To estimate the associated rate constants in (10.33) requires experimental measurements from the PFMP and the dynamic model presented in §10.5.1.

14.10 Complements and Sources

CGMD simulations are used to generate the trajectory of a system of interacting beads in a specific statistical ensemble (temperature, pressure, and energy) and can be used to estimate several parameters of engineered membranes. Key challenges when performing CGMD simulations are the construction of the force field and definition of the atom-to-bead mapping, and validation of the results of the CGMD simulation. Below we discuss these challenges, and how the CGMD simulation results of cholesterol and PGLa were validated.

Statistical Mechanics and Water Density

In this chapter we used the Ornstein–Zernike equation (14.20) with the Percus–Yevick closure (14.21) to construct an analytical expression for the density profile of water at the bioelectronic interface. This was an example of the "integral equation approach" commonly used in statistical mechanics to construct equilibrium distributions of stochastic processes. The general relation between probability distribution functions in molecular dynamics is the Bogoliubov–Born–Green–Kirkwood–Yvon hierarchy. It is possible to construct the Ornstein–Zernike equation (14.20) from this hierarchy using the approximations in [198]. Additionally, the Bogoliubov–Born–Green–Kirkwood–Yvon hierarchy can be used to construct estimates of the density of water at the bioelectronic interface as detailed in [454, 455]. For further details on the use of statistical mechanics to estimate equilibrium distributions refer to [94, 368].

CGMD Simulations of Engineered Membranes

Several methods exist for constructing CGMD force fields and atom-to-bead mappings, including Boltzmann inversion, iterative Boltzmann inversion, inverse Monte Carlo, force matching, and relative entropy matching. In this chapter we used the MARTINI force field [267, 268] to perform all CGMD simulations of the engineered membrane. The MARTINI force field provides a sufficient approximation of the interactions between groups of atoms to allow its use for modeling the dynamics of engineered tethered membranes [160, 165, 245]. We constructed CGMD representations of the gold bioelectronic interface and the tethers as these are not included in the MARTINI force field. The bioelectronic interface was represented by a hydrophobic lattice of beads, and the eight ethylene glycol molecules of the tethers are represented by eight PEG beads. The interaction of the PEG beads is provided in [165, 232, 451]. To ensure the CGMD simulations provide reliable trajectory estimates of the beads, we compare some of the results from the CGMD simulation to the all-atom MD simulation results in [174, 196, 242, 375–378, 428]. For a general comprehensive treatment of CGMD and validation methods refer to [445].

CGMD simulations involve running classical MD simulation on beads (rather than individual atoms). Recall from §14.2.1 that the construction of beads and interactions between beads (the CGMD force field) are based on the results of MD simulations. Therefore, any physical processes neglected by the MD simulation will also be neglected in the associated CGMD simulations. This is important to consider when constructing CGMD models of engineered membranes. For example, MD simulations for modeling the electrode-electrolyte interface is still an active area of research as quantum-mechanical effects are present which result from hydrogen bonding [440, 462]. From the ab initio MD simulations reported in [77], there is a wetting layer of water at the gold surface. However, beyond 3.2 Å, one of the hydrogen molecules in water points toward the gold surface, causing the surface to by hydrophobic. Therefore, in our CGMD simulation model we constructed the bioelectronic interface to be hydrophobic. The dynamics of water at the bioelectronic interface have different density and diffusion compared to that in the bulk solution [19, 147, 484]. From MD simulations, hydrogen-bond breaking and cooperative rearrangement of water in proximity to the bioelectronic interface cause an increase in the lateral diffusion of water compared to the bulk diffusion [363]. This result is consistent with the results from the CGMD simulations of water at the bioelectronic interface.

Control of Engineered Membrane Properties by Sterol Inclusions
Using CGMD simulations, this chapter studied how cholesterol affects the dynamics of lipids, biomechanical properties of the membrane, and electroporation characteristics, as observed experimentally in [175, 308]. Studies of cholesterol span several decades and have included experimental measurements and MD simulations. For a review the reader is referred to [175, 308] and references therein. A substantial amount of work has focused on lipids containing palmitoyl-oleoyl (PO), dioleoyl (DO), distearoyl (DS), and dipalmitoyl (DP) lipid tails. In [37, 64, 119, 340] it is shown that increasing the cholesterol content in POPC, DOPC, DSPC, and DPPC membranes results in an increase in membrane stability, that is, a resistance to membrane defects. Increasing the cholesterol in DOPC membranes reduces the line tension of the membranes [42]. In DPPC membranes, increasing the concentration of cholesterol causes an increase in membrane thickness and a decrease in the lateral diffusion of lipids [37, 159]. CGMD simulations have been applied to study the effects of cholesterol on POPC, DOPC, and DPPC membranes [91]. The CGMD results show that for cholesterol concentrations between 0 and 40 percent, the membrane thickness increases, however, for 50 percent the membrane thickness decreases as a result of the interdigitation between lipid tails.

Studies on the effect of cholesterol on archaebacterial membranes are important for the design of biosensors and engineered membrane platforms. Experimental measurements on black lipid membranes [430] suggest that low concentrations of cholesterol increase archaebacterial membrane stability; however, above 20 percent the membrane stability begins to decrease. Using CGMD simulations and experimental measurements from the EMP, we illustrated how the concentration of cholesterol can be used to control the conductance, thickness, diffusion of lipids, line tension, and surface tension of an archaebacterial membrane.

CGMD Simulations for Resolving the Reaction Mechanism for Peptide Ion Channels
CGMD simulations can be used to resolve the nonidentifiability issue of the reaction mechanism associated with the formation of peptide ion channels as discussed in §10.5.2. This requires combing the results of experimental measurements, MD simulations, and CGMD simulations to estimate the stable states of the monomer(s) leading to the formation of a peptide ion channel. Recall that MD simulations cannot be used in isolation for this task because the simulation time horizon of MD simulations is two to three orders of magnitude (both spatially and temporally) smaller than CGMD simulations [268, 284, 321].

For PGLa pore formation, the CGMD simulation results suggest that there are four stable states for PGLa, namely, the membrane-bound state, transmembrane state, dimer state, and trimer state. How do these results compare with experimental measurements and all-atom MD simulations? Using MD simulations with umbrella sampling for similar-length (19-residue) α-helix peptides, the results in [70] suggest that the PGLa is in a thermodynamically stable configuration when in the membrane-bound and transmembrane-bound states. Additionally, the MD simulation results in [431] illustrate that the PGLa dimer is also a stable configuration. The secondary structure of PGLa monomers is α-helical as measured experimentally from ^{2}H-NMR, ^{15}N-NMR, and ^{19}F-NMR spectroscopy [7, 137, 357, 406, 407]. Additionally, the computed tilt angle of the membrane-bound PGLa monomer from CGMD simulations is 90±5° which is in excellent agreement with the ^{2}H-NMR results [427] and MD simulation [431] of approximately 95°. The transmembrane state of PGLa has not been observed by NMR spectroscopy at physiological temperatures possibly as a result of the PGLa pores being transient [427]. In a gel-phase DMPC/1,2-dimyristoyl-sn-glycero-3-phosphorylglycerol (DMPG) bilayer at temperatures below 15 °C, NMR measurements show that PGLa is in a transmembrane state with a tilt angle of approximately 180° [7]. This is in agreement with the computed tilt angle from the CGMD simulations of 168±5° – the difference in angle is a result of the gel-phase DMPC/DMPG bilayer having a larger membrane thickness than the DphPC membrane, which is in the liquid phase. Additionally, NMR spectroscopy [7, 137, 357, 358, 406, 407] has also been used to experimentally measure the PGLa dimer structure, which is consistent with the CGMD simulation results.

Given that the CGMD simulation results for PGLa are consistent with both MD simulation results and experimental measurements, we conclude that the CGMD simulations can be used to infer the reaction mechanism leading to PGLa pore formation.

14.11 Closing Remarks

This chapter has discussed how coarse-grained molecular dynamics simulations can be used to understand the dynamics of an engineered membrane at a molecular-scale resolution. Specifically we used CGMD to compute key biological parameters such as the diffusion tensor and particle density of water, membrane thickness, lipid diffusion, defect density, free energy of lipid flip-flop, and the interaction of cholesterol and PGLa with the tethered membrane. To link the CGMD simulation results to experimental measurements we provided an atomistic-to-macroscopic model of the tethered membrane and pore formation measurement platform introduced in Chapter 6. From the CGMD simulation results, two important conclusions were found. First, the bioelectronic interface (gold electrode) does not contribute to the membrane dynamics and biomechanics. Therefore, properties of the tethered membranes (membrane thickness, lipid diffusion, defect density, free energy of lipid flip-flop, etc.) are independent of the bioelectronic interface. Second, the density of tethers does contribute to the water and membrane biomechanics. For example, the free energy of lipid flip-flop increases for increasing tether density.

From a pedagogical perspective, we strongly recommend that the reader implement the CGMD simulation code in the Internet supplement. This includes constructing CGMD simulations and computing parameters (diffusion coefficients, density, etc.) from the CGMD simulations that are used in continuum models. For system sizes of a few hundred lipids, a standard desktop computer can be used to perform the CGMD simulations presented in this chapter (except for umbrella sampling, which typically requires running hundreds of CGMD simulations). Given the variety of possible engineered membrane architectures, the ability to construct and run CGMD simulations to evaluate important continuum parameters is an important step in constructing accurate models of engineered tethered membranes.

There still remain several unresolved important questions related to constructing CGMD simulations of engineered membranes which can be the subject of future research. For example, how should the boundary of the electrode-electrolyte interface be defined? This is a nontrivial question since the interaction of water with the electrode surface involves hydrogen-bonding dynamics between water and the gold surface. If a nonzero excitation potential is used then free electrons are present in the gold electrode which attract the anions and cations in the electrolyte solution, causing the formation of electrical double layers. These types of complex charging dynamics are still an open problem in both molecular dynamics and density functional theory. Possible methods for coarse-graining charged electrodes are provided in [132, 208]. Additionally, methods to incorporate polarization effects into the CGMD model are also an active area of research; refer to [277, 443] for details. Although these models account for the polarization effects of water, they cannot be used to study reaction processes that rely on the characteristics of hydrogen bonding, or for studying electron shell dynamics, which are important for estimating the spatially dependent dielectric permittivity, as these are quantum-mechanical effects.

15 All-Atom Molecular Dynamics Simulation Models

15.1 Introduction

This chapter delves into the lowest level of abstraction for modeling engineered membranes considered in this book: namely, all-atom molecular dynamics (MD) simulation models.[1] As in Chapter 14, we are interested in the structure-function relationship of an engineered membrane: that is, given the molecular structure, how can we predict the function of the membrane? To give some perspective, Chapter 14 illustrated how coarse-grained molecular dynamics (CGMD) simulations can be used to estimate important biological parameters such as lipid diffusion and geometric properties of tethered membranes. Although the CGMD model is versatile, it suffers from coarse spatial resolution.[2] To increase the spatial resolution, all-atom MD can be utilized at increased computational cost. In this chapter we describe how all-atom MD can be used to estimate important biological parameters in engineered membranes. The goal is to link the results from all-atom MD to parameters in the fractional-order macroscopic model (8.1) and (11.17) (see Figure 15.1). This allows experimental measurements (discussed in Chapter 6) from the tethered membrane to be used to validate the results from the all-atom MD models. The important biological parameters and processes covered in this chapter include the dynamics of aqueous pore formation, electrostatic potential, intrinsic membrane dipole potential, membrane capacitance, the ion permeation dynamics of the gramicidin ion channel, and the dissociation dynamics of gramicidin dimers.

15.2 Basics of Molecular Dynamics[3]

Molecular dynamics is a computer simulation model for the time evolution of a system of interacting particles based on Newtonian (classical) mechanics. Assuming n

---

[1] MD is a huge area of research; our discussion in this chapter is highly stylized and focuses only on giving insights into MD for engineered membranes. For a general description of MD, please refer to books such as [127, 144, 237, 339] or websites such as www.gromacs.org/, www.charmm-gui.org/, and www.ks.uiuc.edu/Training/Tutorials/namd-index.html.

[2] Recall that CGMD models the dynamics of groups of atoms called beads, not the dynamics of individual atoms. Therefore, CGMD is an approximation of full-atom MD.

[3] This section can be omitted by readers who are familiar with MD simulation models. Readers with no prior experience in MD should consult the references and webpages mentioned at the end of this chapter.

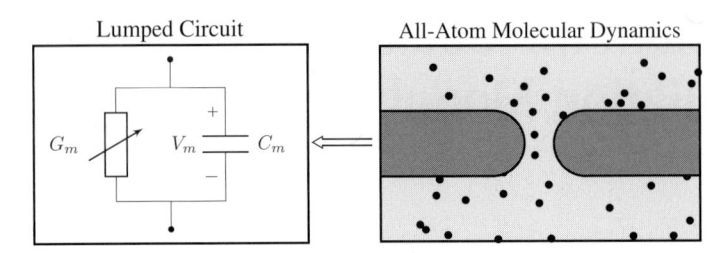

Figure 15.1 Schematic of the relationship between the all-atom molecular dynamics model and lumped circuit parameters in the fractional-order macroscopic model (8.1). The membrane conductance G_m is dependent on the flow of ions through aqueous pores in the tethered membrane, the capacitance C_m of the membrane is dependent on the charge accumulation of ions adjacent to the membrane surface, and the transmembrane potential V_m is dependent on the difference in the number of ionic species on either side of the membrane. The electrolyte is illustrated in light gray, ions by the black dots, and membrane by the dark gray region in the all-atom molecular dynamics schematic.

interacting particles, the equation of motion of the ith particle is given by

$$m_i \ddot{x}_i = F_i, \qquad F_i = -\frac{\partial}{\partial x_i} U(x), \qquad i \in \{1, \ldots, n\}. \tag{15.1}$$

In (15.1), m_i is the mass of the ith particle, x_i is the position of the ith particle in three-dimensional Euclidean space, $\ddot{x}_i$ is the acceleration of the particle, F_i is the total force acting on the particle, $U(x)$ is the potential energy, and n is the total number of particles in the system. The potential energy U is dependent on all n particles in the system, and consequently f_i is also a function of the position of all the particles in the system. Regarding the particle dynamics (15.1) we wish to stress two important aspects:

(i) Equation (15.1) is a nonlinear ordinary differential equation – nonlinear due to the presence of the $dU(x)/dx_i$ term. Therefore, to preserve the total energy (potential plus kinetic) of the overall system, careful numerical implementation is required; see §15.2.3.
(ii) Although the evolution of the particle motions as described by Newton's equation of motion (15.1) is deterministic (nonrandom), the trajectories of the particles are sample paths of random processes since the initial conditions of the particle velocities are variables generated randomly from the Maxwell–Boltzmann distribution.[4] The initial positions of the particles can also be generated

---

[4] The Maxwell–Boltzmann probability density function is

$$f(v) = \sqrt{\left(\frac{m}{2\pi k_B T}\right)^3} 4\pi v^2 \exp\left(-\frac{mv^2}{2k_B T}\right), \tag{15.2}$$

where v is the velocity, m is the mass of the particle, k_B is Boltzmann's constant, and T is the temperature. Generating initial velocities from the Maxwell–Boltzmann distribution can lead to a nonzero total momentum of the system creating a systematic translational drift. To remove this drift it is common to shift the velocities to ensure the net momentum of the system is zero. Formally, the velocity of each particle is shifted by

$$v_i \leftarrow v_i - \frac{\sum_{i=1}^{n} m_i v_i}{n m_i}. \tag{15.3}$$

where n is the total number of particles in the MD simulation.

randomly.[5] §15.2.2 discusses the sources of randomness when performing MD simulations.

When performing MD simulations, it is desirable to mimic real-world experimental conditions. This includes simulations in which macroscopic parameters such as temperature, pressure, volume, particle number, and system energy are held constant. These quantities are not controlled by the potential function $U(x)$ in (15.1). To account for these conserved quantities requires that simulations are performed in specific statistical ensembles.[6] Another important consideration when performing MD simulations is specifying the initial conditions of the particles and the boundary conditions used to solve (15.1).

Performing MD simulations requires (1) the definition of the potential energy function $U(x)$ in (15.1), (2) the conserved macroscopic quantities of the system, and (3) the initial and boundary conditions, and numerical methods for evaluating (15.1). CHARMM, GROMOS, and AMBER are popular force fields used for MD simulations, and we use these force fields for engineered membranes.

In the remainder of this section we briefly discuss the construction of the potential energy function $U(x)$, conserved macroscopic quantities, and initial and boundary conditions necessary for constructing and running MD simulations with a focus on engineered artificial membranes.

15.2.1 Potential Energy Functions

The potential energy function $U(x)$ in (15.1) encodes all the information about the particle-particle interactions in the system. Note that a particle may be a single atom or a collection of atoms. In MD simulations there are two classes of particle-particle interactions that are considered, namely, bonded interactions and nonbonded interactions. Qualitatively, the bonded interactions govern the structure and dynamics of a molecule (multiparticle) structure, while the nonbonded interactions define the interaction of particles with all the other particles in the system. The total potential energy function of the system is given by the sum of nonbonded and bonded interactions,

$$U(x) = U_{\text{nb}}(x) + U_b(x), \tag{15.4}$$

which are defined below.

---

[5] For the engineering and applied math reader: The random placement of molecules into an MD simulation can be considered a "knapsack problem" in which we have a defined simulation cell size, and the goal is to fit as many molecules into the simulation cell while ensuring the ratio between molecules is maintained. For example, if we wish to have a simulation cell of size V that only contains water and lipids with a ratio of 10 water molecules for every lipid, this can be solved using the tools of combinational optimization designed for knapsack and stochastic knapsack problems.

[6] A statistical ensemble is the collection of all possible system states which have different microscopic states (for example, particle position and velocity) but identical macroscopic or thermodynamic properties (for example, temperature or pressure).

Nonbonded interactions. Nonbonded interactions in MD model the van der Waals forces between particles, and also the effects of externally applied potentials. The nonbonded potential is defined by

$$U_{\text{nb}}(x) = \sum_i u(x_i) + \sum_i \sum_{j>i} u(x_i, x_j) + \cdots, \tag{15.5}$$

where $u(x_i)$ is the potential on a particle x_i from an externally applied potential field or other effects resulting from the perimeter of the simulation cell, and potential $u(x_i, x_j)$ accounts for all pairwise interactions between particles. Note that higher-order interactions can also be included; however, for most MD simulations only pairwise interactions are considered for nonbonded interactions.

Bioelectronic interface. In Chapter 14 we discussed how CGMD can model the bioelectronic interface in engineered membranes. Here, we describe how all-atom MD simulations can model the bioelectronic interface. Just as with CGMD simulations, an externally applied potential field $u(x_i)$ can be used to model the gold bioelectronic interface and the application of an externally applied electric field. A popular method for simulating a bioelectronic interface is to define one of the simulation cell walls as the cathode surface, and the opposite cell wall as the anode surface. The cathode and anode are represented by a flat surface of equally spaced atoms with each having a mass and an associated charge density. Since the surface is at a fixed position, the interaction of the entire surface with any particle can be encoded into the potential $u(x_i)$. The potential induced from these charges must be estimated from the particle trajectories. Another method is to dynamically adjust the charges of the surface to maintain a constant potential in the simulation cell [149, 438]. Notice that accurate modeling of the electrode-electrolyte interface is still an active area of research from both modeling and experimental points of view.

A popular method to model van der Waals forces is to use the Lennard-Jones potential and Coulomb's law, both of which provide pairwise interactions between particles. The Lennard-Jones potential $u_{\text{LJ}}(x_i, x_j)$ and Coulomb potential $u_{\text{C}}(x_i, x_j)$ are given by

$$u_{\text{LJ}}(x_i, x_j) = 4\epsilon_{ij} \left[\left(\frac{\sigma_{ij}}{x_{ij}}\right)^{12} - \left(\frac{\sigma_{ij}}{r_{ij}}\right)^{6} \right]$$

$$u_{\text{C}}(x_i, x_j) = \frac{q_i q_j}{4\varepsilon_{\text{eff}}\varepsilon_0 x_{ij}}, \tag{15.6}$$

where $x_{ij} = ||x_i - x_j||$ is the Euclidean distance between particles x_i and x_j, ϵ_{ij} is the well depth of the Lennard-Jones potential, σ_{ij} is the diameter of the Lennard-Jones potential, q_i is the charge, and ε_{eff} is the effective dielectric constant. The effective dielectric constant accounts for the screening effects[7] that water imposes between charged particles. If water molecules are included in the MD simulation then $\varepsilon_{\text{eff}} = 1$ because the water molecules provide the electrostatic screening.

---

[7] Recall from §3.6 that the screening effect results from the reorientation of water dipoles reducing the overall electric field in the system.

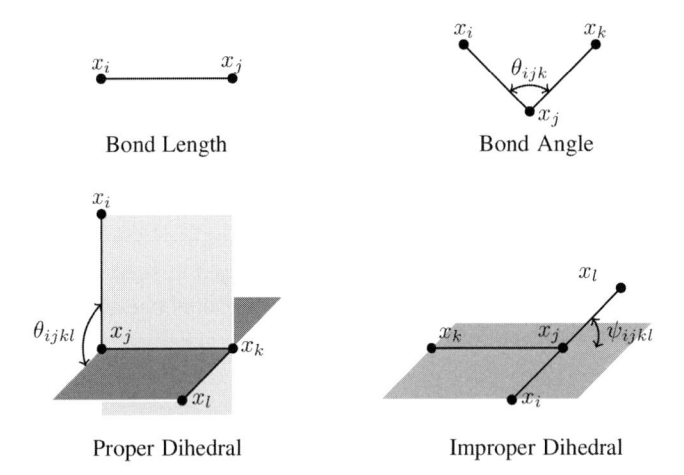

Figure 15.2 Schematic of the four common bonded potentials in MD simulations: bond length, bond angle, proper dihedral, and improper dihedral. The parameters x_i, x_j, x_k, and x_l are the positions of the particles, θ_{ijk} is the bond angle, θ_{ijkl} is the torsion angle, and ψ_{ijkl} is the improper-dihedral angle.

Bonded interactions. The key to bonded interactions is that they preserve the relative positions of small groups of atoms. For example, in water the angle formed between the two hydrogen atoms remains at approximately 104.5°, and the hydrogen and oxygen atoms remain at a distance of approximately 0.95 Å. Larger molecules will typically have several such "shape-preserving" characteristics for preserving the relative positions of atoms in small groups. That is, a large molecule consists of the same geometric features of smaller multiparticle molecules, but it combines these smaller molecules in different ways. There are four main geometric properties of bonds used in MD simulations: bond length u_{bl}, bond angle u_{ba}, proper dihedral u_{pd}, and improper dihedral u_{id}. By superposition, the total potential due to bonding interactions is

$$U_b(x) = \sum_{ij} u_{bl} + \sum_{ijk} u_{ba} + \sum_{ijkl} u_{pd} + u_{id}. \tag{15.7}$$

The above bonding interactions are only considered for particles that are contained in the same molecule. These important geometric properties are illustrated in Figure 15.2.

Bond Length. The bond-length potential energy $u_{bl}(x_{ij})$ is a parameter that characterizes the covalent bond between the two particles x_i and x_j. A commonly used potential function for the bond-length potential is the harmonic potential which is based on Hooke's law. The bond-length potential is given by

$$u_{bl}(x_{ij}) = \frac{1}{2} k_{ij} (x_{ij} - x_{0,ij})^2, \tag{15.8}$$

where k_{ij} is the force of the covalent bond (spring constant), and $x_{0,ij}$ is the equilibrium distance of the covalent bond between particles x_i and x_j.

Bond angle. The bond-angle potential $u_{ba}(x_i, x_j, x_k)$ is designed to reproduce the geometry between the three particles x_i, x_j, and x_k that results from covalent bonds. The bond-angle potential $u_{ba}(x_i, x_j, x_k)$ is commonly modeled using a harmonic potential:

$$u_{ba}(x_i, x_j, x_k) = \frac{1}{2}k_{ijk}(\theta_{ijk} - \theta_{0,ijk})^2, \qquad (15.9)$$

where k_{ijk} is the force constant, and $\theta_{0,ijk}$ is the equilibrium angle between the covalent bonds between atoms x_i, and x_j, and x_j and x_k. An example where a bond-angle potential would be used is with water, which has an equilibrium angle of approximately 104.5°.

Proper dihedral. The proper-dihedral potential is used to account for the energetics of hydrocarbon particles that induce torsional constraints between particles. As an example, consider the butane molecule C_1–C_2=C_3–C_4. The proper-dihedral angle is used to assign a higher energy to the cis conformation (C_1 and C_4 being on the same side). The origin of the proper-dihedral potential is not completely understood, but repulsive interactions between overlapping covalent-bond electron orbitals and steric clashes between atoms (such as C_1 and C_4 in butane) appear to be contributing factors. The proper-dihedral potential $u_{\text{pd}}(x_i, x_j, x_k, x_l)$ is given by

$$u_{\text{pd}}(x_i, x_j, x_k, x_l) = k_{ijkl}\big(1 + \cos(n_{ijkl}\theta_{ijkl} - \theta_{0,ijkl})\big), \qquad (15.10)$$

where k_{ijkl} is the proper-dihedral force constant, n_{ijkl} is the multiplicity, which indicates the number of minima as the bond is rotated through 360°, and $\theta_{0,ijkl}$ is the equilibrium torsion angle between the plane going through the particles x_i, x_j, and x_k and the plane going through the particles x_j, x_k, and x_l.

Improper dihedral. Improper-dihedral potentials are used to ensure that the correct chirality of the particles is selected. Note that a chiral molecule is nonsuperposable on its mirror image. An example of this is a carbon atom attached to four different atoms or molecules. The improper-dihedral potential $u_{\text{id}}(x_i, x_j, x_k, x_l)$ is given by

$$u_{\text{id}}(x_i, x_j, x_k, x_l) = l_{ijkl}(\psi_{ijkl} - \psi_{0,ijkl}), \qquad (15.11)$$

where l_{ijkl} is the force of the improper-dihedral angle, and $\psi_{0,ijkl}$ is the equilibrium improper-dihedral angle between the plane going through the particles x_i, x_j and x_k and the line going through x_j and x_l.

The parameters of the nonbonded potentials (15.6) and bonded potentials in (15.8)–(15.11) are estimated from experimental NMR and X-ray measurements, as well as simulation results from quantum mechanics. Note that only the commonly used potentials for MD have been reported here. In the case for which improved accuracy is required, refined potentials can then be constructed using the tools of quantum mechanics. This is still an active area of research and new force fields are being developed.

15.2.2 Macroscopic Parameters and Statistical Ensembles

Stochasticity of MD. MD simulations are performed by numerically integrating Newton's equations of motion (15.1) over time to obtain the particle trajectories $x(t)$. Even though Newton's equations are deterministic, the trajectories generated by MD are stochastic since the initial conditions for the position and velocity are random variables. In more detail:

- The initial positions of atoms may be chosen randomly via Monte Carlo methods in which molecules (e.g., lipids, water) are randomly inserted, removed, or translated in the MD simulation volume.
- The initial velocities of the atoms must be such that they comprise a desired temperature, and the total linear momentum of the atoms is zero. The initial velocities of the atoms in MD are generated by sampling from a Maxwell–Boltzmann distribution for the desired temperature.
- The thermostat (used to control the macroscopic temperature of MD simulations), such as the Anderson thermostat, also introduces randomness. The Anderson thermostat selects atoms and then sets the selected atom's velocity to the sample generated from a Maxwell–Boltzmann distribution.

Therefore, initializing and running the same MD simulation twice will result in different trajectories for the atoms. The key idea in MD simulations is that, although the trajectories are different, the estimated macroscopic parameters from the MD simulations are identical. That is, the probability distribution associated with the samples generated from different MD simulations converges to an identical stationary joint probability density function of the atom positions and momenta.

Ensemble and time average. The macroscopic parameters of interest are energy E, temperature T, and pressure P. To compute these macroscopic parameters via MD simulations requires that we have a relation between *ensemble averages* (the expected value) and *time averages*, as we now discuss.

Consider a system with n atoms. Then there is a total of $6n$ degrees of freedom that define the kinetic state, with $3n$ values associated with the positions of all the atoms and $3n$ associated with the momenta of all the atoms. Let the column vector $\boldsymbol{x} = [x_1, x_2, \ldots, x_n] \in \mathbb{R}^{3n}$ denote the positions of the n atoms and the column vector $\boldsymbol{p} = [m_1 v_1, m_2 v_2, \ldots, m_n v_n] \in \mathbb{R}^{3n}$ denote the momenta of the n atoms where m_i and v_i are the mass and velocity of atom $i \in \{1, \ldots, n\}$. Each position $\boldsymbol{x}$ and momentum $\boldsymbol{p}$ vector defines a point $(\boldsymbol{x}, \boldsymbol{x})$ in the $6n$-dimensional phase space.

We use the notation $A(\boldsymbol{x}, \boldsymbol{p})$ to denote a generic macroscopic parameter (such as energy, pressure, or temperature) that is dependent on the position and velocity of the particles. Then, evaluating the expectation of A requires computing the multidimensional integral

$$E[A(\boldsymbol{x}, \boldsymbol{p})] = \int_{\mathbb{R}^{3n}} \int_{\mathbb{R}^{3n}} A(\boldsymbol{x}, \boldsymbol{p}) f(\boldsymbol{x}, \boldsymbol{p}) \, d\boldsymbol{x} \, d\boldsymbol{p}. \quad (15.12)$$

Here $f(\boldsymbol{x}, \boldsymbol{p})$ is the stationary joint probability density function of positions and momenta of the n-particle system. The expression (15.12) is the ensemble average of the macroscopic parameter A. There are two issues which make the direct evaluation of (15.12) intractable:

(i) The stationary probability density function $f(x, p)$ may not be known.[8] In some cases, the stationary distribution is specified as follows. The stationary distribution of the NVT (constant number of particles, N, volume V, and temperature T) ensemble is

$$f(x, p) \propto e^{-\frac{1}{k_B T} H(x, p)}, \qquad (15.13)$$

where $H(x, p)$ is the Hamiltonian (total energy) of the system and T is the temperature. The stationary distribution of the NPT (constant number of particles, N, pressure P, and temperature T) ensemble is

$$f(x, p, v) \propto e^{-\frac{1}{k_B T} H(x, p) + Pv}, \qquad (15.14)$$

where T is the temperature, P is the pressure, and $v \in [0, \infty]$ is the volume of the system.

(ii) Even for cases where the stationary distributions are known, computing (15.12) requires the numerical evaluation of a $6n$-dimensional integral, which is numerically intractable if the number of particles, n, is large.

The main idea behind stochastic simulation (and MD) is that, instead of computing $E[A(x, p)]$ using the ensemble average (15.12), we use the time average of the macroscopic parameter $A(x, p)$. The instantaneous value of the macroscopic parameter $A(x(t), p(t))$ at time t can be computed from the microstate $x(t), p(t)$ (position and velocity of all particles at time t). If we generate samples over an interval of time from $t = 0$ to $t = \tau$, then the microstate $x(t), p(t)$ changes as a result of Newton's equations of motion (15.1). Therefore, a *time average* of the macroscopic parameter $A(x(t), p(t))$ can be evaluated as follows:

$$\frac{1}{\tau} \int_{\xi=0}^{\tau} A(x(\xi), p(\xi)) \, d\xi. \qquad (15.15)$$

If we choose τ in (15.15) to be sufficiently large (or infinite), then the microstates $x(t), p(t)$ visited over the time interval τ will traverse all the possible states of the phase space of the system. Therefore, as $\tau \to \infty$, we have

$$E[A(x, p)] = \lim_{t \to \infty} \frac{1}{\tau} \int_{t=0}^{\tau} A(x(t), p(t)) \, dt = \int_{\mathbb{R}^{3n}} \int_{\mathbb{R}^{3n}} A(x, p) f(x, p) \, dx \, dp \qquad (15.16)$$

---

[8] For the reader familiar with Markov processes, the evolution of the positions and momenta of the n particles constitutes a Markov process over a continuum state space. Even though the transition dynamics are known, computing the stationary distribution is intractable. The existence of a stationary distribution requires the Markov process to have stability (recurrence) properties; see [276] for Liapunov-function-type conditions to ensure this.

with probability 1. This useful result is known as the *ergodic hypothesis* in statistical mechanics. The ergodic hypothesis states that, for a sampling time $\tau \to \infty$, the time average (15.15) and ensemble average (15.12) of the macroscopic parameter $A(\boldsymbol{x}, \boldsymbol{p})$ are equal.

Given the ergodic hypothesis (15.16), we can now estimate the macroscopic parameter $A(\boldsymbol{x}, \boldsymbol{p})$ using results from MD simulations. The results of the MD simulations are the trajectories of position and momenta of the particles given by $\{\boldsymbol{x}(t_k), \boldsymbol{p}(t_k)\}_{k=1}^K$ at discrete time points t_k over a simulation time horizon that consists of K discrete time points. The instantaneous value of the macroscopic parameter $A(\boldsymbol{x}, \boldsymbol{p})$ at the t_k is given by $A(\boldsymbol{x}(t_k), \boldsymbol{p}(t_k))$. Therefore, we can approximate the time average of the macroscopic parameter by evaluating[9]

$$\boldsymbol{E}[A(\boldsymbol{x}, \boldsymbol{p})] \approx \frac{1}{K} \sum_{k=1}^{K} A(\boldsymbol{x}(t_k), \boldsymbol{p}(t_k)). \tag{15.17}$$

Computing the temperature, pressure and energy using MD simulations. Given that the time average (15.17) is a useful approximation to the intractable ensemble average (15.12), the idea behind MD is to compute the macroscopic parameters of temperature, pressure, and energy using (15.17).

The total energy of the system is given by the sum of potential and kinetic energy. For a set of n particles with positions $\boldsymbol{x}$ and velocities $\boldsymbol{v}$, the total energy of the system at time t_k is given by

$$E(t_k) = U(\boldsymbol{x}(t_k)) + \frac{1}{2} \sum_{i=1}^{n} m_i |v_i(t_k)|^2, \tag{15.18}$$

where $U(x(t_k))$ is the potential energy defined by (15.18), and the second term is the kinetic energy of the particles with $v_i(t_k)$ the velocity vector of the ith particle and $\|\cdot\|$ the Euclidean norm. In MD simulations the total energy, or just the potential energy, is often used to estimate if a system has been sufficiently equilibrated to begin sampling the trajectory. Note that the initial conditions of nearly all MD simulations have both the particle coordinates and velocities placed randomly. With a random initialization the total energy of the system will be quite high; however, after several time steps the total energy should drop and plateau to an equilibrium value.

The instantaneous temperature of the system is

$$T(t_k) = \frac{1}{k_B N_{\text{df}}} \sum_{i=1}^{n} m_i |v_i(t_k)|^2, \tag{15.19}$$

where k_B is Boltzmann's constant, and N_{df} is the number of degrees of freedom of the system. The instantaneous temperature (15.19) is constructed using the equipartition theorem[10] from statistical mechanics, which relates the average kinetic energy to each

---

[9] The *strong law of large numbers* implies that the time average of $A(\boldsymbol{x}, \boldsymbol{p})$, given by the right-hand side of (15.17), converges with probability 1 to $\boldsymbol{E}[A(\boldsymbol{x}, \boldsymbol{p})]$ as the time horizon $K \to \infty$.
[10] The average kinetic energy of each degree of freedom N_{df} is given by $k_B T/2$, where T is the temperature and k_B is Boltzmann's constant.

degree of freedom of the ith particle. Formally, the equipartition theorem states that

$$\frac{N_{\text{df}} k_B T}{2} = \frac{1}{2} \sum_{i=1}^{n} m_i E\left[|v_i(t_k)|^2\right]. \qquad (15.20)$$

The instantaneous temperature (15.19) is obtained by solving for T and removing the expectation operator in (15.20); that is, the instantaneous velocity is used instead of the ensemble-average velocity. In MD simulations the degrees of freedom are given by $N_{\text{df}} = 3n - N_c - 3$, where N_c accounts for the number of constraints on the particles, and the subtraction of 3 is to account for the fact that the three center-of-mass velocities are constants of the motion and therefore reduce the degrees of freedom. A common constraint imposed when using MD for study engineered artificial membranes is to impose a kinematic constraint that restricts the movement of electrode atoms to a plane (two-dimensional motion). For every particle that is restricted to the plane, this will reduce the degrees of freedom by 1. Notice that the instantaneous temperature $T(t_k)$ can vary between each time step; however, if the system is designed to have a constant temperature, then the time average of the temperature will be a constant T.

The pressure (typically atmospheric) in MD simulations is represented by a pressure tensor $\boldsymbol{P} \in \mathbb{R}^{3 \times 3}$ which defines the pressure resulting from forces in all directions. For example, given a force directed in the x-coordinate direction, the pressure on the plane defined by the constant y coordinate is given by P_{xy}. The collection of all such pressures is contained in the pressure tensor $\boldsymbol{P}$. Note that the diagonal elements of $\boldsymbol{P}$ define the stress, and the off-diagonal elements define the shear. The instantaneous pressure tensor is

$$\boldsymbol{P}(t_k) = \frac{1}{V}\left(\sum_{i=1}^{n} m_i v_i(t_k) \otimes v_i(t_k) + \sum_{i}\sum_{j<i}(x_i(t_k) - x_j(t_k)) \otimes f_{ij}(t_k)\right)$$

$$= \frac{1}{V}\left(\sum_{i=1}^{n} m_i v_i(t_k) \otimes v_i(t_k) + x_i(t_k) \otimes f_i(t_k)\right) \qquad (15.21)$$

$$f_{ij}(t_k) = -\frac{\partial}{\partial x_{ij}}\left(u_{\text{LJ}}(x_{ij}(t_k)) + u_C(x_{ij}(t_k)) + u_{bl}(x_{ij}(t_k))\right), \qquad f_i(t_k) = \sum_{i} f_{ij}(t_k).$$

In (15.21), V is the simulation cell volume, $\otimes$ denotes the Kronecker product, and f_{ij} is the force exerted on particle i by particle j resulting from nonbonded interactions (15.21) and bond-length interactions (15.8). Notice that the pressure tensor is defined using only the pairwise interactions between particles. The pressure tensor can be used to compute important parameters related to engineered artificial cell membranes, including line tension and surface tension, and to construct simulations that match experimental conditions where the pressure may be anisotropic (not equal in all directions). However, in the case of isotropic pressure profiles the scalar instantaneous pressure is given by

$$P(t_k) = \frac{\text{Tr}(\boldsymbol{P}(t_k))}{3}, \qquad (15.22)$$

15.2 Basics of Molecular Dynamics

where $\text{Tr}(\boldsymbol{P}(t_k))$ is the trace (i.e., sum of diagonal elements) of the stress tensor in (15.21). Just as with the temperature, the instantaneous pressure can vary between each time step; however, if the system is designed to have a constant pressure, then the time average of the pressure will be a constant P.

Having defined the macroscopic and instantaneous energy (15.18), temperature (15.19), and pressure (15.21), it is possible to discuss how we can adjust the trajectory dynamics of particles to ensure certain macroscopic parameters remain constant.

Constant energy. To perform MD simulations with constant energy requires that the total energy E (kinetic energy plus potential energy) is known. The simulation is performed such that the system cannot exchange energy with its environment. Therefore – by conservation of energy – the energy of the system remains constant as time evolves. In practical applications, the exact energy of a system is almost never known as a result of uncontrolled experimental factors. Additionally, having a constant energy means that the pressure and/or temperature do not remain constant, which is not desirable when performing most experimental measurements. Although defining a constant energy prior to performing a simulation is very difficult, it is, however, a very useful quantity to ensure an MD simulation is well equilibrated; that is, particle positions and velocities correspond to a stable system. In MD nearly all simulations are performed with a constant energy; however, the energy is not defined prior to performing the simulation.

Constant temperature. Constant-temperature MD simulations are widely used since most real-world experiments are conducted in ambient conditions with constant temperature. Methods to control the temperature of an MD simulation are known as thermostats. Several common types exist, including Berendsen, velocity-rescaling, and Nosé–Hoover thermostats. A common theme with all of these is that the velocity of the particles is adjusted to ensure that the temperature of the system remains constant. This naturally follows from the relation between instantaneous temperature and particle velocity as defined in (15.19). The Berendsen thermostat mimics weak coupling with first-order kinetics to an external heat bath with given temperature T to the system and is the most popular method for constructing a constant-temperature simulation. However, an issue with this method is that this thermostat suppresses the fluctuations of the kinetic energy. Therefore, any macroscopic parameters that require an accurate estimation of the kinetic energy will contain errors when using this thermostat. For example, the heat capacity of the system will contain significant errors if the Berendsen thermostat is used. The velocity-rescaling thermostat is similar to the Berendsen thermostat; however, the thermostat adjusts the kinetic energy by introducing a stochastic term based on a Wiener process. This formulation allows for the correct fluctuations to occur in the kinetic energy while ensuring the temperature of the simulation has a first-order decay to the desired constant temperature with negligible oscillations in the temperature. The Nosé–Hoover thermostat introduces a frictional term into Newton's equation of motion that is proportional to the product of each particle's velocity and a friction parameter. The friction parameter dynamically evolves according to the difference between the kinetic energy and the desired reference temperature. The Nosé–Hoover thermostat, like the velocity-rescaling thermostat, does not suppress the fluctuations of the kinetic energy. However, an issue with this method is that typically it takes approximately four to five times longer to adjust the system temperature to the desired value compared with the Berendsen and velocity-rescaling methods. In practical MD simulations typically the temperature is first relaxed using either the Berendsen or velocity-rescaling thermostat; then the Nosé–Hoover thermostat is used to construct trajectories that will be used to estimate important macroscopic parameters such as diffusion.

Constant pressure. Constant-pressure MD simulations are useful for modeling real-world experiments that are conducted under constant ambient pressure. Methods to control the pressure of an MD simulation are called "barostats." Several common types of barostats exist, including Berendsen and Parrinello–Rahman. The main idea of the barostats is to rescale the particle positions and simulation cell size V in (15.21) in order to adjust the pressure of the system to the desired value. The Berendsen barostat performs this task by scaling the particle positions and simulation cell size relative to the difference in the desired pressure tensor and the instantaneous pressure tensor. An issue with the Berendsen barostat is that the fluctuations in instantaneous pressure or instantaneous volume are not accurately accounted for. It is still not exactly certain what errors the Berendsen barostat may yield during simulations. To accurately account for fluctuations in instantaneous pressure, the Parrinello–Rahman barostat can be utilized. In the Parrinello–Rahman barostat, the simulation cell size is modeled by a second-order differential equation that is dependent on the difference of the desired pressure tensor and instantaneous pressure tensor. Additionally, the equations of motion for the particles are also changed based on the dynamics of the simulation cell size. In typical applications the Parrinello–Rahman barostat requires about four to five times longer to reach the desired pressure compared with the Berendsen barostat. Additionally, if the system instantaneous pressure is very far from the desired pressure then the Parrinello–Rahman barostat can produce large simulation cell size variations that can cause the simulation to crash. In such cases, the Berendsen barostat should be used initially to approximately reach the equilibrium, followed by the Parrinello–Rahman barostat to construct accurate trajectories.

Thermostats and barostats can be combined to construct MD simulations with constant temperature and pressure. Additionally, there exist several other thermostats and barostats that can be used to control the temperature and pressure of an MD simulation.

The control of the macroscopic parameters when performing MD simulations defines what is known as the statistical ensemble of the simulation. For engineered artificial membranes the two most useful ensembles are the NPT ensemble (also known as the isothermal-isobaric ensemble), and the NVT ensemble (also known as the canonical ensemble). The NPT ensemble represents a simulation that is performed at constant pressure and temperature with a constant number of particles. The NVT ensemble represents a simulation that is performed at constant temperature and volume with a constant number of particles. Both the NPT and NVT ensembles require the use of a thermostat, and the NPT requires the additional use of a barostat to control the pressure. If just Newton's equation of motion (15.1) is used to compute the trajectory of the particles then this corresponds to the NVE ensemble (also known as the microcanonical ensemble). In the NVE ensemble, the number of particles, the volume, and the energy of the system all remain constant. However, in practical applications it is very difficult to know the energy of a system. Therefore, both the NPT and NVT ensembles are commonly used. The NVT ensemble is particularly useful when performing MD simulations with engineered artificial membranes because the bioelectronic interface is commonly modeled as a flat surface of equally spaced atoms. If the size of the simulation cell is varied, then the associated number of surface atoms would also vary. However, as long as the membrane, water, and surface are well equilibrated to their correct size, performing the simulation in an NVT ensemble should not introduce significant errors. It is, however, critical to ensure that, prior to performing any NVT ensemble simulations, the system is well equilibrated such that pressure does contribute significantly to the dynamics. Notice that in the NVT ensemble there is no coupling to a pressure bath so the pressure is not guaranteed to remain fixed. Equilibration of the system prior to performing an NVT simulation can be performed by first running the MD simulation in the NPT ensemble to determine the equilibrium volume of the system at the desired pressure value. Then the NVT simulation is performed using the results from the NPT simulation.

Summary. This section has discussed how particle trajectories from MD simulations can be used to estimate a generic macroscopic parameter $A(x, p)$, which is a function of position x and momentum p. The estimate of the macroscopic parameter, which is an ensemble average (mathematical expectation), was estimated as the time average (sample average) of $A(x, p)$ as illustrated in (15.17). Methods to compute the macroscopic temperature, pressure, and total system energy as a function of the position x and momentum p of the atoms in the system were provided. These macroscopic parameters can then be controlled in an MD simulation using barostats (for pressure control) and thermostats (for temperature control). Three important statistical ensembles were presented: the NVE (microcanonical) ensemble, NVT (canonical) ensemble, and NPT (isothermal-isobaric) ensemble. Chapters 14 and 15 both rely on the theory presented in this section to set up both MD and CGMD simulations and to estimate biological parameters from the MD or CGMD simulation results.

15.2.3 Numerical Methods for Molecular Dynamics

MD simulations involve numerically computing the solution of a system of nonlinear ordinary differential equations comprising Newton's equation of motion (15.1) coupled with thermostat and barostat constraints which control the macroscopic temperature and pressure of the simulation. The results of MD simulations are time-evolved trajectories of a set of particles (each with a random initial condition) from which we can estimate macroscopic parameters such as diffusion coefficients and particle density. In this section, we discuss the boundary conditions, initial conditions, and numerical integration techniques used to evaluate this system of nonlinear ordinary differential equations. Careful specification of these initial and boundary conditions is important when setting up MD simulations of engineered membranes and the bioelectronic interface.

Boundary Conditions and Initial Conditions for Molecular Dynamics
Three common boundary conditions used for MD simulations of engineered membranes are fixed, periodic, and mixed boundary conditions. Fixed boundary conditions are

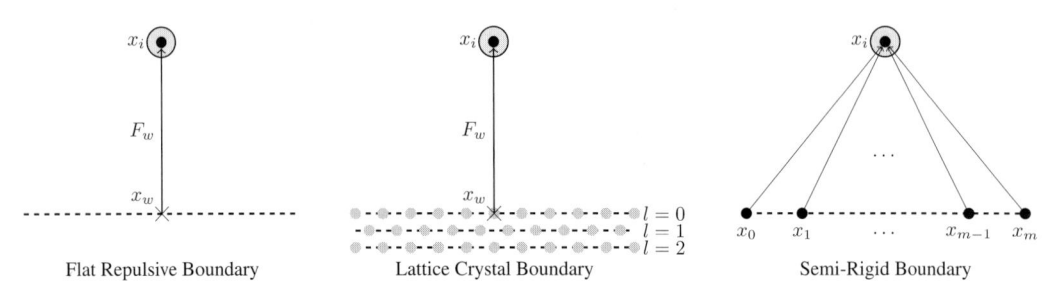

Figure 15.3 Schematic of the flat repulsive, crystal lattice, and semirigid atomistic boundary conditions used in MD simulations. The parameter x_i is the coordinate of the ith atom, the cross indicates the surface coordinate x_w closest to the ith atom, and F_w is the force acting on the atom from the boundary potential or atoms fixed on the boundary. The dashed lines indicate the surface, the grey dots indicate virtual atoms (not explicitly modeled in the MD simulations), and the black dots indicate the coordinate of the atoms. In the lattice crystal boundary $l = 0$ indicates the layer of atoms, and in the semirigid boundary the coordinates $x_0, \ldots, x_m$ indicate the position of the surface atoms. The distance between the ith atom and the surface is given by $z_i = (x_i - x_w) \cdot n_w$, where n_w is the normal vector to the boundary.

useful for modeling the interaction of an electrode surface with an electrolyte solution. Periodic boundary conditions are useful for approximating the macroscale dynamics of large systems. Mixed boundary conditions are useful when modeling engineered membranes where we have periodicity in the directions parallel to the electrode and membrane surface, but not in the direction perpendicular to the electrode surface. Below we discuss each of these important boundary conditions and associated initial conditions of the MD simulations.

Fixed boundaries. Fixed boundary conditions are composed of a virtual potential that accounts for the underlying configuration of atoms that represent a rigid interface. Figure 15.3 illustrates three commonly used fixed boundary conditions: flat repulsive, crystal lattice, and semirigid atomistic.

The flat repulsive boundary is composed of a repulsive potential that prevents electrolyte atoms from crossing the surface. A commonly used flat potential is the Lennard-Jones 12 potential,

$$u_w(z_i) = \varepsilon_{iw} \left(\frac{\sigma_{iw}}{z_i} \right)^{12}, \qquad (15.23)$$

where z_i is the distance of atom i to the boundary, and ε_{iw} and σ_{iw} define the Pauli repulsion[11] of the atom and boundary resulting from short-range overlapping electron orbitals. Note that the flat repulsive boundary can have different Pauli repulsion constants (ε_{iw} and σ_{iw}^{12}) for different atoms $i \in \{1, \ldots, n\}$. This allows the flat repulsive boundary to interact differently with different atoms.

---

[11] Pauli repulsion results from the repulsive net "intermolecular force" that results from two atoms. Note that this Pauli repulsion term is phenomenological because this repulsion between atoms results from quantum-mechanical effects of overlapping electron orbitals of the two atoms. Although the effect of these overlapping electron orbitals can be modeled as a repulsive force between atoms, it is not technically a classical force as no quanta (or force carrier particles) are present.

15.2 Basics of Molecular Dynamics

The crystal lattice boundary condition is commonly used to model the interaction of metal electrodes with an electrolyte solution. Metal electrodes typically have a specific crystal structure (arrangement of metal ions), such as simple cubic (SC), face-centred cubic (FCC), or body-centred cubic (BCC). For example, silver and gold electrodes have the FCC crystal structure. For the FCC crystal structure the fixed boundary potential is given by

$$u_w(z_i) = \sum_{l \in \{1,2,\ldots\}} u_w^l\left(z_i + \frac{l\kappa}{\sqrt{2}}\right) \mathbf{1}\left\{z_i + \frac{l\kappa}{\sqrt{2}} \leq R_c\right\},$$

$$u_w^l(z_i) = \pi \varepsilon_{iw} \left(\frac{\sigma_{iw}}{\kappa}\right)^2 \left[\frac{2}{5}\left(\frac{\sigma_{iw}}{z_i}\right)^{10} - \left(\frac{\sigma_{iw}}{z_i}\right)^4\right] + \frac{\pi V_s z^2}{2\kappa}$$

$$- \pi \varepsilon_{iw} \left(\frac{\sigma_{iw}}{\kappa}\right)^2 \left[\frac{12}{5}\left(\frac{\sigma_{iw}}{R_c}\right)^{10} - 3\left(\frac{\sigma_{iw}}{R_c}\right)^4\right],$$

$$V_s = 4\varepsilon_{iw}\left[\left(\frac{\sigma_{iw}}{R_c}\right)^{12} - \left(\frac{\sigma_{iw}}{R_c}\right)^6\right]. \tag{15.24}$$

In (15.24), the FCC unit cell has a size of $\sqrt{2}\kappa$, ε_{iw} and σ_{iw} define the Pauli repulsion of the ith atom and the FCC boundary, z_i is the distance of the ith atom to the FCC boundary, R_c is the cutoff length of the nonbonded interaction between atoms, V_s is known as the shift function, and $\mathbf{1}\{\cdot\}$ is the indicator function.[12] The FCC boundary potential $u_w(z_i)$ (15.24) accounts for the entire interaction of the ith atom with all the gold electrode atoms. Therefore, $u_w(z_i)$ can be interpreted as a coarse-grained representation of the entire gold electrode. The boundary potential of the SC and BCC crystal structures can be found in [399].

Semirigid atomistic boundary conditions constitute another model for the interaction of an electrolyte with an electrode interface. The semirigid atomistic boundary conditions are imposed on the system by harmonically constraining a set of boundary atoms at specific positions. These atoms interact with other atoms in the MD simulation via nonbonded Lennard-Jones potentials. This is similar to the crystal lattice boundary condition; however, in this case only a single layer of boundary atoms is used to account for the interaction of the electrode surface. Note that semirigid atomistic boundary conditions were used in Chapter 14 to model the gold electrode in the engineered membrane.

Periodic boundary conditions. Our MD simulations of engineered membranes we use periodic boundary conditions that are defined by a *triclinic unit cell*. The triclinic unit cell is defined by three box vectors, denoted by a, b, and c, and the three angles between these box vectors, denoted by $\angle ab$, $\angle ac$, and $\angle ab$. The geometry of the simulation cell is defined by these box vectors and angles. Common triclinic cell shapes include cubic, rhombic dodecahedron, truncated octahedron, and truncated octahedron. When periodic boundary conditions are used, each simulation cell is surrounded by 26 translated copies of the original simulation cell. A cutoff restriction is necessary to

---

[12] The indicator function $\mathbf{1}\{z_i + \frac{l\kappa}{\sqrt{2}} \leq R_c\}$ is equal to 1 if the inequality is satisfied, and 0 otherwise.

ensure that an atom does not interact with itself in any of these 26 translated copies. The reason is that if an atom can interact with its translated copy, it will introduce spurious correlations of the movement of the atom. The *minimum image convention* is used to restrict the interactions of atoms with their translated copies. According to the minimum image convention, the nonbonded interaction between atoms only occurs if the distance between the atoms, denoted by x_{ij}, satisfies

$$x_{ij} \leq R_c < \frac{1}{2} \min(|a|, |b|, |c|), \tag{15.25}$$

where a, b, and c are the three box vectors and R_c is the cutoff distance. The cutoff restriction (15.25) guarantees that atom i interacts with only one image of any atom j. Notice that the cutoff restriction does not guarantee that periodicity effects are not present in the MD simulation results. For example, if the simulation cell size is too small then the estimated diffusion coefficients from MD simulations will scale linearly to changes in the simulation cell size. To mitigate these effects either a large simulation cell size is used, or several MD simulations are performed with different cell sizes to ensure that estimated macroscopic parameters do not change as a function of the cell size. Note that the selection of the cell geometry is nontrivial and can dramatically impact the computational cost of performing MD simulations. For example, using the rhombic dodecahedron simulation cell has a volume that is 71 percent of the volume of a cube having the same image[13] distance. This reduction saves approximately 29 percent of CPU time when simulating flexible molecules such as proteins in a solvent. However, for engineered membrane simulations the rectangular triclinic cell is necessary as the membrane is rectangular in shape.

Mixed boundary conditions. Mixed boundary conditions are a combination of the periodic boundary conditions and fixed boundary conditions. An example of the mixed boundary condition is provided in Chapter 14, where periodic boundary conditions were used in the x- and y-coordinate directions parallel to the membrane and electrode surface; however, a semirigid atomistic boundary condition was used in the z-coordinate direction to model the electrode surface.

Initial conditions. Typically an approximate shape of the system comprised of molecules is known a priori. For example, a membrane is composed of a bilayer of lipids with their hydrophobic tails in contact with each other, and each bilayer has the same number of lipids. However, the positions of the water particles are usually unknown and can be placed in the cell to match the expected density of water in the system. Another parameter that is required is the velocity of the particles. The velocity of the particles can be generated randomly by sampling from a probability distribution. At equilibrium, the speed[14] of the particles will satisfy a Maxwell–Boltzmann distribution given by

$$p(|v_i|) = \left(\frac{m_i}{2\pi k_B T}\right)^{3/2} |v_i|^2 e^{-m_i|v_i|^2/2k_B T}, \tag{15.26}$$

---

[13] An image here is a copy of the particles in the simulation cell.
[14] Given the velocity vector v_i, the speed is the Euclidean norm $|v_i| = \sqrt{v_i^2(1) + v_i^2(2) + v_i^2(3)}$.

where T is the initial system temperature, k_B is Boltzmann's constant, m_i is the mass of the particle, and $|v_i|$ is the speed. Denoting $v_i(1)$, $v_i(2)$, and $v_i(3)$ as the velocities in Euclidean space such that $v_i = [v_i(1), v_i(2), v_i(3)]$, the distribution of each of these velocities is given by the normal probability density function

$$p(v_i(j)) = \sqrt{\frac{m_i}{2\pi k_B T}} e^{-m_i v_i(j)^2 / 2k_B T}, \qquad j \in \{1, 2, 3\}, \tag{15.27}$$

where the generated velocities $v_i(j)$ will have a standard deviation of $\sqrt{k_B T/m_i}$. Note that, even if the speed of the particles satisfies the Maxwell–Boltzmann distribution (15.26), the distribution of the particle speeds will change during simulation, especially if the initial configuration is far from the equilibrium value associated with the selected statistical ensemble. Therefore, an equilibration step is performed which will adjust both the position of the particles to reduce the potential energy of the system, and the velocity of the particles to match the desired initial temperature and pressure of the statistical ensemble defined for the MD simulation. To perform the equilibration a steepest-descent or conjugate-gradients method is used to adjust the particle positions to minimize the potential energy of the system.

Numerical Integration and Approximations

We now discuss numerical methods for evaluating the solution of nonlinear ordinary differential equations comprising Newton's equation of motion (15.1) coupled with a thermostat and barostat, with specified boundary conditions and initial conditions. These numerical methods are the basis for MD simulations.

Several methods can be used for numerically integrating the system of differential equations that comprise an MD simulation including the Velocity Verlet algorithm, leapfrog algorithm, and Beeman algorithm.[15] However, notice that for large systems that can contain more than $n = 10^5$ particles these algorithms are computationally prohibitive because they have a time-step complexity of $O(n^2)$. Several methods have recently been introduced to reduces the number of computations required at every time step, including: cutoff schemes, coarse graining long-range electrostatic interactions, and coarse graining or grouping particles together.

For systems containing large molecules, such as proteins and membranes, there are many more nonbonded interactions than bonded interactions. To reduce the complexity of evaluating all possible interactions, typically a cutoff approximation is used to eliminate the interaction of particles that are far away from each other. Common cutoff schemes include truncation, shift, and switch.

> **Truncation**. In the potential function, the interactions are set to zero for interparticle distances greater than a set cutoff distance. However, this truncation can result in fluctuations in the total energy (kinetic and potential) of the MD simulation and

---

[15] A naive finite-difference implementation will not preserve the total energy of the system. Many of the numerical methods used in MD have deeper group-theoretic interpretations in terms of preserving the total energy.

spurious effects that result from the reduced interaction between atoms. To test if these spurious effects are present requires that multiple MD simulations are run using different levels of truncation and then comparing the values of the estimated macroscopic parameters from the MD simulations. Notice that this cutoff must be less than the minimum image convention cutoff R_c in (15.25).

Shift. This method modifies the entire potential energy surface such that at the cutoff distance the interaction potential is zero. The drawback of this method is that equilibrium distances are slightly decreased.

Switch. This method tapers the interaction potential over a predefined range of distances. The potential takes its usual value up to the first cutoff and is then switched to zero between the first and last cutoff. A limitation of this method is that it suffers from strong forces in the switching region, which can slightly perturb the equilibrium structure. This method is not recommended when using short cutoff regions.

The selection of the cutoff parameters is evaluated based on the accuracy required for computing macroscopic parameters from the simulation, or for studying particular dynamics in the system. A major issue with using cutoff schemes is that long-range electrostatic interactions are suppressed. This will dramatically impact the results of MD simulations that contain surfaces of charged particles. Recall that these can be used to model charged bioelectronic interfaces in engineered artificial membranes. The particle-mesh Ewald summation method can be used to account for these long-range electrostatic interactions. This method essentially constructs a coarse-grained representation of the electrostatic interaction between particles. Obviously care must be taken when selecting how to construct approximate solutions of the original MD equations of motion. However, if the cutoff parameters and long-range electrostatic interactions are sufficiently accounted for, it is possible to perform MD simulations with a time complexity of $O(n \log n)$. This is a significant cost savings compared with the original $O(n^2)$ time complexity.

A viable method to reduce the computational cost of MD simulations is to reduce the number of particles in the system while ensuring important properties of the original system are maintained. Properties that are typically required to be conserved include free energies, distributional properties (e.g., radial distribution function), and forces. Depending on the desired properties, there are different methods to derive such coarse-grained force fields. Note that the MARTINI force field is designed to reproduce the free energies. Additionally, any coarse-grained force field is state dependent and should be recomputed for any change of state. For example, if the coarse-grained force field was constructed at a certain temperature and pressure, then it should only be used for performing MD simulations at this temperature and pressure. Several methods exist for systematically coarse graining all-atom MD simulations: the simplex method, iterative Boltzmann inversion, inverse Monte Carlo, and force matching. A very useful open-source package that implements these coarse-grained methods is the Versatile Object-oriented Toolkit for Coarse-Graining Applications (VOTCA) package available from www.votca.org/.

15.3 MD Simulations for the Dynamics of Engineered Membranes

With the short tutorial on MD in the previous section, we are now ready to consider MD simulations for engineered membranes. This section discusses how MD simulation can be used to estimate important parameters that govern the dynamics of an engineered membrane.[16] These parameters include diffusion coefficients, atomic density and charge density, electrostatic potential, and area per lipid. As will be apparent from the rest of this chapter, MD provides remarkable insight into the workings of the engineered membrane at the atomistic scale.

Let $x(t)$ and $p(t)$ respectively denote the positions and momenta of an ensemble of n particles at time t. Recall from §15.2.2 that a generic macroscopic parameter, denoted as $A(x, p)$, can be estimated using MD simulations by evaluating the time average of the macroscopic parameter using (15.17). In §15.2.2, methods to compute the macroscopic temperature, pressure tensor, and energy were provided. In this section, we provide methods to compute diffusion coefficients, atomic density and charge, and the area of lipid molecules from MD simulations using (15.17).

Diffusion coefficient. The diffusion coefficient of particles is an important macroscopic parameter that is used in continuum theories. There are two possible methods to compute the diffusion coefficient from MD simulations: either use the mean-square displacement or use the velocity-autocorrelation function. The mean-square displacement is used to study the average distance a set of particles moves as a function of time. Formally, the mean-square displacement is

$$\text{MSD}(t_k) = \frac{1}{|\mathcal{A}|} \sum_{i \in \mathcal{A}} ||x_i(t_k) - x_i(0)||^2, \qquad (15.28)$$

where $\mathcal{A}$ are the particle indices of a particular molecule type with $|\mathcal{A}| \leq n$, and $|\mathcal{A}|$ is the number of particles in MD simulation that are contained in the particular molecule type. For example, if the diffusion of water is of interest, $\mathcal{A}$ would contain all the indices of particles associated with the water molecules. The mean-square displacement (MSD) is related to the diffusion coefficient D of the particles via the Einstein equation,

$$D = \lim_{t \to \infty} \left(\frac{1}{2dt} \text{MSD}(t) \right) \approx \frac{1}{2dt_k} \text{MSD}(t_k), \qquad (15.29)$$

where d are the degrees of freedom of the atoms and t_k is the simulation time (assumed to be sufficiently large). In MD simulations, to estimate the diffusion coefficient, typically t_k is chosen larger than 10 ns. The degrees of freedom of an atom define the number of directions in which that atom can move. For example, water can move in all three coordinate directions, whereas lipids can only move in two coordinate directions. Therefore, if the MSD is being estimated for water then $d = 3$, and if the MSD is being estimated for lipids then $d = 2$. Note that the Einstein equation is only valid when subdiffusion or superdiffusion processes are not present. Another method to compute the

---

[16] We emphasize again that since an engineered membrane is a biomimetic approximation to a biological membrane, the MD techniques presented here also apply to biological membranes.

diffusion coefficient is to use the velocity-autocorrelation function:

$$C_V(t) = E\left[\frac{1}{|\mathcal{A}|}\sum_{i\in\mathcal{A}} v_i(t') \cdot v_i(t'+t)\right] \quad (15.30)$$

$$\approx \frac{1}{K}\sum_{k=1}^{K} \frac{1}{|\mathcal{A}|}\sum_{i\in\mathcal{A}} v_i(t_k) \cdot v_i(t_k+t).$$

Using the Green–Kubo relation,[17] the velocity-autocorrelation function can be used to compute the diffusion coefficient as:

$$D = \frac{1}{d}\int_0^\infty C_V(t)dt. \quad (15.31)$$

Since the velocity-autocorrelation function $C_V(t)$ decays to approximately zero for large t, as $t \to \infty$, the integral in (15.31) can be approximated by a finite horizon sum. The decay of $C_V(t)$ is exponential if only binary independent collisions are considered. However, it is possible that collisions between atoms in MD simulations are correlated. For example, a multicollision event such as a lipid colliding with another lipid, then colliding with other lipids, and again colliding with the first lipid this implies there is long-term memory in the system. In such long-term-memory cases, the decay of the velocity autocorrelation function can be slower than an exponential (e.g., $C_V(t)$ may satisfy a power-law decay).

Both methods for computing the diffusion coefficient, using either the mean-square displacement or the velocity-autocorrelation function, can be used and should give approximately the same results.

Atomic density and charge density. The atomic density and charge density are important parameters for computing structural and electrostatic properties of engineered membranes. Suppose we wish to compute the atomic density in the z coordinate of the simulation cell. In MD all particles are represented by points that contain an associated mass and charge. To compute the atomic density we define the spatial resolution of interest as Δz – this is the resolution at which we estimate the atomic density. The MD simulation cell in the z coordinate is then divided into equally sized slabs of thickness Δz. The slabs extend in the xy plane up the boundaries of the simulation cell. Then the atomic density $\rho(z)$ is evaluated as follows:

$$\rho(z(t_k)) = \frac{1}{L_x L_y \Delta z}\sum_{i\in\mathcal{A}} m_i \mathbf{1}\left\{\left\lfloor\frac{z(t_k)}{\Delta z}\right\rfloor \le z_i(t_k) < \left\lceil\frac{z(t_k)}{\Delta z}\right\rceil\right\}$$

$$\rho(z) = \langle\rho(z(t_k))\rangle \approx \frac{1}{K}\sum_{k=1}^{K}\rho(z(t_k)). \quad (15.32)$$

In (15.32), the parameters L_x and L_y are the length of the simulation cell size in the x- and y-coordinate directions, $\mathcal{A}$ is the set of particles associated with a particular molecule

---

[17] A Green–Kubo relation is an expression that relates the diffusion coefficient to the integral of a time-correlation function.

type (e.g., water) for which the density is to be computed, m_i is the mass of the particles, $\mathbf{1}\{\cdot\}$ is the indicator function, $\lfloor \cdot \rfloor$ is the floor function (rounded off to the nearest smallest integer), and $\lceil \cdot \rceil$ is the ceiling function (rounded off to the nearest largest integer).

Intuitively, (15.32) represents the time-averaged density of particles contained in a rectangular slab of volume $L_x L_y \Delta z$. The particle density $\rho(z)$ is useful for estimating membrane thickness as well as the spatially dependent density of water in proximity to the bioelectronic interface and membrane surface. The charge density can be evaluated using an analogous formula; however, we substitute the particle charge q_i in (15.32) in place of the mass m_i, and typically $\mathcal{A}$ is then defined as all particles in the simulation cell. A very useful macroscopic parameter to compute using the charge density is the electrostatic potential of the system. Using Poisson's equation, the electrostatic potential of the simulation cell is given by

$$\phi(z) = \frac{1}{\varepsilon_0} \int_0^z \int_0^y q(x)dxdy + \phi(0), \qquad (15.33)$$

where $\phi(0)$ is the electrostatic potential at $z = 0$ nm, and $q(z)$ is the computed charge density. The potential can be used to estimate the capacitance of engineered tethered membranes directly using the results of MD simulations.

Area per lipid. The area per lipid in engineered membranes, denoted by A_L, is an important parameter that can be used to validate properties of the MD force field (or potentials used to construct the MD force field that describes the interaction between particles). By using NMR spectroscopy, it is possible to estimate A_L directly from experimental measurements. If the membrane is composed of a homogeneous mixture of lipids (i.e., the same lipids), then the A_L can be computed as the following time average for large K:

$$A_L(t_k) = \frac{L_x(t_k)L_y(t_k)}{N_L},$$

$$A_L = \langle A_L(t_k) \rangle \approx \frac{1}{K} \sum_{k=1}^{K} A_L(t_k). \qquad (15.34)$$

In (15.34), N_L is the number of lipids in one leaflet of the membrane. For homogeneous membranes N_L is equal to half the total number of lipids in the simulation cell. In the case of multiple lipids, or lipid and protein mixtures, it is still possible to estimate their area. To estimate the area of multiple lipids, first an equally spaced grid is selected for the surface parallel to the membrane surface. Then, each grid point is selected as being associated with a particular lipid or protein. The area occupied by each lipid is estimated by summing the areas of all the cells assigned to the lipid of interest. Note that, in the case of significant membrane undulations (bending of the membrane), more sophisticated methods are required to estimate the area per lipid. This, however, is not necessary for engineered artificial membranes because the tethers suppress significant membrane undulations.

15.4 Aqueous Pore Formation Dynamics in Tethered Membranes

Recall from Chapter 11 that electroporation is the process of formation of aqueous pores in a tethered membrane as a result of applying a transmembrane potential. This section illustrates that, even at zero transmembrane potential, it is still possible for aqueous pores to exist in the membrane as a result of random thermal fluctuations or from the diffusion of ions. Recall that the transmembrane potential is a macroscopic parameter that describes the potential drop across the entire membrane; however, it is still possible that nonzero potential gradients exist in small regions of the membrane, which can lead to aqueous pore formation.

What are the dynamics of the aqueous pore formation process? The aqueous pore formation process in cell membranes can result from several mechanisms: membrane stretching, electrochemical gradients, pore-forming proteins and peptides, vesicle fusion, and lipid flip-flop. Using all-atom MD it is possible to study how transient aqueous pores form in membranes. The process of aqueous pore formation is complex and dependent on the membrane composition, the mechanism of pore formation, and environmental parameters such as an electrolyte solution concentration and temperature. For example, using electrochemical gradients, it is illustrated in [324] that the electroporation process of *Escherichia coli* and *Staphylococcus aureus* have subtle differences: in *S. aureus* a membrane defect (aqueous pore) can be formed by either the outer or inner lipids; however, for *E. coli* the defect is preferentially formed by the inner leaflet phospholipids (e.g., phosphatidylethanolamine) irrespective of the applied potential gradient. In this section we present the aqueous pore formation process of transient hydrophilic pores in engineered membranes resulting from electrochemical gradients using all-atom MD simulations. The simulation results provide a mechanism for hydrophilic pore formation that can result in the formation of hydrophobic conducting pores. Recall that the Smoluchowski–Einstein equation (11.1) provides a probabilistic characterization of the population and size of these pores; in comparison, here we provide a mechanism for the formation of such pores.

Figure 15.4 illustrates the formation of an aqueous pore in a DphPC membrane using all-atom MD simulations when the transmembrane potential $V_m = 4.0$ V. The simulation details are provided in §C.5.1. Let us describe Figure 15.4 in more detail. The process of aqueous pore creation begins with water fingers penetrating the hydrophobic portion of the intact membrane. The formation of these water fingers occurs on both sides of the DphPC membrane and is independent of the direction of the applied transmembrane potential. These results are in agreement with the initial pore formation dynamics observed for other types of lipid membranes. The water fingers then penetrate through the hydrophilic portion of the membrane to form a water bridge. Note that the formation of these water bridges in DphPC membranes appears to be cooperative, requiring a water finger from each side of the membrane to come into contact as illustrated in Figure 15.4. A nonconducting hydrophilic pore forms when a water bridge connects both electrolyte solutions across the membrane. This hydrophilic pore expands until ions can cross, at which point the pore becomes a hydrophilic pore.

15.4 Aqueous Pore Formation Dynamics in Tethered Membranes

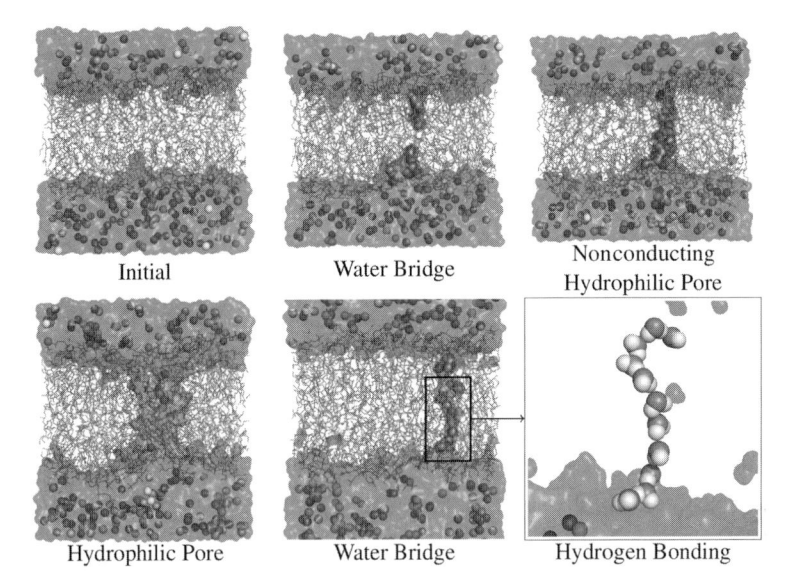

Figure 15.4 A color version of this figure can be found online at www.cambridge.org/engineered-artificial-membranes. Schematic of the all-atom aqueous pore formation and closure resulting from a charge of $+7q$ C in one bathing solution, and $-7q$ C in the other bathing solution initially. The lipids are illustrated in green, the water is represented by the blue translucent surface, the chloride ions by the yellow spheres, the sodium ions by the orange spheres, white spheres are hydrogen atoms, and red spheres are the oxygen atoms. Details of the simulation setup are provided in §C.5.1.

Given sufficient time the lipid head groups would line the pore, creating a hydrophobic conducting pore; however, in our simulation setup we did not observe the formation of a hydrophilic conducting pore. This results is because ions quickly move between the electrolyte baths causing a decrease in the transmembrane potential V_m which promotes the closure of the hydrophilic conducting pore prior to the lipid head groups being able to diffuse to the pore interior. As the pore closes, the water bridge once again forms, which is composed of a well-ordered set of water molecules arranged by the hydrogen-bonding structure. This illustrates the importance of hydrogen bonding in the electroporation process for dynamics of pore formation and destruction. The water molecules in the water bridge eventually exit the hydrophilic region of the membrane, leaving an intact membrane. In the simulation, the final transmembrane potential was approximately $V_m = 0.5$ V which is above the threshold voltage of electroporation, V_{ep}, observed from experimental measurements. This discrepancy likely results from the quantum-mechanical effects of water (e.g., hydrogen bonding), and the water-lipid interaction. State-of-the-art methods for correlated electronic structure calculations such as ab initio electron-correlated theory [460], or moving beyond the pairwise interactions of all-atom MD to many-body potential energy functions [79, 313] may be used to account for the hydrogen bonding in water; however, the construction of accurate models of water is still an active area of research.

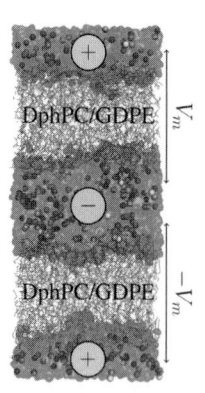

Figure 15.5 A color version of this figure can be found online at www.cambridge.org/engineered-artificial-membranes. Schematic of the all-atom MD simulation where a transmembrane potential V_m is applied across the membrane. The potential V_m results from different concentrations of chloride ions (yellow spheres) and sodium ions (orange spheres) in each electrolyte bath. The lipids are illustrated in green, and the water is represented by the blue translucent surface. Details of the MD simulation setup are provided in §C.5.1.

15.5 Capacitance and Dipole Potential of Tethered Membranes

In this section we use MD simulations to compute the electrostatic potential resulting from a difference in ions on either side of the membrane. Computing this electrostatic potential is essential for evaluating important biological parameters such as the membrane capacitance C_m and transmembrane potential V_m resulting from charge imbalances across the membrane. Also, the dipole potential, denoted by V_d, can be estimated. The dipole potential provides a measure of the forces acting on water and the polar lipid head groups as a result of the charge imbalance across the membrane.

To compute the electrostatic potential via MD simulation, we use Poisson's equation (10.3) with periodic boundary conditions described in Chapter 10. Given the average charge density along the membrane normal, the resulting electrostatic potential is given by

$$\phi(z) = \frac{1}{\varepsilon_0} \int_0^z \int_0^y \rho(x) dx dy + \phi(0), \qquad (15.35)$$

where $\phi(0)$ is the potential at $z = 0$ nm. To construct the MD simulation with a potential gradient across the membrane, we place two membranes in the simulation with a different number of chloride and sodium ions in each electrolyte bath as illustrated in Figure 15.5.

Given (15.35) and the results from the molecular dynamics simulation of (15.5), how can the capacitance C_m of the membrane be estimated? The capacitance of the membrane is related to the MD parameters by

$$C_m = \frac{q_m}{V_m} = \frac{q_{\text{sim}} A_m}{V_m A_{\text{sim}}}, \qquad (15.36)$$

15.5 Capacitance and Dipole Potential of Tethered Membranes

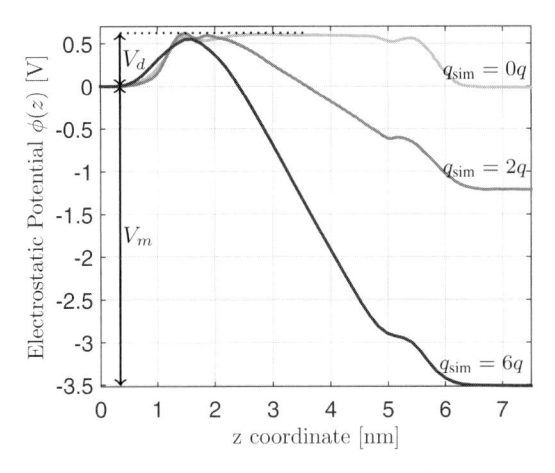

Figure 15.6 Computed electrostatic potential $\phi(z)$ normal to the membrane surface. V_d is the intrinsic membrane dipole potential which is the voltage change across the lipid-water interface and center of the lipid bilayer with $q_{\text{sim}} = 0$, and V_m is the transmembrane potential. The potential is computed for three cases where $q_{\text{sim}} = 0q, 2q$, and $6q$ using (15.35) with the simulation setup provided in Figure 15.5.

where q_{sim} is the excess charge on either side of the membrane in the MD simulation (Figure 15.5), and A_{sim} is the surface area of the membrane in the MD simulation (Figure 15.5). The transmembrane potential V_m in (15.36) is obtained by evaluating the difference in $\phi(z)$ (15.35) at the surface of the membrane as illustrated in Figure 15.5. Three MD simulations where performed, each with a different number of sodium and chloride ions in the baths, which resulted in $q_{\text{sim}} = 0q, 2q$, and $6q$, where q is the elementary charge of an electron. The resulting potential $\phi(z)$ evaluated using (15.35) is provided in Figure 15.6. The estimated capacitance of the zwittrionic C20 diphytanyl-ether-glycero-phosphatidylcholine (DpcPC) membrane evaluated using (15.36) is $C_m = 15$ nF, which is in excellent agreement with the experimentally measured capacitance of 1 percent tethered DphPC membranes, which is in the range of 15.0 to 17.5 nF (Table C.6 on page 418). Assuming that the membrane has a thickness of $h_m = 3.6$ nm, then the average relative dielectric permittivity of the membrane is approximately $\varepsilon_m = 2.9$, which is in good agreement with the relative dielectric permittivity $\varepsilon = 2$ of pure hydrocarbons. The membrane dipole potential V_d is the difference in the voltage potential $\phi(z)$ (15.35) between the surface of the membrane and the interior of the membrane with $q_{\text{sim}} = 0$. V_d is an important parameter because it provides insight into ion and water transport across the membrane surface. The estimated membrane dipole potential for the DphPC membrane is $V_d = 520$ mV, which is in excellent agreement with the experimentally measured dipole potential of 510 mV obtained using cryogenic electron microscopy [450]. If conductance of ions through the membrane was a result of permeability through the membrane, and not through aqueous pores, then the DphPC membrane would have a much larger conductance for anions compared to that of cations. This difference in conductance results because the force acting on a charge q near the membrane surface is approximately given by $-qV_d$. However, in all our electroporation measurements the polarity of the excitation does not appear to impact the conductance of the DphPC membrane, which suggests that ions primarily cross the membrane via conducting aqueous pores.

15.6 Modeling Ion Permeation and Channel Conductance

Ion permeation refers to the process of how ions traverse the channel pore through the cell membrane. The trajectory of a permeating ion gives insight into the structure-function relationship of an ion channel, that is, how the full atom structure of an ion channel determines its function. For example, typically ion channels have one or more binding sites where the ion remains trapped before jumping to other binding sites and finally exiting the ion channel. Ion permeation is a widely studied area [80]. In this section, we make only some brief remarks in the context of engineered membranes and permeation of gramicidin channels (since they are an integral component of the ion-channel switch (ICS) biosensor).

Using experimental measurements and the fractional-order macroscopic model (Figure 13.1) it is possible to estimate the ensemble conductance of all conducting pores in a tethered membrane. However, it is useful to have an estimate of the conductance of a single ion channel, G_c. Such information would allow the experimental measurement of the conductance of the tethered membrane to be related to the number of conducting ion channels in the membrane. This is, however, a nontrivial task because it requires modeling the permeation of ions through small geometries (a few angstroms) that may be charged. In addition, the ion channel may itself have gating mechanisms that need to be modeled in combination with the ion permeation dynamics. In this section we provide a brief account of the mathematical tools that are used to model ion permeation.

15.6.1 Models for Ion Permeation: From Ab Initio to Reaction Rate

The models used for ion permeation include levels of abstraction ranging from fundamental to phenomenological: ab initio quantum mechanics, classical molecular dynamics, coarse-grained molecular dynamics, Brownian dynamics, continuum (Poisson–Nernst–Planck) theories, and reaction-rate theory. Each of these approaches has its strengths and limitations and involves a degree of approximation as discussed below. Such models link channel structure (which is typically defined in terms of an atomic model at the subnanoscale) to channel function (which is observed via experimental measurements at the macroscopic time scale). Any high-resolution dynamic model for ion permeation needs to consider three ingredients: the ions, water molecules, and the atoms of the protein that forms the ion channel. It is essential for such dynamic models to be based on physical principles and to result in computationally tractable simulation algorithms. Such models also need to elucidate the detailed mechanisms of ion permeation – where the binding sites are in the channel, how fast an ion moves from one binding site to another, and where the rate-limiting steps are in conduction. Finally, it should make predictions that can be confirmed or refuted experimentally.

Ab initio quantum mechanics is the lowest level of abstraction: the interactions between the ions, water, lipids, and protein atoms are computed by solving a multibody Schrödinger equation. This approach offers the ultimate tool for the modeling of biomolecular systems as there are no free parameters present in the multibody Schrödinger equation. However, solving such an equation is computationally intractable. Even with simplifying assumptions, one can only numerically solve very small systems over very short simulation time horizons. Given the high computational cost of solving the multibody Schrödinger equation, if electron-shell interactions play an important role in the conductance of an ion channel, then multiphysics models can be employed where the multibody Schrödinger equation is coupled with a higher level of abstraction model such as MD.

A higher level of abstraction compared to ab initio quantum mechanics is to assume a phenomenological form for the potential energy between the atoms in the water, lipids, proteins, and water. This is the main idea of MD – to replace the multibody Schrödinger equation that describes the interaction between atoms with a phenomenological potential energy function. As we discussed in §15.2, in MD simulations this potential energy function accounts for all interactions between the atoms (nonbonded and bonded interactions). It is possible to simulate the permeation of ions through the ion channel using MD; however, the maximum time horizon for computationally tractable ion-channel simulations is approximately $0.1~\mu s$. This simulation time horizon is too short to observe the permeation of ions across an ion channel and to determine its conductance G_c, which is the most important channel property. Although MD cannot be used to directly observe the ion permeations, the potential of mean force (PMF) can be computed for an ion traveling through the ion channel. This information is then utilized in the Fokker–Planck equation to estimate the conductance G_c of the ion channel. Recall that in Chapter 14 we utilized the Fokker–Planck equation to compute the diffusion tensor of water at the bioelectronic interface. If we choose F_w as the PMF of the ion, and we assume the diffusion of ions in the channel is known, then substituting these into the Fokker–Planck equation allows the computation of the ionic flux (e.g., current) of ions in the channel. This can be used to compute the channel conductance G_c. Coarse-grained MD can be used to estimate the channel conductance in an equivalent way as in MD simulations; that is, we compute the PMF of the ion and then use the Fokker–Planck equation to estimate the associated conductance. Notice, however, that the results from molecular dynamics are likely to give a higher-accuracy estimate of the PMF compared to the results from coarse-grained MD simulations.

In between the all-atom and continuum levels of abstraction there exist Brownian dynamics, which can be utilized to estimate the conductance of ion channels. In Brownian dynamics, two assumptions are made in addition to the simplifying assumptions used in MD. First, water is treated as a continuum and only the ions are treated as point-like particles. The net effect of water on the ions is treated as frictional and a result of random forces – random because water molecules are constantly moving due to thermal fluctuations. These random forces model the incessant collisions between ions and water. This treatment of explicit water molecules by implicit water can be viewed as a functional central limit theorem.[18] The second approximation when using Brownian dynamics is that the atoms in the protein and lipids are fixed (stationary over time) and do not undergo any thermal fluctuations. In the case that ion permeation occurs on the slow time scale, and these random thermal fluctuations occur on the fast time scale, then we can use stochastic averaging theory[19] to justify this assumption of a fixed protein and lipid structure. The main idea of stochastic averaging theory is that the protein and lipid atoms moving on the fast time scale will perceive the slowly moving ions as fixed, and the slowly moving ions will only see a time-averaged effect from the fast-moving lipid and protein atoms. Therefore, the time-averaged position of the protein and lipid atoms can be used. With these two assumptions, Brownian dynamics model the movement of the ions through the channel via a set of stochastic differential equations known as the Langevin dynamics equations. Brownian dynamics simulations can be used to model the permeation of hundreds of ions on the time scale of tens of microseconds. This is sufficiently long to estimate the conductance G_c of an ion channel.

The next higher level of abstraction uses continuum theories to estimate the ion-channel conductance G_c. One of the most common continuum theories to estimate ion-channel conductance is to utilize the Poisson–Nernst–Planck (PNP) system of equations, which we derived in Chapter 10. The main approximation used to move from an atomistic description to a continuum description of ion movement is to impose the mean-field approximation. In the mean-field approximation, the space-time average of the motion of several ions is used in place of the specific ion positions. That is, the expected values of the space-time average ion positions are used in place of specific ion positions. Using the mean-field approximation of ion interactions and continuum descriptions of concentration and electrostatic potential, the dynamics of the ions in an electrolyte solution can be modeled via the PNP set of equations. Notice, however, that since the PNP operates on the space-time average, it cannot capture physical phenomena that occur on the atomic scale such as ion-ion correlated movement. Note that this effect can play an important role in the conductance of ion channels, which requires more the correlated movement of multiple ions to open the gating mechanism of the ion channel.

At the highest level of abstraction we have the reaction-rate model described in Chapter 8. The typical application of the reaction-rate model is to use experimental measurements of an ion channel and then construct a set of ordinary differential equations (ODEs) that can describe the conductance dynamics of the ion channel. Although the reaction-rate model parameters have no direct physical relation to the channel structure, many useful insights have been gleaned in the past about ion permeation using this approach. This approach is also very useful from an engineering perspective as the

---

[18] The classical central theorem says: Suppose $\{X_k\}$ is an independent and identically distributed sequence of random variables with zero mean and unit variance. Then $\lim_{n \to \infty} \frac{1}{\sqrt{n}} \sum_{k=1}^{n} X_n$ converges in distribution to the Gaussian random variable with zero mean and unit variance. The functional central limit theorem is an infinite-dimensional extension and deals with a sequence of random processes converging weakly to a Gaussian random process on a function space; see [45] for details.

[19] Stochastic averaging theory is widely used to analyze the convergence of adaptive algorithms in the areas of signal processing, control, and machine learning.

ODEs can be used to construct control algorithms suitable for controlling the conductance dynamics of the ions channels.

To summarize, as new analytical methods have been developed and the available computational power increases, theoretical models of ion permeation have become increasingly sophisticated. It has now become possible to relate the atomic structure of an ion channel to its function through the fundamental laws of physics operating in electrolyte solutions. Many aspects of macroscopic observable properties of ion channels are being addressed by molecular and stochastic dynamics simulations. Quantitative statements based on rigorous physical laws are replacing qualitative explanations of how ions permeate across narrow pores formed by the protein wall and how ion channels allow one ionic species to pass while blocking others. The computational methods of solving complex biological problems, such as permeation, selectivity, and gating mechanisms of ion channels, will increasingly play prominent roles as the speed of computers increases and theoretical approaches that are currently under development become refined further.

15.6.2 Gramicidin Channel Conductance Estimation Using Distributional Molecular Dynamics

In this section we illustrate how the conductance of a gramicidin A (gA) channel can be estimated using MD in combination with a generalized Langevin dynamics equation. This combination of MD and Langevin dynamics allows the computation of the distribution of ion trajectories without requiring the results from large-time-horizon MD simulations.[20]

The main idea of the method is a two-tiered approach: MD is used to measure various properties of the system, and the results are fed into a higher-level system which is able to extend the simulation time horizon beyond what is currently practical for an MD simulation alone. There are three steps to the modeling procedure. First, a stochastic physical model such as the generalized Langevin equation is assumed for the system. The second step involves using MD simulations to estimate parameters that govern the evolution of the system. The last step involves numerically solving a stochastic dynamics equation using the estimated parameters. Here we provide the basis of the stochastic model. Interested readers are referred to [141] for the algorithms used to compute the MD simulation and experimental analysis of this method.

Distributional MD assumes that the evolution of the system variables (i.e., position and momentum) can be described using the generalized Langevin equation consisting of Newton's laws of motion, a frictional force term, and a random force term which is related to the frictional force through a fluctuation-dissipation theorem. The generalized Langevin equation used to describe the motion of the particles is given by

$$\frac{\partial \mathbf{x}(t)}{\partial t} = m^{-1}\mathbf{p}(t),$$

$$\frac{\partial \mathbf{p}(t)}{\partial t} = \mathbf{F}_D(\mathbf{q}(t)) - \int_0^t K(t')\mathbf{p}(t-t')dt' + \mathbf{F}_R(t), \quad (15.37)$$

---

[20] Permeation in gA channels using molecular dynamics is now a well-studied area; see, for example, [12].

where m is an $N \times N$ diagonal matrix with the mass of particle i specified by the matrix element m_{ii}, $\mathbf{p} = [p_1, p_2, \ldots, p_N]'$ is a column vector with p_i denoting the momentum of particle i, $\mathbf{x} = [x_1, x_2, \ldots, x_N]'$ is a column vector with x_i denoting the position of particle i, $\mathbf{F}_D(\mathbf{x}(t))$ corresponds to a deterministic force term, $\mathbf{F}_R(t)$ is a random force term, and $K(t)$ is a friction kernel intrinsic to the system. The friction term in (15.37) contains a negative sign because it retards the velocity; that is, friction causes a negative contribution to the change in momentum $\mathbf{p}(t)$.

The deterministic force $\mathbf{F}_D(\mathbf{x}(t))$ is equal to the sum of the system force ($\mathbf{F}_s$) and the gradient of the potential mean force ($\mathbf{F}_{PMF}$). The system force $\mathbf{F}_s$ is the sum of forces exerted by the system particles, and $\mathbf{F}_{PMF}$ accounts for the equilibrium average force exerted on the system by the system and bath particles. The value of $K(t)$ is obtained by relating $K(t)$ to the momentum autocorrelation function $C(t) = \langle \mathbf{p}(t)\mathbf{p}'(0) \rangle$ and making the assumption that $\mathbf{F}_D(\mathbf{x}(t))$ can be approximated by a harmonic potential $U(\mathbf{x}) = k(\mathbf{x} - \mathbf{x}_0)^2/2$ [141]. Using these assumptions, we obtain the following relation between the friction kernel $K(t)$ and autocorrelation $C(t)$:

$$\frac{\partial C}{\partial t} = -\int_0^t \left(K(t-\tau) + \frac{k}{m} \right) C(\tau) d\tau, \tag{15.38}$$

where k is the coefficient of the harmonic potential $U(\mathbf{x})$. Note that $K(t) \to 0$ as $t \to \infty$. As such, the coefficient k can be estimated using (15.38) for sufficiently large t given $C(t)$ from MD simulations.

A further assumption made for the system is that the fluctuation-dissipation theorem[21] holds. That is,

$$\langle \mathbf{F}_R(0)'\mathbf{F}_R(t) \rangle = \lim_{\Delta \to \infty} \frac{1}{\Delta} \int_0^\Delta \mathbf{F}_R(0+\tau)'\mathbf{F}_R(t+\tau) d\tau = k_B T K(t) m \tag{15.39}$$

with k_B being Boltzmann's constant, T the temperature, and $\langle \cdot \rangle$ the time average. The random force $\mathbf{F}_R(t)$ is assumed to be a Gaussian random process. The validity of this assumption is empirically verified in [141] using MD simulations with the statistical goodness-of-fit tests such as Kolmogorov–Smirnov and Andersen–Darling tests.

The generalized Langevin equation (15.37) describing ion permeation can now be fully characterized as follows. The momentum-autocorrelation function $C(t)$ is evaluated using the particle trajectories from the MD simulation. Specifically, we multiply each particle velocity term in (15.30) by the associated particle mass to obtain $C(t)$. Substituting $C(t)$ into (15.38) results in a Volterra integral equation that relates the friction kernel $K(t)$ to $C(t)$. $K(t)$ is evaluated by numerically solving the Volterra integral equation. The deterministic force $F_D(x(t))$ is computed using the potential of mean force

---

[21] The fluctuation-dissipation theorem asserts that the response of a system to a small perturbation force is the same as its response to a spontaneous fluctuation. For example, in an electrolyte the fluctuation-dissipation theorem can be used to relate the diffusion coefficient D (property of the unperturbed system) to the electric mobility u (average velocity of the molecule) resulting from a small perturbation force from an applied electric field. Note that this is known as the *Einstein relation between diffusion and electric mobility*. In this section, the fluctuation-dissipation theorem is used to relate the correlation function of the fluctuating force, denoted by $\langle \mathbf{F}_R(0)'\mathbf{F}_R(t) \rangle$, to the friction coefficient $K(t)$, that is, the dissipation.

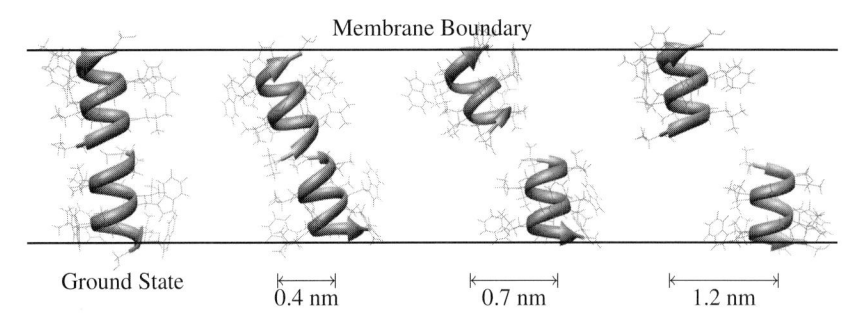

Figure 15.7 Snapshots of the gA dimer dissociation dynamics for variations in the lateral displacement, $R_{\text{lat}} = \{0 \text{ nm}, 0.4 \text{ nm}, 0.7 \text{ nm}, 1.2 \text{ nm}\}$, obtained using MD simulations with umbrella sampling [448].

(PMF) that is estimated using umbrella sampling techniques. Using the above procedure, it is illustrated in [141] that the conductance of a gramicidin A channel is $G_c = 23$ pS, which is in excellent agreement with the experimentally measured conductance of 22.5 pS.

15.7 Gramicidin A (gA) Dimer Dissociation and Reaction-Rate Estimation

In this section, all-atom MD is used to study the dynamics of gA channels in the ICS biosensor: specifically how gA dimers move and split into monomers in the engineered membrane.[22] By constructing MD simulations for this dimer dissociation, we have a first-principles methodology for understanding how the ICS biosensor operates at an atomic spatial scale. Recall from Chapter 4 that gA is an integral component of the ICS biosensor; it is the controlled switching between gA monomers and dimers that results in a selective sensing device. Also, from §10.3, the continuum models of the ICS biosensor are dependent on several parameters such as conductance, diffusivity, permittivity, and chemical reaction rates. These parameters were traditionally obtained experimentally; however, an increase in computational capabilities linked with sophisticated mathematical algorithms is allowing these physical parameters to be modeled from all-atom MD. Characteristics of ion channels such as ion binding, permeation pathways, ion conductance, selectivity, and gating are required to compute important macrolevel parameters of the ICS biosensor.

The main outcomes of this section are twofold:

(i) §15.7.1 uses MD simulations to show that when a gA dimer splits into monomers, the dimer first tilts within the membrane and then one-half (monomer) shears off; see Figure 15.7.

(ii) §15.7.2 shows how MD can be used to estimate the chemical reaction rate of gA dimer dissociation, denoted as f_6 in (10.12). This gives a complete characterization from atoms to macroscopic device.

---
[22] This is in contrast to §15.6, where we studied how individual gA dimers conduct ions (permeation problem).

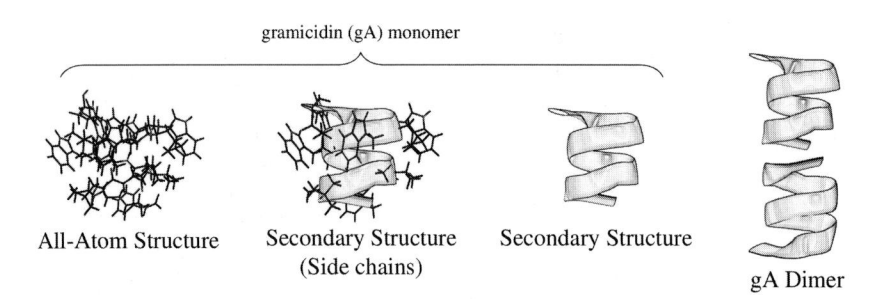

Figure 15.8 Schematic of the secondary structure of the gramicidin A peptide. The three representations of the gA monomer (all-atom structure, secondary structure with side chains, and secondary structure with no side chains), and the gA dimer secondary structure. The gA dimer is a conducting ion channel that allows the permeation of ions through the biological membrane.

15.7.1 Molecular Reaction Dynamics of Gramicidin Channel Dissociation

Molecular reaction dynamics (MRD) studies chemical reactions at the molecular level. Examples of MRD include the study of atomic-level events such as the coupling of antibody and analyte or the dissociation of the gA monomers in the ICS biosensor. As MRD develops, ever more sophisticated chemical processes are understood at the atomic scale, allowing the understanding of molecular level reactions and also allowing macroscale reaction rates of processes to be computed from all-atom MD simulation results.

MRD relies on having an accurate structural representation of the chemical reactants present. For example, for gA dissociation this requires a high spatial resolution (few angstroms) of the coordinates of atoms in gramicidin. The structure of ion channels can be determined due to remarkable advances in X-ray crystallography. This allows the construction of computer simulation models and MRD that accurately account for the atomic structure present in ion channels. For example, Lomize et al. successfully mapped the static structure of the gA ion channel [251]. Recall that gA is a polypeptide that consists of 15 amino acid residues with two coupled gA monomers. Figure 15.8 represents the structure of the gA ion channel (gA dimer) and the associated gA monomers obtained from X-ray crystallographic techniques. Current X-ray crystallographic techniques are unable to detect hydrogen bonds; however, the coupling of the two gA monomers is likely a result of six intermolecular hydrogen bonds, as predicted using MD simulations [448].

Previous experimental studies [112, 343, 359] of the gating characteristics of the gA dimer show that the dissociation-association rate is dependent on voltage, ion occupancy, and hydrocarbon thickness of the membrane. Additionally, the gA dimer can exist in multiple conformational states. Using MRD, a combination of MD simulation with umbrella sampling (see footnote 26 on page 322), it is possible to capture the dynamic behavior of the dissociation process of gA dimers. Umbrella sampling is used to compute the PMF of a system when it is perturbed along a specific reaction coordinate. To perform umbrella sampling, we define the pulling coordinate as the lateral distance between the center of mass – the average position of all the atoms weighted relative to the mass of each atom – of each gA monomer. The results of performing an MD

15.7 Gramicidin A (gA) Dimer Dissociation and Reaction-Rate Estimation

simulation with umbrella sampling for the defined pulling coordinate allows estimation of the dynamics of the dissociation event.

Dissociation pathway. The dissociation event of the gA dimer occurs via a lateral displacement of the gA monomers followed by tilting of monomers with respect to the lipid bilayer normal. Graphically this dissociation event is illustrated in Figure 15.7. The entire dissociation event occurs over a few milliseconds and requires approximately equal amounts of energy to break each of the six intermolecular hydrogen bonds. The dissociation event illustrated in Figure 15.7 provides an approximation of the dissociation event that takes place in the ICS biosensor when an antibody tethered to the mobile gA monomer is attracted to a target molecule in the analyte solution.

As illustrated in Figure 15.7, detailed MD studies [448] reveal the following gA dimer disassociation pathway: First, lateral displacement (tilting) of the dimer occurs within membrane. This tilting allows for increasing lateral displacement without breaking noncovalent bonds between gA monomers. Initially both monomers tilt at the same rate. As the gA dimer tilts, the lipids adjacent to it shift and the dimer begins to experience a lateral resistance from the surrounding lipid, which reduces the degrees of freedom (flexibility) of the dimer. When the tilting is no longer able to overcome the increasing lateral resistance, the noncovalent bonds between the monomers start to break and one monomer tilts more compared to the other monomer. This difference in tilts between the two monomers further enhances the breaking of noncovalent bonds between them. As further lateral displacement occurs, the dimer completely disassociates.

The above disassociation of a gA dimer by lateral displacement is an incremental process. In comparison, MD simulations show that if gA dimers were to disassociate by axial separation (that is, to be pulled apart orthogonal to the membrane surface), it would be a rapid process that would require an instant supply of a large disassociation energy; see [448] for the PMF plots for lateral versus axial separation. Thus from MD simulations one can conclude that the gA dimer dissociates with lateral displacement, as displayed in Figure 15.7, rather than as a direct axial separation.

15.7.2 Gramicidin A Reaction Rates

Umbrella sampling (see footnote 26 on page 322) yields an estimate of the potential mean force for the dissociation dynamics of gA; however, the potential mean force alone cannot be used to estimate the actual reaction rate of gA dissociation. This is because the potential of mean force is dependent on the selected reaction coordinate, which may be functionally independent of the reaction-path bottleneck taken for the dissociation to occur. Assuming that the reaction coordinate used to construct the potential mean force is defined to correctly describe the state transition, the energy barrier computed from the potential mean force can be used to construct the rate constant of the dissociation. The potential mean force of the gA dissociation event in Figure 15.7 is given in Figure 15.9. From Figure 15.9 the energy gap between the ground-state gA dimer and dissociated monomers is approximately 14 kcal/mol. To estimate the dissociation reaction rate, we use the Smoluchowski–Einstein equation with the energy defined by the PMF from the MD simulation, and the position-dependent diffusion coefficient estimated from MD simulation results. For dimer dissociation we are interested in

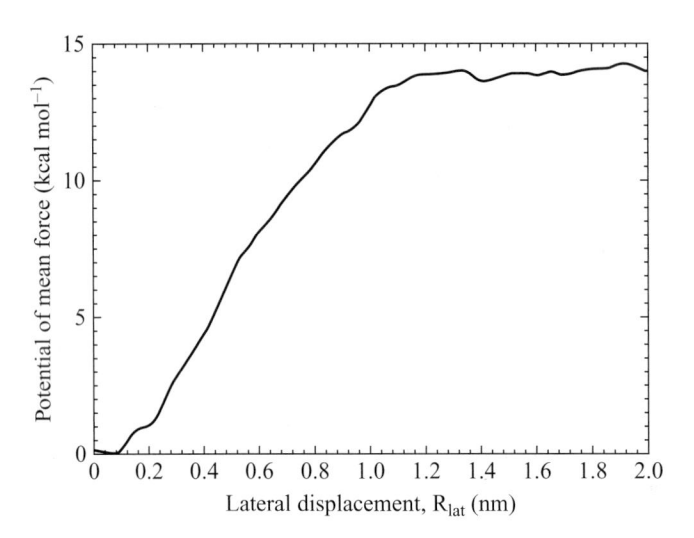

Figure 15.9 Potential of mean force of the gA ion channel with lateral displacement of the two gA monomers as illustrated in Figure 15.7.

computing the rate constant k_{AB} where the system transitions from the dimer, denoted by r_A, to the two monomers r_B. From the plot of potential of mean force, these reaction-coordinate positions are given by $r_A = 0$ nm and $r_B = 1.4$ nm. Let us assume that the reaction coordinate r is a stochastic variable with a probability density function $p(r, t)$ given by the Smoluchowski–Einstein equation (recall Chapter 11):

$$\frac{\partial p}{\partial t} = \frac{\partial}{\partial r}\left[D(r)\left(\frac{p}{k_B T}\frac{\partial F}{\partial r} + \frac{\partial p}{\partial r}\right)\right],$$

$$D(r) = \left(\frac{k_B T}{k}\right)^2 \left[\int_0^\infty \boldsymbol{E}[r(\xi)r(0)]\,d\xi\right]^{-1}, \tag{15.40}$$

where $D(r)$ is the position-dependent diffusion coefficient, k_B is Boltzmann's constant, T is the temperature, k is the coefficient of the harmonic potential used to constrain the gA monomers, F is the free energy, and $\boldsymbol{E}[\cdot]$ denotes the expectation operator. Note that the relation between potential mean force and free energy is still an open problem in the literature [310]. In this analysis we will assume the differences are negligible, with the potential mean force and free energy being approximately equal. The initial condition of (15.40) is given by $p(r, 0) = \delta(r_A)$, and the boundary conditions are no-flux boundary conditions. To estimate the rate constant k_{AB} for dimer dissociation, we assume it is given by the reciprocal of the mean first-passage time of the transition from state r_A to state r_B. From (15.40), the inverse of the rate constant k_{AB} is given by

$$k_{AB}^{-1} = \int_{r_A}^{r_B}\int_0^y \frac{1}{D(y)} e^{F(y)/k_B T} e^{-F(z)/k_B T}\,dz\,dy. \tag{15.41}$$

As seen from (15.41), the reaction rate of transitioning from state r_A to state r_B can be estimated by computing the free energy of the chemical reaction and integrating the position autocorrelation function. Both of these can be estimated from MD simulations. Setting r_A as the state of a gA dimer, and r_B as the state of two gA monomers, then (15.41) is equivalent to estimating the reaction rate f_6^{-1} in (10.12).

Remark: Estimating the rate constant k_{AB} in (15.41) requires knowledge of the transition path between the two states r_A and r_B. For example, to estimate k_{AB} for dimer dissociation, this transition path was specified as the distance between the center of mass of the two gA monomers. However, for more general reactions the transition path between the two states may not be known. Is there a method that allows estimation of k_{AB} without defining the reaction coordinate a priori? *Transition path sampling* is a powerful method to perform this task, and it does not require the reaction coordinate to be defined a priori. Using transition path sampling, in which the system evolves from one stable state A to another stable state B, it is possible to recover the rate constant of the reaction from A to B.

15.8 Complements and Sources

Molecular dynamics. MD simulation has been widely studied over several decades. For a general description of MD, refer to books such as [127, 144, 237, 339] or websites such as www.gromacs.org/, www.charmm-gui.org/, and www.ks.uiuc.edu/Training/Tutorials/namd-index.html.

All-atom MD simulations are used to model the trajectory of a system of interacting atoms in a specific statistical ensemble and can be used to estimate several important parameters of engineered membranes. The statistical ensemble (temperature, pressure, and energy) of an MD simulation are controlled using thermostats (Berendsen [40], velocity rescaling [59], and Nosé–Hoover [166, 303]) and barostats (Berendsen [40] and Parrinello–Rahman [304, 315]). MD is a sampling method such that, at each time t, the position and velocity of each atom is a sample from the phase space of the MD simulation. The phase space is the $6N$-dimensional space of all the possible positions x_i and momenta p_i of the N atoms. To estimate macroscopic parameters requires us to take statistical averages of these samples from the phase space. Therefore, given the results of an MD simulation (trajectory of atoms), time and spatial averages (mean-field approximations) are taken to estimate continuum parameters such as atom density, diffusion, and temperature. In addition, the tools of statistical mechanics (e.g., the Smoluchowski–Einstein equation [17, 311, 395] and Langevin dynamics [141]) and statistical sampling theory (umbrella sampling [448] and transition path sampling [17, 311, 395]) can be combined with MD to estimate the energy of a chemical reaction, the rate of chemical reactions, and the conductance of ion channels. Note that the field of studying reaction dynamics using MD is known as molecular reaction dynamics (MRD) [237].

Aqueous pores. Although the formation of aqueous pores is still not an experimentally observable process, MD can be used to simulate such an event. The aqueous pore formation process in cell membranes can result from several mechanisms: membrane stretching, electrochemical gradients, pore-forming proteins and peptides, vesicle fusion, and lipid flip-flop [35]. Using all-atom MD simulation it is possible to study how transient aqueous pores form in membranes [35, 48, 98, 118, 145, 213, 262, 324, 330, 425, 426]. The formation of these water fingers occurs on both sides of the DphPC membrane and is independent of the direction of the applied transmembrane potential.

Permeation. MD simulation also provides a method for modeling the dynamics of ion channels at short time scales (several nanoseconds). The three-dimensional structure of ion channels can be obtained using X-ray crystallography. This structure facilitates the construction of simulation models (such as Brownian dynamics, CGMD, and MD) to relate the atomic structure of the ion channels [141, 156, 218, 261] to their function. An introduction and review of several types of ion channel models is presented in [156, 218, 261]. The book [39] gives a lucid description of multiphysics models at various levels of abstraction.

The PNP theory is a continuum approach for modeling ion permeation and has been pioneered by Eisenberg and coworkers [80, 167, 176, 224, 265, 280, 371]. An issue with the classic PNP theory is that it neglects ion-ion interactions, nonelectrostatic effects between ions, steric effects, and screening effects of ions and water. The PNP theory may not be accurate for narrow ion channels such as gramicidin A, where the discrete ion effects cannot be neglected [218, 261]. This is because the mean-field approximation necessary for the PNP theory to be applicable breaks down when it is applied to narrow ion channels.

One way of accommodating for the shortcoming of PNP is to adopt a two-tiered approach utilizing MD coupled with Brownian or Lagrangian dynamics. MD simulation is used to compute various parameters of the system such as the potential of mean force [12, 352]; then these parameters are fed into the higher-level abstraction models, allowing simulation time horizons that are comparable to experimental data which is on the scale of 10 to 100 μs. This two-tiered method has been used by a number of research groups to model ion permeation. For example, Brownian dynamics (see the seminal paper [437]), together with the potential of mean force computed from MD, have been used to model permeation in various ion channels in [74, 75, 169, 444]. In [142, 385], the combination of MD and Markovian state-transition properties is combined to model cardiac action potentials. The combination of MD and random walks has also been used to model ion-channel permeation [43, 370]. Methods incorporating PNP theory with PMF approximations are active areas of research [12, 156, 264, 352]. Other methods to model ion permeation include velocity-autocorrelation functions, the second fluctuation-dissipation theorem, mean-square displacement, and the generalized Langevin equation for a harmonic oscillator [265]. In [141], distributional MD is used to model the ion permeation using a non-Markovian generalized Langevin dynamics model in combination with MD. Notice, however, that the results for ion permeation are highly dependent on the conformational dynamics of the ion channel and the environment in which the ion channel resides. Consider the gA channel, for example. Experimental studies of the gating characteristics of the gA dimer show that the dissociation-association rate is dependent on voltage, ion occupancy, and hydrocarbon thickness of the membrane [112, 343, 359]. Additionally, experimental studies have indicated that the gA dimer can exist in multiple conformational states [148, 256]. All these properties of the channel must be taken into account to construct an accurate estimate of the ion permeation and therefore the conductance of the ion channel.

Disassociation dynamics of gramicidin. The disassociation pathway of gA dimers was studied in [278] using Monte Carlo methods. They predicted that the disassociation of a gA dimer involves intermonomer hydrogen-bond breaking, backbone realignment, and relative monomer tilt at the intermonomer junction. The detailed MD study for the disassociation dynamics of gA dimers reported in §15.7 is from [448], where the potential of mean force of the gA dimer is computed as a function of the distances between centres of mass of the two gA monomers. By comparing PMF profiles obtained for lateral displacement and axial separation, [448] determined the dissociation energy for the gA dimer. The conclusion is that the gA dimer dissociates with lateral displacement as displayed in Figure 15.7, rather than as direct axial separation.

15.9 Closing Remarks

In this chapter, all-atom MD simulations were used to study the structure-function relationship of engineered membranes. We used MD to give insight into various aspects of the engineered membrane at the atomic scale. Also using MD simulation, we computed important biological parameters (membrane capacitance, dipole potential, and gramicidin A conductance) to study the dynamics of aqueous pore formation, and finally how gramicidin A dimers dissociate into monomers in the membrane. The results from all-atom MD simulation yield parameter values at the macroscopic level that are in excellent agreement with the experimentally measured current response from tethered membranes. This in turn yields a complete atomistic-to-device-level model of the engineered membrane.

A key future challenge to the development of MD simulation models is the incorporation of electron-shell interactions. As we saw from MD simulations of aqueous pore formation, hydrogen bonding plays a central role in the formation and destruction of aqueous pores. Methods that can account for correlated electronic structures include ab initio electron-correlated theory [460], or moving beyond the pairwise interactions of all-atom MD to many-body potential energy functions [79, 313]. However, these models have not been extended to the interaction dynamics of water and biomolecules such as peptides, proteins, and lipids. Additionally, the electrode-electrolyte dynamics also require the incorporation of electron-shell interactions. Methods for modeling the electrode-electrolyte interface is an active area of research in which hydrogen bonding plays an important role [440, 462].

16 Closing Summary for Part III: From Atoms to Device

Part III dealt with models for going from structure to function for engineered membranes and with four engineered membrane devices (ion-channel switch, pore formation measurement platform, electroporation measurement platform, and electrophysiological response platform). For the reader's convenience, Figure 16.1 summarizes the different levels of modeling abstractions. In Part III we dealt with multiphysics models ranging from classical molecular dynamics to macroscopic reaction-rate models.

With the detailed descriptions of the various levels of model abstraction, we are now in a position to summarize the complete atom-to-device model for the engineered membrane. Abstractly, the interaction of the macroscopic, mesoscopic, microscopic, and quantum-mechanical models is given by

$$I(t) = \text{macroscopic}\,(V(t), G_m, C_m, C_{dl}, p, R_e),$$
$$G_m, \ldots, R_e = \text{mesoscopic}\,(D, \gamma, \sigma, \ldots, h_m, h_r),$$
$$D, \gamma, \sigma, \ldots, h_m, h_r = \text{microscopic}\,(U, T, P),$$
$$U = \text{quantum mechanics},$$

where U is the interaction potential between atoms, T is the temperature, P is the pressure. The interaction potentials U between atoms can be estimated from quantum mechanics, which has no free parameters. This potential is then used in the microscopic model (all-atom and coarse-grained molecular dynamics) to estimate important parameters for the mesoscopic model. Details on the estimation of these mesoscopic model parameters from the microscopic models are provided in Chapters 14 and 15. Given the mesoscopic model parameters estimated from the microscopic models, the mesoscopic models in Chapters 10 and 11 are then used to estimate the parameters in the macroscopic models in Chapters 8 and 9. The macroscopic models are then used to estimate

Levels of Abstraction				
Ab Initio Molecular Dynamics	Classical Molecular Dynamics	Coarse-Grained Molecular Dynamics	Continuum Theory	Macroscopic Theory
nm fs	nm ns	nm μs	μm μs	m s

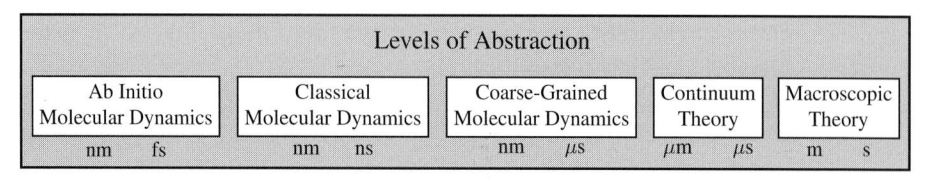

Figure 16.1 Schematic of the levels of abstraction for models of engineered artificial membranes. Ab initio is the lowest level of abstraction while macroscopic is the highest level.

Closing Summary for Part III: From Atoms to Device

the current response $I(t)$ in the engineered membrane that results from an excitation voltage $V(t)$. The complete system of models from molecular dynamics to macroscopic provides the atom-to-device model of the engineered membrane.

Conceptually, Part III can be viewed as a two-step approximation to real-life biological membranes. The first approximation step involves constructing an artificial membrane to mimic a biological membrane (as detailed in Part II). The second approximation is to mathematically model the dynamics of this artificial membrane from atom to device. This two-level modeling procedure provides substantial insight into the properties of living membranes.

Finally, §3.7 described future technologies involving engineered membranes such as implantable medical devices, diagnostics, and therapeutics; we believe that the multiphysics models detailed in Part III will become increasingly relevant to the engineering of these futuristic membrane-based technologies.

Appendices

Appendix A Elementary Primer on Partial Differential Equations (PDEs)

Chapters 10–12 described continuum models of engineered membranes that involved partial differential equations (PDEs). This appendix gives a brief introduction to the classification of PDEs, linear PDEs, nondimensionalization of PDEs, and numerical methods for solving PDEs. Uniqueness and existence results are not discussed here since they involve advanced results in functional analysis that are outside the scope of this book. For a comprehensive advanced treatment of PDEs at a graduate mathematics level see [115].

A.1 Linear, Semilinear, and Nonlinear Partial Differential Equations

PDEs are typically classified into linear, semilinear, and nonlinear. Linear PDEs are important as several methods exist to obtain closed-form solutions of these PDEs (under simple boundary conditions), including separation of variables, superposition, Fourier series, Laplace transform, and Fourier transform. The solution to some semilinear PDEs can be obtained using the symmetry method or method of characteristics. When a linear or semilinear PDE has more sophisticated boundary conditions (as is the case with engineered membranes), the PDE needs to be solved numerically. Similarly, nonlinear PDEs typically do not have an exact solution and must also be solved numerically.

Consider a generic PDE of two scalar variables (space x and time t), where $u(x, t)$ is a function of the variables x and t. A linear PDE has the form

$$\frac{\partial u}{\partial t} = a\frac{\partial^2 u}{\partial x^2} + b\frac{\partial u}{\partial x} + cu + d, \tag{A.1}$$

where $a(x, t)$, $b(x, t)$, $c(x, t)$, and $d(x, t)$ are generic functions of x and t but not u. A PDE is semilinear if the coefficient function $a(x, t)$ of the highest partial derivative is dependent on x and t but not u. Therefore, (A.1) is a semilinear PDE if the coefficient functions can be expressed as $a(x, t)$, $b(x, t, u)$, $c(x, t, u)$, and $d(x, t, u)$. If the PDE is neither linear nor semilinear, it is nonlinear. Linear, semilinear, and nonlinear PDEs are all used to model the dynamics of engineered membranes. For example:

Linear. Poisson's equation ((10.3) on page 180), the Nernst–Planck equation ((10.9) on page 182), the Poisson–Fermi equation ((11.23) on page 233), and the Fokker–Planck equation ((14.28) on page 318).

Semilinear. Reaction-diffusion equation ((10.36) on page 199).
Nonlinear. Smoulochowski–Einstein equation ((11.1) on page 217), and the generalized Nernst–Planck equation ((11.18a) on page 229).

Additionally, nonlinear systems of PDEs are also used to model the dynamics in engineered membranes.

A.2 Linear Partial Differential Equations and Boundary Conditions

Linear PDEs model several phenomena in engineered membranes, including the spatially dependent electric potential and diffusion of electrolyte ions.

Consider the linear PDE

$$\sum_{i,j=1}^{4} \partial_i(\partial_j a_{ij} u) + \sum_{i=1}^{4} \partial_i(b_i u) + \sum_{i=1}^{4} c_i u + d = 0, \qquad (A.2)$$

where a_{ij}, b_i, c_i, and d are generic coefficient functions that are not dependent on u. Note the indices $i \in \{1, 2, 3, 4\}$ represent the three spatial coordinates (x, y, z) and time t and ∂_i denotes the partial derivative with respect to this variable. Let us construct the symmetric matrix A with elements $A_{ij} = a_{ij}$. The matrix A plays an important role in classifying the solution characteristics of the linear PDE. If the matrix A has no zero eigenvalues, then the PDE (A.2) is classified as *hyperbolic*. Hyperbolic PDEs have the unique property that their solution can contain discontinuities. Therefore, these PDEs are typically used to model convection-driven transport problems, for example, in the advection transport equation of ions. If A is positive semidefinite and has exactly one eigenvalue at zero, then the PDE (A.2) is classified as *parabolic*. Parabolic PDEs are useful for modeling time-varying systems such as ions undergoing diffusion.[1] An example of a parabolic PDE is the diffusion equation. The advection-diffusion equation is an important PDE which combines both parabolic and hyperbolic partial differential equations to describe the movement of ions in an electrolyte. An example of an advection-diffusion equation is the Nernst–Planck equation ((10.9) on page 182) if no electric field is present. If A is positive definite then the PDE (A.2) is classified as *elliptic*. The solution of elliptic equations contains no discontinuities. Therefore, elliptic equations are useful for modeling systems that have reached equilibrium. Note that since the elliptic equation can describe the equilibrium state of a system, the solution of the elliptic equation is affected by the full domain, denoted by Ω, and boundary conditions defined on the boundary $\partial \Omega$ of the PDE. An example of an elliptic PDE is Poisson's equation ((10.3) on page 180).

To completely specify a PDE requires defining the domain Ω of the PDE, the boundary conditions $\partial \Omega$, initial conditions, and terminal conditions. The initial and terminal conditions of the PDE specify properties of u at specific times, whereas the boundary conditions specify properties of u at specific spatial locations. The domain Ω specifies

---

[1] Diffusion in fluids is also called hydrodynamic dispersion.

the spatial geometry and time where the PDE applies. The boundary condition $\partial \Omega$ defines a constraint for u on the spatial boundary of Ω. Four commonly used boundary conditions are the Dirichlet, Neumann, Robin, and mixed conditions.

Dirichlet condition. The Dirichlet condition specifies the u on the boundary of the domain and is given by

$$u(x, t) = g(x, t) \text{ in } \partial \Omega, \tag{A.3}$$

where $g(x, t)$ is a generic function. For Poisson's equation, the Dirichlet boundary condition is used to specify the electric potential on the boundary representing the electrode.

Neumann condition. The Neumann condition specifies the value of the normal derivative of u on the boundary of the domain. The Neumann boundary condition is

$$\nabla u(x, t) \cdot n = g(x, t) \text{ in } \partial \Omega, \tag{A.4}$$

where n is the normal derivative on the boundary and $g(x, t)$ is a generic function. For Poisson's equation, the Neumann boundary condition is used to specify that no change in electric potential occurs at the boundary.

Robin condition. The Robin boundary is a linear combination of the Dirichlet condition and the Neumann condition in which the linear combination of the value of u and the normal derivative of u (denoted by n) are defined on the boundary. The Robin condition is given by

$$g(x, t)u(x, t) + h(x, t)\nabla u(x, t) \cdot n = l(x, t) \text{ in } \partial \Omega, \tag{A.5}$$

where g, h, and l are generic functions. For Poisson's equation, the Robin boundary condition is used to account for the Stern double-layer capacitance present at the electrode surface.

Mixed condition. A mixed boundary condition represents a boundary-value problem in which the boundary of the domain Ω is composed of disjoint subsets with each having a unique boundary condition.

Additionally, other types of boundary conditions can be specified to account for surface reaction dynamics that are dependent on the value of u.

A.3 Nondimensionalization of Partial Differential Equations

From a practical point of view, nondimensionalization of PDEs (and also ODEs) is useful for reducing the number of parameters in the PDE, analyzing the behavior of the PDE, and also rescaling parameters of the PDE. Scaling the parameters of PDEs is a key step to performing perturbation analysis and was used in Chapter 9 for modeling the response of the ion-channel switch (ICS) biosensor in the reaction-rate-limited regime.

The procedure for nondimensionalization of a PDE (and ordinary differential equations (ODEs)) is given by the following three steps:

(i) Define dimensionless variables.
(ii) Define nondimensionalizing constants.
(iii) Construct the new dimensionless PDE (or ODE) model.

Below we apply the nondimensionalization method for an advection-diffusion PDE and a two–time scale ODE.

Example 1: Advection-Diffusion Equation with Langmuir Binding Kinetics

An important application of nondimensionalization is with the advection-diffusion equation coupled with surface reaction boundary conditions. Recall that this PDE was used for modeling the ICS biosensor and the pore formation measurement platform in Chapter 10. Below we illustrate how to apply the nondimensionalization technique to the advection-diffusion PDE.

Consider the advection-diffusion equation with surface reaction boundary conditions given by

$$\frac{\partial c}{\partial t} = D\nabla^2 c - \nabla \cdot (vc) = D\nabla^2 c - \frac{\partial}{\partial x}\left[\frac{6Qy}{Lh^3}(h-y)c\right] \quad \text{in } \Omega,$$

$$\frac{\partial b}{\partial t} = R_s = k_f c(b_{\max} - b) - k_r b \quad \text{on } \partial\Omega_{\text{surf}},$$

$$n \cdot D\nabla c = \begin{cases} -R_s & \text{on } \partial\Omega_{\text{surf}} \\ 0 & \text{otherwise} \end{cases}, \quad\quad\quad\quad (A.6)$$

$$c = c_o \quad \text{on } \partial\Omega_{\text{in}},$$

$$b(0) = 0 \quad \text{on } \partial\Omega_{\text{surf}}.$$

Here, c is the analyte concentration, D is the analyte diffusion coefficient, b is the surface-bound analyte concentration, Q is the input flow rate, and the parameters k_f, k_r, and $b_{\max}$ are the forward binding rate, reverse binding rate, and maximum allowable surface concentration of bound analyte, respectively. The domain Ω and boundary conditions $\partial\Omega_{\text{surf}}$ and $\partial\Omega_{\text{in}}$ are defined in Figure 10.3 on page 179. The goal is to nondimensionalize (A.6).

Let us select the scaling parameters:

$$\bar{c} = \frac{c}{c_o}, \bar{x} = \frac{x}{W}, \bar{y} = \frac{y}{h}, \tau = \frac{Dt}{h^2}, \text{ and } \bar{b} = \frac{b}{b_{\max}}, \quad\quad\quad\quad (A.7)$$

A.3 Nondimensionalization of Partial Differential Equations

where W and h are the width and height of the flow chamber, respectively. Substituting (A.7) into the PDE (A.6) results in

$$\frac{\partial \bar{c}}{\partial \tau} = \gamma^2 \frac{\partial^2 \bar{c}}{\partial \bar{x}^2} + \frac{\partial^2 \bar{c}}{\partial \bar{y}^2} - 6\lambda P_e \bar{y}(1-\bar{y})\frac{\partial \bar{c}}{\partial \bar{x}} \quad \text{in } \Omega,$$

$$\frac{\partial \bar{b}}{\partial \tau} = \bar{R}_s = D_a(\bar{c}(1-\bar{b}) - \tilde{c}\bar{b}) \quad \text{on } \partial\Omega_{\text{surf}},$$

$$n \cdot \bar{\nabla}\bar{c} = \begin{cases} -\bar{R}_s & \text{on } \partial\Omega_{\text{surf}} \\ 0 & \text{otherwise} \end{cases}, \qquad (A.8)$$

$$\bar{c} = 1 \quad \text{on } \partial\Omega_{\text{in}},$$

$$\bar{b}(0) = 0 \quad \text{on } \partial\Omega_{\text{surf}}.$$

The parameters D_a, P_e, γ, λ, and $\tilde{c}$ in (A.8) are related to the dimensional quantities in (A.6) as follows:

$$D_a = \frac{k_f c_o h^2}{D}, \quad P_e = \frac{Q}{DW}, \quad \lambda = \frac{h}{L}, \quad \gamma = \frac{h}{W}, \quad \tilde{c} = \frac{k_r}{c_o k_f}. \qquad (A.9)$$

The parameters in (A.9) play an important role in the dynamics of the biosensor. D_a is the Damköhler number and P_e is the Péclet number (which is the product of the Reynolds number and the Prandtl number). Here, D_a is defined as the ratio of the characteristic mixing time $\tau_D = h^2/D$ and the characteristic surface reaction time $\tau_R = 1/k_f c_o$. If $D_a \gg 1$, then $\tau_D \gg \tau_R$ and the biosensor operates in the diffusion-limited regime as the diffusion time scale of the analyte molecules is significantly larger than the surface reaction-rate time scale. If $D_a \ll 1$ then the biosensor operates in the reaction-rate-limited regime because diffusion occurs on a significantly faster time scale than the chemical reactions. Therefore, the Damköhler number D_a provides a measure of the relative influence of the mass-transport effects and the surface chemical reactions on the binding kinetics. Here, the Péclet number is defined as the ratio of the characteristic time for the analyte to diffuse across the channel, $\tau_D = h^2/D$, and the characteristic time $\tau_C = h^2 W/Q$ for the analyte to travel the same distance h perpendicular to the membrane surface as a result of convection. If $P_e \ll 1$, then $\tau_D \ll \tau_C$ and the mass-transport dynamics of the analyte molecules in the flow chamber are governed by diffusion. However, if $P_e \gg 1$ then the mass-transport dynamics of the analyte molecules in the flow chamber are governed by the convection term. Therefore, the Péclet number provides a measure of the mass-transport conditions of the analyte molecules in the flow chamber of the biosensor. The parameter $\tilde{c}$ gives the equilibrium concentration of bound analyte species. The parameters in (A.9) can be used to infer important properties of the biosensor. For example, if $P_e \ll 1$ and $\lambda \ll 1$ (e.g., for a very slow flow rate Q and thin flow chamber $h \ll L$) then nearly all analyte molecules in the electrolyte will bind to the surface of the biosensor. However, if $P_e \gg 1$ and $\lambda \gg 1$ (e.g., for a very fast flow rate Q and thick flow chamber $h \gg L$) then only a few analyte molecules will bind to the biosensor surface. If $\lambda \ll 1$ (very thin flow chamber $h \ll L$), $c_o h \ll b_{\text{max}}$ (significantly larger number of binding sites compared to analyte molecules), and $D_a \gg 1$ (diffusion-limited surface binding kinetics), then a traveling wave of equilibration will occur on the surface of the biosensor because the target molecules remain in solution until they reach an unbound section of the biosensor surface.

Example 2: Two–Time Scale Pore Formation Reaction Mechanism

Chemical reaction kinetics can have slow reactions and fast reactions; that is, the reaction system has two time scales. Examples include the chemical reactions in the ICS biosensor in Chapter 9 and the pore formation process in Chapter 10. Singular perturbation theory [201, 207] can be used to approximate the solution of the two–time scale system. This technique was applied to the ICS biosensor to construct the reaction-limited equation ((9.10) on page 166). But before we can apply singular perturbation we need to identify which variables are slow and which variables are fast. It is here that nondimensionalization is crucial. Below we illustrate how to apply nondimensionalization and singular perturbation theory to an ODE that models the dynamics of the pore formation reaction mechanism.

Consider the following pore formation reaction mechanism ODE:

$$\frac{dp_1}{dt} = k_f a(p_{\max} - p_1 - np_n) - k_r p_1,$$

$$\frac{dp_n}{dt} = k_p p_1^n, \tag{A.10}$$

where p_1 is the protomer concentration, p_n is the conducting pore concentration, k_f, k_r, and k_p are reaction rates, n is the number of protomers in each conducting pore, a is the analyte concentration of peptides, and $p_{\max}$ is the maximum concentration of protomers in the membrane. Notice that (A.10) is similar to the surface reaction mechanism proposed for peptidyl-glycine-leucine-carboxyamide (PGLa) in (10.33) on page 197. Typically, the rate constants in (A.10) satisfy the relations $k_f > k_r$ and $k_f \gg k_p$.

Let us assume we are in the reaction-rate-limited regime such that a in (A.10) is a constant. The first task is to nondimensionalize (A.10). Here we select the scaling parameters:

$$\bar{p}_1 = \frac{p_1}{p_{\max}}, \quad \bar{p}_n = \frac{p_n}{p_{\max}}, \quad \tau = \varepsilon t, \quad \varepsilon = k_p. \tag{A.11}$$

The parameter ε was selected as the smallest time constant of the ODEs (A.10). Substituting (A.11) into the PDE (A.10) results in the two–time scale system

$$\varepsilon \frac{d\bar{p}_1}{d\tau} = k_f a(1 - \bar{p}_1 - n\bar{p}_n) - k_r \bar{p}_1,$$

$$\frac{d\bar{p}_n}{dt} = \bar{p}_1^n. \tag{A.12}$$

In (A.12) the fast variable is $\bar{p}_1$ and the slow variable is $\bar{p}_n$. Using Tikhonov's theorem and the initial condition $\bar{p}_n(0) = 0$, it is possible to approximate the solution to (A.12) as $\varepsilon \to 0$. The approximation solution to (A.12) is

$$\frac{d\hat{\bar{p}}_n}{d\tau} = \left(\frac{k_f a}{k_f a + k_r}\right)^n (1 - n\hat{\bar{p}}_n)^n, \quad \hat{\bar{p}}_n(0) = 0, \tag{A.13}$$

where $\hat{\bar{p}}$ denotes the approximate solution of $\bar{p}_n$. Tikhonov's theorem states that if $\bar{p}_n(0)$ is within an $O(\varepsilon)$ neighborhood of

$$\bar{p}_1 = \frac{k_f a(1 - n\bar{p}_n)}{k_f a + k_r},$$

then for all time $\tau \in [0, T]$, $|\bar{p}_n(\tau) - \hat{\bar{p}}_n| = O(\varepsilon)$, where τ denotes a finite time horizon.

A.4 Solutions of Partial Differential Equations

Numerical methods for solving PDEs are the subjects of numerous books and papers. In this section we briefly discuss numerical methods for solving PDEs, which include the weak form, discretization methods, and time-differencing methods. All the PDEs in this book were solved numerically using COMSOL.[2] Our aim is to give a short listing of the methods COMSOL uses to solve PDEs.

The strong form of the PDE along with the boundary conditions states the conditions at every point $x \in \Omega$ over a domain Ω and boundary $\partial\Omega$ that a solution must satisfy.[3] As such, the strong form imposes continuity and differentiability requirements on the solution. An issue with using the strong form is that typically PDEs do not have a strong solution. However, it is possible to construct weak solutions of the PDEs using the weak form of the PDE. The weak form represents the PDE and boundary conditions as an integral expression. The solution of the integral expression satisfies the conditions of the original PDE and boundary conditions in an integral sense (e.g., on an average). A weak formulation of the PDE and boundary conditions can be constructed by applying the following steps:

- Multiply each equation by a generic function $w(x)$.
- Integrate this product over the spatial domain Ω.
- Use integration by parts to reduce the order of the partial derivative operators to a minimum.
- Introduce the boundary conditions $\partial\Omega$ into the integral expression if possible.

The result of applying the above steps to the linear PDE (A.2) will yield the weak form of the PDE,

$$\int_\Omega \left(\sum_{i,j=1}^4 \partial_i(\partial_j a_{ij} u) + \sum_{i=1}^4 \partial_i(b_i u) + \sum_{i=1}^4 c_i u + d \right) dV(x) = 0,$$

$$\int_\Omega (L(u) + d)\, dV(x) = 0, \tag{A.14}$$

---

[2] The multiphysics finite-element solver COMSOL can be obtained from COMSOL Inc. (founded in Stockholm, Sweden, in 1986) or from the website www.comsol.com/.
[3] This is known as the strong solution or classical solution to the PDE.

where $L(u)$ is a differential operator, and $dV(x)$ is the spatial volume element. Note that the integral is only over the spatial domain Ω and not time t. The main advantage of the weak formulation (A.14) is that the integral expression can be evaluated using numerical techniques.

To numerically evaluate u using the weak form (A.14) with the associated boundary conditions, we approximate u using a set of N basis functions,[4] denoted by $\Phi_n(x)$, as follows:

$$u \approx \hat{u} = \sum_{n=1}^{N} u_n(t)\Phi_n(x), \tag{A.15}$$

where $u_n(t)$ are unknown coefficients that approximate the solution of u. Substituting (A.15) into (A.14) results in

$$\int_{\Omega} \left[\sum_{n=1}^{N} u_n(t) L(\Phi_n(x)) + d(x,t) \right] dV(x) = 0. \tag{A.16}$$

The term contained in $[\cdot]$ of (A.16) can be viewed as the residual error between $\hat{u}$ and u. Expression (A.16) states that the mean residual error between $\hat{u}$ and u is minimized. Note, however, that this is not desired because it does not restrict the magnitude of the residual in the domain Ω – it only restricts the mean residual error. Instead, we desire the solution to have minimal residual error over the entire domain Ω. To accomplish this requires introducing a test function $w(x)$ into (A.16), which results in

$$\int_{\Omega} w(x) \left[\sum_{n=1}^{N} u_n(t) L(\Phi_n(x)) + d(x,t) \right] dV(x) = 0. \tag{A.17}$$

For the solution of $f_n(t)$ in (A.17) to have minimal residual error over the domain Ω requires the test function $w(x)$ to exist in a Sobolev space (informally, a function space where weak derivatives up to order k exist that are L_p integrable for specified positive integers k and p). A common method to construct the test function $w(x)$ is via the basis function expansion

$$w(x) = \sum_{m=1}^{M} w_m \Psi_m(x), \tag{A.18}$$

where w_m are unknown scalar coefficients, and $\Psi_m(x)$ is a smooth function with compact support. Substituting (A.18) into (A.17) results in

$$\sum_{m=1}^{M} \int_{\Omega} w_m \Psi_m(x) \left[\sum_{n=1}^{N} u_n(t) L(\Phi_n(x)) + d(x,t) \right] dV(x) = 0. \tag{A.19}$$

The solution of (A.16) for the unknown coefficients $u_n(t)$ and w_m provides the approximate solution to u where the residual error is minimized over the domain Ω.

---

[4] The basis functions are also commonly referred to as shape functions and interpolation functions.

A.4 Solutions of Partial Differential Equations

Given (A.19), the main idea is to select the basis functions Φ and Ψ such that $\hat{u}$ can be evaluated by solving a system of algebraic equations (if u is not time dependent) or a system of ordinary differential equations (if u is time dependent). Several methods can be used for this task including the finite-difference, finite-volume, and finite-element method. The only difference between these methods is how the basis functions Φ and Ψ are selected in (A.19). For a detailed discussion of numerical methods for PDEs refer to [34, 171, 187].

All numerically computed solutions to the PDEs in this book are evaluated using the commercial finite element analysis (FEA) solver COMSOL. In FEA, the coordinate space Ω is partitioned into a set of M discrete elements Ω_e such that

$$\Omega = \cup_{e=1}^{M} \Omega_e. \qquad (A.20)$$

This is known as meshing the domain Ω. Each element in Ω_e is associated with a set of vertices (element nodes). The vertices that result from meshing are used to define the properties of the basis functions $\Psi_m(x)$ for the test function $w(x)$ (A.18) and $\Phi_n(x)$ for the trial function $\hat{u}$ (A.15). These basis functions are then used to transform (A.19) into a system of ordinary differential equations. There exist several methods to solve systems of ordinary differential equations. In COMSOL, we utilize a variable-order variable-step-size backward-differentiation formula to evaluate the system of ordinary differential equations.

Appendix B Tutorial on Coarse-Grained Molecular Dynamics with Peptides

In this appendix we use GROMACS 5.1, PyMOL 1.7x, and VMD 1.9.3, together with the MARTINI force field to outline how to set up coarse-grained molecular dynamics (CGMD) simulations of a peptide. Our description includes how to construct the all-atom structure of the peptide, how to build all-atom models of bilayer lipid membranes, how to insert the peptides into the membrane, and how to construct publication-quality figures. The material in this appendix is relevant to Chapter 14 for constructing coarse-grained molecular dynamics simulations using the MARTINI force field. Note that performing CGMD simulations and MD simulations are identical except for the definitions of the bead-particle interactions and simulation time steps used. All the files associated with these tutorials are provided at www.cambridge.org/engineered-artificial-membranes.

B.1 Constructing the All-Atom and Coarse-Grained Structure of a Peptide

This section illustrates how to construct the all-atom and coarse-grained structure of the antimicrobial PGLa peptide using VMD and PyMOL. Recall that molecular dynamics (MD) and PyMOL were discussed in Chapter 2, and PGLa was discussed in Chapter 10. The goal is to construct the all-atom and coarse-grained structures of PGLa illustrated in Figure B.1.

To construct the all-atom coordinate structure of the PGLa peptide the following steps are used.

  (i) Obtain the amino-acid sequence of interest. For PGLa the amino acid sequence is:

  GMASKAGAIAGKIAKVALKAL-NH_2.

  (ii) Open VMD, and proceed to Extensions → Modeling → Molfacture. Click the "Start Molefacture", leaving the entry field blank.
  (iii) In Molefacture, Build → Protein Builder. Enter the amino-acid sequence of PGLa, and select the α-helix secondary structure (this is the expected secondary structure of PGLa in both the surface and transmembrane configuration as discussed in Chapter 14). Now click build.
  (iv) After pressing build, the all-atom structure of PGLa will appear in the VMD display. In VMD Main, select File → Save Coordinates, and save the all-atom PGLa

B.1 Constructing the All-Atom and Coarse-Grained Structure of a Peptide

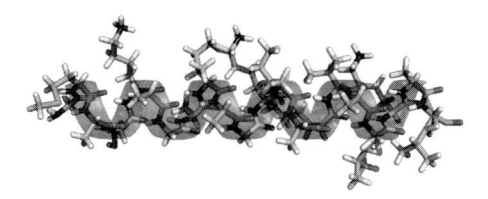

(a) All-atom structure of PGLa:
GMASKAGAIAGKIAKVALKAL-NH$_2$.

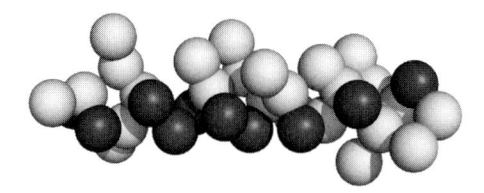

(b) Coarse-grained representation of the PGLa peptide.

Figure B.1 All-atom and coarse-grained structure of the PGLa antimicrobial peptide.

structure as PGLa.pdb. The file PGLa.pdb contains the all-atom representation of the PGLa peptide in the protein data bank format.

Notice that this PGLa.pdb file can be viewed in both VMD and PyMOL.

Having constructed the PGLa.pdb coordinate file, we now map the all-atom structure into a coarse-grained structure for use with the MARTINI force field. This mapping operation is performed using the *martinize.py* python script (available from http://md.chem.rug.nl/cgmartini/index.php/home). The detailed steps to perform this mapping from the all-atom structure of PGLa to the coarse-grained structure of PGLa are provided below.

(i) To use the martinize.py python script requires that a PGLa.ssd file be constructed that defines the secondary structure of PGLa. To construct the PGLAs.ssd file we use the program *dssp-2.0.4-win32.exe* (available from www.mmnt.net/db/0/0/ftp.cmbi.kun.nl/pub/molbio/software/dssp-2). The letters contained in the PGLa.ssd file define the secondary structure of PGLa (e.g., H represents an α-helix). A detailed description of the format of the PGLa.ssd file is provided at www.cmbi.ru.nl/dssp.html.

(ii) Now use martinize.py to produce three files: cg_PGLa.gro (coordinate file), PGLa.top (topology file), and PGLa.itp (structure file).
Command:
./martinize.py -f PGLa.pdb -o PGLa.top -x cg_PGLa.gro -ss PGLa.ssd -p backbone

(iii) Open the file PGLa.gro using VMD, and check that the structure is identical to that seen in Figure B.1.

Figure B.2 Coarse-grained 128-lipid DphPC bilayer with water removed.

B.2 Construction of Coarse-Grained Lipid Bilayer

There are two primary methods for constructing coarse-grained (MARTINI) lipid bilayers using VMD, PyMOL, and GROMACS. The first is by a self-assembly process starting from a single lipid topology and coordinate file. The outline of the steps necessary to perform the self-assembly can be found at http://cgmartini.nl/index.php/tutorials-general-introduction/bilayers. Here we provide a second possible approach to construct a CGMD simulation of a DphPC bilayer from a preequilibrated CGMD DPPC lipid bilayer. Note that DPPC and DphPC have an equivalent coarse-grained representation in the MARTINI force field. Additionally, various membranes can be found at http://cgmartini.nl/index.php/force-field-parameters/lipids. Here we focus on the construction of the coarse-grained DphPC bilayer membrane. The steps to construct the DphPC bilayer are provided below.

(i) Obtain the files martini_v2.0_lipids.itp, martini_v2.2.itp, dppc_bilayer.gro, and dppc_bilayer.top. The *.itp files contain the MARTINI force-field definition of the bead interactions, the *.gro file is the coordinate file of the DPPC bilayer, and the *.top file defines the number of each type of molecule in the *.gro file. In addition to these files, we also require the water-1bar-303K.gro which will be used to solvate the coarse-grained membrane.

(ii) Open the dppc_bilayer.gro file and remove all the water ions (W). In the dppc_bilayer.top file, remove the W molecules as these are not contained in the *.gro file after removal. The resulting dry_dppc_bilayer.gro topology after water removal is given in Figure B.2.

(iii) Before we construct a larger lipid bilayer from the dry_dppc_bilayer.gro, we must remove the periodicity which allows subsections of the molecules to reside outside the dimensions of the simulation cell. GROMACS automatically recognizes that these are from the same molecule using the simulation cell dimensions. However, since we are going to adjust the simulation cell dimensions we must remove this periodicity. Once the periodicity is removed, then we can replicate

B.2 Construction of Coarse-Grained Lipid Bilayer

Figure B.3 Coarse-grained 512-lipid DphPC bilayer.

the bilayer structure contained in the dry_dppc_bilayer.gro file. The commands to do this are included below. The final result is provided in Figure B.3.

Command:
```
gmx grompp -f em1.mdp -c dry_dppc_bilayer.gro -p dppc_bilayer.top -maxwarn 10 -o em1.tpr
```
gmx trjconv -s em1.tpr -f dry_dppc_bilayer.gro -p dppc_bilayer.top -o dry_dppc_nper.gro -pbc mol -ur compact
```
gmx genconf -f dry_dppc_nper.gro -o dry_dppc_ext.gro -nbox 2 2 1
```
Note: The topology file dppc_bilayer.top must be updated with the new DPPC lipids.

(iv) Do a short energy minimization with em2.mdp.
Command:
```
gmx grompp -f em2.mdp -c dry_dppc_ext.gro -p dppc_bilayer.top -maxwarn 10 -o em2.tpr
mdrun -v -deffnm em2
```

(v) Expand the simulation box for inclusion of PGLa and water.
Command:
```
gmx editconf -f em2.gro -o dppc_box.gro -center 6.31915 6.46099 6 -box 12.63830 12.92198 12
```

(vi) Add 3 coarse-grained PGLa to the simulation structure. The commands are given below. The final result is presented in Figure B.4.
Command:
```
gmx editconf -f cg_PGLa.gro -o PGLa_box1.gro -center 6.31915 6.46099 8.5 -box 12.63830 12.92198 12
gmx editconf -f cg_PGLa.gro -o PGLa_box2.gro -center 4.31915 4.46099 8.5 -box 12.63830 12.92198 12
gmx editconf -f cg_PGLa.gro -o PGLa_box3.gro -center 8.31915 8.46099 8.5 -box 12.63830 12.92198 12
cat PGLa_box1.gro PGLa_box2.gro PGLa_box3.gro em2_ext.gro > system.gro
```
Note: Edit the system.gro coordinate file using a standard text editor to have 6252 beads which correspond to the 3 PGLa molecules and 512 DphPC molecules.

Figure B.4 Coarse-grained 512-lipid DphPC bilayer with 3 PGLa.

Additionally, remove all coordinate file lines that result from the cat operation as these will cause an error when reading the system.gro file. Check the system.gro structure using VMD, and write a system.top file for the system.gro coordinates. (This will have a very similar structure to the dppc_bilayer.top file except now we include the 3 PGLa molecules.) Remember to add #include "PGLa.itp" to the system.top file that defines the interactions and bonds of beads in the PGLa molecule for GROMACS.

(vii) Do a short energy minimization. Note that "defin=-DPOSRES" should be added to the equilibration file em3.mdp to constrain the PGLa from moving substantially. The commands are given below.
Command:
```
gmx grompp -f em3.mdp -c system.gro -p system.top -maxwarn 10 -o em3.tpr
gmx mdrun -v -deffnm em3
```

(viii) Check the energy minimization by analysis of the potential and density to ensure convergence. The commands are given below.
Command:
```
gmx energy -f em3.edr -o potential.xvg
xmgrace potential.xvg
```

(ix) Solvate the system with the equilibrated water system water-1bar-303K.gro.
Command:
```
gmx solvate -cp em3.gro -cs water-1bar-303K.gro -p system.top -radius 0.25 -o wet_system.gro
```
Note: Remember to update the system.top file to include the added water molecules.

(x) Perform another energy minimization.
Command:
```
gmx grompp -f em4.mdp -c wet_system.gro -p system.top -maxwarn 10 -o em4.tpr
gmx mdrun -v -deffnm em4
```

(xi) Conduct production MD using md.mdp.
Note: In the md.mdp file, tc-grp= W DPPC Protein, for the groups, with tau-t and ref-t set accordingly. Additionally, this production run is extremely short and not suitable for sampling. Typically hundreds of nanoseconds of simulation time

are required for the estimation of biological parameters such as the diffusion coefficient of lipids.
Command:
```
gmx grompp -f md.mdp -c em4.gro -p system.top -maxwarn 10 -o prod.tpr
mdrun -v -deffnm prod
```
(xii) Analysis of the potential energy and density of molecules from the production run. The program xmgrace is used to plot the information in the *.xvg files generated from the GROMACS.
Command:
```
gmx energy -f prod.edr -o potential.xvg
xmgrace potential.xvg
gmx energy -f prod.edr -o density.xvg
xmgrace density.xvg
```
(xiii) Extending simulations
Open the md.mdp file, and adjust the number of steps to extend the simulation.
Command:
```
gmx grompp -f md.mdp -c prod.tpr -o contprod1.tpr -t prod.cpt -p system.top -maxwarn 10
gmx mdrun -v -deffnm contprod1
```

B.3 How to Insert PGLa Peptides in the Transmembrane State

In this section we use GROMACS, the MARTINI force field, and PyMOL to insert the antimicrobial peptide PGLa into an equilibrated and solvated DphPC membrane. Recall that this method was used to construct the CGMD results presented in Chapter 14. The PGLa peptides are placed in the interior of the membrane parallel to the lipid tail groups prior to equilibration to allow us to sample relatively rare transmembrane configurations efficiently.

The command line steps to run the simulation are provided below.

(i) Prepare the membrane topology file by centering the center of mass of the molecules in the box and use the unit cell representation such that all atoms are put closest to the center of the box. Convert both dry_dppc_bilayer.gro and cg_PGLa.gro files to *.pdb files for input to PyMOL.
Command:
```
gmx editconf -f cg_PGLa.gro -o cg_PGLa.pdb
gmx editconf -f dppc_bilayer.gro -o dppc_bilayer.pdb
```
(ii) We now open the files cg_PGLa.pdb and dppc_bilayer.pdb in PyMOL to position the PGLa in the transmembrane position. To begin editing molecules in PyMOL go to "Mouse→3 Button Editing". Ensure that PGLa peptide is deprotected and the DphPC membrane state is frozen. This can be done in the graphical user interface by pressing A at the top right beside the associated molecule group and selecting A→state→freeze, A→movement→deprotect, respectively. To copy the PGLa molecule: A→duplicate object, then rename the created object. Position the

PGLa to the desired position in the membrane, and then output the membrane as merged_dppc.pdb and PGLa as merged_pgla.pdb.

To prevent atom reordering in PyMOL, input the following lines in the PyMOL terminal:

Command:
```
set retain_order, 1
set pdb_retain_ids, 1
```

(iii) Create the files dppcM.gro and PGLaM proteins.gro to be used with genbox. This can be done by converting merged_dppc.pdb and merged_pgla.pdb files to dppcM.gro and PGLaM.gro using the GROMACS function editconf. Change the dimensions of the simulation cell of both the dppcM.gro and PGLaM.gro files to [12.6380 12.92198 10.05482] (the original dimensions of the dppc_bilayer.gro file).

(iv) Place the peptides in the membrane and remove any lipids and/or solvents that overlap.

Command:
```
gmx genbox -cp PGLaM.gro -cs dppcM.gro -radius 0.21 -o system.gro
```

(v) Create a system.top file for the system.gro coordinate file. At this point you can also add further water molecules to the simulation cell by first solvating the system and performing a short energy minimization. This is only necessary if too many water molecules were removed during the PGLa insertion, or if the original dppcM.gro file contains no water molecules.

(vi) Perform an energy minimization on the system.

Command:
```
gmx grompp -f em2.mdp -c system.gro -p system.top -maxwarn 10 -o em2.tpr
mdrun -v -deffnm em2
```

(vii) Check the energy minimization to ensure it was successfully completed.

Command:
```
gmx energy -f em2.edr -o potential.xvg
xmgrace potential.xvg
```

(viii) Perform a full production run (the first 1 ns or so is part of the equilibration to allow the lipids to surround the peptides).

Command:
```
gmx grompp -f md.mdp -c em2.gro -p system.top -maxwarn 10 -o prod.tpr
mdrun -v -deffnm prod
```

B.4 Note on Publication-Quality Figures

PyMOL is an excellent tool for making movies and publication-quality figures; however, for poster-sized images POV-Ray[1] is a dedicated 3D ray-tracing program that can read

---

[1] www.povray.org/

B.4 Note on Publication-Quality Figures

Figure B.5 Generated image from PyMOL and POV-Ray that is comprised of 512 POPC lipids and 9 PGLa antimicrobial peptides.

files generated by PyMOL and is suitable for large poster-sized or cover photos for journals. Figure B.5 provides a rough illustration of the capabilities of using PyMOL in combination with POV-Ray.

Appendix C Experimental Setup and Numerical Methods

This section details the precise ambient conditions and hardware specifications under which all experiments reported in this book were conducted. §C.1 provides the parameters for the ion-channel switch (ICS) biosensor experiments. §C.2 provides the experimental parameters estimated for the pore formation measurement platform (PFMP) for the dynamics of peptidyl-glycine-leucine-carboxyamide (PGLa) pore formation. §C.3 provides the experimental parameters of the electroporation measurement platform (EMP) model. Finally, §C.4 through §C.5.1 discuss coarse-grained molecular dynamics (CGMD) and molecular dynamics (MD) simulation parameters.

C.1 Ion-Channel Switch Biosensor

All experimental measurements, unless otherwise stated, were conducted at 27°C in a phosphate-buffered solution with a pH of 7.2, and a 0.15 M saline solution composed of Na^+, K^+, and Cl^-. At this temperature the tethered membrane is in the liquid phase. A pH of 7.2 was selected to match that typically found in the cellular cytosol of real cells. The forward and reverse reaction rates in Table C.1 are obtained from [220, 221, 283]. In Table C.1, notice that the computed diffusion coefficient for ferritin is 80 $\mu m^2/s$ and that for hCG is 250 $\mu m^2/s$; this is an expected result because the ferritin has a molecular weight of 450 kDa and hCG a molecular weight of 25.7 kDa.

The governing equations (10.10), (10.12), (10.13), and (10.15) with the boundary conditions (10.11) are solved numerically with the commercially available finite-element solver COMSOL 4.3a (Comsol Multiphysics, Burlington, MA). To solve the advection-diffusion equations (10.10) and (10.15) the COMSOL module *Transport of Diluted Species* is used. The surface reaction diffusion equation (10.13) is solved using the *Weak Form Boundary PDE* module. The simulation domain is meshed with approximately 46,000 triangular elements constructed using an advancing-front meshing algorithm. The generalized Poisson–Nernst–Planck and Poisson–Nernst–Planck are numerically solved using the *multifrontal massively parallel sparse direct solver* [14] with a variable-order variable-step-size backward-differential formula [53]. The Levenberg–Marquardt algorithm (10.17) is implemented using the MATLAB function *lsqnonlin* with (10.16) computed from the results of the COMSOL simulations.

Table C.1 Model parameters for ICS biosensor.

Symbol	Definition	Value
L_w	Width of flow chamber	3.0 mm
L	Length of flow chamber	0.7 mm
h	Height of flow chamber	100 μm
D_s^j	Surface bound diffusion	0–12 μm^2/s
$c_s^w(0), c_s^x(0), c_s^y(0), c_s^z(0)$	initial surface concentration	0 mol/m^2

Symbol	Definition	Streptavidin	Ferritin	TSH	hCG
c_o^A	Inlet analyte concentration	10 pM–1 nM	50–600 pM	100 fM–312 pM	10–353 nM
$c_s^c(0)$	Mobile gA monomers	16 pmol/m^2	16 pmol/m^2	16 pmol/m^2	1.6 pmol/m^2
$c_s^s(0)$	Tethered gA monomers	166 pmol/m^2	166 pmol/m^2	16 pmol/m^2	16 pmol/m^2
$c_s^b(0)$	Tethered binding sites	16 pmol/m^2	16 pmol/m^2	33 pmol/m^2	33 pmol/m^2
$c_s^d(0)$	gA dimers	16 pmol/m^2	16 pmol/m^2	33 pmol/m^2	33 pmol/m^2
$f_1 = f_2 = f_6$	Forward reaction rate	4×10^3 m^3/smol	4×10^3 m^3/smol	8×10^3 m^3/smol	5×10^2 m^3/smol
$f_3 = f_4$	Forward reaction rate	3×10^{11} m^2/smol	3×10^{11} m^2/smol	3×10^{11} m^2/smol	3×10^9 m^2/smol
$f_5 = f_7$	Forward reaction rate	6×10^9 m^2/smol	6×10^9 m^2/smol	6×10^8 m^2/smol	3×10^{11} m^2/smol
$r_1 = r_2 = r_6$	Reverse reaction rate	10^{-6} s^{-1}	10^{-6} s^{-1}	10^{-6} s^{-1}	10^{-4} s^{-1}
$r_3 = r_4$	Reverse reaction rate	10^{-6} s^{-1}	10^{-6} s^{-1}	10^{-6} s^{-1}	10^{-4} s^{-1}
$r_5 = r_7$	Reverse reaction rate	1.5×10^{-2} s^{-1}	1.5×10^{-2} s^{-1}	1.5×10^{-2} s^{-1}	1.0×10^{-2} s^{-1}
D^a	Diffusivity of analyte a	150 μm^2/s	80 μm^2/s	150 μm^2/s	250 μm^2/s
Q	Flow rate	100 μL/min	10 μL/min	100 μL/min	100 μL/min

Table C.2 Model parameter for varying PGLa (Figure 10.10).

Symbol	Definition	Value			
r_a	Effective radius of PGLa	1.33 nm			
p_{max}	Saturated surface concentration	5×10^{-10}			
k_a	Adsorption rate constant	5000 m^2/smol			
k_d	Desorption rate constant	0 1/s			
k_d	Desorption rate constant	0 1/s			
k_p	Rate of protomer formation	0.5 1/s			
k_1	Rate of protomer binding	0.5 1/s			
n	n in (10.33)	1			
D_a	Analyte diffusion coefficient	2–5 nm^2/ns			
PGLa concentration a_o:		10 μM	20 μM	30 μM	40 μM
m	m in (10.33)	2	2	3	3
k_c	Rate of closing	1×10^2 m^2/smol	1.2×10^6 m^2/smol	26×10^{15} m^4/smol^2	1.1×10^{15} m^4/smol^2

C.2 Pore Formation Measurement Platform: PGLa

The maximum concentration of PGLa in a solution can be computed by estimating the average volume occupied by 1 mol of PGLa, then computing the associated average concentration of PGLa that would reside in a 1 m^3 volume if only PGLa was present. The average volume of PGLa is 1438 cm^3/mol which is computed by multiplying the molecular weight of PGLa (1970 g/mol) by the average protein specific volume of PGLa (0.73 cm^3/g) [122]. The maximum concentration of PGLa in solution is 695 mol/m^3 or 695 mM; this corresponds to $r_a = 1.33$ nm in (10.35). The maximum surface concentration of membrane-bound PGLa and protomers is taken as corresponding to 1 percent of the total molar concentration of the tethered membrane lipids. Each lipid in the tethered membrane has a surface area of 0.68 nm^2; therefore, for a 2.1 mm^2 membrane there are approximately 3×10^{12} lipids in the surface layer. Therefore, the maximum surface concentration of membrane-bound PGLa and protomers is 2.5×10^{-8} mol/m^2. This corresponds to an effective surface radius for each membrane-bound PGLa and protomer (i.e., r_m and r_i in (10.36)) to be 8 nm. The maximum surface concentration for PGLa, m_{max}, defined following (10.33), was selected to match that of the maximum surface concentration of cytolysin A [432]. The reaction-diffusion parameters for numerical results presented in Figures 10.10 and 10.12 are provided in Tables C.2 and C.3, respectively.

The governing equations (10.35) and (10.36) with boundary conditions (10.37) and initial conditions (10.38) are solved numerically with the commercially available finite-element solver COMSOL 4.3a (Comsol Multiphysics, Burlington, MA). The simulation domain is meshed with approximately 28,199 triangular elements constructed using an advancing-front meshing algorithm. Equations (10.35) and (10.36) are numerically solved using the *multifrontal massively parallel sparse direct solver* [14] with a variable-order variable-step-size backward-differential formula [53]. Equation (10.39)

Table C.3 Model parameter for varying POPG (Figure 10.12).

Symbol	Definition	Value
r_a	Effective radius of PGLa	1.33 nm
$m_{\max}$	Saturated surface concentration	5×10^{-10}
k_a	Adsorption rate constant	$1 \times 10^5 - 2 \times 10^5$ m^2/smol
k_d	Desorption rate constant	0 1/s
k_p	Rate of protomer formation	10 1/s
n	n in (10.33)	3
k_1	Rate of protomer binding	$0.6 \times 10^{15} - 1.5 \times 10^{15}$ m^4/smol^2
m	m in (10.33)	1
k_c	Rate of closing	$1 \times 10^{-3} - 1.3 \times 10^{-3}$ 1/s
D_a	Analyte diffusion coefficient	3 nm^2/ns

is used to compute the pore conductance with the integration done in the region $\partial \Omega_{\text{surf}}$. The computational domain of the continuum model is provided in Figure 10.9. A two-dimensional simulation domain is considered because the chamber width of the PFMP is $W = 3$ mm and the chamber height is $h_e = 0.1$ mm. As shown in [51], for $h_e/W < 0.1$ the variation in concentration along the width of the chamber is negligible. For the PFMP the aspect ratio is $h_e/W = 0.03$; therefore, a two-dimensional domain can be used to model the reaction-diffusion dynamics in the PFMP. See Table C.4 for model parameters of the PFMP.

C.3 Tethered-Membrane Parameters: Pore Conductance and Electrical Energy

To compute ϕ and J^i that are required for evaluating the pore conductance G_p and the electrical energy W_{es} required to form a pore, the following assumptions are made. First, the diagonal diffusion coefficients in $\mathbf{D}^i$ are assumed to be equal (i.e., isotropic diffusion); however, $\mathbf{D}^i \neq \mathbf{D}^j$ for $i \neq j$. Second, the diffusion of Na$^+$, K$^+$, and Cl$^-$ is assumed to change proportionally with how the diffusion of water varies. The third assumption is that the free energies of Na$^+$, K$^+$, and Cl$^-$ are all constant such that $\nabla F_w^i = 0$. With these assumptions, the material parameters in Table C.5 are used to

Table C.4 Model parameter for PFMP.

Symbol	Definition	Value
h_c	Inlet chamber height	4 mm
L_{in}	Inlet chamber length	2 mm
L	Channel buffer length	1 mm
h_e	Channel height	0.1 mm
L_e	Electrode length	0.7 mm
L_{out}	Outlet chamber length	20 mm
h_{out}	Outlet chamber height	4 mm

solve equation (11.18) in the simulation domain defined in Figure 11.5 with diffusion coefficient given by

$$D^i(x) = \begin{cases} D_r^i & \text{if } x \in \Omega_r, \\ D_w^i & \text{if } x \in \Omega_w. \end{cases} \quad (C.1)$$

The dielectric permittivity in (11.18) is spatially dependent as defined below:

$$\varepsilon(x) = \begin{cases} \varepsilon_w & \text{if } x \in \Omega_r \cup \Omega_w, \\ \varepsilon_m & \text{if } x \in \Omega_m. \end{cases} \quad (C.2)$$

In Table C.5, the concentrations match those used in the experimental measurements of the EMP. The selection of effective ion size (i.e., solvated ionic radius) is based on the mobility measurements reported in [178]. The diffusion coefficients of the ions and electrical permittivities of water and the biological membrane are provided in [157] and from the CGMD simulations. The geometric parameters h_r and h_m are selected to match the experimentally measured results obtained from neutron-reflectometry measurements of similar tethered membranes reported in [152] and the CGMD simulation results. The parameters G_o, C_m, C_{dl}, and R_e in Table C.6 are estimated using a single impedance measurement for each tethered membrane. The electroporation parameters C, D, and r_m are obtained from [126, 136, 214, 297, 393]. The parameters σ and γ are computed from the CGMD simulations. Since α and q are not dependent on the tether density, only a single current measurement was used to estimate these parameters, and it was found to be consistent with those reported in [136].

The pore density ρ_o can be estimated using

$$\rho_o = \frac{G_o}{A_m G_p(r_m)}, \quad (C.3)$$

with $A_m = 2.1$ mm^2 the area of the membrane and G_o and $G_p(r_m)$ defined in Table C.6.

C.4 Coarse-Grained Molecular Dynamics (CGMD) Simulations

Simulation details for cholesterol. The molecular dynamics simulations were performed using GROMACS [154] version 4.6.2 with the MARTINI force field [267, 268]. The interactions of the CGMD beads are defined by the Lennard-Jones (LJ) potential, and harmonic potentials are utilized for bond and angle interactions. A shift function is added to the Coulombic force to smoothly and continuously decay to zero from 0 to 1.2 nm. The LJ interactions were treated likewise except that the shift function was turned on between 0.9 and 1.4 nm. The grid-type neighbor searching algorithm is utilized for the simulation; that is, atoms in the neighboring grid were updated every ten time steps. The equations of motion are integrated using the leapfrog algorithm with a time step of 20 fs. Periodic boundary conditions are implemented in the x and y directions (Figure 14.4). Simulations are performed in the NAP$_z$T ensemble using a temperature of

Table C.5 Parameters for pore conductance and electrical energy required for aqueous pore formation predictions.

Symbol	Definition	Value
$c^{Na}\|_{t=0}$	Initial Na$^+$ concentration	321.45 mol/m^3
$c^{K}\|_{t=0}$	Initial K$^+$ concentration	13.39 mol/m^3
$c^{Cl}\|_{t=0}$	Initial Cl$^-$ concentration	334.84 mol/m^3
a_{Na}	Na$^+$ effective ion size	4 Å
a_K	K$^+$ effective ion size	5 Å
a_{Cl}	Cl$^-$ effective ion size	4 Å
D_w^{Na}	Na$^+$ diffusion coefficient in Ω_w	1.33×10^{-9} m^2/s
D_w^{K}	K$^+$ diffusion coefficient in Ω_w	1.96×10^{-9} m^2/s
D_w^{Cl}	Cl$^-$ diffusion coefficient in Ω_w	2.07×10^{-9} m^2/s
D_r^{Na}	Na$^+$ diffusion coefficient in Ω_r	0.92×10^{-9} m^2/s
D_r^{K}	K$^+$ diffusion coefficient in Ω_r	1.37×10^{-9} m^2/s
D_r^{Cl}	Cl$^-$ diffusion coefficient in Ω_r	1.43×10^{-9} m^2/s
ε_w	Electrolyte electrical permittivity	7.083×10^{-10} F/m
ε_m	Membrane electrical permittivity	2.65×10^{-11} F/m
F	Faraday constant	9.6485×10^4 C/mol
C_s	Stern layer capacitance	1 pF
k_B	Boltzmann constant	$1.3806488 \times 10^{-23}$ J/K
T	Temperature	300 K
ϕ_e	Electrode potential	100-500 mV
ϕ_{ec}	Counter electrode potential	0 mV
l_r	Tether reservoir length	400 nm
h_r	Tether reservoir height	4 nm
h_m	Membrane thickness	3.5 nm
h_e	Electrolyte height	60 nm

350 K to match that used in [245] for similar membrane structures. The temperature is held constant using a velocity-rescaling algorithm [59] with a time constant of 0.5 ps. Furthermore, a semi-isotropic Berendsen barostat was applied for the pressure coupling. The lipid and water molecules are coupled separately for temperature and pressure control. The gold surface is modeled using the *walls* option in GROMACS. Note that CGMD simulation times are reported as effective time – that is, four times the actual simulation time. The effective time is introduced to account for the speed-up in the CGMD model [267, 268].

Simulation details for DphPC. The CGMD simulations were performed using GROMACS [154] version 4.6.2 (double precision) with the MARTINI force field [267, 268]. Unless otherwise stated, the CGMD simulation parameters are provided in [245, 451]. The interaction of the CGMD beads is defined by the potential, and harmonic potentials are utilized for bond and angle interactions. A shift function is added to the Coulombic force to smoothly and continuously decay to zero from 0 to 1.2 nm. The LJ interactions were treated likewise except that the shift function was turned on between 0.9 nm and 1.2 nm. The grid-type neighbor searching algorithm is utilized for the simulation; that is, atoms in the neighboring grid were updated every ten time steps. The equations of

Table C.6 Parameters for CED current predictions.

Symbol	Definition	Value		
γ	Edge energy	1.2×10^{-11} J/m		
σ	Surface tension	15×10^{-3} J/m^2		
α	Creation-rate coefficient	1×10^9 s^{-1}		
q	$q = (r_m/r_*)^2$	2.46		
C	Steric repulsion constant	9.67×10^{-15} J$^{1/4}$ m		
D	Radial diffusion coefficient	1×10^{-14} m^2/s		
r_m	Equilibrium pore radius	0.8 nm		
$G_p(r_m)$	Equilibrium pore conductance	1.56 nS		
R_e	Electrolyte resistance	100–800 Ω		
Spacer Surface:				
p	Fractional order parameter	0.83		
C_{dl}	Double-layer capacitance	230 nF*		
DphPC Membrane	**Tether Density:**	**1%**	**10%**	**100%**
G_o	Initial membrane conductance	1.00 μS	0.66 μS	0.33 μS
C_m	Membrane capacitance	15.0–17.5 nF	12.5–16.0 nF	12.4 nF
p	Fractional-order parameter	0.90–0.95	0.90–0.95	0.93
C_{dl}	Double-layer capacitance	100–180 nF*	100–180 nF*	120–180 nF*
V_{ep}	Voltage of electroporation	350–415 mV	480–560 mV	650 mV
K_t	Spring constant	0 N/m	0 N/m	20 mN/m
S. cerevisiae Membrane	**Tether Density:**	**1%**	**10%**	
G_o	Initial membrane conductance	5.00 μS	1.11–1.66 μS	
C_m	Membrane capacitance	16.0–18.0 nF	14.0 nF	
p	Fractional-order parameter	0.90	0.90–0.92	
C_{dl}	Double-layer capacitance	180 nF*	180 nF*	
V_{ep}	Voltage of electroporation	330–350 mV	410–430 mV	
K_t	Spring constant	0 N/m	0 N/m	
E. coli Membrane	**Tether Density:**	**1%**	**10%**	
G_o	Initial membrane conductance	1.00–2.00 μS	0.66 μS	
C_m	Membrane capacitance	14.0 nF	15.0–17.0 nF	
p	Fractional-order parameter	0.90–0.91	0.90–0.91	
C_{dl}	Double-layer capacitance	180 nF*	180 nF*	
V_{ep}	Voltage of electroporation	360–380 mV	400–450 mV	
K_t	Spring constant	0 N/m	0 N/m	

motion are integrated using the leapfrog algorithm with a time step of 20 fs. Periodic boundary conditions are implemented in three dimensions. Simulations are performed in the NVT ensemble using a temperature of 320 K. The temperature is held constant using a velocity-rescaling algorithm [59] with a time constant of 0.5 ps. The lipids, tethers, spacers, and water molecules are coupled separately for temperature control. All the systems studied here were first energy minimized using the steepest-descent method in GROMACS. A 50-ns equilibration run was performed prior to the production run. Production runs were performed for a simulation time of 1.5 μs in a simulation with cell size $L_x = L_y = 5.4$ nm and $L_z = 16.0$ nm. Visualization of the results was reported using Visual Molecular Dynamics (VMD) and PyMOL. The numerical method used to evaluate D_{zz} (14.31) is provided in [372].

C.5 CGMD Simulation Setup for PGLa

All CGMD simulations were implemented in GROMACS version 4.6.2 [41, 154, 401]. For all production runs, the Berensden temperature coupling was used with a temperature of 323 K and a time constant of 0.3 ps. A temperature of 323 K was selected to ensure that the CGMD water does not freeze [268, 321]. Berendsen semi-isotropic pressure coupling was used with a time constant of 3.0 ps, compressibility of 3×10^{-5} 1/bar, and a reference pressure of 1.0 bar [40]. The time step of the simulation is 20 fs with the electrostatic interactions smoothly shifted from zero at 12 Å and Lennard-Jones interaction from 9–12 Å. The membrane is modeled using 512 1,2-dipalmitoyl-sn-glycero-3-phosphocholine (DPPC) molecules. Note that DPPC has an identical structure to zwittrionic C20 diphytanyl-ether-glycero-phosphatidylcholine (DphPC) and a similar structure to C20 diphytanyl-diglyceride ether (GDPE) when using the CGMD representation because mesoscopic details such as the phytanyl tails in the DphPC and GDPE are equivalent to the palmitoyl tails in DPPC. The 512-lipid CGMD DphPC membrane is constructed by replicating the equilibrated 128 DPPC bilayer, from the MARTINI website,[1] twice in the X and Y directions. The 512 DphPC membrane is solvated using CGMD water beads and energy minimized followed by an equilibration in NPT for 200 ns to produce the equilibrated membrane structure. The dimensions of the simulation cell containing the membrane are $126 \times 129 \times 150$ Å^3 corresponding to $X \times Y \times Z$ coordinate axis. The solvent solution surrounding the peptide and membrane surface is composed of water molecules and Na$^+$ and Cl$^-$ ions to make the solvent a 0.15 M NaCl solution and also to neutralize the charge on the peptides.

Surface binding of PGLa. To study the monomer binding of PGLa to the membrane surface (Figure 14.18), a single PGLa peptide is placed 17 Å above the surface of the membrane. After energy minimization, the production run was carried out for a simulation time horizon of 1 μs.

Transmembrane insertion of PGLa. To construct the transient aqueous pores in Figure 14.18 which allow the PGLa to translocate from the membrane surface to the transmembrane state, we employ the method outlined in [236]. The Berendsen semi-isotropic pressure coupling is used with a time constant of 3.0 ps, compressibility of 3×10^{-5} 1/bar, and a reference pressure of 1.0 bar in the direction normal to the membrane surface. The lateral pressure is held at -50 bar until a transient pore has formed. The negative pressure promotes the formation of transient aqueous pores in the membrane. After a transient pore has formed and the PGLa has diffused into the aqueous pore, the lateral pressure is set to 1.0 bar, allowing the aqueous pore to close. The production run for pore closure has a simulation time horizon of 500 ns.

Oligomerization of transmembrane PGLa. To study the oligomerization process of PGLa we place four PGLa monomers in the transmembrane state, as illustrated in Figure 14.18, using the method outlined in [423]. PyMOL (PyMOL Molecular Graphics System, Version 1.3, Schrödinger, LLC) is used to place the peptides in the

---

[1] http://md.chem.rug.nl/cgmartini/index.php/downloads

transmembrane state in the membrane. The system is equilibrated in the NPT ensemble for 20 ns. After energy minimization the production run is carried out for 4 μs.

C.5.1 Simulation Setup of All-Atom Molecular Dynamics

The MD simulation illustrated in Figure 15.5 is composed of two bilayer lipid membranes separated by a distance of 4 nm. The MD simulations were carried out in GROMACS. The MD simulation consisted of 180 DphPC, 76 HPC (modification of DphPC lipid by deleting elements in the tail, that is, deleting 11 atoms starting from tail (N)) and 8120 water molecules surrounding the membranes and 150 Na ions and 150 Cl ions. Each of the two membranes has 90 DphPC and 38 HPC. The box dimensions are $6.2 \times 6.5 \times 15.1$ nm^3. The GROMACS ff53a6 force field and 'tip3p' water model are used for the simulation.

Bibliography

[1] Abidor, I., Arakelyan, V., Chernomordik, L., Chizmadzhev, Y., Pastushenko, V., and Tarasevich, M. 1979. Electric breakdown of bilayer lipid membranes: I. The main experimental facts and their qualitative discussion. *Journal of Electroanalytical Chemistry and Interfacial Electrochemistry*, **104**, 37–52.

[2] Abolfath-Beygi, M., and Krishnamurthy, V. 2014. Biosensor arrays for estimating molecular concentration in fluid flows. *IEEE Transactions on Signal Processing*, **62**(1), 239–251.

[3] Abolfath-Beygi, M., Krishnamurthy, V., and Cornell, B. 2013. Multiple surface-based biosensors for enhanced molecular detection in fluid flow systems. *IEEE Sensors Journal*, **13**(4), 1265–1273.

[4] Ackerberg, R., Patel, R., and Gupta, S. 1978. The heat/mass transfer to a finite strip at small Péclet numbers. *Journal of Fluid Mechanics*, **86**(1), 49–65.

[5] Adamczyk, Z., and Warszyński, P. 1996. Role of electrostatic interactions in particle adsorption. *Advances in Colloid and Interface Science*, **63**, 41–149.

[6] Aernouts, J., Couckuyt, I., Crombecq, K., and Dirckx, J. 2010. Elastic characterization of membranes with a complex shape using point indentation measurements and inverse modelling. *International Journal of Engineering Science*, **48**(6), 599–611.

[7] Afonin, S., Grage, S., Ieronimo, M., Wadhwani, P., and Ulrich, A. 2008. Temperature-dependent transmembrane insertion of the amphiphilic peptide PGLa in lipid bilayers observed by solid state 19F NMR spectroscopy. *Journal of the American Chemical Society*, **130**(49), 16512–16514.

[8] Aguilella-Arzo, M., Aguilella, V., and Eisenberg, R. 2005. Computing numerically the access resistance of a pore. *European Biophysics Journal*, **34**(4), 314–322.

[9] Al-Sakere, B., André, F., Bernat, C., et al. 2007. Tumor ablation with irreversible electroporation. *PLoS One*, **2**(11), e1135.

[10] Alberts, B. 2008. *Molecular Biology of the Cell*. Garland Science.

[11] Allagui, A., Freeborn, T., Elwakil, A., and Maundy, B. 2016. Reevaluation of performance of electric double layer capacitors from constant-current charge/discharge and cyclic voltammetry. *Scientific Reports*, **6**.

[12] Allen, T., Andersen, O., and Roux, B. 2006. Molecular dynamics – potential of mean force calculations as a tool for understanding ion permeation and selectivity in narrow channels. *Biophysical Chemistry*, **124**(3), 251–267.

[13] Als-Nielsen, J., and Kjaer, K. 1989. X-ray reflectivity and diffraction studies of liquid surfaces and surfactant monolayers. Pages 113–138 in *Phase Transitions in Soft Condensed Matter*. Springer.

[14] Amestoy, P., Duff, I., L'Excellent, J., and Koster, J. 2001. A fully asynchronous multifrontal solver using distributed dynamic scheduling. *SIAM Journal on Matrix Analysis and Applications*, **23**(1), 15–41.

[15] Anderson, N., Richter, L., Stephenson, J., and Briggman, K. 2007. Characterization and control of lipid layer fluidity in hybrid bilayer membranes. *Journal of the American Chemical Society*, **129**(7), 2094–2100.

[16] Andronesi, O., Pfeifer, J., Al-Momani, L., et al. 2004. Probing membrane protein orientation and structure using fast magic-angle-spinning solid-state NMR. *Journal of Biomolecular NMR*, **30**(3), 253–265.

[17] Ansari, A. 2000. Mean first passage time solution of the Smoluchowski equation: Application to relaxation dynamics in myoglobin. *Journal of Chemical Physics*, **112**(5), 2516–2522.

[18] Appleby, A. 2005. Electron transfer reactions with and without ion transfer. Pages 175–301 in *Modern Aspects of Electrochemistry*. Springer.

[19] Archer, A., and Evans, R. 2013. Relationship between local molecular field theory and density functional theory for non-uniform liquids. *Journal of Chemical Physics*, **138**(1), 014502.

[20] Ashcroft, F. 1999. *Ion Channels and Disease*. Academic Press.

[21] Asphahani, F., and Zhang, M. 2007. Cellular impedance biosensors for drug screening and toxin detection. *Analyst*, **132**, 835–841.

[22] Asphahani, F., Thein, M., Veiseh, O., et al. 2008. Influence of cell adhesion and spreading on impedance characteristics of cell-based sensors. *Biosensors and Bioelectronics*, **23**(8), 1307–1313.

[23] Atkins, A., Wyborn, N., Wallace, A., et al. 2000. Structure-function relationships of a novel bacterial toxin, hemolysin E: The role of αG. *Journal of Biological Chemistry*, **275**(52), 41150–41155.

[24] Baba, T., Minamikawa, H., Hato, M., and Handa, T. 2001. Hydration and molecular motions in synthetic phytanyl-chained glycolipid vesicle membranes. *Biophysical Journal*, **81**(6), 3377–3386.

[25] Baker, M., Maloy, W., Zasloff, M., and Jacob, L. 1993. Anticancer efficacy of Magainin2 and analogue peptides. *Cancer Research*, **53**(13), 3052–3057.

[26] Barnett, A. 1990. The current-voltage relation of an aqueous pore in a lipid bilayer membrane. *Biochimica et Biophysica Acta, Biomembranes*, **1025**(1), 10–14.

[27] Barnett, A., and Weaver, J. 1991. Electroporation: A unified, quantitative theory of reversible electrical breakdown and mechanical rupture in artificial planar bilayer membranes. *Bioelectrochemistry and Bioenergetics*, **25**(2), 163–182.

[28] Barroso, M., De-Los-Santos-Álvarez, N., Delerue-Matos, C., and Oliveira, M. 2011. Towards a reliable technology for antioxidant capacity and oxidative damage evaluation: Electrochemical (bio) sensors. *Biosensors and Bioelectronics*, **30**(1), 1–12.

[29] Bayley, H., and Cremer, P. 2001. Stochastic sensors inspired by biology. *Nature*, **413**(6852), 226.

[30] Bazant, M., Kilic, M., Storey, B., and Ajdari, A. 2009. Towards an understanding of induced-charge electrokinetics at large applied voltages in concentrated solutions. *Advances in Colloid and Interface Science*, **152**, 48–88.

[31] Bazant, M., Storey, B., and Kornyshev, A. 2011. Double layer in ionic liquids: Overscreening versus crowding. *Physical Review Letters*, **106**(4), 046102.

[32] Beard, D., and Qian, H. 2008. *Chemical Biophysics: Quantitative Analysis of Cellular Systems*. Cambridge University Press.

[33] Beebe, S., Fox, P., Rec, L., Willis, L., and Schoenbach, K. 2003. Nanosecond, high-intensity pulsed electric fields induce apoptosis in human cells. *FASEB Journal*, **17**(11), 1493–1495.

[34] Belytschko, T., Liu, W., Moran, B., and Elkhodary, K. 2013. *Nonlinear Finite Elements for Continua and Structures.* John Wiley & Sons.

[35] Bennett, D., and Tieleman, P. 2014. The importance of membrane defects? Lessons from simulations. *Accounts of Chemical Research,* **47**(8), 2244–2251.

[36] Bennett, W., and Tieleman, D. 2011. Water defect and pore formation in atomistic and coarse-grained lipid membranes: Pushing the limits of coarse graining. *Journal of Chemical Theory and Computation,* **7**, 2981–2988.

[37] Bennett, W., MacCallum, J., and Tieleman, D. 2009. Thermodynamic Analysis of the effect of cholesterol on dipalmitoylphosphatidylcholine lipid membranes. *Journal of the American Chemical Society,* **131**, 1972–1978.

[38] Bennett, W., Sapay, N., and Tieleman, D. 2014. Atomistic simulations of pore formation and closure in lipid bilayers. *Biophysical Journal,* **106**, 210–214.

[39] Berendsen, H. 2007. *Simulating the Physical World: Hierarchical Modeling from Quantum Mechanics to Fluid Dynamics.* Cambridge University Press.

[40] Berendsen, H., Postma, J., Gunsteren, W., DiNola, A., and Haak, J. 1984. Molecular dynamics with coupling to an external bath. *Journal of Chemical Physics,* **81**(8), 3684–3690.

[41] Berendsen, H., Spoel, D., and Drunen, R. 1995. GROMACS: A message-passing parallel molecular dynamics implementation. *Computer Physics Communications,* **91**(1–3), 43–56.

[42] Berkowitz, M. 2009. Detailed molecular dynamics simulations of model biological membranes containing cholesterol. *Biochimica et Biophysica Acta, Biomembranes,* **1788**(1), 86–96.

[43] Berneche, S., and Roux, B. 2003. A microscopic view of ion conduction through the K+ channel. *Proceedings of the National Academy of Sciences of the United States of America,* **100**(15), 8644–8648.

[44] Bertsekas, D. 1999. *Nonlinear Programming.* Athena Scientific.

[45] Billingsley, P. 2013. *Convergence of Probability Measures.* John Wiley & Sons.

[46] Blau, A. 2013. Cell adhesion promotion strategies for signal transduction enhancement in microelectrode array in vitro electrophysiology: An introductory overview and critical discussion. *Current Opinion in Colloid & Interface Science,* **18**(5), 481–492.

[47] Blum, L., and Stell, G. 1976. Solution of Ornstein–Zernike equation for wall-particle distribution function. *Journal of Statistical Physics,* **15**(6), 439–449.

[48] Böckmann, R., Groot, B., Kakorin, S., Neumann, E., and Grubmüller, H. 2008. Kinetics, statistics, and energetics of lipid membrane electroporation studied by molecular dynamics simulations. *Biophysical Journal,* **95**(4), 1837–1850.

[49] Bohner, M., and Kästner, J. 2012. An algorithm to find minimum free-energy paths using umbrella integration. *Journal of Chemical Physics,* **137**(3), 034105.

[50] Brian, A., and McConnell, H. 1984. Allogeneic stimulation of cytotoxic T cells by supported planar membranes. *Proceedings of the National Academy of Sciences of the United States of America,* **81**(19), 6159–6163.

[51] Brody, J., Yager, P., Goldstein, R., and Austin, R. 1996. Biotechnology at low Reynolds numbers. *Biophysical Journal,* **71**(6), 3430–3441.

[52] Bronzino, J., and Peterson, D. 2014. *Biomedical Signals, Imaging, and Informatics.* Boca Raton, FL: CRC Press.

[53] Brown, P., Hindmarsh, A., and Petzold, L. 1994. Using Krylov methods in the solution of large-scale differential-algebraic systems. *SIAM Journal on Scientific Computing,* **15**(6), 1467–1488.

[54] Budvytyte, R., Valincius, G., Niaura, G., Voiciuk, V., Mickevicius, M., Chapman, H., Goh, H., Shekhar, P., Heinrich, F., and Shenoy, S. 2013. Structure and properties of tethered bilayer lipid membranes with unsaturated anchor molecules. *Langmuir*, **29**(27), 8645–8656.

[55] Bufler, J., Kahlert, S., Tzartos, S., Maelicke, A., and Franke, C. 1996. Activation and blockade of mouse muscle nicotinic channels by antibodies directed against the binding site of the acetylcholine receptor. *Journal of Physiology*, **492**, 107–114.

[56] Bunjes, N., Schmidt, E., Jonczyk, A., et al. 1997. Thiopeptide-supported lipid layers on solid substrates. *Langmuir*, **13**(23), 6188–6194.

[57] Burger, M. 2011. Inverse problems in ion channel modelling. *Inverse Problems*, **27**, 083001.

[58] Burgess, J., Rhoten, M., and Hawkridge, F. 1998. Cytochrome C oxidase immobilized in stable supported lipid bilayer membranes. *Langmuir*, **14**(9), 2467–2475.

[59] Bussi, G., Donadio, D., and Parrinello, M. 2007. Canonical sampling through velocity rescaling. *Journal of Chemical Physics*, **126**(1), 014101.

[60] Cafiso, D. 1994. Alamethicin: A peptide model for voltage gating and protein-membrane interactions. *Annual Review of Biophysics and Biomolecular Structure*, **23**(1), 141–165.

[61] Campbell, I. 2012. *Biophysical Techniques*. Oxford University Press.

[62] Caponetto, R. 2010. *Fractional Order Systems: Modeling and Control Applications*. Vol. 72. World Scientific.

[63] Cappé, O., Moulines, E., and Rydén, T. 2005. *Inference in Hidden Markov Models*. Springer-Verlag.

[64] Casciola, M., Bonhenry, D., Liberti, M., Apollonio, F., and Tarek, M. 2014. A molecular dynamic study of cholesterol rich lipid membranes: Comparison of electroporation protocols. *Bioelectrochemistry*, **100**, 11–17.

[65] Catterall, W. 2012. Voltage-gated sodium channels at 60: Structure, function and pathophysiology. *Journal of Physiology*, **590**(11), 2577–2589.

[66] Cavanagh, J., Fairbrother, W., Palmer, A., and Skelton, N. 1995. *Protein NMR Spectroscopy: Principles and Practice*. Academic Press.

[67] Chang, D., and Reese, T. 1990. Changes in membrane structure induced by electroporation as revealed by rapid-freezing electron microscopy. *Biophysical Journal*, **58**(1), 1.

[68] Charalambous, K., and Wallace, B. 2011. NaChBac: The long lost sodium channel ancestor. *Biochemistry*, **50**(32), 6742–6752.

[69] Chee, C., Lee, H., and Lu, C. 2008. Using 3D fluid–structure interaction model to analyse the biomechanical properties of erythrocyte. *Physics Letters A*, **372**(9), 1357–1362.

[70] Chetwynd, A., Wee, C., Hall, B., and Sansom, M. 2010. The energetics of transmembrane helix insertion into a lipid bilayer. *Biophysical Journal*, **99**(8), 2534 – 2540.

[71] Chinchar, V., Bryan, L., Silphadaung, U., et al. 2004. Inactivation of viruses infecting ectothermic animals by amphibian and piscine antimicrobial peptides. *Virology*, **323**(2), 268–275.

[72] Chong, S., Aichele, M., Meyer, H., Fuchs, M., and Baschnagel, J. 2007. Structural and conformational dynamics of supercooled polymer melts: Insights from first-principles theory and simulations. *Physical Review E*, **76**(5), 051806.

[73] Chung, S.H., Krishnamurthy, V., and Moore, J.B. 1991. Adaptive processing techniques based on hidden Markov models for characterising very small channel currents buried in noise and deterministic interferences. *Philosophical Transactions of the Royal Society B: Biological Sciences*, **334**, 357–384.

[74] Chung, S.H., Allen, T.W., and Kuyucak, S. 2002a. Conducting-state properties of the KcsA potassium channel from molecular and Brownian dynamics simulations. *Biophysical Journal*, **82**, 628–645.

[75] Chung, S.H., Allen, T.W., and Kuyucak, S. 2002b. Modeling diverse range of potassium channels with Brownian dynamics. *Biophysical Journal*, **83**, 263–277.

[76] Chung, S.H., Andersen, O., and Krishnamurthy, V. (eds). 2007. *Biological Membrane Ion Channels: Dynamics, Structure and Applications*. Springer-Verlag.

[77] Cicero, G., Calzolari, A., Corni, S., and Catellani, A. 2011. Anomalous wetting layer at the Au (111) surface. *Journal of Physical Chemistry Letters*, **2**(20), 2582–2586.

[78] Cirac, A., Moiset, G., Mika, J., et al. 2011. The molecular basis for antimicrobial activity of pore forming cyclic peptides. *Biophysical Journal*, **100**(10), 2422 – 2431.

[79] Cisneros, G., Wikfeldt, K., Ojamaäe, L., et al. 2016. Modeling Molecular Interactions in Water: From Pairwise to Many-Body Potential Energy Functions. *Chemical Reviews*, **116**(13), 7501–7528.

[80] Coalson, R., and Kurnikova, M. 2007. Poisson–Nernst–Planck theory of ion permeation through biological channels. Pages 449–484 in Chung, S. H., Andersen, O., and Krishnamurthy, V. (eds.), *Biological Membrane Ion Channels*. Springer-Verlag.

[81] Cole, K. 1968. *A Quantitiative Description of Membrane Current and Its Application to Conductance and Excitation in Nerve*. Berkeley: University of California Press.

[82] Conlon, J., Mechkarska, M., et al. 2012. Host-defense peptides in skin secretions of the tetraploid frog *Silurana epitropicalis* with potent activity against methicillin-resistant *Staphylococcus aureus* (MRSA). *Peptides*, **37**(1), 113–119.

[83] Cornell, B. 2002. Membrane-based biosensors. Page 457 in Ligler, F.S., and Taitt, C.A.R. (eds), *Optical Biosensors: Present and Future*. Elsevier.

[84] Cornell, B., Braach-Maksvytis, V., King, L., et al. 1997. A biosensor that uses ion-channel switches. *Nature*, **387**, 580–583.

[85] Cornell, B., Krishna, G., Osman, P., Pace, R., and Wieczorek, L. 2001. Tethered bilayer lipid membranes as a support for membrane-active peptides. *Biochemical Society Transactions*, **29**(4), 613–617.

[86] Cornell, B., Scolan, G., Powl, A., Carnie, S., and Wallace, B. 2012. Comparative study of the bacterial sodium channel NaChBac function using patch clamp and AC impedance spectroscopy in a tethered membrane. *Biophysical Journal*, **102**(3), 338a.

[87] Coster, H., Chilcott, T., and Coster, A. 1996. Impedance spectroscopy of interfaces, membranes and ultrastructures. *Bioelectrochemistry and Bioenergetics*, **40**(2), 79–98.

[88] Cranfield, C., Cornell, B., Grage, S., et al. 2014. Transient potential gradients and impedance measures of tethered bilayer lipid membranes: Pore-forming peptide insertion and the effect of electroporation. *Biophysical Journal*, **106**(1), 182–189.

[89] Cruickshank, C., Minchin, R., Dain, A., and Martinac, B. 1997. Estimation of the pore size of the large-conductance mechanosensitive ion channel of *Escherichia coli*. *Biophysical Journal*, **73**(4), 1925–1931.

[90] Daillant, J., and Gibaud, A. 2008. *X-Ray and Neutron Reflectivity: Principles and Applications*. Vol. 770. Springer.

[91] Daily, M., Olsen, B., Schlesinger, P., Ory, D., and Baker, N. 2014. Improved coarse-grained modeling of cholesterol-containing lipid bilayers. *Journal of Chemical Theory and Computation*, **10**(5), 2137–2150.

[92] Das, S. 2011. *Functional Fractional Calculus*. Springer Science & Business Media.

[93] De Rosa, M., Gamacorta, M.A., Nicolaus, B., Chappe, B., and Albrecht, P. 1983. Isoprenoid ethers; backbone of complex lipids of the archaebacterium *Sulfolobus solfataricus*. *Biochimica et Biophysica Acta*, **753**, 249–256.

[94] de With, G. 2013. *Liquid-State Physical Chemistry: Fundamentals, Modeling, and Applications*. Vol. 1. Wiley.

[95] DeBruin, K., and Krassowska, W. 1999a. Modeling electroporation in a single cell. I. Effects of field strength and rest potential. *Biophysical Journal*, **77**(3), 1213–1224.

[96] DeBruin, K., and Krassowska, W. 1999b. Modeling electroporation in a single cell. II. Effects of ionic concentrations. *Biophysical Journal*, **77**(3), 1225–1233.

[97] Dehghan, M. 2004. Numerical solution of the three-dimensional advection-diffusion equation. *Applied Mathematics and Computation*, **150**, 5–19.

[98] Delemotte, L., and Tarek, M. 2012. Molecular dynamics simulations of lipid membrane electroporation. *Journal of Membrane Biology*, **245**(9), 531–543.

[99] Deminsky, M., Eletskii, A., Kniznik, A., et al. 2013. Molecular dynamic simulation of transmembrane pore growth. *Journal of Membrane Biology*, **246**(11), 821–831.

[100] Deng, P., Lee, Y., Lin, R., and Zhang, T. 2012. Nonlinear electro-mechanobiological behavior of cell membrane during electroporation. *Applied Physics Letters*, **101**(5), 053702.

[101] Deniaud, A., Rossi, C., Berquand, A., et al. 2007. Voltage-dependent anion channel transports calcium ions through biomimetic membranes. *Langmuir*, **23**(7), 3898–3905.

[102] Devasahayam, S. 2012. *Signals and Systems in Biomedical Engineering: Signal Processing and Physiological Systems Modeling*. Springer Science & Business Media.

[103] Diaz, A., Albertorio, F., Daniel, S., and Cremer, P. 2008. Double cushions preserve transmembrane protein mobility in supported bilayer systems. *Langmuir*, **24**(13), 6820–6826.

[104] Diethelm, K., and Freed, A. 1998. The FracPECE subroutine for the numerical solution of differential equations of fractional order. *Forschung und wissenschaftliches Rechnen*, **1999**, 57–71.

[105] Dorairaj, S., and Allen, T. 2007. On the thermodynamic stability of a charged arginine side chain in a transmembrane helix. *Proceedings of the National Academy of Sciences of the United States of America*, **104**, 4943–4948.

[106] DuMont, A., and Torres, V. 2014. Cell targeting by the *Staphylococcus aureus* pore-forming toxins: It's not just about lipids. *Trends in Microbiology*, **22**(1), 21–27.

[107] Dunlop, J., Bowlby, M., Peri, R., Vasilyev, D., and Arias, R. 2008. High-throughput electrophysiology: An emerging paradigm for ion-channel screening and physiology. *Nature Reviews Drug Discovery*, **7**(4), 358–368.

[108] Dura, J., Pierce, D., Majkrzak, C., et al. 2006. AND/R: Advanced neutron diffractometer/reflectometer for investigation of thin films and multilayers for the life sciences. *Review of Scientific Instruments*, **77**(7), 074301.

[109] Edwards, David A. 1999. Estimating rate constants in a convection-diffusion system with a boundary reaction. *IMA Journal of Applied Mathematics*, **63**, 89–112.

[110] Eifler, N., Vetsch, M., Gregorini, M., et al. 2006. Cytotoxin ClyA from *Escherichia coli* assembles to a 13-meric pore independent of its redox-state. *EMBO Journal*, **25**(11), 2652–2661.

[111] El-Andaloussi, S., Lee, Y., Lakhal-Littleton, S. 2012. Exosome-mediated delivery of siRNA in vitro and in vivo. *Nature Protocols*, **7**(12), 2112–2126.

[112] Elliott, J., Needham, D., Dilger, J., and Haydon, D. 1983. The effects of bilayer thickness and tension on gramicidin single-channel lifetime. *Biochimica et Biophysica Acta, Biomembranes*, **735**(1), 95–103.

[113] Ernst, R., Bodenhausen, G., and Wokaun, A. 1987. *Principles of Nuclear Magnetic Resonance in One and Two Dimensions*. Clarendon Press Oxford.

[114] Esfandyarpour, R., Javanmard, M., Koochak, Z., et al. 2013. Label-free electronic probing of nucleic acids and proteins at the nanoscale using the nanoneedle biosensor. *Biomicrofluidics*, **7**(4), 044114.

[115] Evans, L. 2010. *Partial Differential Equations*. American Mathematical Society.

[116] Fang, Y., Persson, B., Löfås, S., and Knoll, W. 2004. *Protein Microarray Technology*. Wiley InterScience. Chapter 6: Surface Plasmon Fluorescence Spectroscopy for Protein Binding Studies, pp. 131–151.

[117] Ferguson, T., and Bazant, M. 2012. Nonequilibrium thermodynamics of porous electrodes. *Journal of the Electrochemical Society*, **159**(12), A1967–A1985.

[118] Fernández, L., and Reigada, R. 2014. Effects of dimethyl sulfoxide on lipid membrane electroporation. *Journal of Physical Chemistry B*, **118**(31), 9306–9312.

[119] Fernández, L., Marshall, G., Sagués, F., and Reigada, R. 2010. Structural and kinetic molecular dynamics study of electroporation in cholesterol-containing bilayer. *Journal of Physical Chemistry B*, **114**(20), 6855–6865.

[120] Fertig, N., Blick, R.H., and Behrends, J.C. 2002. The cell patch clamp recording performed on a planar glass chip. *Biophysical Journal*, **82**(June), 3056–3062.

[121] Filippov, A., Orädd, G., and Lindblom, G. 2007. Domain formation in model membranes studied by pulsed-field gradient-NMR: The role of lipid polyunsaturation. *Biophysical Journal*, **93**(9), 3182–3190.

[122] Fischer, H., Polikarpov, I., and Craievich, A. 2004. Average protein density is a molecular-weight-dependent function. *Protein Science*, **13**(10), 2825–2828.

[123] Fitter, J., Gutberlet, T., and Katsaras, J. 2006. *Neutron Scattering in Biology: Techniques and Applications*. Springer Science & Business Media.

[124] Flenner, E., Das, J., Rheinstädter, M., and Kosztin, I. 2009. Subdiffusion and lateral diffusion coefficient of lipid atoms and molecules in phospholipid bilayers. *Physical Review E*, **79**(1), 011907.

[125] Franceschetti, D., Macdonald, R., and Buck, R. 1991. Interpretation of finite-length-Warburg-type impedances in supported and unsupported electrochemical cells with kinetically reversible electrodes. *Journal of the Electrochemical Society*, **138**(5), 1368–1371.

[126] Freeman, S., Wang, M., and Weaver, J. 1994. Theory of electroporation of planar bilayer membranes: Predictions of the aqueous area, change in capacitance, and pore-pore separation. *Biophysical Journal*, **67**(1), 42–56.

[127] Frenkel, D., and Smit, B. 2001. *Understanding Molecular Simulation: From Algorithms to Applications*. Vol. 1. Academic Press.

[128] Fromherz, P. 2003. Neuroelectronics interfacing: Semiconductor chips with ion channels, nerve cells and brain. Pages 781–810 in Weise, R. (ed), *Nanoelectronics and Information Technology*. Wiley-VCH.

[129] Garett, R., and Grisham, C. 2016. *Biochemistry*. Brooks Cole.

[130] Ge, L., Bernasconi, L., and Hunt, P. 2013. Linking electronic and molecular structure: Insight into aqueous chloride solvation. *Physical Chemistry Chemical Physics*, **15**(31), 13169–13183.

[131] Georganopoulou, D. 2009 (April). Reagentless electrochemical biosensors for clinical diagnostics. In: *41st Annual Oak Ridge Conference*. Frontiers in Clinical Diagnostics, Baltimore, MD.

[132] Giera, B., Henson, N., Kober, E., Shell, S., and Squires, T. 2015. Electric double-layer structure in primitive model electrolytes: Comparing molecular dynamics with local-density approximations. *Langmuir*, **31**(11), 3553–3562.

[133] Giess, F., Friedrich, M., Heberle, J., Naumann, R., and Knoll, W. 2004. The protein-tethered lipid bilayer: A novel mimic of the biological membrane. *Biophysical Journal*, **87**(5), 3213–3220.

[134] Gillespie, D. 2010. Analytic theory for dilute colloids in a charged slit. *Journal of Physical Chemistry B*, **114**(12), 4302–4309.

[135] Gillespie, D. 2015. A review of steric interactions of ions: Why some theories succeed and others fail to account for ion size. *Microfluidics and Nanofluidics*, **18**(5–6), 717–738.

[136] Glaser, R., Leikin, S., Chernomordik, L., Pastushenko, V., and Sokirko, A. 1988. Reversible electrical breakdown of lipid bilayers: Formation and evolution of pores. *Biochimica et Biophysica Acta, Biomembranes*, **940**(2), 275–287.

[137] Glaser, R., Sachse, C., Dürr, U., et al. 2005. Concentration-dependent realignment of the antimicrobial peptide PGLa in lipid membranes observed by solid-state ^{19}F-NMR. *Biophysical Journal*, **88**(5), 3392–3397.

[138] Glaser, R.W. 1993. Antigen-antibody binding and mass transport by convection and diffusion to a surface: A two-dimensional computer model of binding and dissociation kinetics. *Analytical Biochemistry*, **213**, 152–161.

[139] Gliozzi, A., Rolandi, R., De Rosa, M., and Gamacorta, A. 1983. Monolayer black membranes from bipolar lipids of archaebacteria and their temperature-induced structural changes. *Journal of Membrane Biology*, **75**(1), 45–56.

[140] Goldstein, B., Coombs, D., He, X., Pineda, A.R., and Wofsy, C. 1999. The influence of trasnport on the kinetics of binding to surface receptors: Application to cell and BIAcore. *Journal of Molecular Recognition*, **12**, 293–299.

[141] Gordon, D., Krishnamurthy, V., and Chung, S. 2009. Generalized Langevin models of molecular dynamics simulations, with applications to ion channels. *Journal of Chemical Physics*, **131**, 111.

[142] Graf, P., Kurnikova, M., Coalson, R., and Nitzan, A. 2004. Comparison of dynamic lattice Monte Carlo simulations and the dielectric self-energy Poisson–Nernst–Planck continuum theory for model ion channels. *Journal of Physical Chemistry B*, **108**(6), 2006–2015.

[143] Grenier, E., Louvet, V., and Vigneaux, P. 2014. Parameter estimation in non-linear mixed effects models with SAEM algorithm: Extension from ODE to PDE. *ESAIM: Mathematical Modelling and Numerical Analysis*, **48**(5), 1303–1329.

[144] Griebel, M., Knapek, S., and Zumbusch, G. 2007. *Numerical Simulation in Molecular Dynamics: Numerics, Algorithms, Parallelization, Applications*. Vol. 5. Springer Science & Business Media.

[145] Gurtovenko, A., and Lyulina, A. 2014. Electroporation of asymmetric phospholipid membranes. *Journal of Physical Chemistry B*, **118**(33), 9909–9918.

[146] Hänggi, P., Talkner, P., and Borkovec, M. 1990. Reaction-rate theory: Fifty years after Kramers. *Reviews of Modern Physics*, **62**(2), 251.

[147] Hansen, J., and McDonald, I. 1990. *Theory of Simple Liquids*. Elsevier.

[148] Harms, G., Orr, G., Montal, M., et al. 2003. Probing conformational changes of gramicidin ion channels by single-molecule patch-clamp fluorescence microscopy. *Biophysical Journal*, **85**(3), 1826–1838.

[149] Haskins, J., and Lawson, J. 2016. Evaluation of molecular dynamics simulation methods for ionic liquid electric double layers. *Journal of Chemical Physics*, **144**(18), 184707.
[150] Hatlo, M., Van Roij, R., and Lue, L. 2012. The electric double layer at high surface potentials: The influence of excess ion polarizability. *EPL*, **97**(2), 28010.
[151] Heinrich, F. 2016. Deuteration in biological neutron reflectometry. *Methods in Enzymology*, **566**, 211–230.
[152] Heinrich, F., Ng, T., Vanderah, D., et al. 2009. A new lipid anchor for sparsely tethered bilayer lipid membranes. *Langmuir*, **25**(7), 4219–4229.
[153] Henderson, D., Abraham, F., and Barker, J. 1976. The Ornstein–Zernike equation for a fluid in contact with a surface. *Molecular Physics*, **31**(4), 1291–1295.
[154] Hess, B., Kutzner, C., Spoel, D., and Lindahl, E. 2008. GROMACS 4: Algorithms for highly efficient, load-balanced, and scalable molecular simulation. *Journal of Chemical Theory Computation*, **4**(3), 435–447.
[155] Heysel, S., Vogel, H., Sanger, M., and Sigrist, H. 1995. Covalent attachment of functionalized lipid bilayers to planar waveguides for measuring protein binding to biomimetic membranes. *Protein Science*, **4**(12), 2532–2544.
[156] Hille, B. 2001. *Ionic Channels of Excitable Membranes*. 3rd edition. Sinauer Associates, Inc.
[157] Hobbie, R., and Roth, B. 2007. *Intermediate Physics for Medicine and Biology*. Springer.
[158] Hoffmann, J., and Gillespie, D. 2013. Ion correlations in nanofluidic channels: Effects of ion size, valence, and concentration on voltage- and pressure-driven currents. *Langmuir*, **29**(4), 1303–1317.
[159] Hofsäß, C., Lindahl, E., and Edholm, O. 2003. Molecular dynamics simulations of phospholipid bilayers with cholesterol. *Biophysical Journal*, **84**(4), 2192–2206.
[160] Hoiles, W., and Krishnamurthy, V. 2015. Dynamic modeling of antimicrobial pore formation in engineered tethered membranes. *IEEE Transactions on Molecular, Biological and Multi-Scale Communications*, **1**(3), 265–276.
[161] Hoiles, W., Krishnamurthy, V., and Cornell, B. 2012. Mathematical models for sensing devices constructed out of artificial cell membranes. *Nanoscale Systems: Mathematical Modeling, Theory and Applications*, **1**, 143–171.
[162] Hoiles, W., Krishnamurthy, V., Cranfield, C., and Cornell, B. 2014. An engineered membrane to measure electroporation: Effect of tethers and bioelectronic interface. *Biophysical Journal*, **107**(6), 1339–1351.
[163] Hoiles, W., Krishnamurthy, V., and Cornell, B. 2014. Modelling the bioelectronic interface in engineered tethered membranes: From biosensing to electroporation. *IEEE Transactions on Biomedical Circuits and Systems*, **PP**(99), 1–13.
[164] Hoiles, W., Krishnamurthy, V., and Cornell, B. 2015. Membrane bound molecular machines for sensing. *Journal of Analytical & Bioanalytical Techniques*, 1.
[165] Hoiles, W., Gupta, R., Cornell, B., Cranfield, C., and Krishnamurthy, V. 2016. The effect of tethers on artificial cell membranes: A coarse-grained molecular dynamics study. *PLoS One*, **11**(10), e0162790.
[166] Hoover, W. 1985. Canonical dynamics: Equilibrium phase-space distributions. *Physical Review A*, **31**(3), 1695.
[167] Horng, T., Lin, T., Liu, C., and Eisenberg, B. 2012. PNP Equations with steric effects: A model of ion flow through channels. *Journal of Physical Chemistry B*, **116**(37), 11422–11441.

[168] Howorka, S., Nam, J., Bayley, H., and Kahne, D. 2004. Stochastic detection of monovalent and bivalent protein-ligand interactions. *Angewandte Chemie, International Edition*, **43**, 842–846.

[169] Hoyles, M., Krishnamurthy, V., Siksik, M., and Chung, S.H. 2008. Brownian Dynamics theory for predicting internal and external blockages of tetraethylammonium in the KcsA potassium channel. *Biophysical Journal*, **94**(January), 366–378.

[170] Hu, Q., and Joshi, R. 2009. Transmembrane voltage analyses in spheroidal cells in response to an intense ultrashort electrical pulse. *Physical Review E*, **79**, 011901.

[171] Hughes, T. 2012. *The Finite Element Method: Linear Static and Dynamic Finite Element Analysis*. Courier Corporation.

[172] Humphrey, W., Dalke, A., and Schulten, K. 1996. VMD: Visual Molecular Dynamics. *Journal of Molecular Graphics*, **14**(1), 33–38.

[173] Hung, W., Lee, M., Chen, F., and Huang, H. 2007. The condensing effect of cholesterol in lipid bilayers. *Biophysical Journal*, **92**(11), 3960–3967.

[174] Husslein, T., Newns, D., Pattnaik, P., et al. 1998. Constant pressure and temperature molecular-dynamics simulation of the hydrated diphytanolphosphatidylcholine lipid bilayer. *Journal of Chemical Physics*, **109**(7), 2826–2832.

[175] Ikonen, E. 2008. Cellular cholesterol trafficking and compartmentalization. *Nature Reviews Molecular Cell Biology*, **9**(2), 125–138.

[176] Im, W., and Roux, B. 2002. Ion permeation and selectivity of OmpF porin: A theoretical study based on molecular dynamics, Brownian dynamics, and continuum electrodiffusion theory. *Journal of Molecular Biology*, **322**(4), 851–869.

[177] Ingólfsson, H., and Andersen, O. 2010. Screening for small molecules' bilayer-modifying potential using a gramicidin-based fluorescence assay. *Assay and Drug Development Technologies*, **8**(4), 427–436.

[178] Israelachvili, J. 2011. *Intermolecular and Surface Forces: Revised Third Edition*. Academic Press.

[179] Jafari, H., and Daftardar-Gejji, V. 2006. Solving a system of nonlinear fractional differential equations using Adomian decomposition. *Journal of Computational and Applied Mathematics*, **196**(2), 644–651.

[180] Jeon, J., Monne, H., Javanainen, M., and Metzler, R. 2012. Anomalous diffusion of phospholipids and cholesterols in a lipid bilayer and its origins. *Physical Review Letters*, **109**(18), 188103.

[181] Jesus, I., and Machado, T. 2009. Development of fractional order capacitors based on electrolyte processes. *Nonlinear Dynamics*, **56**(1-2), 45–55.

[182] Jeuken, L. 2009. Electrodes for integral membrane enzymes. *Natural Product Reports*, **26**(10), 1234–1240.

[183] Jiang, F., Bouret, Y., and Kindt, J. 2004. Molecular dynamics simulations of the lipid bilayer edge. *Biophysical Journal*, **87**, 182–192.

[184] Jiang, Y., and Kindt, J. 2007. Simulations of edge behaviour in a mixed bilayer: Fluctuation analysis. *Journal of Chemical Physics*, **126**, 045105–9.

[185] Jo, S., Lim, J., Klauda, J., and Im, W. 2009. CHARMM-GUI membrane builder for mixed bilayers and its application to yeast membranes. *Biophysical Journal*, **97**(1), 50–58.

[186] Joannis, J., Jiang, F., and Kindt, J. 2006. Coarse-grained model simulations of mixed-lipid systems: Composition and line tension of a stabilized bilayer edge. *Langmuir*, **22**, 998–1005.

[187] Johnson, C. 2012. *Numerical Solution of Partial Differential Equations by the Finite Element Method*. Courier Corporation.

[188] Joshi, R., and Schoenbach, K. 2000. Electroporation dynamics in biological cells subjected to ultrafast electrical pulses: A numerical simulation study. *Physical Review E*, **62**, 1025–1033.

[189] Joshi, R., Hu, Q., Schoenbach, K., and Hjalmarson, H. 2002. Improved energy model for membrane electroporation in biological cells subjected to electrical pulses. *Physical Review E*, **65**, 041920.

[190] Junghans, A., and Koöper, I. 2010. Structural analysis of tethered bilayer lipid membranes. *Langmuir*, **26**(13), 11035–11040.

[191] Junghans, A., Watkins, E., Barker, R., et al. 2015. Analysis of biosurfaces by neutron reflectometry: From simple to complex interfaces. *Biointerphases*, **10**(1), 019014.

[192] Kakorin, S., Brinkmann, U., and Neumann, E. 2005. Cholesterol reduces membrane electroporation and electric deformation of small bilayer vesicles. *Biophysical Chemistry*, **117**(2), 155–171.

[193] Kalinowski, S., Ibron, G., Bryl, K., and Figaszewski, Z. 1998. Chronopotentiometric studies of electroporation of bilayer lipid membranes. *Biochimica et Biophysica Acta, Biomembranes*, **1369**(2), 204–212.

[194] Kanthou, C., Kranjc, S., Sersa, et al. 2006. The endothelial cytoskeleton as a target of electroporation-based therapies. *Molecular Cancer Therapeutics*, **5**(12), 3145–3152.

[195] Karatzas, I., and Shreve, S. 1991. *Brownian Motion and Stochastic Calculus*. Second edition. Springer.

[196] Karjiban, R., Shaari, N., Gunasakaran, U., and Basri, M. 2013. A coarse-grained molecular dynamics study of DLPC, DMPC, DPPC, and DSPC mixtures in aqueous solution. *Journal of Chemistry*, **2013**, 931051.

[197] Kästner, J. 2011. Umbrella sampling. *Wiley Interdisciplinary Reviews: Computational Molecular Science*, **1**(6), 932–942.

[198] Kayser, R., and Raveché, H. 1982. Derivation of the Ornstein–Zernike differential equation from the Bogoliubov–Born–Green–Kirkwood–Yvon hierarchy. *Physical Review A*, **26**(4), 2123.

[199] Kendall, J., Johnson, B., Symonds, P., et al. 2010. Effect of the structure of cholesterol-based tethered bilayer lipid membranes on ionophore activity. *ChemPhysChem*, **11**(10), 2191–2198.

[200] Ketchem, R., Hu, W., and Cross, T. 1993. High-resolution conformation of gramicidin A in a lipid bilayer by solid-state NMR. *Science*, **261**, 1457–1457.

[201] Khalil, H. 2002. *Nonlinear Systems*. Prentice Hall.

[202] Kim, S., Cho, H., et al. 2009. Nanogap biosensors for electrical and label-free detection of biomolecular interactions. *Nanotechnology*, **20**(45), 455502.

[203] Kirchner, B., Dio, P., and Hutter, J. 2012. Real-world predictions from ab initio molecular dynamics simulations. Pages 109–153 in *Multiscale Molecular Methods in Applied Chemistry*. Springer.

[204] Knight, J., Lerner, M., Marcano-Velázquez, J., Pastor, R., and Falke, J. 2010. Single molecule diffusion of membrane-bound proteins: Window into lipid contacts and bilayer dynamics. *Biophysical Journal*, **99**(9), 2879–2887.

[205] Knoll, W., Bender, K., Förch, R., et al. 2010. Polymer-tethered bimolecular lipid membranes. Pages 197–233 in *Polymer Membranes/Biomembranes*. Springer.

[206] Kok, T., Mickan, L., and Burrell, C. 1994. Routine diagnosis of seven respiratory viruses and *Mycoplasma pneumoniae* by enzyme immunoassay. *Journal of Virological Methods*, **50**(1–3), 87–100.

[207] Kokotovic, P., Khalil, H., and O'Reilly, J. 1999. *Singular Perturbation Methods in Control: Analysis and Design*. SIAM.

[208] Kong, X., Lu, D., Liu, Z., and Wu, J. 2015. Molecular dynamics for the charging behavior of nanostructured electric double layer capacitors containing room temperature ionic liquids. *Nano Research*, **8**(3), 931–940.

[209] Kornyshev, A. 2013. The simplest model of charge storage in single file metallic nanopores. *Faraday Discussions*, **164**, 117–133.

[210] Koronkiewicz, S., and Kalinowski, S. 2004. Influence of cholesterol on electroporation of bilayer lipid membranes: Chronopotentiometric studies. *Biochimica et Biophysica Acta, Biomembranes*, **1661**(2), 196–203.

[211] Kotulska, M., Basalyga, J., Derylo, M., and Sadowski, P. 2010. Metastable pores at the onset of constant-current electroporation. *Journal of Membrane Biology*, **236**(1), 37–41.

[212] Kozuch, J., Weichbrodt, C., Millo, D., et al. 2014. Voltage-dependent structural changes of the membrane-bound anion channel hVDAC1 probed by SEIRA and electrochemical impedance spectroscopy. *Physical Chemistry Chemical Physics*, **16**(20), 9546–9555.

[213] Kramar, P., Delemotte, L., Lebar, A., et al. 2012. Molecular-level characterization of lipid membrane electroporation using linearly rising current. *Journal of Membrane Biology*, **245**(10), 651–659.

[214] Krassowska, W., and Filev, P. 2007. Modeling electroporation in a single cell. *Biophysical Journal*, **92**(2), 404–417.

[215] Kresák, S., Hianik, T., and Naumann, R. 2009. Giga-seal solvent-free bilayer lipid membranes: From single nanopores to nanopore arrays. *Soft Matter*, **5**(20), 4021–4032.

[216] Krishnamurthy, V. 2016. *Partially Observed Markov Decision Processes: From Filtering to Controlled Sensing*. Cambridge University Press.

[217] Krishnamurthy, V., and Chung, S.H. 2007. Large-Scale Dynamical Models and Estimation for Permeation in Biological Membrane Ion Channels. *Proceedings of the IEEE*, **95**(5), 853–880.

[218] Krishnamurthy, V., and Cornell, B. 2012. Engineering aspects of biological ion channels: from biosensors to computational models for permeation. *Protoplasma*, **249**, 3–9.

[219] Krishnamurthy, V., Luk, K., Cornell, B., and Martin, D. 2007. Gramicidin ion channel based nano-biosensors: Construction, stochastic dynamical models and statistical detection algorithms. *IEEE Sensors Journal*, **7**(9), 1281–1288.

[220] Krishnamurthy, V., Monfared, S., and Cornell, B. 2010. Ion-channel biosensors – part I: Construction, operation and clinical studies. *IEEE Transactions on Nanotechnology (Special Issue on Nanoelectronic Interface to Biomolecules and Cells)*, **9**(3), 303–312.

[221] Krishnamurthy, V., Monfared, S., and Cornell, B. 2010. Ion-channel biosensors – part II: Dynamic modeling, analysis and statistical signal processing. *IEEE Transactions on Nanotechnology (Special Issue on Nanoelectronic Interface to Biomolecules and Cells)*, **9**(3), 313–321.

[222] Kruskal, P., Jiang, Z., Gao, T., and Lieber, C. 2015. Beyond the patch clamp: Nanotechnologies for intracellular recording. *Neuron*, **86**(1), 21–24.

[223] Kučerka, N., Nagle, J., Sachs, J., et al. 2008. Lipid bilayer structure determined by the simultaneous analysis of neutron and X-ray scattering data. *Biophysical Journal*, **95**(5), 2356–2367.

[224] Kurnikova, M., Coalson, R., Graf, P., and Nitzan, A. 1999. A lattice relaxation algorithm for three-dimensional Poisson–Nernst–Planck theory with application to ion transport through the gramicidin A channel. *Biophysical Journal*, **76**(2), 642–656.
[225] Kushner, H.J. 1990. *Weak Convergence and Singularly Perturbed Stochastic Control and Filtering Problems*. Birkhauser.
[226] Kushwaha, S.C., Kates, M., Sprott, G.D., and Smith, I.C. 1981. Novel complex polar lipids from the methanogenic archaebacterium *Methanospirillum hungatei*. *Science*, **211**, 1163–1164.
[227] Kuzmenkin, A., Bezanilla, F., and Correa, A. 2004. Gating of the bacterial sodium channel, NaChBac: Voltage-dependent charge movement and gating currents. *Journal of General Physiology*, **124**(4), 349–356.
[228] Latz, A., and Zausch, J. 2013. Thermodynamic derivation of a Butler–Volmer model for intercalation in Li-ion batteries. *Electrochimica Acta*, **110**, 358–362.
[229] Lauffenburger, D., and Linderman, J. 1993. *Receptors: Models for Binding, Tracking, and Signaling*. Oxford University Press.
[230] Leake, M. 2016. *Biophysics: Tools and Techniques*. CRC Press.
[231] Lee, S., Cascão-Pereira, L., Sala, R., Holmes, S., Ryan, K., and Becker, T. 2005. Ion channel switch array: A biosensor for detecting multiple pathogens. *Industrial Biotechnology*, **1**(1), 26–31.
[232] Lee, H., Vries, A., Marrink, S., and Pastor, R. 2009. A coarse-grained model for polyethylene oxide and polyethylene glycol: Conformation and hydrodynamics. *Journal of Physical Chemistry B*, **113**(40), 13186–13194.
[233] Lee, YiKuen, and Deng, PeiGang. 2012. Review of micro/nano technologies and theories for electroporation of biological cells. *Science China Physics, Mechanics & Astronomy*, **55**(6), 996–1003.
[234] Lenzi, E., Zola, R., Rossato, R., et al. 2017. Asymptotic behaviors of the Poisson–Nernst–Planck model, generalizations and best adjust of experimental data. *Electrochimica Acta*, **226**, 40–45.
[235] Lenzi, K., de Paula, L., Silva, R., and Evangelista, R. 2013. A connection between anomalous Poisson–Nernst–Planck model and equivalent circuits with constant phase elements. *The Journal of Physical Chemistry C*, **117**(45), 23685–23690.
[236] Leontiadou, H., Mark, A., and Marrink, S. 2004. Molecular dynamics simulations of hydrophilic pores in lipid bilayers. *Biophysical Journal*, **86**(4), 2156–2164.
[237] Levine, R. 2005. *Molecular Reaction Dynamics*. Cambridge University Press.
[238] Li, J., and Lin, H. 2010. The current-voltage relation for electropores with conductivity gradients. *Biomicrofluidics*, **4**(1), 013206.
[239] Li, S., Cutrera, J., Heller, R., and Teissie, J. 2014. *Electroporation Protocols: Preclinical and Clinical Gene Medicine*. Humana.
[240] Li-Fries, J. 2007. *Ion Channels in Mixed Tethered Bilayer Lipid Membranes*. Ph.D. thesis, Max Planck Institut für Polymerforschung.
[241] Ligler, F., Fare, T., Seib, E., et al. 1988. Fabrication of key components of a receptor-based biosensor. *Medical Instrumentation*, **22**(5), 247–256.
[242] Lim, J., and Klauda, J. 2011. Lipid chain branching at the iso- and anteiso-positions in complex chlamydia membranes: A molecular dynamics study. *Biochimica et Biophysica Acta, Biomembranes*, **1808**(1), 323–331.
[243] Lin, Y., Minner, D., Herring, V.L., and Naumann, C. 2012. Physisorbed polymer-tethered lipid bilayer with lipopolymer gradient. *Materials*, **5**(11), 2243–2257.

[244] Lipinski, C. 2004. Lead- and drug-like compounds: The rule-of-five revolution. *Drug Discovery Today: Technologies*, **1**(4), 337–341.

[245] Liu, C., and Faller, R. 2012. Conformational, dynamical, and tensional study of tethered bilayer lipid membranes in coarse-grained molecular simulations. *Langmuir*, **28**, 15907–15915.

[246] Liu, J., and Eisenberg, B. 2013. Correlated ions in a calcium channel model: A Poisson–Fermi theory. *Journal of Physical Chemistry B*, **117**(40), 12051–12058.

[247] Liu, J., and Eisenberg, B. 2014. Poisson–Nernst–Planck–Fermi theory for modeling biological ion channels. *Journal of Chemical Physics*, **141**(22), 12B640_1.

[248] Liu, J., and Eisenberg, B. 2015. Numerical methods for a Poisson–Nernst–Planck–Fermi model of biological ion channels. *Physical Review E*, **92**(1), 012711.

[249] Liu, P., Harder, E., and Berne, B. 2004. On the calculation of diffusion coefficients in confined fluids and interfaces with an application to the liquid–vapor interface of water. *Journal of Physical Chemistry B*, **108**(21), 6595–6602.

[250] Ljung, L. 1999. *System Identification: Theory for the User*. Wiley Online Library.

[251] Lomize, A., Orekhov, V., and Arsen'ev, A. 1992. Refinement of the spatial structure of the gramicidin A ion channel. *Bioorganicheskaya Khimiya*, **18**(2), 182–200.

[252] Lopatin, A.N., Makhina, E.N., and Nichols, C.G. 1995. The mechanism of inward rectification of potassium channels: Long-pore plugging by cytoplasmic polyamines. *Journal of General Physiology*, **106**(November), 923–955.

[253] Lopreore, C., Bartol, T., Coggan, J., et al. 2008. Computational modeling of three-dimensional electrodiusion in biological systems: Application to the node of Ranvier. *Biophysical Journal*, **95**(6), 2624–2635.

[254] Lu, B., Zhou, Y., Huber, G., et al. 2007. Electrodiffusion: A continuum modeling framework for biomolecular systems with realistic spatiotemporal resolution. *Journal of Chemical Physics*, **127**, 135102.

[255] Lu, B., Holst, M., McCammon, J., and Zhou, Y. 2010. Poisson–Nernst–Planck equations for simulating biomolecular diffusion-reaction processes I: Finite element solutions. *Journal of Computational Physics*, **229**(19), 6979–6994.

[256] Lu, H. 2005. Probing single-molecule protein conformational dynamics. *Accounts of Chemical Research*, **38**(7), 557–565.

[257] Lu, X.D., Ottova, A.L., and Tien, H.T. 1996. Biophysical aspects of agar-gel supported bilayer lipid nembranes: A new method for forming and studying planar bilayer lipid membranes. *Bioelectrochemistry and Bioenergetics*, **39**(2), 285–289.

[258] Lück, J., and Latz, A. 2016. Theory of reactions at electrified interfaces. *Physical Chemistry Chemical Physics*, **18**, 17799–17804.

[259] Ludwig, A., Völkerink, G., and Rhein, C., et al. 2010. Mutations affecting export and activity of cytolysin A from *Escherichia coli*. *Journal of Bacteriology*, **192**(15), 4001–4011.

[260] Léobon, B., Garcin, I., Menasché, P., et al. 2003. Myoblasts transplanted into rat infarcted myocardium are functionally isolated from their host. *Proceedings of the National Academy of Sciences of the United States of America*, **100**(13), 7808–7811.

[261] Maffeo, C., Bhattacharya, S., Yoo, J., Wells, D., and Aksimentiev, A. 2012. Modeling and simulation of ion channels. *Chemical Reviews*, **112**(12), 6250–6284.

[262] Majhi, A., Kanchi, S., Venkataraman, V., Ayappa, K., and Maiti, P. 2015. Estimation of activation energy for electroporation and pore growth rate in liquid crystalline and gel phases of lipid bilayers using molecular dynamics simulations. *Soft Matter*, **11**(44), 8632–8640.

[263] Majkrzak, C., Carpenter, E., Heinrich, F., and Berk, N. 2011. When beauty is only skin deep: Optimizing the sensitivity of specular neutron reflectivity for probing structure beneath the surface of thin films. *Journal of Applied Physics*, **110**(10), 102212.

[264] Mamonov, A., Coalson, R., Nitzan, A., and Kurnikova, M. 2003. The role of the dielectric barrier in narrow biological channels: A novel composite approach to modeling single-channel currents. *Biophysical Journal*, **84**(6), 3646–3661.

[265] Mamonov, A., Kurnikova, M., and Coalson, R. 2006. Diffusion constant of K+ inside gramicidin A: A comparative study of four computational methods. *Biophysical Chemistry*, **124**(3), 268–278.

[266] Markel, V. 2016. Introduction to the Maxwell Garnett approximation: Tutorial. *Journal of the Optical Society of America A*, **33**(7), 1244–1256.

[267] Marrink, S., and Tieleman, P. 2013. Perspective on the MARTINI model. *Chemical Society Reviews*, **42**, 6801–6822.

[268] Marrink, S., Risselada, H., Yefimov, S., Tieleman, D., and Vries, A. 2007. The MARTINI force field: A coarse grained model for biomolecular simulations. *Journal of Physical Chemistry B*, **111**(27), 7812–7824.

[269] Martinac, B., Nomura, T., Chi, G., et al. 2014. Bacterial mechanosensitive channels: Models for studying mechanosensory transduction. *Antioxidants & Redox Signaling*, **20**(6), 952–969.

[270] Masel, R. 1996. *Principles of Adsorption and Reaction on Solid Surfaces*. Vol. 3. John Wiley & Sons.

[271] Mason, T., Pineda, A.R., Wofsy, C., and Goldstein, B. 1999. Effective rate models for the analysis of transport-dependent biosensor data. *Mathematical Bioscience*, **159**, 123–144.

[272] Maynard, J., Lindquist, N., Sutherland, J., et al. 2009. Surface plasmon resonance for high-throughput ligand screening of membrane-bound proteins. *Biotechnology Journal*, **4**(11), 1542–1558.

[273] McGillivray, D., Valincius, G., Vanderah, D., et al. 2007. Molecular-scale structural and functional characterization of sparsely tethered bilayer lipid membranes. *Biointerphases*, **2**(1), 21–33.

[274] McGillivray, D., Valincius, G., Heinrich, F., et al. 2009. Structure of functional *Staphylococcus aureus* α-hemolysin channels in tethered bilayer lipid membranes. *Biophysical Journal*, **96**(4), 1547–1553.

[275] Melikov, K., Frolov, V., Shcherbakov, A., et al. 2001. Voltage-induced nonconductive pre-pores and metastable single pores in unmodified planar lipid bilayer. *Biophysical Journal*, **80**(4), 1829–1836.

[276] Meyn, S., and Tweedie, R. 2012. *Markov Chains and Stochastic Stability*. Springer Science & Business Media.

[277] Miguel, V., Perillo, M., and Villarreal, M. 2016. Improved prediction of bilayer and monolayer properties using a refined BMW-MARTINI force field. *Biochimica et Biophysica Acta, Biomembranes*, **1858**(11), 2903–2910.

[278] Miloshevsky, G., and Jordan, P. 2006. The open state gating mechanism of gramicidin A requires relative opposed monomer rotation and simultaneous lateral displacement. *Structure*, **14**(8), 1241–1249.

[279] Mocenni, C., Madeo, D., and Sparacino, E. 2011. Linear least squares parameter estimation of nonlinear reaction diffusion equations. *Mathematics and Computers in Simulation*, **81**(10), 2244–2257.

[280] Modi, N., Winterhalter, M., and Kleinekathofer, U. 2012. Computational modeling of ion transport through nanopores. *Nanoscale*, **4**, 6166–6180.

[281] Molleman, A. 2003. *Patch Clamping: An Introductory Guide to Patch Clamp Electrophysiology.* John Wiley & Sons.

[282] Momani, S., and Odibat, Z. 2007. Numerical approach to differential equations of fractional order. *Journal of Computational and Applied Mathematics*, **207**(1), 96–110.

[283] Monfared, S., Krishnamurthy, V., and Cornell, B. 2012. A molecular machine biosensor: Construction, predictive models, and experimental studies. *Biosensors and Bioelectronics*, **34**, 261–266.

[284] Monticelli, L., Kandasamy, S., Periole, X., Larson, R., Tieleman, P., and Marrink, S. 2008. The MARTINI coarse-grained force field: Extension to proteins. *Journal of Chemical Theory and Computation*, **4**(5), 819–834.

[285] Motesharei, K., and Ghadiri, M.R. 1997. Diffusion-limited size-selective ion sensing based on SAM-supported peptide nanotubes. *Journal of the American Chemical Society*, **119**, 11306–11312.

[286] Movahed, S., and Li, D. 2013. A theoretical study of single-cell electroporation in a microchannel. *Journal of Membrane Biology*, **246**, 151–160.

[287] Mueller, P., Rudin, D., Tien, T., and Wescott, W. 1962. Reconstitution of cell membrane structure in vitro and its transformation into an excitable system. *Nature*, **194**(4832), 979–980.

[288] Myszka, D.G., Morton, T.A., Doyle, M.L., and Chaiken, I.M. 1997. Kinetic analysis of a protein antigen–antibody interaction limited by mass transport on an optical biosensor. *Biophysical Chemistry*, **64**, 127–137.

[289] Naumann, C., Prucker, O., Lehmann, T., et al. 2001a. The polymer-supported phospholipid bilayer: Tethering as a new approach to substrate-membrane stabilization. *Biomacromolecules*, **3**, 27–35.

[290] Naumann, C.A., Knoll, W., and Frank, C.W. 2001b. Hindered diffusion in polymer-tethered membranes: A monolayer study at the air–water interface. *Biomacromolecules*, **2**(4), 1097–1103.

[291] Naumann, R., Schmidt, E.K., Jonczyk, A., et al. 1996. Self-assembly in natural and unnatural systems. *Angewandte Chemie, International Edition*, **35**(11), 1154–1196.

[292] Naumann, R., Schmidt, E., Jonczyk, A., et al. 1999. The peptide-tethered lipid membrane as a biomimetic system to incorporate cytochrome C oxidase in a functionally active form. *Biosensors and Bioelectronics*, **14**(7), 651–662.

[293] Naumann, R., Baumgart, T., Gräber, P., et al. 2002. Proton transport through a peptide-tethered bilayer lipid membrane by the H+-ATP synthase from chloroplasts measured by impedance spectroscopy. *Biosensors and Bioelectronics*, **17**(1), 25–34.

[294] Naumann, R., Schiller, S., Giess, F., et al. 2003. Tethered lipid bilayers on ultraflat gold surfaces. *Langmuir*, **19**(13), 5435–5443.

[295] Neher, E. 2001. Molecular biology meets microelectronics. *Nature Biotechnology*, **19**(2), 114.

[296] Nelson, D., Lehninger, A., and Cox, M. 2008. *Principles of Biochemistry.* Macmillan.

[297] Neu, J., and Krassowska, W. 1999. Asymptotic model of electroporation. *Physical Review E*, **59**, 3471–3482.

[298] Neu, J., and Krassowska, W. 2003. Modeling postshock evolution of large electropores. *Physical Review E*, **67**, 021915.

[299] Neu, J., and Krassowska, W. 2006. Singular perturbation analysis of the pore creation transient. *Physical Review E*, **74**, 031917.

[300] Neu, J., Smith, K., and Krassowska, W. 2003. Electrical energy required to form large conducting pores. *Bioelectrochemistry*, **60**(1), 107–114.

[301] Newman, J. 1966. Resistance for flow of current to a disk. *Journal of the Electrochemical Society*, **113**(5), 501–502.
[302] Nölting, B. 2009. *Methods in Modern Biophysics*. Springer Science & Business Media.
[303] Nosé, S. 1984. A molecular dynamics method for simulations in the canonical ensemble. *Molecular Physics*, **52**(2), 255–268.
[304] Nosé, S., and Klein, M. 1983. Constant pressure molecular dynamics for molecular systems. *Molecular Physics*, **50**(5), 1055–1076.
[305] Obergrussberger, A., Stölzle-Feix, S., Becker, N., Brüggemann, A., Fertig, N., and Möller, C. 2015. Novel screening techniques for ion channel targeting drugs. *Channels*, **9**(6), 367–375.
[306] Ogier, S., Bushby, R., Cheng, Y., et al. 2000. Suspended planar phospholipid bilayers on micromachined supports. *Langmuir*, **16**(13), 5696–5701.
[307] Oh, S., Cornell, B., Smith, D., et al. 2008. Rapid detection of influenza A virus in clinical samples using an ion channel switch biosensor. *Biosensors and Bioelectronics*, **23**(7), 1161–1165.
[308] Ohvo-Rekilä, H., Ramstedt, B., Leppimäki, P., and Slotte, P. 2002. Cholesterol interactions with phospholipids in membranes. *Progress in Lipid Research*, **41**(1), 66–97.
[309] Oscarsson, J., Mizunoe, Y., Li, L., et al. 1999. Molecular analysis of the cytolytic protein ClyA (SheA) from *Escherichia coli*. *Molecular Microbiology*, **32**(6), 1226–1238.
[310] Otter, W. 2013. Revisiting the exact relation between potential of mean force and free-energy profile. *Journal of Chemical Theory and Computation*, **9**(9), 3861–3865.
[311] Ovchinnikov, V., Nam, K., and Karplus, M. 2016. A simple and accurate method to calculate free energy profiles and reaction rates from restrained molecular simulations of diffusive processes. *Journal of Physical Chemistry B*, **120**(33), 8457–8472.
[312] Pabst, M., Wrobel, G., Ingebrandt, S., Sommerhage, F., and Offenhäusser, A. 2007. Solution of the Poisson–Nernst–Planck equations in the cell–substrate interface. *European Physical Journal E: Soft Matter and Biological Physics*, **24**, 1–8.
[313] Paesani, F. 2016. Getting the right answers for the right reasons: Toward predictive molecular simulations of water with many-body potential energy functions. *Accounts of Chemical Research*, **49**(9), 1844–1851.
[314] Pakhomov, A., Miklavcic, D., and Markov, M. 2010. *Advanced Electroporation Techniques in Biology and Medicine*. CRC Press.
[315] Parrinello, M., and Rahman, A. 1981. Polymorphic transitions in single crystals: A new molecular dynamics method. *Journal of Applied Physics*, **52**(12), 7182–7190.
[316] Pastor, R., and Karplus, M. 1988. Parametrization of the friction constant for stochastic simulations of polymers. *Journal of Physical Chemistry*, **92**(9), 2636–2641.
[317] Pastushenko, V., and Chizmadzhev, Y. 1982. Stabilization of conducting pores in BLM by electric current. *General Physiology and Biophysics*, **1**, 43–52.
[318] Pastushenko, V., Chizmadzhev, Yu., and Arakelyan, V. 1979. Electric breakdown of bilayer lipid membranes: II. Calculation of the membrane lifetime in the steady-state diffusion approximation. *Journal of Electroanalytical Chemistry and Interfacial Electrochemistry*, **104**, 53–62.
[319] Pedrotty, D., Koh, J., Davis, B., et al. 2005. Engineering skeletal myoblasts: Roles of three-dimensional culture and electrical stimulation. *American Journal of Physiology: Heart and Circulatory Physiology*, **288**(4), H1620–H1626.
[320] Peng, Z., Tang, J., Han, X., Wang, E., and Dong, S. 2002. Formation of a supported hybrid bilayer membrane on gold: A sterically enhanced hydrophobic effect. *Langmuir*, **18**(12), 4834–4839.

[321] Periole, X., and Marrink, S. 2012. The Martini coarse-grained force field. Pp. 533–565 in *Biomolecular Simulations*. Methods in Molecular Biology, vol. 924. Springer.

[322] Peterman, M.C., Ziebarth, J.M., Braha, O., et al. 2002. Ion channels and lipid bilayer membranes under high potentials using microfabricated apertures. *Biomedical Microdevices*, **4**, 236–236.

[323] Petrov, E., Rohde, P., Cornell, B., and Martinac, B. 2012. The protective effect of osmoprotectant TMAO on bacterial mechanosensitive channels of small conductance MscS/MscK under high hydrostatic pressure. *Channels*, **6**(4), 262–271.

[324] Piggot, T., Holdbrook, D., and Khalid, S. 2011. Electroporation of the *E. coli* and *S. aureus* membranes: Molecular dynamics simulations of complex bacterial membranes. *Journal of Physical Chemistry B*, **115**(45), 13381–13388.

[325] Pintelon, R., and Schoukens, J. 2012. *System Identification: A Frequency Domain Approach*. John Wiley & Sons.

[326] Plant, A. 1993. Self-assembled phospholipid/alkanethiol biomimetic bilayers on gold. *Langmuir*, **9**(11), 2764–2767.

[327] Plant, A.L. 1999. Supported hybrid bilayer membranes as rugged cell membrane mimics. *Langmuir*, **15**, 5128–5135.

[328] Podlubny, I. 1998. *Fractional Differential Equations: An Introduction to Fractional Derivatives, Fractional Differential Equations, to Methods of Their Solution and Some of Their Applications*. Vol. 198. Academic Press.

[329] Polak, A., Bonhenry, D., Dehez, F., Kramar, P., Miklavčič, D., and Tarek, M. 2013. On the electroporation thresholds of lipid bilayers: Molecular dynamics simulation investigations. *Journal of Membrane Biology*, **246**(11), 843–850.

[330] Polak, A., Tarek, M., Tomšič, M., et al. 2014. Electroporation of archaeal lipid membranes using MD simulations. *Bioelectrochemistry*, **100**, 18–26.

[331] Prashar, J., Sharp, P., Scarffe, M., and Cornell, B. 2007. Making lipid membranes even tougher. *Journal of Materials Research*, **22**(08), 2189–2194.

[332] Qian, S., and Bau, H.H. 2003. A mathematical model of lateral flow bioreactions applied to sandwich assays. *Analytical Biochemistry*, **322**, 89–98.

[333] Qiao, Y., Liu, X., Chen, M., and Lu, B. 2016. A local approximation of fundamental measure theory incorporated into three dimensional Poisson–Nernst–Planck equations to account for hard sphere repulsion among ions. *Journal of Statistical Physics*, **163**(1), 156–174.

[334] Rabiner, L. 1989. A tutorial on hidden Markov models and selected applications in speech recognition. *Proceedings of the IEEE*, **77**(2), 257–286.

[335] Ragaliauskas, T., Mickevicius, M., Rakovska, B., et al. 2017. Fast formation of low-defect-density tethered bilayers by fusion of multilamellar vesicles. *Biochimica et Biophysica Acta, Biomembranes*, **1859**(5), 669–678.

[336] Raguse, B., Braach-Maksvytis, V., Cornell, B., et al. 1998. Tethered lipid bilayer membranes: Formation and ionic reservoir characterization. *Langmuir*, **14**(3), 648–659.

[337] Raicu, V., and Popescu, A. 2008. *Integrated Molecular and Cellular Biophysics*. Springer.

[338] Redick, S., Settles, D., Briscoe, G., and Erickson, H. 2000. Defining Fibronectin9s Cell Adhesion Synergy Site by Site-Directed Mutagenesis. *The Journal of cell biology*, **149**(2), 521–527.

[339] Rapaport, D. 2002. *The Art of Molecular Dynamics Simulation*. Vol. 2. Cambridge University Press.

[340] Reigada, R. 2014. Electroporation of heterogeneous lipid membranes. *Biochimica et Biophysica Acta, Biomembranes*, **1838**(3), 814–821.

[341] Ren, D., Navarro, B., Xu, H., Yue, L., Shi, Q., and Clapham, D. 2001. A prokaryotic voltage-gated sodium channel. *Science*, **294**(5550), 2372–2375.

[342] Richter, R., Bérat, R., and Brisson, A. 2006. Formation of solid-supported lipid bilayers: An integrated view. *Langmuir*, **22**(8), 3497–3505.

[343] Ring, A. 1992. Influence of ion occupancy and membrane deformation on gramicidin A channel stability in lipid membranes. *Biophysical Journal*, **61**(5), 1306–1315.

[344] Risken, H. 1984. Fokker–Planck equation. Pages 63–95 in *The Fokker-Planck Equation*. Springer.

[345] Robelek, R., Lemker, E., Wiltschi, B., et al. 2007. Incorporation of in vitro synthesized GPCR into a tethered artificial lipid membrane system. *Angewandte Chemie, International Edition*, **46**(4), 605–608.

[346] Robinson, J. 2001. *Infinite-Dimensional Dynamical Systems*. Cambridge University Press.

[347] Rollins-Smith, L., Doersam, J., Longcore, J., et al. 2002. Antimicrobial peptide defenses against pathogens associated with global amphibian declines. *Developmental & Comparative Immunology*, **26**(1), 63–72.

[348] Rols, M., and Teissié, J. 1992. Experimental evidence for the involvement of the cytoskeleton in mammalian cell electropermeabilization. *Biochimica et Biophysica Acta, Biomembranes*, **1111**(1), 45–50.

[349] Römer, W., and Steinem, C. 2004. Impedance analysis and single-channel recordings on nano-black lipid membranes based on porous alumina. *Biophysical Journal*, **86**(2), 955–965.

[350] Römer, W., Lam, Y., Fischer, D., et al. 2004. Channel activity of a viral transmembrane peptide in micro-BLMs: Vpu1-32 from HIV-1. *Journal of the American Chemical Society*, **126**(49), 16267–16274.

[351] Rosazza, C., Escoffre, J., Zumbusch, A., and Rols, M. 2011. The actin cytoskeleton has an active role in the electrotransfer of plasmid DNA in mammalian cells. *Molecular Therapy*, **19**(5), 913–921.

[352] Roux, B., Allen, T., Bemeche, S., and Im, W. 2004. Theoretical and computational models of biological ion channels. *Quarterly Reviews of Biophysics*, **37**(1), 15–103.

[353] Rubinstein, I. 1990. *Electrodiffusion of Ions*. SIAM Studies in Applied Mathematics.

[354] Sackmann, E. 1996. Supported membranes: Scientific and practical applications. *Science*, **271**(5245), 43–48.

[355] Sackmann, E., and Tanaka, M. 2000. Supported membranes on soft polymer cushions: Fabrication, characterization and applications. *Trends in Biotechnology*, **18**(2), 58–64.

[356] Sakmann, B. 2013. *Single-Channel Recording*. Springer Science & Business Media.

[357] Salnikov, E., and Bechinger, B. 2011. Lipid-controlled peptide topology and interactions in bilayers: Structural insights into the synergistic enhancement of the antimicrobial activities of PGLa and magainin 2. *Biophysical Journal*, **100**(6), 1473–1480.

[358] Salnikov, E., Aisenbrey, C., Aussenac, F., et al. 2016. Membrane topologies of the PGLa antimicrobial peptide and a transmembrane anchor sequence by dynamic nuclear polarization/solid-state NMR spectroscopy. *Scientific Reports*, **6**, 20895.

[359] Sandblom, J., Galvanovskis, J., and Jilderos, B. 2001. Voltage-dependent formation of gramicidin channels in lipid bilayers. *Biophysical Journal*, **81**(2), 827–837.

[360] Sansom, M. 1991. The biophysics of peptide models of ion channels. *Progress in Biophysics and Molecular Biology*, **55**(3), 139–235.

[361] Santangelo, C. 2006. Computing counterion densities at intermediate coupling. *Physical Review E*, **73**(4), 041512.

[362] Santo, K., and Berkowitz, M. 2012. Difference between magainin-2 and melittin assemblies in phosphatidylcholine bilayers: Results from coarse-grained simulations. *Journal of Physical Chemistry B*, **116**(9), 3021–3030.

[363] Santos, F., and Franzese, G. 2012. Relations between the diffusion anomaly and cooperative rearranging regions in a hydrophobically nanoconfined water monolayer. *Physical Review E*, **85**(1), 010602.

[364] Scheibler, L., Dumy, P., Boncheva, H., et al. 1999. Functional molecular thin films: Topological templates for the chemoselective ligation of antigenic peptides to self-assembled monolayers. *Angewandte Chemie, International Edition*, **38**, 696–699.

[365] Scheidt, H., Huster, D., and Gawrisch, K. 2005. Diffusion of cholesterol and its precursors in lipid membranes studied by 1H pulsed field gradient magic angle spinning NMR. *Biophysical Journal*, **89**(4), 2504–2512.

[366] Schepetiuk, S.K., and Kok, T. 1993. The use of MDCK, MEK and LLC-MK2 cell lines with enzyme immunoassay for the isolation of influenza and parainfluenza viruses from clinical specimens. *Journal of Virological Methods*, **42**, 241–50.

[367] Schick, S., Chen, L., Li, E., et al. 2010. Assembly of the M2 tetramer is strongly modulated by lipid chain length. *Biophysical Journal*, **99**(6), 1810–1817.

[368] Schirmacher, W. 2015. *Theory of Liquids and Other Disordered Media. Lecture Notes in Physics*. Vol. 887. Springer.

[369] Schneider, G. 2010. Virtual screening: An endless staircase? *Nature Reviews Drug Discovery*, **9**(4), 273.

[370] Schumaker, M., Pomes, R., and Roux, B. 2001. Framework model for single proton conduction through gramicidin. *Biophysical Journal*, **80**(1), 12–30.

[371] Schuss, Z., Nadler, B., and Eisenberg, R.S. 2001. Derivation of Poisson and Nernst–Planck equations in a bath and channel from a molecular model. *Physical Review E*, **64**, 036116.

[372] Sedlmeier, F., Hansen, Y., Mengyu, L., Horinek, D., and Netz, R. 2011. Water dynamics at interfaces and solutes: Disentangling free energy and diffusivity contributions. *Journal of Statistical Physics*, **145**(2), 240–252.

[373] Selberherr, S. 1984. *Analysis and Simulation of Semiconductor Devices*. Springer-Verlag.

[374] Separovic, F., and Cornell, B. 2007. Gated ion channel-based biosensor device. Pages 595–621 in Chung, S.H., Andersen, O., and Krishnamurthy, V. (eds), *Biological Membrane Ion Channels*. Springer-Verlag.

[375] Shinoda, K., Shinoda, W., Baba, T., and Mikami, M. 2004. Comparative molecular dynamics study of ether- and ester-linked phospholipid bilayers. *Journal of Chemical Physics*, **121**(19), 9648–9654.

[376] Shinoda, K., Shinoda, W., and Mikami, M. 2007. Molecular dynamics simulation of an archaeal lipid bilayer with sodium chloride. *Physical Chemistry Chemical Physics*, **9**(5), 643–650.

[377] Shinoda, W., Mikami, M., Baba, T., and Hato, M. 2003. Molecular dynamics study on the effect of chain branching on the physical properties of lipid bilayers: Structural stability. *Journal of Physical Chemistry B*, **107**(50), 14030–14035.

[378] Shinoda, W., Mikami, M., Baba, T., and Hato, M. 2004. Dynamics of a highly branched lipid bilayer: A molecular dynamics study. *Chemical Physics Letters*, **390**(1), 35–40.

[379] Siegel, A., Murcia, M., Johnson, M., Reif, M., et al. 2010. Compartmentalizing a lipid bilayer by tuning lateral stress in a physisorbed polymer-tethered membrane. *Soft Matter*, **6**(12), 2723–2732.

[380] Siekmann, I., Sneyd, J., and Crampin, E. 2014. Statistical analysis of modal gating in ion channels. *Proceedings of the Royal Society A: Mathematical, Physical and Engineering Sciences*, **470**(2166), 20140030.

[381] Siekmann, I., Fackrell, M., Crampin, E., and Taylor, P. 2016. Modelling modal gating of ion channels with hierarchical Markov models. *Proceedings of the Royal Society A: Mathematical, Physical and Engineering Sciences*, **472**(2192), 20160122.

[382] Sigworth, F.J., and Klemic, K.G. 2002. Patch clamp on a chip. *Biophysical Journal*, **82**(June), 2831–2832.

[383] Silin, V., Wieder, H., Woodward, J., et al. 2002. The role of surface free energy on the formation of hybrid bilayer membranes. *Journal of the American Chemical Society*, **124**(49), 14676–14683.

[384] Silin, V., Kasianowicz, J., Michelman-Ribeiro, A., et al. 2016. Biochip for the detection of *Bacillus anthracis* lethal factor and therapeutic agents against anthrax toxins. *Membranes*, **6**(3), 36.

[385] Silva, J., Pan, H., Wu, D., et al. 2009. A multiscale model linking ion-channel molecular dynamics and electrostatics to the cardiac action potential. *Proceedings of the National Academy of Sciences of the United States of America*, **106**(27), 11102–11106.

[386] Simon, A., Gounou, C., Tan, S., Tiefenauer, L., Berardino, M., and Brisson, A. 2013. Free-standing lipid films stabilized by Annexin-A5. *Biochimica et Biophysica Acta, Biomembranes*, **1828**(11), 2739–2744.

[387] Singer, J., and Nicolson, G. 1972. The fluid mosaic model of the structure of cell membranes. *Science*, **175**(4023), 720–731.

[388] Singh, M., and Kant, R. 2014. Theory of anomalous dynamics of electric double layer at heterogeneous and rough electrodes. *Journal of Physical Chemistry C*, **118**(10), 5122–5133.

[389] Sinner, E., and Knoll, W. 2001. Functional tethered membranes. *Current Opinion in Chemical Biology*, **5**(6), 705–711.

[390] Sinner, E., Reuning, U., Kök et al. 2004. Incorporation of integrins into artificial planar lipid membranes: Characterization by plasmon-enhanced fluorescence spectroscopy. *Analytical Biochemistry*, **333**(2), 216–224.

[391] Sinner, E., Ritz, S., Naumann, R., Schiller, S., and Knoll, W. 2009. Self-assembled tethered bimolecular lipid membranes. *Advances in Clinical Chemistry*, **49**, 159–179.

[392] Smadbeck, P., and Kaznessis, Y. 2012. Stochastic model reduction using a modified Hill-type kinetic rate law. *Journal of Chemical Physics*, **137**(23), 234109.

[393] Smith, K., Neu, J., and Krassowska, W. 2004. Model of creation and evolution of stable electropores for DNA delivery. *Biophysical Journal*, **86**(5), 2813–2826.

[394] Smondyrev, A., and Berkowitz, M. 1999. Structure of dipalmitoylphosphatidylcholine/cholesterol bilayer at low and high cholesterol concentrations: Molecular dynamics simulation. *Biophysical Journal*, **77**(4), 2075–2089.

[395] Socci, N., Onuchic, J., and Wolynes, P. 1996. Diffusive dynamics of the reaction coordinate for protein folding funnels. *Journal of Chemical Physics*, **104**(15), 5860–5868.

[396] Somorjai, G., and Li, Y. 2010. *Introduction to Surface Chemistry and Catalysis*. John Wiley & Sons.

[397] Song, L., Hobaugh, M., Shustak, C., Cheley, S., Bayley, H., and Gouaux, J. 1996. Structure of staphylococcal α-hemolysin, a heptameric transmembrane pore. *Science*, **274**(5294), 1859–1865.

[398] Soumpasis, D. 1983. Theoretical analysis of fluorescence photobleaching recovery experiments. *Biophysical Journal*, **41**(1), 95–97.

[399] Spijker, P., Eikelder, H., Markvoort, A., Nedea, S., and Hilbers, P. 2008. Implicit particle wall boundary condition in molecular dynamics. *Proceedings of the Institution of Mechanical Engineers, Part C: Journal of Mechanical Engineering Science*, **222**(5), 855–864.

[400] Spira, M., and Hai, A. 2013. Multi-electrode array technologies for neuroscience and cardiology. *Nature Nanotechnology*, **8**(2), 83–94.

[401] Spoel, D., Lindahl, E., Hess, B., et al. 2005. GROMACS: Fast, flexible and free. *Journal of Computing Chemistry*, **26**, 1701–1718.

[402] Squires, T., Messinger, R., and Manalis, S. 2008. Making it stick: Convection, reaction and diffusion in surface-based biosensors. *Nature Biotechnology*, **26**(4), 417–426.

[403] Steinem, C., Janshoff, A., Ulrich, W.P., Sieber, M., and Galla, H.J. 1996. Impedance analysis of supported lipid bilayer membranes: A scrutiny of different preparation techniques. *Biochimica et Biophysica Acta.*, **1279**, 169–180.

[404] Stone, B., Burrows, J., Schepetiuk, S., et al. 2004. Rapid detection and simultaneous subtype differentiation of influenza A viruses by real time PCR. *Journal of Virological Methods*, **117**, 103–112.

[405] Stora, T., Lakey, J.H., and Vogel, H. 1999. Ion-channel gating in transmembrane receptor proteins: Functional activity in tethered lipid membranes. *Angewandte Chemie, International Edition*, **38**, 389–392.

[406] Strandberg, E., Wadhwani, P., Tremouilhac, P., Dürr, U., and Ulrich, A. 2006. Solid-state NMR analysis of the PGLa peptide orientation in DMPC bilayers: Structural fidelity of ^{2}H-Labels versus high sensitivity of ^{19}F-NMR. *Biophysical Journal*, **90**(5), 1676–1686.

[407] Strandberg, E., Tremouilhac, P., Wadhwani, P., and Ulrich, A. 2009. Synergistic transmembrane insertion of the heterodimeric PGLa/magainin 2 complex studied by solid-state NMR. *Biochimica et Biophysica Acta, Biomembranes*, **1788**(8), 1667–1679.

[408] Stroeve, P., and Miller, I. 1975. Lateral diffusion of cholesterol in monolayers. *Biochimica et Biophysica Acta, Biomembranes*, **401**(2), 157–167.

[409] Sumino, A., Dewa, T., Takeuchi, T., et al. 2011. Construction and structural analysis of tethered lipid bilayer containing photosynthetic antenna proteins for functional analysis. *Biomacromolecules*, **12**(7), 2850–2858.

[410] Sundararajan, R. 2014. *Electroporation-Based Therapies for Cancer: From Basics to Clinical Applications*. Elsevier.

[411] Sung, W., and Park, P. 1997. Dynamics of pore growth in membranes and membrane stability. *Biophysical Journal*, **73**, 1797–1804.

[412] Tabaei, S., Jackman, J., Kim, S., et al. 2014. Formation of cholesterol-rich supported membranes using solvent-assisted lipid self-assembly. *Langmuir*, **30**(44), 13345–13352.

[413] Talele, S., Gaynor, P., Cree, M., and Ekeran, J. 2010. Modelling single cell electroporation with bipolar pulse parameters and dynamic pore radii. *Journal of Electrostatics*, **68**(3), 261–274.

[414] Tamm, L., and McConnell, H. 1985. Supported phospholipid bilayers. *Biophysical Journal*, **47**(1), 105.

[415] Tanaka, M., and Sackmann, E. 2005. Polymer-supported membranes as models of the cell surface. *Nature*, **437**(7059), 656–663.

[416] Taylor, J., Phillips, S., and Cheng, Q. 2007. Microfluidic fabrication of addressable tethered lipid bilayer arrays and optimization using SPR with silane-derivatized nanoglassy substrates. *Lab on a Chip*, **7**(7), 927–930.

[417] Taylor, J., Linman, M., Wilkop, T., and Cheng, Q. 2009. Regenerable tethered bilayer lipid membrane arrays for multiplexed label-free analysis of lipid protein interactions on poly (dimethylsiloxane) microchips using SPR imaging. *Analytical chemistry*, **81**(3), 1146–1153.

[418] Taylor, S., and Gileadi, E. 1995. Physical interpretation of the Warburg impedance. *Corrosion*, **51**(9), 664–671.

[419] Teissie, J., and Rols, M. 1994. Manipulation of cell cytoskeleton affects the lifetime of cell membrane electropermeabilization. *Annals of the New York Academy of Sciences*, **720**(1), 98–110.

[420] Teixeira, S., Zaccai, G., Ankner, J., et al. 2008. New sources and instrumentation for neutrons in biology. *Chemical Physics*, **345**(2), 133–151.

[421] Thakore, V., Molnar, P., and Hickman, J. 2012. An optimization-based study of equivalent circuit models for representing recordings at the neuron electrode interface. *IEEE Transactions on Biomedical Engineering*, **59**(8), 2338–2347.

[422] Thein, M., Asphahani, F., Cheng, A., et al. 2010. Response characteristics of single-cell impedance sensors employed with surface-modified microelectrodes. *Biosensors and Bioelectronics*, **25**(8), 1963–1969.

[423] Thøgersen, L., Schiøtt, B., Vosegaard, T., Nielsen, N., and Tajkhorshid, E. 2008. Peptide aggregation and pore formation in a lipid bilayer: A combined coarse-grained and all atom molecular dynamics study. *Biophysical Journal*, **95**(9), 4337–4347.

[424] Thompson, J., Cronin, B., Bayley, H., and Wallace, M. 2011. Rapid assembly of a multimeric membrane protein pore. *Biophysical Journal*, **101**(11), 2679–2683.

[425] Tieleman, P. 2004. The molecular basis of electroporation. *BMC Biochemistry*, **5**(1), 10.

[426] Tokman, M., Hyojin, J., Levine, Z., et al. 2012. Electric field-driven water dipoles: Nanoscale architecture of electroporation. *Biophysical Journal*, **102**(1), 401a.

[427] Tremouilhac, P., Strandberg, E., Wadhwani, P., and Ulrich, A. 2006. Synergistic transmembrane alignment of the antimicrobial heterodimer PGLa/magainin. *Journal of Biological Chemistry*, **281**(43), 32089–32094.

[428] Tristram-Nagle, S., Kim, D., Akhunzada, N., et al. 2010. Structure and water permeability of fully hydrated diphytanoylPC. *Chemistry and Physics of Lipids*, **163**(6), 630–637.

[429] Tu, K., Klein, M., and Tobias, D. 1998. Constant-pressure molecular dynamics investigation of cholesterol effects in a dipalmitoylphosphatidylcholine bilayer. *Biophysical Journal*, **75**(5), 2147–2156.

[430] Uitert, I., Gac, S., and Berg, A. 2010. The influence of different membrane components on the electrical stability of bilayer lipid membranes. *Biochimica et Biophysica Acta, Biomembranes*, **1798**(1), 21–31.

[431] Ulmschneider, J., Smith, J., Ulmschneider, M., Ulrich, A., and Strandberg, E. 2012. Reorientation and dimerization of the membrane-bound antimicrobial peptide PGLa from microsecond all-atom MD simulations. *Biophysical Journal*, **103**(3), 472–482.

[432] Vaidyanathan, S., Sathyanarayana, P., Maiti, P., Visweswariah, S., and Ayappa, K. 2014. Lysis dynamics and membrane oligomerization pathways for Cytolysin A (ClyA) pore-forming toxin. *RSC Advances*, **4**, 4930–4942.

[433] Vaknin, D., Kjaer, K., Als-Nielsen, J., and Lösche, M. 1991. Structural properties of phosphatidylcholine in a monolayer at the air/water interface: Neutron reflection study and reexamination of x-ray reflection measurements. *Biophysical Journal*, **59**(6), 1325.

[434] Valenzuela, S., Alkhamici, H., Brown, L., et al. 2013. Regulation of the membrane insertion and conductance activity of the metamorphic chloride intracellular channel protein CLIC1 by cholesterol. *PLoS One*, **8**(2), e56948.

[435] Valincius, G., Meškauskas, T., and Ivanauskas, F. 2011. Electrochemical impedance spectroscopy of tethered bilayer membranes. *Langmuir*, **28**(1), 977–990.

[436] Valincius, G., Mickevicius, M., Penkauskas, T., and Jankunec, M. 2016. Electrochemical impedance spectroscopy of tethered bilayer membranes: An effect of heterogeneous distribution of defects in membranes. *Electrochimica Acta*, **222**, 904–913.

[437] van Gunsteren, W.F., Berendsen, H.J., and Rullmann, J.A.C. 1981. Stochastic dynamics for molecules with constraints: Brownian dynamics of n-alkalines. *Molecular Physics*, **44**(1), 69–95.

[438] Vatamanu, J., Borodin, O., and Smith, G. 2010. Molecular dynamics simulations of atomically flat and nanoporous electrodes with a molten salt electrolyte. *Physical Chemistry Chemical Physics*, **12**(1), 170–182.

[439] Vela, M., Martin, H., Vericat, C., et al. 2000. Electrodesorption kinetics and molecular interactions in well-ordered thiol adlayers on Au (111). *Journal of Physical Chemistry B*, **104**(50), 11878–11882.

[440] Velasco-Velez, J., Wu, H., Pascal, T., et al. 2014. The structure of interfacial water on gold electrodes studied by X-ray absorption spectroscopy. *Science*, **346**(6211), 831–834.

[441] Vijayendran, R.A., Ligler, F.S., and Leckband, D.E. 1999. A computational reaction-diffusion model for analysis of transport-limited kinetics. *Journal of Analytical Chemistry*, **71**, 5405–5412.

[442] Vockenroth, I., Atanasova, P., Jenkins, A., and Koeper, I. 2008. Incorporation of α-hemolysin in different tethered bilayer lipid membrane architectures. *Langmuir*, **24**(2), 496–502.

[443] Vögele, M., Holm, C., and Smiatek, J. 2015. Properties of the polarizable MARTINI water model: A comparative study for aqueous electrolyte solutions. *Journal of Molecular Liquids*, **212**, 103–110.

[444] Vora, T., Corry, B., and Chung, S.H. 2004. A model of sodium channels. *Biochimica et Biophysica Acta, Biomembranes*, **1668**, 106–116.

[445] Voth, G. 2008. *Coarse-Graining of Condensed Phase and Biomolecular Systems*. CRC Press.

[446] Wagner, M., and Tamm, L. 2000. Tethered polymer-supported planar lipid bilayers for reconstitution of integral membrane proteins: Silane-polyethyleneglycol-lipid as a cushion and covalent linker. *Biophysical Journal*, **79**(3), 1400–1414.

[447] Waisman, E., Henderson, D., and Lebowitz, J. 1976. Solution of the mean spherical approximation for the density profile of a hard-sphere fluid near a wall. *Molecular Physics*, **32**(5), 1373–1381.

[448] Wanasundara, S., Krishnamurthy, V., and Chung, S. 2011. Free energy calculations of gramicidin dimer dissociation. *The Journal of Physical Chemistry B*, **115**(46), 13765–13770.

[449] Wang, H., Thiele, A., and Pilon, L. 2013. Simulations of cyclic voltammetry for electric double layers in asymmetric electrolytes: A generalized modified Poisson–Nernst–Planck model. *Journal of Physical Chemistry C*, **117**(36), 18286–18297.

[450] Wang, L., Bose, P., and Sigworth, F. 2006. Using cryo-EM to measure the dipole potential of a lipid membrane. *Proceedings of the National Academy of Sciences of the United States of America*, **103**(49), 18528–18533.

[451] Wang, S., and Larson, R.G. 2013. Coarse-grained molecular dynamics simulation of tethered lipid assemblies. *Soft Matter*, **9**, 480–486.

[452] Weaver, J., and Chizmadzhev, Y. 1996. Theory of electroporation: A review. *Bioelectrochemistry and Bioenergetics*, **41**(2), 135–160.

[453] Weaver, J., and Mintzer, R. 1981. Decreased bilayer stability due to transmembrane potentials. *Physics Letters A*, **86**(1), 57–59.

[454] Weeks, J., Vollmayr, K., and Katsov, K. 1997. Intermolecular forces and the structure of uniform and nonuniform fluids. *Physica A: Statistical Mechanics and Its Applications*, **244**(1–4), 461–475.

[455] Weeks, J., Katsov, K., and Vollmayr, K. 1998. Roles of repulsive and attractive forces in determining the structure of nonuniform liquids: Generalized mean field theory. *Physical Review Letters*, **81**(20), 4400.

[456] Wertheim, M. 1963. Exact solution of the Percus–Yevick integral equation for hard spheres. *Physical Review Letters*, **10**(8), 321–323.

[457] Westerhoff, H., Zasloff, M., Rosner, J., et al. 1995. Functional synergism of the magainins PGLa and magainin-2 in Escherichia coli, tumor cells and liposomes. *European Journal of Biochemistry*, **228**(2), 257–264.

[458] Wiener, M., and White, S. 1991. Fluid bilayer structure determination by the combined use of X-ray and neutron diffraction. I. Fluid bilayer models and the limits of resolution. *Biophysical Journal*, **59**(1), 162.

[459] Wiener, M., and White, S. 1991. Fluid bilayer structure determination by the combined use of X-ray and neutron diffraction. II. "Composition-space" refinement method. *Biophysical Journal*, **59**(1), 174–185.

[460] Willow, S., Salim, M., Kim, K., and Hirata, S. 2015. Ab initio molecular dynamics of liquid water using embedded-fragment second-order many-body perturbation theory towards its accurate property prediction. *Scientific Reports*, **5**.

[461] Wiltschi, B., Knoll, W., and Sinner, E. 2006. Binding assays with artificial tethered membranes using surface plasmon resonance. *Methods*, **39**(2), 134–146.

[462] Winter, B. 2015. Interfaces: Scientists strike wet gold. *Nature Chemistry*, **7**(3), 192–194.

[463] Wohlert, J., Otter, W., Edholm, O., and Briels, W. 2006. Free energy of a trans-membrane pore calculated from atomistic molecular dynamics simulations. *The Journal of Chemical Physics*, **124**(15), 154905.

[464] Woodhouse, G., King, L., Wieczorek, L., Osman, P., and Cornell, B. 1999. The ion channel switch biosensor. *Journal of Molecular Recognition*, **12**(5), 328–334.

[465] Woodhouse, G., King, L., Wieczorek, L., Osman, P., and Cornell, B. 1999. The ion channel switch biosensor. *Journal of Molecular Recognition*, **12**(5), 328–334.

[466] Woodhouse, G., King, L., Wieczorek, L., and Cornell, B. 1999c. Kinetics of the competitive response of receptors immobilised to ion-channels which have been incorporated into a tethered bilayer. *Faraday Discussions*, **111**, 247–258.

[467] Woolley, A., and Wallace, B. 1992. Model ion channels: Gramicidin and alamethicin. *Journal of Membrane Biology*, **129**(2), 109–136.

[468] Wu, Y., He, K., Ludtke, S., and Huang, H. 1995. X-ray diffraction study of lipid bilayer membranes interacting with amphiphilic helical peptides: Diphytanoyl phosphatidylcholine with alamethicin at low concentrations. *Biophysical Journal*, **68**(6), 2361.

[469] Xun, X., Cao, J., Mallick, B., Maity, A., and Carroll, R. 2013. Parameter estimation of partial differential equation models. *Journal of the American Statistical Association*, **108**(503), 1009–1020.

[470] Yamashita, K., Kawai, Y., Tanaka, Y., et al. 2011. Crystal structure of the octameric pore of staphylococcal γ-hemolysin reveals the β-barrel pore formation mechanism by two components. *Proceedings of the National Academy of Sciences of the United States of America*, **108**(42), 17314–17319.

[471] Yeh, I., and Berkowitz, M. 1999. Dielectric constant of water at high electric fields: Molecular dynamics study. *Journal of Chemical Physics*, **110**(16), 7935–7942.

[472] Yildiz, A., Kang, C., and Sinner, E. 2013a. Biomimetic membrane platform containing hERG potassium channel and its application to drug screening. *Analyst*, **138**(7), 2007–2012.

[473] Yildiz, A., Yildiz, U., Liedberg, B., and Sinner, E. 2013b. Biomimetic membrane platform: Fabrication, characterization and applications. *Colloids and Surfaces B: Biointerfaces*, **103**, 510–516.

[474] Yin, P., Burns, C., Osman, P.D., and Cornell, B. 2003. A tethered bilayer sensor containing alamethicin channels and its detection of amiloride based inhibitors. *Biosensors and Bioelectronics*, **18**(4), 389–397.

[475] Yu, L., Sheng, Y., and Chiou, A. 2013. Three-dimensional light-scattering and deformation of individual biconcave human blood cells in optical tweezers. *Optics Express*, **21**(10), 12174–12184.

[476] Yuan, H., Leitmannova-Ottova, A., and Tien, T. 1996. An agarose-stabilized BLM: A new method for forming bilayer lipid membranes. *Materials Science and Engineering: C*, **4**(1), 35–38.

[477] Zasloff, M. 1987. Magainins, a class of antimicrobial peptides from *Xenopus* skin: Isolation, characterization of two active forms, and partial cDNA sequence of a precursor. *Proceedings of the National Academy of Sciences of the United States of America*, **84**(15), 5449–5453.

[478] Zeng, X., and Li, S. 2011. Multiscale modeling and simulation of soft adhesion and contact of stem cells. *Journal of the Mechanical Behavior of Biomedical Materials*, **4**(2), 180–189.

[479] Zeng, Y., Yip, A., Teo, S., and Chiam, K. 2012. A three-dimensional random network model of the cytoskeleton and its role in mechanotransduction and nucleus deformation. *Biomechanics and Modeling in Mechanobiology*, **11**(1–2), 49–59.

[480] Zhang, J., Johnson, P., and Popel, A. 2008. Red blood cell aggregation and dissociation in shear flows simulated by lattice Boltzmann method. *Journal of Biomechanics*, **41**(1), 47–55.

[481] Zhelev, D., and Needham, D. 1993. Tension-stabilized pores in giant vesicles: Determination of pore size and pore line tension. *Biochimica et Biophysica Acta*, **1147**, 89–104.

[482] Zheng, Q., and Wei, G. 2011. Poisson–Boltzmann–Nernst–Planck model. *Journal of Chemical Physics*, **134**(19), 194101.

[483] Zheng, Q., Chen, D., and Wei, G. 2011. Second-order Poisson–Nernst–Planck solver for ion transport. *Journal of Computational Physics*, **230**(13), 5239–5262.

[484] Zhou, S., and Jamnik, A. 2006. Structure of inhomogeneous Lennard-Jones fluid near the critical region and close to the vapor-liquid coexistence curve: Monte Carlo and density-functional theory studies. *Physical Review E*, **73**(1), 011202.

Index

absolute permittivity, *see* dielectric constant
activity coefficient, 230
advection-diffusion equation, *see* Nernst–Planck equation
   nondimensionalization, 397
amino acids
   polypeptide, 16
   side-chain, 17
anomalous diffusion, 330, 339, 340
antimicrobial peptide
   gramicidin A, 22
antimicrobial peptides, 196
   alamethicin, 123
   coarse-grained molecular dynamics, 341–348
   diffusion, 344–345
   dynamics, 196
   generalized reaction-diffusion equation, 197–200
   oligomerization, 196
   PGLa, 104
   surface binding, 196
   transmembrane orientation, 198
apolar molecules, 303
archaebacteria, 14
archaebacteria lipids, 14
area per lipid, 324
Arrhenius reaction rate, 308
artificial neural networks, 55
autoregressive Gaussian process, 169

barostats, 364
   Berendsen, 364
   Parrinello–Rahman, 364
bioelectronic interface, 26,
   coarse-grained molecular dynamics, 310
   density profile of water, 315
   hydration ion size, 269
   Percus–Yevick equation, 311
   capacitance, 275–279
   constant-phase element, 143–144
   dielectric constant, 238–242, 271–275
   double-layer impedance, 242, 243
   faradic impedance, 242
   faradic reactions, 242–247
   generalized Poisson–Nernst–Planck, 229–232, 238
   ionic correlation effects, 238–242
   kinetic impedance, 242
   Poisson–Fermi–Nernst–Planck, 232–235, 238
biological neural networks, 54
biological parameters
   aliphatic chain deuterium order parameter, 298
   area per lipid, 324, 373
   defect density, 325
   diffusion tensor, 317
   intrinsic membrane dipole potential, 377
   line tension, 326
   lipid diffusion, 329
      ballistic, 339
      Fickian diffusion, 339
      subdiffusion, 339
   lipid flip-flop energy, 324
   membrane capacitance, 376
   membrane thickness, 330
   number of aqueous pores, 324
   radial distribution function, 300
   surface tension, 327
biomimetic, 31
biomolecules
   amino acids, 16
      nonpolar, 18
      polar, 18
      weak acid, 18
      weak base, 18
   C-terminus, 17*f*
   N-terminus, 17*f*
   peptides, 16
   proteins, 16
      integral monotopic, 19
      integral polytopic, 19
      peripheral membrane, 19
   structure, 19
      primary structure, 19
      quaternary structure, 20
      secondary structure, 19
      tertiary structure, 20

biosensor
  analyte detection, 87
  arrays, 97–100, 187–195
  design, 89–91
  hidden Markov Model, 168–171
  ICS biosensor, 83–102
  influenza A, 95–97
  macroscopic model, 162–165
  membrane conductance, 163–165
  mesoscopic model, 178–185
  multianalyte, 97–100
  nanomachine conductance, 378–383
  nanomachine dynamics, 383–387
  singular perturbation approximation, 165–166
  specificity, 94–95
  surface reactions, 163–165
Bjerrum length, 233, 241
black-box model, 48
Boltzmann distribution, 312
Brownian dynamics, 311, 379
Brownian motion, 305
Butler–Volmer equation, 206

canonical ensemble, 312
cell lysis, 74
cell membrane, 30
  cytoskeletal filaments, 31
  cytosol, 30
CGMD, see coarse-grained molecular dynamics
charge accumulation, 50
charge distribution, 11
chemical bond, 9
  coordinate covalent bond, 10
  covalent bond, 9
  ionic bond, 11
  van der Waals, 11
chemical potential
  Widom insertion method, 231
chirality, 20
cholesterol
  effect on membrane, 333
  coarse-grained molecule dynamics, 332–335
  coarse-grained representation, 304
  electric measurements, 267–269
  electroporation, 267–269
  fractional-order macroscopic model, 154–156
  heterogeneous membranes, 267–269
  impedance measurements, 156–157
  lateral diffusion, 333–334
  line tension, 335
  membrane thickness, 334–335
  surface tension, 335
coarse-grained molecular dynamics, 295–352
  bioelectronic interface, 304
coarse graining
  Boltzmann inversion, 300

force matching, 301
inverse Monte Carlo, 300
relative entropy matching, 301
COMSOL, 401, 403
configurational phase space, 301
construction, see coarse graining
  deuterium order parameter, 327, 328
  diffusion tensor, 317–321
  DphPC lipids, 304
  electrode, 304
  line tension, 325–327
  lipid desorption, 323
  lipid diffusion, 328–330
  lipid energetics, 322–325
  lipid flip-flop, 322
  MARTINI force field, 303
  membrane structure, 330–331
  PGLa pore formation, 341–348
  round-trip time, 319
  setup, 299–300
  sterol effects, 332–335
  sterols, 304
  surface tension, 325–327
  tethers and spacers, 304
  water density, 315–317
continuum models
  Fokker–Planck equation, 305–310, 317–319
  generalized Poisson–Nernst–Planck, 229–232
  generalized reaction-diffusion equation, 197–200
  Navier–Stokes equation, 182
  Nernst–Planck equation, 181, 182
  Poisson's equation, 180
  Poisson–Boltzmann equation, 240–242
  Poisson–Fermi–Nernst–Planck, 232–235
  Poisson–Nernst–Planck, 182, 185
  Smoluchowski–Einstein equation, 217–222
Coulomb correlation length, 233
Coulomb correlations, 49
cubic packing structure, 230, 239

Damköhler number, 161, 177–178, 399
  mass-transport influenced regime, 177
  reaction-rate-influenced regime, 177
de Broglie wavelength, 230
Debye length, 205, 241
defect density, 325
density profile, 311
depolarization, 290
Deuterium order parameter, 328
dielectric constant
  bioelectronic interface, 271–275
  coarse-grained molecular dynamics, 324
  effective dielectric constant, 356
  ionic correlation effects, 238–242
  Maxwell–Garnett equation, 239–240

molecular dynamics, 356, 375–378
Poisson–Boltzmann equation, 240–242
Poisson–Fermi–Nernst–Planck, 232–235
pure hydrocarbon, 154
diffusion process, 305
    Arrhenius reaction rate, 308
    Fokker–Planck equation, 306
    Kolmogorov equation, 306
    mean first passage time, 308
    stationary distribution, 307
diffusion tensor
    bioelectronic interface, 320–321
    coarse-grained molecular dynamics, 317–321
    Fokker–Planck equation, 317–319
    round-trip time, 319
    water, 320–321
distal layer, 330
double-layer charging, 224
    Coulomb correlations, 228
    polarization, 228
    screening effects, 223
    Stern and diffuse layers, 224
double-layer impedance, 242, 243
DphPC bilayers, 14, 262
    aqueous pore formation, 373–375
    archaebacteria lipids, 14
    area per lipid, 128
    capacitance, 375–378
    coarse-grained representation, 304
    deuterium order parameter, 327–328
    fluorescence recovery after photobleaching, 129–131
    line tension, 325–327
    lipid diffusion, 129–131, 328–330
    lipid energetics, 322–325
    membrane structure, 330–331
    molecular dynamics, 373–378
    neutron reflectometry, 131–133
    reaction-rate model, 142
    surface tension, 325–327
    tether density, 262
    thickness, 131–133
    X-ray reflectometry, 128

electric permittivity, *see* dielectric constant
electrochemotherapy, 213
electrodesorption, 71
electrolyte dynamics
    charge accumulation, 50
    Coulomb correlations, 49
    diffusion-limited charge transfer, 50
    ionic adsorption dynamics, 50
    polarization effects, 50
    reaction-limited charge transfer, 50
    screening effects, 50
    steric effect, 49

electroneutrality, 142
electrophysiological response , 282
electrophysiological response platform, *see* ERP
electroporation, 212
    aqueous pore energy model, 220
    aqueous pores, 214
        toroidal pore, 231
    continuum model, 227–235
    hydrophilic pore, 214
    hydrophobic pore, 214
    irreversible, 213
    mesoscopic model, 214–215, 222–227
    multiphysics model, 222
    reversible, 213
electroporation measurement platform, *see* EMP
electrostatic pressure, *see* Maxwell stress tensor
ELISA, 53, 87, 95, 97
embedded ion channel, 103
EMP, 107–110
    aqueous pore capacitance, 259
    aqueous pore conductance, 254–256
    cholesterol, 267–269
    continuum model, 227–235
    formation, 108–109
    heterogeneous membrane, 265–267
    mesoscopic model, 214–215, 222–227
    operation, 109–110
    tether density, 262–265
energy minimization principle, 226
engineered artificial membranes, 32
    bioelectronic interface, 26
    cushioned membranes, 34
    freestanding lipid bilayer, 34
    hybrid bilayer lipid membrane, 33
    laboratory construction, 76–81
    spacers, 25, 67
    supported lipid bilayer, 33
    tethered bilayer lipid membranes, 35
    tethers, 25, 67
engineered tethered membrane, 39–45
    bioelectronic interface, 26
    electrophysiological response platform, 44
    electroporation measurement platform, 43
    ion-channel switch biosensor, 41
    laboratory construction, 76–81
    multiphysics models, 45–49
    pore formation measurement platform, 42
ensemble average, 359
equilibrium electrode potential, 206
equipartition theorem, 361
ergodic hypothesis (law of large numbers), 361
ERP, 110–115
    formation, 112–114
    fractional-order macroscopic model, 284–285, 286–290

ERP (*cont.*)
    noninvasive screening of cells, 293
    operation, 114–115
    skeletal myoblasts, 290–291
    systems biology, 283
    voltage-gated ion channel, 285–286
excess chemical potential, 230

faradic flux, 244
faradic impedance, 242
faradic reactions, 242–247
    absolute chemical activity, 246
    Butler–Volmer equation, 206
    frequency prefactor, 245
    Frumkin–Butler–Volmer equation, 244
Fick's laws
    Fick's first law, 181
    Fick's second law, 319
finite-length Warburg impedance, 207
fluctuation-dissipation theorem, 311, 337
fluorescence recovery after photobleaching, 129
Fokker–Planck equation, 305–310, 317–319
fractional Fokker–Planck equation, 340
fractional Langevin equations, 340
fractional-order differential equation, 143
fractional order macroscopic model, 140–144
    charge accumulation, 143
    constant-phase element, 143–144
    diffusion-limited charge transfer, 143
    electrode double-layer capacitance, 143–144
    electrolyte resistance, 142
    fractional derivatives, 144–147
    ionic adsorption dynamics, 143
    membrane capacitance, 142
    membrane conductance, 142
    reaction-limited charge transfer, 143
    transmembrane potential, 140*f*
fractional transport processes, 340
frequency prefactor, 245
friction kernel, 382
Frumkin–Butler–Volmer equation, 244

gamma function, 144
gene electrotransfer, 213
generalized Poisson–Nernst–Planck, 229–232
    chemical potential, 230
    equilibrium concentration, 241
    steric constraint, 230
gramicidin A, 22,
    conductance, 381–383
    dimer dissociation, 383–384
    ICS biosensor, 87
    ion permeation models, 378–381
    molecular dynamics, 383–387
    nuclear magnetic resonance, 128–129
    spontaneous insertion, 76–81

hard-sphere potential, 313
hard-wall potential, 315
hidden Markov Model, 168
    maximum likelihood estimate, 170
    parameter estimator, 170
    state estimation filter, 170
Hill-type approximation, 198
Hodgkin–Huxley models, 291
hydration ion size, 269
hydrophilic, 12
hydrophobic effect, 15
hyperpolarization, 290

ICS biosensor, 83–102
    arrays, 97–100, 187–195
    clinical evaluation, 95–97
    construction and formation, 85–87
    ferritin, 186–187
    hCG, 186–187
    hidden Markov model, 168
    least-squares estimation, 185
    macroscopic model, 162–165
    mass-transport regime, 177
    membrane conductance, 163–165
    mesoscopic model, 178–185
    microscopic model, 378–387
    multianalyte, 97–100
    multicompartment model, 208
    nanomachine conductance, 378–383
    nanomachine dynamics, 383–387
    reaction-rate-limited regime, 177
    singular perturbation approximation, 165–166
    stochastic model, 168
    streptavidin, 186–187
    surface reactions, 163–165
    TSH, 186–187
implantable devices, 51
    cochlear and retinal, 51
in vitro medical diagnostics, 52
instantaneous temperature, 361
ion-channels, 21–24
    ligand-gated, 23
    light-gated, 24
    mechanosensitive, 24
    peptide, 22
    protomers, 22
    transient receptor potential, 24
    voltage-gated, 22
ion permeation, 378–381
    Brownian dynamics, 379
    continuum theories, 380
    Fokker–Planck equation, 379
    gramicidin, 381
    reaction-rate model, 380
ion-channel gating models, 282
ion-channel shot noise, 168

ionic conductance noise, 168
ionic transport
  diffusive flux, 181
  electrical-migration flux, 181
  velocity field flux, 181
ionophores, 116

Johnson–Nyquist noise, 168

kinetic impedance, 242
Kolmogorov equation, 306
  Arrhenius reaction rate, 308
  mean first-passage time, 308

laminar flow, 185
Langevin equation, 318, 336–339, 381
  Brownian dynamics, 336, 380
  large molecules, 336, 380
  lipid diffusion, 336
Langmuir binding kinetics, 398
Langmuir–Hinshelwood equation, 198
lateral diffusion, 330
law of large numbers, 336, 361
least-squares estimation, 185
Lennard-Jones potential, 314
Levenberg–Marquardt algorithm, 186
line tension, 322, 325, 326
linear Poisson–Boltzmann equation, 241
lipid bilayers, 14–16
  DphPC, see DphPC bilayers
  molecules, see lipids
  phospholipids, 12–13
lipid desorption, 323, 332
lipid flip-flop, 322, 324, 332
lipid structures
  lipid bilayer, 12
  liposomes, 12
  micelles, 12
  vesicles, 12
lipids, 12
  cholesterol, 12
  glycolipids, 12
  lipid domain, 155
  lipid rafts, 13
  phospholipids, 12
  tethered, 67
local packing fraction, 230
Lorentz force, 180

macroscopic models, see reaction-rate models
macroscopic reaction-rate model
  tethered transmembrane potential, 163f
MARTINI force field, see coarse-grained
  bioelectronic interface, 304
  electrode, 304
  lipids, 304

sterol, 304
tethers and spacers, 304
mass conservations equations, 230
  chemical potential, 230
  excess chemical potential, 230
mass-transport regime, 177
maximum likelihood estimate, 170
Maxwell stress tensor, 237
Maxwell's equations, 180
Maxwell–Boltzmann distribution, 354
Maxwell–Garnett equation, 239
mean first-passage time, 305
  Arrhenius reaction rate, 308
  partial differential equation, 308
mean-square displacement, 329
Measurement techniques
  electrical response, 118–127
  fluorescence recovery after photobleaching, 129–131
  impedance, 120
  neutron reflectometry, 131–133
  nuclear magnetic resonance, 129
  X-ray reflectometry, 128
membrane
  geometric properties, 330
membrane permanent surface charge density, 202
mesoscopic models, see continuum models
microscopic model, see coarse-grained
mode-coupling theory, 339
Models
  ab initio molecular dynamics, 46
  black-box, 48
  coarse-grained molecular dynamics, 47
  continuum theories, 48
  molecular dynamics, 47
  reaction-rate theory, 48
molecular dynamics
  aqueous pore formation, 373–375
  bioelectronic interface, 356, 366–367
  bond angle, 357
  bond length, 357
  bonded interactions, 356–358
  boundary conditions, 367
  degrees of freedom, 362
  energy, 361
  improper dihedral, 358
  initial conditions, 368
  membrane capacitance, 375–378
  membrane dipole potential, 375–378
  non-bonded interactions, 355
  NPT, 364
  numerical integration, 369–370
  NVE, 364
  NVT, 364
  potential energy function, 355

Index

molecular dynamics (*cont.*)
  pressure, 362
  proper dihedral, 358
  temperature, 361
  triclinic unit cell, 367
  umbrella sampling, 322
molecular therapeutics, 54
multicompartment model, 208
multicellular communication, 55
multiphysics model, 45

Navier–Stokes equation, 182
Nernst–Planck equation, 181, 182
  boundary conditions, 183–184
  Dirichlet boundary conditions, 183–184
  mobility, 182
neutron reflectometry, 131
noise
  1/f, 169
nuclear magnetic resonance, 129
numerical integration
  cutoff approximation
    shift, 370
    switch, 370
    truncation, 369
  particle-mesh Ewald summation, 370

Ohmic contacts, 52
organism
  eukaryotes, 13
  prokaryotes, 13
Ornstein–Zernike equation, 313, 314
overdamped Langevin equation, 318
oxidation potential, 243

packing coefficient, 239
pair-correlation function, 312
patch microelectrodes, 282
patch-clamp electrophysiology, 282
Pauli repulsion force, 316
PDEs
  nondimensionalization, 397
  Dirichlet boundary condition, 397
  Fokker–Planck equation, 305–310, 317–319
  generalized Poisson–Nernst–Planck, 229–232
  generalized reaction-diffusion equation, 197–200
  mixed boundary condition, 397
  Navier–Stokes equation, 182
  Nernst–Planck equation, 181, 182
  Neumann boundary condition, 397
  Poisson's equation, 180
  Poisson–Boltzmann equation, 240–242
  Poisson–Fermi–Nernst–Planck, 232–235
  Poisson–Nernst–Planck, 182, 185
  Robin boundary condition, 397
  Smoluchowski–Einstein equation, 217–222

Percus–Yevick equation, 311, 313
  density profile, 311
  Fredholm integral equation, 313
perturbation analysis, 203, 224
  double-layer impedance, 203
  Smoluchowski–Einstein equation, 224
PFMP, 104–107
  construction, 105
  generalized reaction-diffusion equation, 197–200
  operation, 106–107
  PGLa dynamics, 200–203
PGLa
  coarse-grained molecular dynamics, 345–348
  coarse-grained representation, 344
  diffusion, 344–345
  experimental measurements, 105, 200–203
  generalized reaction-diffusion equation, 197–200
  oligomerization, 347
  reaction mechanism, 348
  surface binding, 346
  translocation, 347
point-of-care diagnostic, 53
Poisson's equation, 180
  boundary conditions, 183–184
  dielectric permittivity, 180
  Neumann boundary conditions, 183–184
  polarization charge density, 240
Poisson–Boltzmann equation, 240–242
Poisson–Fermi–Nernst–Planck, 232–235
  Bjerrum length, 241
  Coulomb correlation length, 233
  dielectric permittivity operator, 233
  equilibrium concentration, 241
  Poisson–Fermi equation, 233
Poisson–Nernst–Planck, 182, 185
  asymptotic model, 203, 207
  double-layer capacitance, 206, 207
polar molecule, 303
polarization, 50
polarization charge density, 240
pore density, 322
pore formation measurement platform, *see* PFMP
potential energy function
  bonded interactions, 357
    bond angle, 357
    bond length, 357
    improper dihedral, 358
    proper dihedral, 358
  nonbonded interactions, 356
    Coulomb's potential, 356
    Lennard-Jones potential, 356
    van der Waals forces, 356
potential of mean force, 322
Prandtl number, 399
protein
  chirality, 20
  conformation change, 18

gramicidin A, 22
ion channels, 21–24
structure, 19–21
visualization, 27–29
proximal layer, 330
Péclet number, 161, 177–178, 399
  advection-transport-influenced regime, 177
  mass-transport-influenced regime, 178

radial distribution function, 312
Ramachandran plot, 344
reaction rates, 161
reaction-rate-limited regime, 177
reaction-rate models
  fractional derivatives, 144–147
  fractional-order macroscopic model, 140–144
repolarization, 290
resting potential, 290
Reynolds number, 399
round-trip time, 319

Singer–Nicolson model, 81
singular perturbation approximation
  ICS biosensor, 165
  nondimensionalization, 400
  Smoluchowski–Einstein equation, 225
Smoluchowski–Einstein equation, 217–222
  boundary conditions, 218
  homogeneous, 218
  nonhomogenous, 218
  source term, 219
solvent interaction
  amphiphilic, 12
  hydrophilic, 12
  hydrophobic, 12
spreading conductance, 254
stationary distribution, 307
statistical ensemble, 355
  canonical ensemble (NVT), 364
  constant energy, 363
  constant pressure, 364
  constant temperature, 363
  isothermal–isobaric ensemble (NPT), 364
  microcanonical ensemble (NVE), 364
  NVT, 312
steric effects, 49

steric interactions, 230
Stern layer, 244
sterol, *see* cholesterol
stochastic averaging theory, 380
stochastic differential equation, 380
stochastic model, 168, 336
stoichiometry matrix, 164
substrate-integrated microelectrode arrays, 282
surface tension, 322, 325, 327

tethered bilayer lipid membranes, 35–39
  avidin/biotin tethered membrane, 38
  cholesterol tethered membrane, 38
  engineered tethered membrane, 38
  peptide tethered membrane, 37
  polymer tethered membrane, 38
  protein tethered membrane, 37
thermostats, 363
  Berendsen, 363
  Nosé–Hoover, 363
  velocity rescaling, 363
Tikhonov's theorem, 166, 400
transmembrane potential, 213
triclinic unit cell, 367
two-compartment model, 208

umbrella sampling, 322

voltage-gated ion channel, 285
voltage-gated ion channels, 22–23
  depolarization, 290
  Hodgkin–Huxley, 291
  hyperpolarization, 290
  repolarization, 290
  resting potential, 290

Warburg impedance, 144, 207
water
  density profile, 311
  diffusion tensor, 320
  spreading conductance, 254, 256
weighted histogram analysis method, 322
Widom insertion method, 231

X-ray reflectometry, 128

zwitterions, 202